INTRODUCTORY FLUID MECHANICS

The primary objective of this introductory text is to familiarize students exposed to only one course on fluids with the basic elements of fluid mechanics so that, should their future work rely on occasional numerical solutions, they will be familiar with the jargon of the discipline and the expected results. This book also can serve as a long-term reference text, in contrast to the oversimplified approach often used for such introductory courses. Additionally, it provides a comprehensive foundation for more advanced courses in fluid mechanics within disciplines such as mechanical or aerospace engineering. To avoid confusing students, the governing equations are introduced early and the assumptions leading to the various models are presented. This provides a logical hierarchy and explains the interconnectivity between the various models. Supporting examples demonstrate the principles and provide engineering analysis tools for many engineering calculations.

Dr. Joseph Katz is a professor of Aerospace Engineering and Engineering Mechanics at San Diego State University. His rich and diverse academic and engineering background covers typical aerospace and automotive disciplines such as computational and experimental aerodynamics, vehicle dynamics, race-car aerodynamics, and engine cooling. As a race-car designer for the past 25 years, Dr. Katz has participated in a large number of projects involving open wheel (Formula One and Indy), prototype (IMSA), hill-climb, and NASCAR. His fluid mechanics research interests include unsteady aerodynamics and incompressible flow with a strong emphasis on developing numerical techniques.

Before assuming his academic position, he spent four years at the NASA Ames Research Center full-scale wind-tunnel facility. For his work in developing the PMARC computational tool, he and his team received the 1997 NASA Space Act Award. In recent years, he has been active in the aerodynamic development of unmanned aerial vehicles such as the Global Hawk and E-Hunter.

Professor Katz has received numerous awards for being a most influential teacher and outstanding educator as recognized by the American Institute of Aeronautics and Astronautics. He is the author of several books and more than 100 other publications. His book on race-car aerodynamics can be found on the desks of most race-car designers around the world.

Introductory Fluid Mechanics

Joseph Katz
San Diego State University

CAMBRIDGE UNIVERSITY PRESS
Cambridge, New York, Melbourne, Madrid, Cape Town,
Singapore, São Paulo, Delhi, Mexico City

Cambridge University Press
32 Avenue of the Americas, New York NY 10013-2473, USA

Published in the United States of America by Cambridge University Press, New York

www.cambridge.org
Information on this title: www.cambridge.org/9781107617131

First published 2010
First paperback edition 2013

A catalogue record for this publication is available from the British Library

Library of Congress Cataloguing in Publication Data

Katz, Joseph, 1947–
Introductory fluid mechanics / Joseph Katz.
p. cm.
Includes bibliographical references and index.
ISBN 978-0-521-19245-3 (hardback)
1. Fluid mechanics. 2. Fluid mechanics – Mathematics. I. Title.
TA357.K376 2010
620.1′06–dc22 2010024916

ISBN 978-0-521-19245-3 Hardback
ISBN 978-1-107-61713-1 Paperback

Contents

Preface

Fluid mechanics is a fascinating but complex science and some problems cannot be solved by simple intuition. The reason behind this is the complex nonlinear differential equations, which cannot be solved analytically. The approach that evolved over recent centuries is to develop simple models for specific flow regions so that engineering calculations and predictions become possible. Unfortunately, even these simple models rely on complex mathematics, which makes introductory courses on this subject extremely difficult and sometimes confusing to students.

On the other hand, numerical solutions have matured recently and generating a solution for a given geometry can be achieved by a simple "run" command. The approach of many users is to run a large number of cases and develop their own "learning curve" of the problem, exactly as is done by experiments. The ease of generating attractive, colorful solutions creates the illusion (for many students) that further study of the subject is unnecessary.

The first objective of this introductory text is to familiarize students (and many will be exposed to only one course on fluids) with the basic elements of fluid mechanics. Therefore, if their future work relies on occasional numerical solutions, they will be familiar with the jargon of the discipline and with the expected results. At the same time, this book can serve as a long-term reference text, contrary to the oversimplified approach occasionally used for such introductory courses. The second objective is to provide a comprehensive foundation for more advanced courses in fluid mechanics (in areas such as mechanical or aerospace engineering disciplines). In order not to confuse the students, the governing equations are introduced early and the assumptions leading to the various models are clearly presented. This provides a logical hierarchy for the material that follows and explains the interconnectivity between the various models. Subsequent topics are then logically developed from the early chapter (Chapter 2) and the discussions are simple, brief, nonconfusing, and accompanied by useful examples (e.g., make it easy to understand to stimulate students' interest in the subject). Supporting examples demonstrate and explain the underlying principles and provide engineering analysis tools for various practical engineering calculations.

Also, emphasis is placed on providing complete solutions (e.g., flow in pipes, boundary-layer flow, etc.) for simple laminar flow cases (mostly one-dimensional) to explain the method and to capture student interest. Therefore, the approach is self-containing and readers are not directed elsewhere for more detailed formulations (e.g., the boundary layer problem is solved by using an integral method, and all details are presented here). Once the solution for a simple case is explained, extrapolation to more realistic and more complex cases (as in turbulent flow in pipes or in boundary layers) is provided.

A Word to the Instructor

Teaching a first course in fluid mechanics is always challenging because of the numerous new concepts that were not used in previous engineering courses. The source of the problem is in the highly nonlinear nature of the governing equations, which cannot be solved analytically for a general case. Over the years, various flow regimes were treated by neglecting large portions of the governing equations, leading to approximate solutions, which are unique to that limited problem. Consequently, many introductory fluid mechanics courses focus on surveying those localized solutions and not dealing directly with the governing equations. This approach perhaps is less mathematically intense, but it may confuse the novice student who cannot connect the various partial solutions to construct a more comprehensive understanding of the fluid mechanics discipline.

The present approach is aimed at easing the learning process by providing an early overview of the field and identifying a roadmap for understanding the different flow regimes. This is accomplished by presenting the governing equations (in their simplest laminar form) early on. In each of the following chapters, students are reminded how the particular subject of the chapter relates to the original roadmap (i.e., hydrostatics is a case in which the velocity in the governing equations is zero). Another benefit is that some similar cases are combined and students do not feel that each case needs a new and different mathematical approach. Bearing in mind that, for some students this is the only fluid dynamic course they'll take, providing a comprehensive survey of the field also becomes a major objective. Thus, delivery efficiency (of contents per time) becomes of paramount importance, demanding certain sacrifices such as pushing the discussion on dimensional analysis to the second half of the semester. Consequently, the introduction of the Reynolds number is delayed to Chapter 5, where it pops up automatically. This realignment of the topics is done mainly to allow early discussions (and assignment of homework problems) on more practical fluid-related problems such as fluid static or one-dimensional flows. As a result of these simplifications, *more* material can be covered during a given teaching period.

As noted, this text is proposed for the first introductory course on fluid mechanics. Therefore, the discussion is focused on cases in which simple laminar flow models can be effectively presented (whereas complex subjects such as surface waves, turbulence modeling, or transition can be discussed in more advanced courses). A one-semester introductory course may cover the following sections (in the order presented):

Chapter 1 is basically an introduction. A survey of engineering units is recommended, and Section 1.5 can be skipped.

Chapter 2 introduces the fluid dynamic equations. The integral form is easily developed in class, but Section 2.7 and on can be skipped (reading these sections at home, for information only, is recommended).

Chapter 3 focuses on the pressure term in the equations and concepts such as the pressure distribution and the center of pressure are introduced.

Chapter 4 focuses on the inertia term of the general equations. All one-dimensional tools are easily incorporated, including the Bernoulli equation.

Chapter 5 introduces the effects of viscosity through laminar flow models. Some exact solutions, such as the pipe flow, are developed. Once the flow parameters are identified (e.g., the Reynolds number), nondimensional numbers such as the friction coefficient enable the solution of a wider range of engineering problems (including turbulent flow). In a one-semester course, Subsection 5.3.4 and the sections beyond Subsection 5.9.3 are not taught.

Chapter 6 is a short introduction to the next two chapters dealing mainly with high-Reynolds-number flows. It is important because it provides the connection between the models for surface friction and external aerodynamics and hydrodynamics. It also justifies why potential flow models are capable of estimating the pressure field in high-Reynolds-number flows.

Chapter 7 discusses the role of the boundary layer in a manner similar to that of flow in pipes (e.g., after a simple solution, parameters such as the friction coefficient are extrapolated into the turbulent flow region). Section 7.6 and subsequent sections usually are not taught during an introductory course.

Chapter 8 introduces the concept of ideal flow. Note that only the velocity potential is presented, as in more advanced courses it can be extended to three-dimensional flows. The approach used in the previous chapter continues by solving a limited analytical case (e.g., the flow over a cylinder) and then developing coefficients so that various practical problems for both laminar and turbulent flows can be solved (e.g., the lift and drag of various bodies). Students usually enjoy this chapter, but the end of the semester is already on the horizon.

Chapters 9–11 deal with more advanced topics and portions can be briefly discussed, if time allows, or can be used in a second, more advanced course. In practice, the computational fluid dynamics of Chapter 9 can be discussed during the last week of the semester (but should not be included in the final test). A presentation of colorful, three-dimensional computational fluid dynamic solutions of the day is highly recommended.

1 Basic Concepts and Fluid Properties

1.1 Introduction

The science of fluid mechanics has matured over the last 200 years, but even today we do not have complete and exact solutions to all possible engineering problems. Although the governing equations (called the Navier–Stokes equations) were established by the mid-1800s, solutions did not follow immediately. The main reason is that it is close to impossible to analytically solve these nonlinear partial differential equations for an arbitrary case. Consequently, the science of fluid mechanics has focused on simplifying this complex mathematical model and on providing partial solutions for more restricted conditions. Therefore the different chapters on classical fluid mechanics are based on retaining different portions of the general equation while neglecting other lower-order terms. This approach allows the solution of the simplified equation, yet preserves the dominant physical effects (relevant to that particular flow regime). Finally, with the enormous development of computational power in the 21st century, numerical solutions of the fluid mechanic equations have become a reality. However, in spite of these advances, elements of modeling are still used in these solutions, and the understanding of the "classical" but limited models is essential for successfully using these modern tools.

This first chapter provides a short introduction on the historical evolution of fluid mechanics and a brief survey of fluid properties. After this introduction, the fluid dynamic equations are developed in the next chapter.

1.2 A Brief History

The science of fluid mechanics is neither new nor biblical; however, most of the progress in this field was made in the 20th century. Therefore it is appropriate to open this text with a brief history of the discipline, with only a very few names mentioned.

As far as we can document history, fluid dynamics and related engineering were always integral parts of human evolution. Ancient civilizations built ships, sails, irrigation systems, and flood-management structures, all requiring some basic understanding of fluid flow. Perhaps the best known early scientist in this field is

Archimedes of Syracuse (287–212 B.C.E.), founder of the field now we call "fluid statics," whose laws on buoyancy and flotation are used to this day.

A major leap in understanding fluid mechanics began with the European Renaissance of the 14th–17th centuries. The famous Italian painter-sculptor, Leonardo da Vinci (1452–1519), was one of the first to document basic laws such as the conservation of mass. He sketched complex flow fields, suggested feasible configurations for airplanes, parachutes, and even helicopters, and introduced the principle of streamlining to reduce drag.

During the next couple of hundred years, the sciences were gradually developed and then suddenly accelerated by the rational mathematical approach of an Englishman, Sir Isaac Newton (1642–1727), to physics. Apart from the basic laws of mechanics, and particularly the second law connecting acceleration with force, the concepts for drag and shear in a moving fluid were developed by Newton, and his principles are widely used today.

The foundations of fluid mechanics really crystallized in the 18th century. One of the more famous scientists, Daniel Bernoulli (1700–1782, Dutch-Swiss), pointed out the relation between velocity and pressure in a moving fluid, an equation that bears his name appears in every textbook. However, his friend Leonhard Euler (1707–1783, Swiss born), a real giant in this field, is the one who actually formulated the Bernoulli equations in the form known today. In addition, Euler, using Newton's principles, developed the continuity and momentum equations for fluid flow. These differential equations, the Euler equations, are the basis for modern fluid dynamics and perhaps the most significant contribution to the process of understanding fluid flows. Although Euler derived the mathematical formulation, he did not provide solutions to his equations. (Note that Euler is pronounced "oiler," not "yuler"; hence we have "*an* Euler equation.")

Science and experimentation in the field increased, but it was only in the 19th century that the governing equations were finalized in the form known today. A Frenchman, Claude-Louis-Marie-Henri Navier (1785–1836), understood that friction in a flowing fluid must be added to the force balance. He incorporated these terms into the Euler equations and published the first version of the complete set of equations in 1822. These equations are known today as the Navier–Stokes equations. Communications and information transfers were not well developed in those days. For example, Sir George Gabriel Stokes (1819–1903) lived on the English side of the English Channel but did not communicate directly with Navier. Independently, he also added the viscosity term to the Euler equations. Hence the glory is shared by both scientists for these equations. Euler can be also considered the first to solve the equations for the motion of a sphere in a viscous flow, which is now called "Stokes flow."

Although the theoretical basis for the governing equation had been laid down by now, it was clear that the solution was far out of reach. Therefore scientists focused on "approximate models," using only portions of the equation that could be solved. Experimental fluid mechanics also gained momentum, with important discoveries by Englishman Osborne Reynolds (1842–1912) about turbulence and transition from laminar to turbulent flow. This brings us to the 20th century, when science and technology grew at an explosive rate, particularly after the first powered flight of the Wright brothers in the United States (December 1903). Fluid mechanics attracted not only the greatest talent but also investments from governments as the potential of flying machines was recognized. If only one name is mentioned per century, then Ludwig Prandtl (1874–1953) of Göttingen, Germany, deserves the glory for the 20th century. He made tremendous progress in developing simple models for problems such as the flow in boundary layers and over airplane wings.

This trend of solving models and not the complex Navier–Stokes equations continued well into the mid-1990s, until the tremendous growth in computer power finally allowed numerical solutions of these equations. Physical modeling is still required, but the numerical approach allows the solutions of nonlinear partial differential equations, an impossible task from the pure analytical point of view. Nowadays, the flow over complex shapes and the resulting forces can be computed by commercial computer codes, but without being exposed to simple models, our ability to analyze the results would be incomplete.

1.3 Dimensions and Units

The magnitude (or dimensions) of physical variables is expressed in engineering units. In this book we follow the metric system, which was accepted by most professional societies in the mid-1970s. This International system, (SI for Systeme International) of units is based on the decimal system and is much easier to use than other (e.g., British) systems of units. For example the basic length is measured in meters (m): 1000 m is a kilometer (km) and 1/100 of a meter is a centimeter (cm). Along the same line, 1/1000 m is a millimeter (mm).

Mass is measured in grams (g), which is the mass of one cubic centimeter (1 cm^3) of water. One thousand grams are one kilogram (kg), and 1000 kg are one metric ton. Time is still measured the old-fashioned way, in hours (h), 1/60 of an hour is a minute (min), and 1/60 of a minute is a second (s).

For this book, velocity is one of the most important variables, and its basic measure therefore is meters per second (m/s). Vehicle speeds are usually measured in kilometers per hour (km/h) and clearly $1\,\text{km/h} = 1000/3600 = 1/3.6\,\text{m/s}$. Acceleration is the rate of change of velocity and therefore it is measured in meters per second squared (m/s^2).

Newton's second law defines the units for the force F when a mass m is accelerated at a rate of a:

$$F = ma = \text{kg}\frac{\text{m}}{\text{s}^2}.$$

Therefore this unit is called a newton ($\text{N} = \text{kg}\frac{\text{m}}{\text{s}^2}$). Sometimes the unit *kilogram-force* (kg_f) is used because the gravitational pull of 1-kg mass at sea level is 1 kg_f. If we approximate the gravitational acceleration as $g = 9.8\ \text{m/s}^2$, then

$$1\,\text{kg}_\text{f} = 9.8\,\text{N}.$$

The pressure, which is the force per unit area, is measured with the previous units,

$$p = \frac{F}{S} = \frac{\text{kg} \cdot \dfrac{\text{m}}{\text{s}^2}}{\text{m}^2} = \frac{\text{N}}{\text{m}^2} = 1\ \text{Pascal (Pa)};$$

this unit is named for the French scientist Blaise Pascal (1623–1662). Sometimes an atmosphere (atm) is used to measure pressure, and this unit is about 1 kg_f/cm^2, or, more accurately,

$$1 \text{ atm} = 1.013 \times 10^5 \text{ N/m}^2.$$

In the following sections we discuss some of the more important fluid properties along with the units used to quantify them. In reality, there are a large number of engineering units, and a list of the most common ones is provided in Appendix A.

1.4 Fluid Dynamics and Fluid Properties

Fluid dynamics is the science dealing with the motion of fluids. Fluids, unlike solids, cannot assume a fixed shape under load and will immediately deform. For example, if we place a brick in the backyard pool it will sink because the fluid below is not rigid enough to hold it.

Both gases and liquids behave similarly under load and both are considered fluids. A typical engineering question that we'll try to answer here is this: What are the forces that are due to fluid motion? Examples could focus on estimating the forces required for propelling a ship or for calculating the size and shape of a wing required for lifting an airplane. So let us start with the first question: What is a fluid?

As noted, in general, we refer to liquids and gases as fluids, but we can treat the flow of grain in agricultural machines, a crowd of people leaving a large stadium, or the flow of cars by using the principles of fluid mechanics. Therefore one of the basic features is that we can look at the fluid as a continuum and not analyze each element or molecule (hence the analogy to grain or seeds). The second important feature of fluids is that they deform easily, unlike solids. For example, a static fluid cannot resist a shear force and the particles will simply move. Therefore, to generate shear force, the fluid must be in motion. This is clarified in the following subsections.

1.4.1 Continuum

Most of us are acquainted with Newtonian mechanics, and therefore it would be natural for us to look at particle (or group of particles) motion and discuss their dynamics by using the same approach used in courses such as dynamics. Although this approach has some followers, let us first look at some basics.

Consideration a: The number of molecules is very large and it would be difficult to apply the laws of dynamics, even when a statistical approach is used. For example, the number of molecules in one gram-mole (1 g mole) is called the Avogadro number (after the Italian scientist, Amadeo Avogadro, 1776–1856). 1 g mole is the molecular weight multiplied by 1 g. For example, for a hydrogen molecule (H_2) the molecular weight is 2; therefore 2 g of hydrogen are 1 g mole. The Avogadro number N_A is

$$N_A = 6.02 \times 10^{23} \text{molecules/g mole}. \tag{1.1}$$

Because the number of molecules is very large, it is easier for us to assume a continuous fluid rather than to discuss the dynamics of each molecule or even their dynamics by using a statistical approach.

Consideration b: In gases, which we can view as the least condensed fluids, the particles are far from each other, but as Brown (Robert Brown, botanist, 1773–1858) observed in 1827, the molecules are constantly moving, and hence this phenomenon is called Brownian motion. The particles move at various speeds and in arbitrary directions, and the average distance between particle collisions is called the mean free path λ, which for standard air is about 6×10^{-6} cm. Now, suppose that a pressure disturbance (or a jump in the particle velocity) is introduced; this effect will be communicated to the rest of the fluid by the preceding interparticle collisions. The speed that this disturbance spreads in the fluid is called the speed of sound, and this gives us an estimate about the order of molecular speeds (the speed of sound is about 340 m/s in air at 288 K). Of course, many particles must move faster than this speed because of the three-dimensional (3D) nature of the collisions (see Section 1.6). It is only logical that the speed of sound depends on temperature because temperature is related to the internal energy of the fluid. If this molecular mean-free-path distance λ is much smaller than the characteristic length L in the flow of interest (e.g., $L \sim$ the chord of an airplane's wing) then, for example, we can consider the air (fluid) as a *continuum*! In fact, a nondimensional number, called the Knudsen number (after the Danish scientist Martin Knudsen, 1871–1949), exists based on this relation:

$$\mathrm{Kn} = \frac{\lambda}{L}. \tag{1.2}$$

Thus, if $\mathrm{Kn} < 0.01$, meaning that the characteristic length is 100 times larger than the mean free path, then the continuum assumption may be used. Exceptions for this assumption of course would be when the gas is very rare ($\mathrm{Kn} > 1$), e.g., in vacuum or at very high altitudes in the atmosphere.

Therefore, if we agree on the concept of a continuum, we do not need to trace individual molecules (or groups of molecules) in the fluid but rather we should observe the changes in the average properties. Apart from properties such as density or viscosity, the fluid flow may have certain features that must be clarified early on. Let us first briefly discuss frequently used terms such as laminar and turbulent and attached and separated flows, and then focus on the properties of the fluid material itself.

1.4.2 Laminar and Turbulent Flows

Now that, by means of the continuum assumption, we have eliminated the discussion about arbitrary molecular motion, a somewhat similar but much larger-scale phenomenon must be discussed. For the discussion let us assume a free-stream flow along the x axis with uniform velocity U. If we follow the traces made by several particles in the fluid we would expect to see parallel lines, as shown in the upper part of Fig. 1.1. If, indeed, these lines are parallel and follow the direction of the average velocity and the motion of the fluid seems to be "well organized," then this flow is called laminar. If we consider a velocity vector in a Cartesian system,

$$\vec{q} = (u, v, w), \tag{1.3}$$

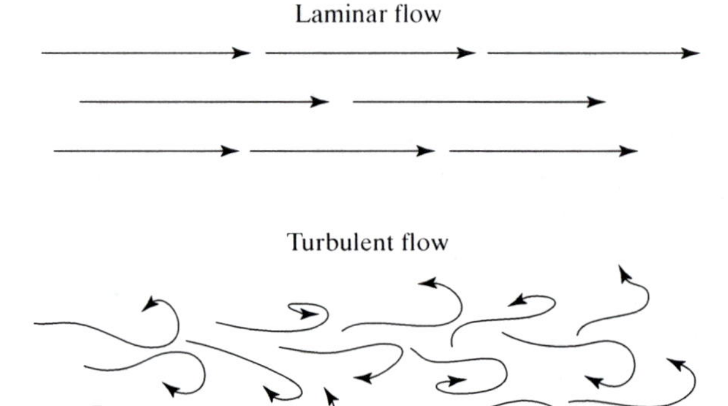

Figure 1.1. Schematic description of laminar and turbulent flows having the same average velocity.

then for this steady-state flow the velocity vector will be

$$\vec{q} = (U, 0, 0), \tag{1.3a}$$

and here U is the velocity in the x direction.

On the other hand, it is possible to have the same average speed in the flow, but in addition to this average speed the fluid particles will momentarily move in the other directions (lower part of Fig. 1.1). The fluid is then called turbulent (even though the average velocity U_{av} could be the same for both the laminar and turbulent flows). In this two-dimensional (2D) case the flow is time dependent everywhere, and the velocity vector then becomes

$$\vec{q} = (U_{av} + u', v', w'), \tag{1.4}$$

where u', v', and w' are the perturbations in the x, y, and z directions. Also, it is clear that the average velocities in the other directions are zero:

$$V_{av} = W_{av} = 0.$$

So if a simple one-dimensional (1D) laminar flow transitions into a turbulent flow, then it also becomes 3D (not to mention time dependent). Knowing whether the flow is laminar or turbulent is very important for most engineering problems because features such as friction and momentum exchange can change significantly between these two types of flow. The fluid flow can become turbulent in numerous situations such as inside long pipes or near the surface of high-speed vehicles.

1.4.3 Attached and Separated Flows

By observing several streamline traces in the flow (by injecting smoke, for example), we can see if the flow follows the shape of an object (e.g., a vehicle's body) close to its surface. When the streamlines near the solid surface follow exactly the shape of the body [as in Fig. 1.2(a)], the flow is considered to be attached. If the flow does not follow the shape of the surface [as seen behind the vehicle in Fig. 1.2(b)], then the flow is considered detached or separated. Usually such separated flows behind the vehicle will result in an unsteady wake flow, which can be felt up to large distances behind the vehicle. Also, in Fig. 1.2(b) the flow is attached on the upper surface and is separated only behind the vehicle. As we shall see later, having attached flow fields is extremely important because the vehicle with the larger areas of flow separation

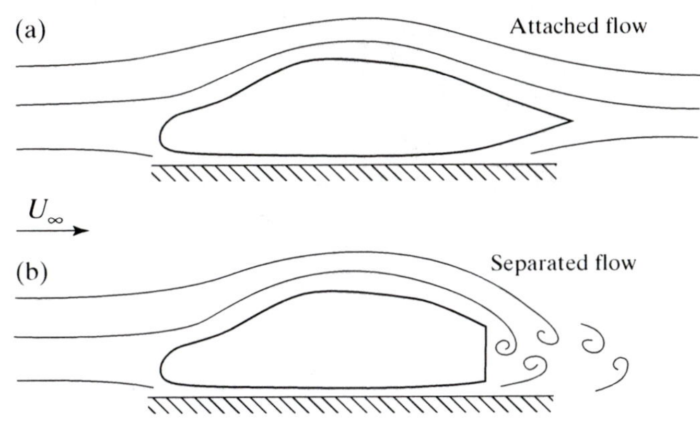

Figure 1.2. (a) Attached flow over a streamlined car and (b) the locally separated flow behind a more realistic automobile shape.

is likely to experience higher resistance (drag). Now, to complicate matters, if the flow above this model is turbulent then, because of the momentum influx from the outer fluid layers, the flow separation can be delayed.

1.5 Properties of Fluids

Fluids, in general, may have many properties related to thermodynamics, mechanics, or other fields of science. In the following subsections, only a few, which are used in introductory fluid mechanics, are mentioned.

1.5.1 Density

Density, by definition, is mass per unit volume. In the case of fluids, we can define the density (with the aid of Fig. 1.3) as the limit of this ratio when a measuring volume V shrinks to zero. We need to use this definition because density can change from one point to the other. Also in this picture, we can relate to a volume element in space that we can call "control volume," which moves with the fluid or can be stationary (in any case it is better to place this control volume in inertial frames of reference).

Therefore the definition of density at a point is

$$\rho = \lim_{V \to 0} \left(\frac{m}{V} \right). \tag{1.5}$$

Typical units are kilograms per cubic meter (kg/m^3) or grams per cubic centimeter (g/cm^3).

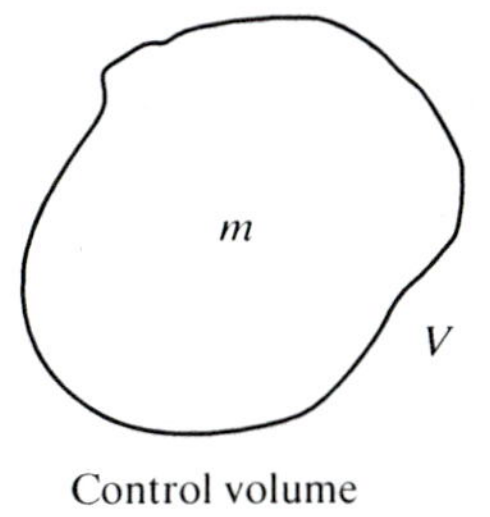

Figure 1.3. Mass m in a control volume V. Density is the ratio of m/V.

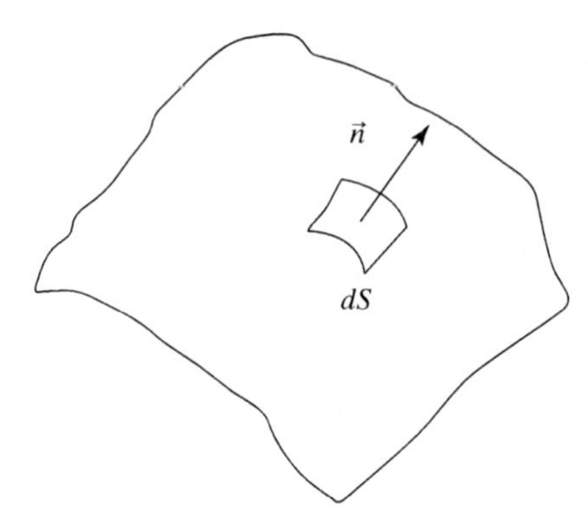

Figure 1.4. Pressure acts normal to the surface dS ($\vec{n}$ is the unit vector normal to the surface).

1.5.2 Pressure

We can describe the pressure p as the normal force F per unit area acting on a surface S. Again, we use the limit process to define pressure at a point, as it may vary on a surface:

$$p = \lim_{S \to 0} \left(\frac{F}{S} \right). \tag{1.6}$$

Bernoulli pictured the pressure as being a result of molecules impinging on a surface (so this force per area is a result of the continuous bombardment of the molecules). Therefore, the fluid pressure acting on a solid surface is normal to the surface, as shown in Fig. 1.4. Consequently the force direction is obtained by multiplying with the unit vector $\vec{n}$ normal to the surface. Because the pressure acts *normal* to a surface the resulting ΔF force is

$$\Delta F = -p\vec{n}\, ds. \tag{1.7}$$

Here the minus sign is a result of the normal unit vector pointing outside the surface while the force that is due to pressure points inward. Also note that the pressure at a point inside a fluid is the same in all directions. This property of the pressure is called *isetropic*. The observation about the fluid pressure at a point acting equally in any arbitrary direction was documented first by Blaise Pascal (1623–1662).

The units used for pressure were introduced in Section 1.3. However, the pascal is a small unit; the units used more often are the kilopascal (kP), the atmosphere (atm), or the bar (bar has no abbreviation; hence the correct use is: 1 bar or 5 bars):

$$1\,\text{kP} = 1000\frac{\text{N}}{\text{m}^2}, \qquad 1\,\text{atm} = 101{,}300\frac{\text{N}}{\text{m}^2}, \qquad 1\,\text{bar} = 100{,}000\frac{\text{N}}{\text{m}^2}.$$

1.5.3 Temperature

Temperature is a measure of the internal energy at a point in the fluid. Over the years different methods have evolved to measure temperature; for example, the freezing point of water is considered zero in the Celsius system and the boiling temperature of water under standard conditions is 100 °C. Kelvin units (K) are similar to Celsius; however, they measure the temperature from absolute zero, a temperature found in space, and they represent a condition when molecular motion will stop. The relation between the two temperature-measuring systems is

$$K = 273.16 + {}^{\circ}C. \tag{1.8}$$

The Celsius system is widely used in European countries whereas in the United States, Fahrenheit units are still used. In this case, 100 °F was set to be close to the human body's temperature. The conversion between these temperature-measuring systems is

$$^{\circ}C = 5/9(^{\circ}F - 32), \tag{1.9}$$

which indicates that $0\,^{\circ}C = 32\,^{\circ}F$. The absolute temperature in these units is in Rankine units (°R) and this scale is higher by 459.69°:

$$^{\circ}R = 459.69 + {}^{\circ}F. \tag{1.10}$$

Now that we have introduced density, pressure, and temperature, it is important to recall the ideal-gas relation, in which these properties are linked together by the gas constant R:

$$p/\rho = RT. \tag{1.11}$$

If we define v as the volume per unit mass then $v = 1/\rho$, and we can write

$$pv = RT. \tag{1.12}$$

However, R is different for various gases or for their mixtures, but it can be easily calculated with the universal gas constant $\mathcal{R}$ ($\mathcal{R} = 8314.3$ J/mol K). Then we can find R by dividing this universal $\mathcal{R}$ by the average molecular weight M of the mixture of gases.

EXAMPLE 1.1. THE IDEAL-GAS FORMULA. As an example, for air we can assume $M = 29$ and therefore

$$R = \mathcal{R}/M = 8314.3/29 = 286.7\,\text{m}^2/(\text{s}^2\ \text{K}) \text{ for air.} \tag{1.13}$$

Suppose we want to calculate the density of air when the temperature is 300 K and the pressure is 1 kg_f/cm^2:

$$\rho = p/RT = 1 \times 9.8 \times 10^4/286.7 \times 300 = 1.139\,\text{kg/m}^3.$$

Here we used $1\,kg_f/cm^2 = 9.8 \times 10^4\,N/m^2$ and $g = 9.8\,m/s^2$.

Another interesting use of the universal gas constant is when we can calculate the volume (V) of 1 g mole of gas in the following conditions (e.g., $T = 300$ K and $p = 1$ atm $= 101{,}300\,N/m^2$). For air we should take 29 g because $M = 29$ and therefore $\mathcal{R}$ is multiplied by 10^{-3} because we considered 1 g mole and not 1 kg mole:

$$V = \mathcal{R}T/p = 8314.3 \times 10^{-3} \times 300/101{,}300 = 24.62 \times 10^{-3}\,\text{m}^3 = 24.62\,\text{L}.$$

Note that 1 g mole of any gas will occupy the same volume because we have the same number of molecules (as postulated by Avogadro). Also, L is one liter ($= 0.001\,m^3$).

1.5.4 Viscosity

The viscosity is a very important property of fluids, particularly when fluid motion is discussed. In fact, the schematic diagram of Fig. 1.5 is often used to demonstrate the

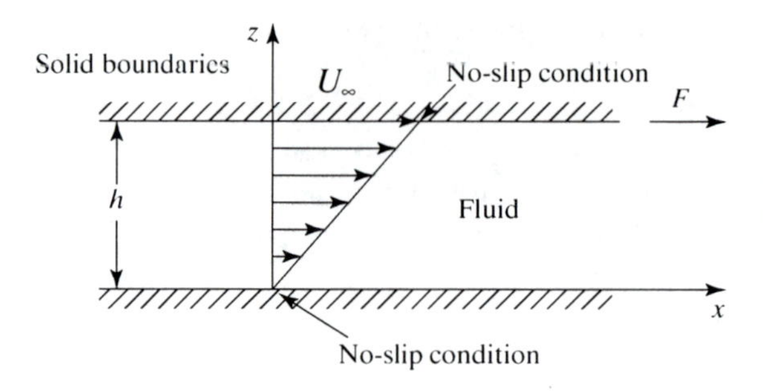

Figure 1.5. The flow between two parallel plates. The lower is stationary while the upper moves at a velocity of U_∞.

difference between solids and fluids. A fluid must be in motion in order to generate a shear force, whereas a solid can support shear forces in a stationary condition.

In this figure the upper plate moves at a velocity of U_∞ while the lower surface is at rest. A fluid is placed between these parallel plates, and when the upper plate is pulled, a force F is needed. At this point we can make another important observation. The fluid particles in immediate contact with the plates will not move relative to the plates (as if they were glued to it). This is called the *no-slip boundary condition*, and we will use this in later chapters. Consequently we can expect the upper particles to move at the upper plate's speed while the lowest fluid particles attached to the lower plate will be at rest. Newton's law of friction states that

$$\tau = \mu \frac{dU}{dz}. \tag{1.14}$$

Here τ is the shear force per unit area (shear stress) and μ is the fluid viscosity. In this case the resulting velocity distribution is linear and the shear will be constant inside the fluid (for $h > z > 0$). For this particular case we can write

$$\tau = \mu \frac{U_\infty}{h}. \tag{1.15}$$

A fluid that behaves like this is called a Newtonian fluid, indicating a linear relation between the stress and the strain. As noted earlier, this is an important property of fluids because without motion there is no shear force.

The units used for τ are force per unit area, and the units for the viscosity μ are defined by Eq. (1.14). Some frequently used properties of some common fluids are provided in Table 1.1.

Table 1.1. *Approximate properties of some common fluids at 20 °C (ρ = density, μ = viscosity, σ = surface tension)*

Fluid	ρ (kg/m^3)	μ (N s/m^2)	σ (N/m)
Air	1.22	1.8×10^{-5}	
Helium	0.179	1.9×10^{-5}	
Gasoline	680	3.1×10^{-4}	2.2×10^{-2}
Kerosene	814	1.9×10^{-3}	2.8×10^{-2}
Water	1000	1.0×10^{-3}	7.3×10^{-2}
Sea water	1030	1.2×10^{-3}	7.3×10^{-2}
Motor oil (SAE 30)	919	0.29	3.6×10^{-2}
Glycerin	1254	0.62	6.3×10^{-2}
Mercury	13600	1.6×10^{-3}	4.7×10^{-1}

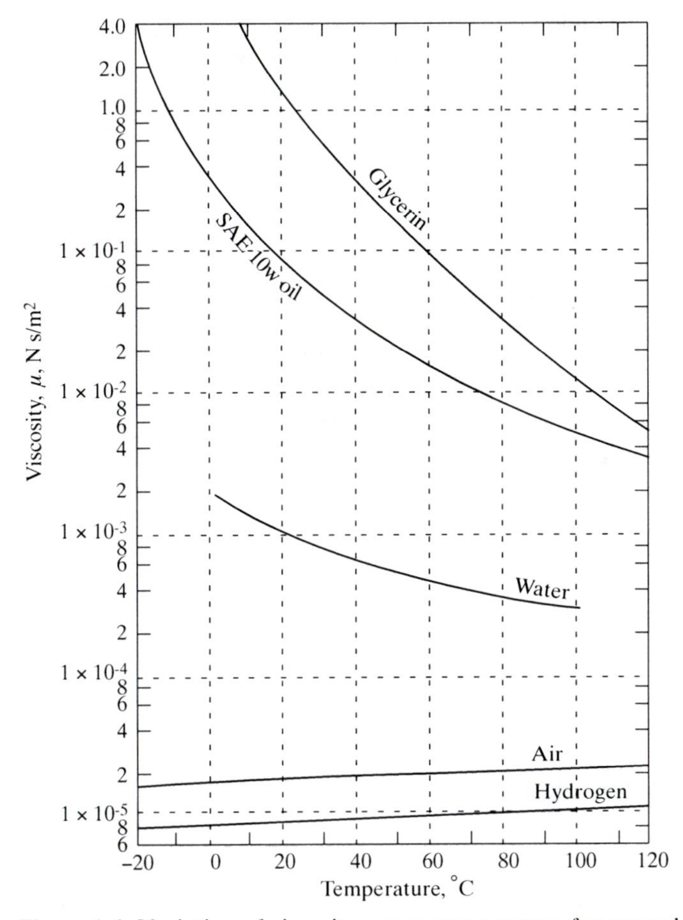

Figure 1.6. Variation of viscosity versus temperature for several fluids.

Also note that the viscosity of most fluids depends on the temperature, and this is shown for several common fluids in Fig. 1.6.

EXAMPLE 1.2. THE UNITS OF SHEAR. To demonstrate the units of shear let us calculate the force required to pull a plate floating on a 2-cm-thick layer of SAE 30 oil at $U_\infty = 3$ m/s.

Taking the value of the viscosity from Table 1.1, we have $\mu = 0.29$ kg/m s at 20 °C. Thus,

$$\tau = 0.293 \frac{3}{0.02} = 43 \,\text{kg/m s}^2 = 43 \,\text{N/m}^2.$$

Sometimes the ratio between viscosity and the density is denoted as ν, the *kinematic viscosity*. Its definition is

$$\nu = \frac{\mu}{\rho}. \tag{1.16}$$

1.5.5 Specific Heat

Fluids have several thermodynamic properties, and only two, related to heat exchange, are mentioned here. For example, if heat Q is added in a constant-pressure process to a mass m, then the relation between temperature change and heat is stated by the simple formula

$$Q = mc_p \Delta T. \tag{1.17}$$

Here c_p is the specific-heat coefficient used in a constant-pressure process. However, if the fluid is not changing its volume during the process, then c_v is used for the specific heat in this constant-volume process:

$$Q = mc_v \Delta T. \tag{1.18}$$

The ratio between these two specific-heat coefficients is denoted by

$$\gamma = \frac{c_p}{c_v}. \tag{1.19}$$

The heat (energy) required for raising the temperature of 1 g of water by 1 °C is called a *calorie* (cal). Therefore, the units for c_p or c_v are $\frac{\text{cal}}{\text{kg}\,^\circ\text{C}}$ and 1 cal = 4.2 J (J = joule). Work in mechanics is force times distance, and therefore units of 1 J are

$$1\ \text{J} = \text{kg}\frac{\text{m}}{\text{s}^2}\text{m} = \text{kg}\frac{\text{m}^2}{\text{s}^2}.$$

Also, for an ideal gas undergoing an adiabatic process, the two heat capacities relate to the gas constant R (see [1, p. 90]):

$$c_p - c_v = R. \tag{1.20}$$

1.5.6 Heat Transfer Coefficient *k*

Heat transfer can take several forms, such as conduction, convection, or radiation. Because this introductory text does not include heat transfer only one basic mode of heat transfer called conduction is mentioned. The elementary 1D model is depicted by Fig. 1.7, where the temperature on one side of the wall is higher than that on the other side. The basic heat transfer equation for this case, called the Fourier equation, states that the heat flux is proportional to the area A, the temperature gradient, and the coefficient k, which depends on the material through which the heat is conducted:

$$Q = -kA\frac{dT}{dx}. \tag{1.21}$$

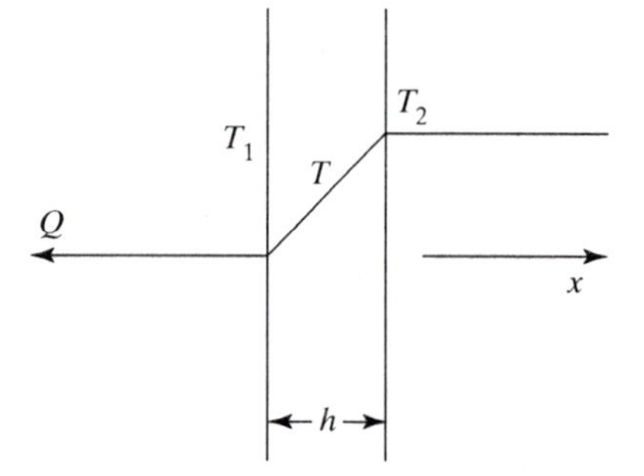

Figure 1.7. Conductive heat transfer through a wall of thickness h.

For the case in Fig. 1.7, we could state that the heat flux is

$$Q = -kA\frac{T_2 - T_1}{h}. \tag{1.21a}$$

Here T_2 is larger that T_1, and the minus sign indicates that the heat flux is in the left direction. Also note that the temperature distribution in the wall is linear. The units for k are defined by Eq. (1.21) as $\frac{\text{cal}}{\text{m}^2\,{}^\circ\text{C}}$.

1.5.7 Surface Tension σ

Molecules in a liquid attract each other equally, but near the surface this equilibrium is changed, resulting in the surface tension σ. A simplistic visualization of this effect is provided in Fig. 1.8. The molecule, deep in the fluid, is attracted equally by its neighbors and therefore is in equilibrium. However, on the surface (as shown), the vertical force is not balanced and this inward force is balanced by the surface tension, very much like a thin cover stretched over the surface of a pool. Near the wall this surface tension may change direction and pull the surface up or down, depending on the wetting angle.

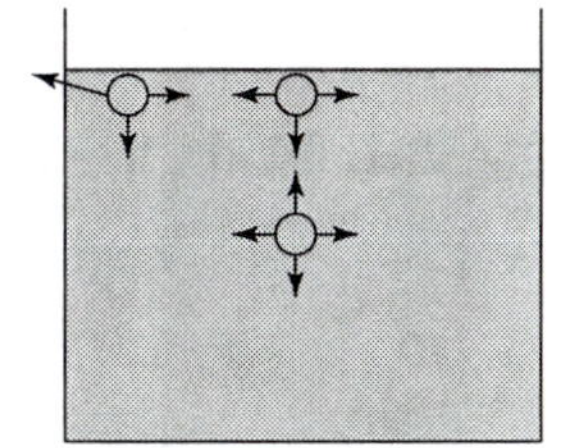

Figure 1.8. Schematic description of the surface tension in a liquid.

This wetting angle or the *contact angle* θ is used to define the direction of the force. An example to demonstrate surface tension is shown in Fig. 1.9. Here a movable link is closing the open end of a U-shaped wire. After dipping it, say in soapy water, a film is visible. A load F may be applied, as shown, to measure the magnitude of the surface tension σ, which has units of force per unit length.

We can therefore estimate the total force resulting from the surface tension by multiplying the contact length L by the surface tension σ:

$$F = \sigma L. \tag{1.22}$$

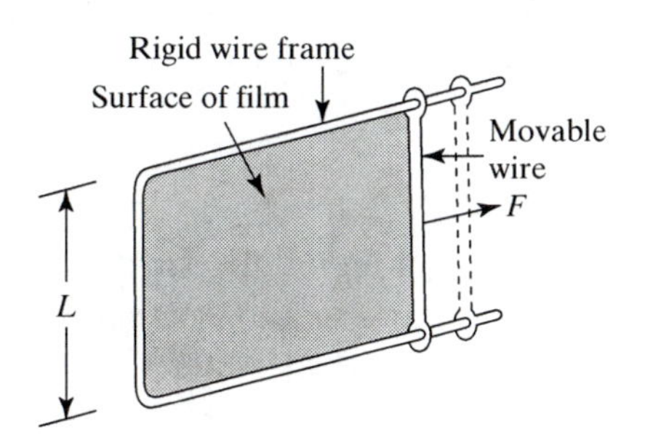

Figure 1.9. A film of liquid forming inside a wire frame.

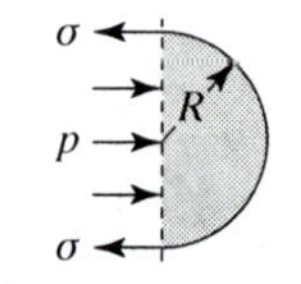

(a) Spherical droplet

(b) Spherical bubble

Figure 1.10. The forces on one half of a bubble or droplet.

Typical values for the surface tension for an air–water surface at standard conditions are $\sigma = 0.073$ N/m and for an air–mercury surface at standard conditions $\sigma = 0.48$ N/m. The surface tension for several other fluids is listed in Table 1.1.

EXAMPLE 1.3. PRESSURE INSIDE A LIQUID BUBBLE OR DROPLET. Let us estimate the pressure (difference) inside a liquid bubble. We do this by imaginarily splitting the droplet into two halves and estimating the forces that must be in equilibrium. The surface tension force around the circular opening, pulling the hemisphere of radius R to the left (Fig. 1.10), is

$$\sigma 2\pi R.$$

The force that is due to pressure inside the hemisphere, pushing to the right, is

$$\pi R^2 \Delta p.$$

Because the hemisphere is in equilibrium, those two forces are equal:

$$\sigma 2\pi R = \pi R^2 \Delta p,$$

or

$$\Delta p = 2\sigma / R. \tag{1.23}$$

Therefore the pressure difference depends on the radius R when σ is given. In the case of a droplet, the smaller ones will have higher pressure inside, and as the radius is increased the droplet becomes less stable, so there is a limit to the size of possible droplets. In the case of the bubble, if the pressure inside increases, the bubble stretches and becomes larger; again, for very large bubbles, the stress holding them together is too low and they will disintegrate.

EXAMPLE 1.4. PRESSURE DIFFERENCE IN A SOAP-WATER BUBBLE. As an example, calculate the pressure difference for a 10-cm-diameter spherical water and soap bubble. The surface tension of the water and soap mixture is actually much less than for clean water; let us assume $\sigma = 2.5 \times 10^{-2}$ N/m. By the way, the long molecules of the soap stabilize the bubble and reduce evaporation. Using Eq. (1.23), we get

$$\Delta p = \frac{2\sigma}{R} = \frac{2 \times 2.5 \times 10^{-2}\,\text{N/m}}{5 \times 10^{-2}\,\text{m}} = 1.0 \frac{\text{N}}{\text{m}^2}.$$

Another important parameter related to the surface tension is the contact angle (or wetting angle) θ, as shown in Fig. 1.11.

For a glass–water surface in air and at standard conditions, $\theta = 0°$. For a glass–mercury surface in air and at standard conditions, $\theta = 130°$.

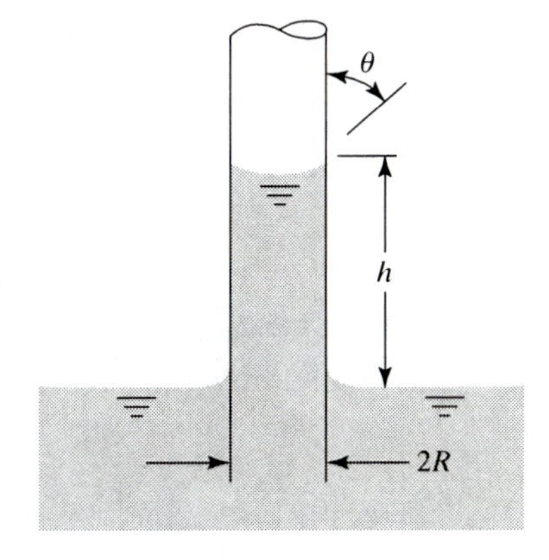

Figure 1.11. The effect of surface tension when a capillary tube is dipped in a fluid container. θ is the contact angle.

To demonstrate the effect of surface tension and wetting angle, let us consider a case in which a capillary tube is inserted vertically into a container filled with a liquid (Fig. 1.11) and the parameters θ and σ are known.

Let us observe the vertical force balance. This force that is due to the surface tension on the upper, circular contact area is

$$\sigma 2\pi R \cos\theta$$

The weight of the raised liquid column is

$$\rho g \pi R^2 h.$$

Because the liquid in the column is in equilibrium, we can write

$$\sigma 2\pi R = \rho\pi R^2 h,$$

and by solving for h we get

$$h = \frac{2\sigma \cos\theta}{\rho g R}. \tag{1.24}$$

This means, for example, that for tubes with smaller radius R the liquid will rise higher (in the case of $\theta < 90°$). Also note that for $\theta < 90°$ the liquid inside the tube will rise while a capillary depression will be observed (the level inside the tube will be lower then in the outer container) for $\theta > 90°$. If a liquid droplet is positioned on a surface, then this last condition is called a *nonwetted surface*, as illustrated in Fig. 1.12b. Along the same line, for $\theta < 90°$ we may call the surface a *wetted*.

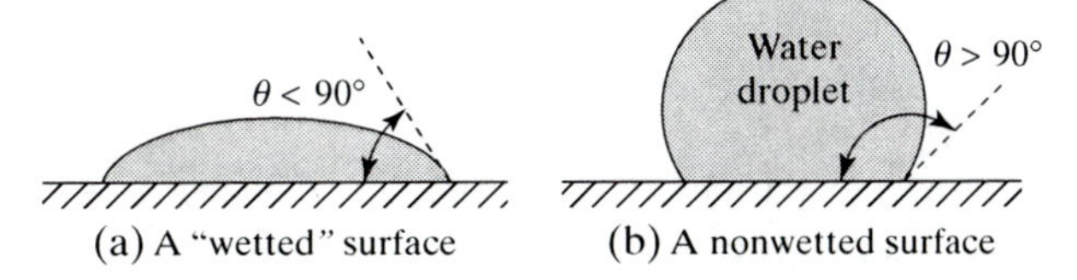

Figure 1.12. Effect of the wetting angle on the shape of a droplet placed on a solid surface.

EXAMPLE 1.5. CAPILLARY EFFECT. For a numerical example, calculate the rise of water in a capillary tube. Consider $R = 1$ mm and from Table 1.1 we get $\sigma = 0.073$ N/m, $\theta = 0°$, and water density $\rho = 1000$ kg/m^3. Then, using Eq. (1.24), we get

$$h = \frac{2 \times 0.073 \times \cos 0}{1000 \times 9.8 \times 0.001} = 0.015\,\text{m},$$

or 15 mm. Now we continue the same example but with a nonwetted condition and assume mercury is the fluid. Then from Table 1.1 we get $\sigma = 0.48$ N/m, $\theta = 130°$, and density $\rho = 13{,}600$ kg/m^3. Then, using Eq. (1.24), we get

$$h = \frac{2 \times 0.48 \times \cos 130}{13{,}600 \times 9.8 \times 0.001} = -0.0046\,\text{m},$$

or 4.6 mm. Note that the negative sign means that the liquid in the tube is below the liquid level outside the tube.

These effects seem small, but when a liquid column is used for pressure measurements (as in some manometers) then these small differences that are due to surface tension must be accounted for.

1.5.8 Modulus of Elasticity *E*

The modulus of elasticity E is a measure of compressibility. It can be defined as

$$E = dp/(dV/V) \quad \text{or} \quad dp = E(dV/V). \tag{1.25}$$

The second form indicates how much pressure is needed to compress a material having a modulus of E. Also, the change in volume is directly related to the change in density, and we can write

$$d\rho/\rho = dV/V. \tag{1.26}$$

And by substituting dV/V for $d\rho/\rho$ in Eq. (1.25) we get

$$E = dp/(d\rho/\rho). \tag{1.27}$$

Most liquids are not very compressible, but gases are easily compressed and for an ideal gas we already introduced this relation [in Eq. (1.11)]:

$$dp/d\rho = RT. \tag{1.28}$$

Therefore, substituting Eq. (1.28) into Eq. (1.27) results in (for an ideal gas)

$$E = \rho RT. \tag{1.29}$$

The units for E, based on Eq. (1.25), are $\dfrac{\text{N/m}^2}{\text{m}^3/\text{m}^3} = \text{N/m}^2$.

EXAMPLE 1.6. COMPRESSIBILITY OF A LIQUID. For this example, let us consider the compressibility of seawater. The modulus of elasticity is $E = 2.34 \times 10^9$ N/m^2, and let us evaluate the change in volume at a depth of 1 km. The change in pressure at a 1000-m depth is

$$dp = \rho g h = 1000 \times 9.8 \times 1000\,\text{N/m}^2,$$
$$dV/V = dp/E = 1000 \times 9.8 \times 1000/2.34 \times 10^9 = 4.188 \times 10^{-3}\,(0.42\%),$$

which is less than half a percent. This shows that water is really incompressible.

It is interesting to point out that compressibility relates to the speed of sound in a fluid. If we use the letter a to denote the speed of sound, then later (Section 10.2) we shall see that

$$a^2 = dp/d\rho. \tag{1.30}$$

For liquids we can use Eq. (1.27) to show that

$$a^2 = dp/d\rho = E/\rho. \tag{1.31}$$

For ideal gases undergoing an adiabatic process (thermally isolated), we get

$$dp/d\rho = \gamma RT. \tag{1.32}$$

Therefore

$$a = \sqrt{\gamma RT}, \tag{1.33}$$

indicating that the speed of sound is a function of the temperature.

EXAMPLE 1.7. THE SPEED OF SOUND. Let us calculate the speed of sound in air at 300 K. Taking the value of R from Eq. (1.13) and assuming $\gamma = 1.4$, we get

$$a = \sqrt{1.4 \times 286.6 \times 300} = 346.9 \text{ m/s for air.}$$

Now, to calculate the speed of sound in water we must use Eq. (1.31). Based on the modulus of elasticity of seawater,

$$a = \sqrt{\frac{E}{\rho}} = \sqrt{\frac{2.34 \times 10^9}{1000}} = 1529 \text{ m/s},$$

and the resulting speed of sound is significantly higher.

1.5.9 Vapor Pressure

Vapor pressure is a property related to the phase change of fluids. One way to describe it is to observe the interface between the liquid and the gas phase of a particular fluid; the vapor pressure indicates that there is an equilibrium between the molecules leaving and joining the liquid phase. The best example is to examine the vapor pressure of water as shown in Fig. 1.13. Because molecular energy is a function of the temperature, it is clear that vapor pressure will increase with temperature. The vapor pressure is zero at 0 °C, and of course is equal to 1 atm at 100 °C, which is the standard boiling point of water.

In later chapters we shall see that the pressure can change in a moving fluid. So even if there is no temperature change, there could be a situation in which the pressure in the fluid falls below the vapor pressure. The result is the formation of bubbles, because at this condition the liquid will evaporate locally. This phenomenon is called *cavitation*, and Fig. 1.14 shows an example of the bubble trail created near the tip of a propeller (where a tip vortex creates a high local velocity). Cavitation in pumps or fast-moving objects through fluid can result in mechanical damage that is

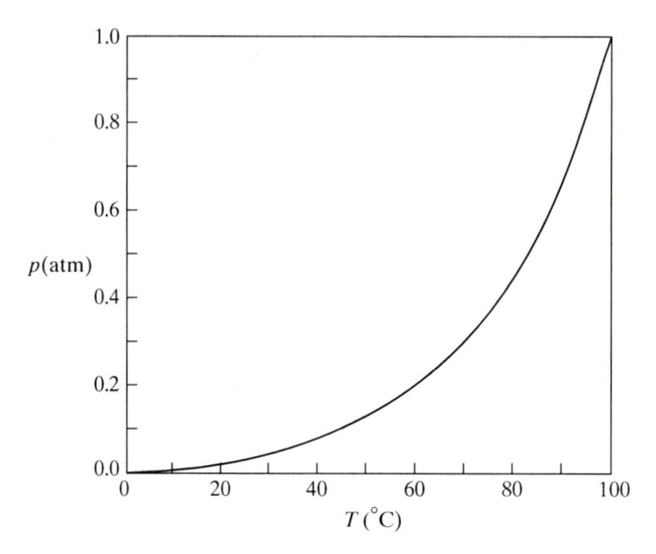

Figure 1.13. Vapor pressure of water versus temperature.

due to pitting of the surface and vibration. In the case of liquid pumps, cavitation will reduce flow rate and performance and will affect the pump efficiency.

Fluids have many more properties such as enthalpy, entropy, internal energy, etc., but they are mostly not used in this text.

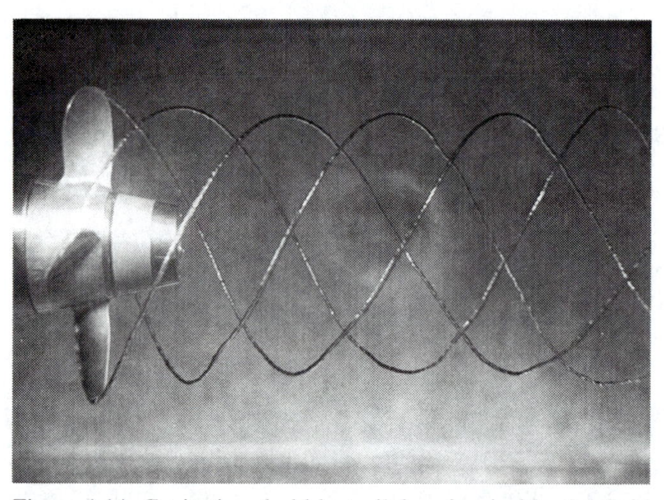

Figure 1.14. Cavitation: bubble trail forming inside the fluid when the local pressure falls below the vapor pressure. In this case the bubbles originate near the propeller tip where the pressure is the lowest. (Courtesy of Garfield Thomas Water Tunnel, Pensylvania State University.)

1.6 Advanced Topics: Fluid Properties and the Kinetic Theory of Gases

Gases were defined earlier as fluids in which the molecules can move freely and are far from each other, occasionally colliding with each other. This model led Daniel Bernoulli in 1738 to explain pressure in gases based on this type of molecular motion. Bernoulli considered a cylindrical container, filled with gas, as shown in Fig. 1.15. As the molecules move inside the container, they also impinge on the walls, as shown in the figure. Now we may neglect the intermolecular collisions and

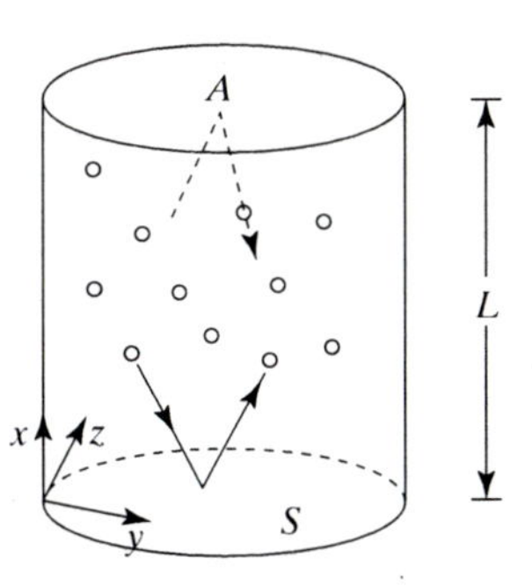

Figure 1.15. Gas molecules moving randomly inside a container.

assume that when a molecule hits the wall it will bounce back without losses (elastic collision). This assumption also includes pure elastic collisions with the sides of the cylinder.

Therefore the total forces that are due to these collisions must produce the pressure on the container's walls. For example, the particle in Fig. 1.15 (marked by the letter A) hitting the top has a velocity of

$$q = (u, v, w), \tag{1.34}$$

and when it hits the top, the change in its linear momentum in the x direction is

$$2mu,$$

where the 2 is a result of the elastic collision, m is the mass of the molecule, and u is the velocity component in the x direction. Because the particle is contained inside the cylinder and is continuously bouncing back and forth, we can estimate the time Δt between these collisions on the upper wall by

$$\Delta t = 2L/u,$$

where L is the length of the cylinder. The force that is due to the collisions of this particle, based on Newton's momentum theory, is

$$F = \frac{\Delta(mu)}{\Delta t} = \frac{2mu}{2L/u} = mu^2/L. \tag{1.35}$$

Now recall that the particles are likely to move at the same speed in any direction and

$$q^2 = u^2 + v^2 + w^2;$$

if all directions are of the same order of magnitude we can assume that

$$q^2 \approx 3u^2.$$

Now suppose that there are N particles in the container; therefore the force that is due to the inner gas is

$$F = N\frac{mq^2/3}{L}, \tag{1.36}$$

and the pressure is simply the force per unit area,

$$p = \frac{F}{S} = N\frac{mq^2/3}{LS} = \frac{N}{3V}mq^2 \tag{1.37}$$

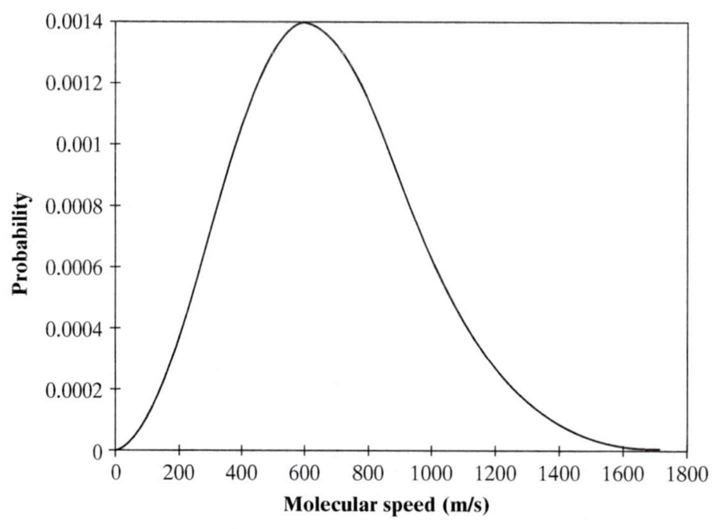

Figure 1.16. Maxwell's universal velocity distribution for the molecules of air ($M = 29$), at 300 K.

where the volume $V = LS$ and S is the cylinder top (or bottom) area. This is a surprisingly simple approach that connects the pressure to the molecular kinetic energy. Now we recall the ideal-gas equation,

$$pV = nRT = \frac{N}{N_A} RT, \tag{1.38}$$

where $n = N/N_A$ is the number of moles in the cylinder [recall that N_A is the Avogadro number in Eq. (1.1)]. By equating these two equations, we solve for the temperature:

$$T = \frac{N_A}{3R} mq^2. \tag{1.39}$$

This simple model shows that for an ideal gas the molecular kinetic energy is proportional to the absolute temperature. This means that at absolute zero the molecular motion will stop, a concept that was not received well in Bernoulli's era. About 100 years later, a Scottish physicist, James Maxwell (1831–1879), revived this theory and introduced a statistical approach. He suggested a universal velocity distribution (Fig. 1.16) that shows the velocity range of molecular motion. Our interest at this point is to demonstrate the magnitude of the molecular velocity, which mainly depends on temperature and molecular weight. The probability is depicted on the ordinate (the y axis) and the probable velocity is on the abscissa (the x axis). Of course, the total area under the curve is always 1 because all particles in the container are included. Note that the average velocity is a bit over the top to the right (468 m/s for air), which is somewhat higher than the speed of sound, mentioned earlier.

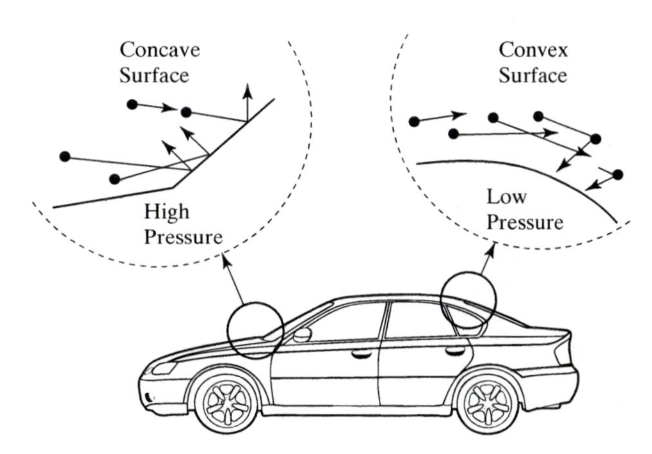

Figure 1.17. Using the kinetic theory of gases, we can explain the high pressure near a concave curvature and the lower pressure near a convex curvature.

Another interesting aspect of this molecular model is that, for flow over bodies, it can explain intuitively the effect of curvature on the pressure distribution. For example, Fig. 1.17 shows a generic automobile that is moving forward at a constant speed. The air molecules are moving toward the car at an average velocity, in addition to their Brownian motion (see Fig. 1.17). At the base of the windshield the number of collisions will increase because the incoming molecules will hit head on and some may even bounce back again because of intermolecular collisions. On the other hand, when observing the flow over a concave surface as shown in the figure (behind the rooftop), we see that the particles will not hit the rear window head on. They will fill the void mainly due to intermolecular collisions. Hence a lower pressure is expected there. We can also guess that the velocity at the base of the windshield (concave surface) slows down whereas the undisturbed particles at the back (convex surface) will accelerate to cover the additional distance created by the void. This generic discussion suggests that the pressure is lower if the velocity is increased in such flows. We shall see later that this observation led to the formulation of the well-known Bernoulli equation.

1.7 Summary and Concluding Remarks

In this introductory chapter the properties of fluids were discussed. The reader must have seen those during earlier studies and the only ones worth mentioning are related to the forces in fluids. The first is of course the pressure, which acts *normal* to a surface and the second is the shear force. The shear stress in a fluid exists only when the fluid is in motion, contrary to solids that can resist shear under static conditions. This situation is created by the *no-slip boundary condition*, which postulates that the fluid particles in contact with a surface will have zero relative velocity at the contact area.

REFERENCE

[1] Karlekar, B. V., *Thermodynamics for Engineers*, Prentice-Hall, Englewood Cliffs, NJ, 1983.

PROBLEMS

1.1. A uniform pressure is acting on a plate 0.5 m tall and 3.0 m wide. Assuming the pressure difference between the two sides of the plate is 0.05 atm, calculate the resultant force.

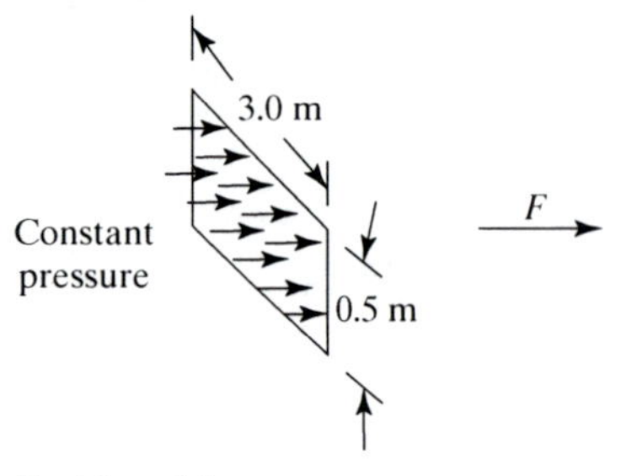

Problem 1.1.

1.2. Identical bricks 0.1 m wide, and weighing 2 kg are placed on a plate (assume the plate has negligible weight). Calculate the total weight and force F required for balancing the plate. How far from $x = 0$ should F be placed so that the plate will not tip over?

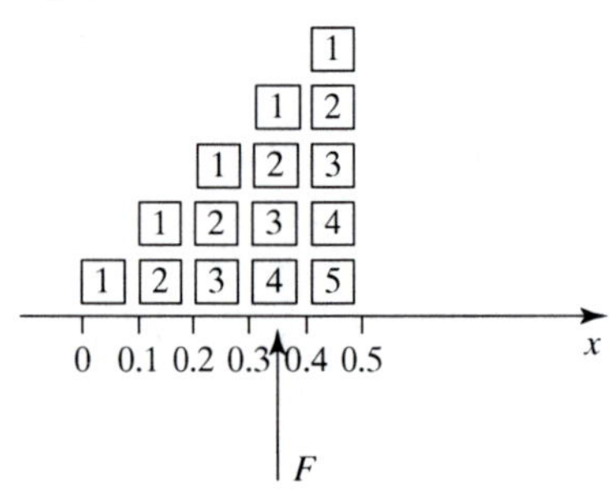

Problem 1.2.

1.3. A linearly varying pressure $[p(x) = P_{\max}/\mathrm{L} \times x]$ is acting on a plate. Calculate the total force (resultant) and how far it acts from the origin. Later we call this the center of pressure.

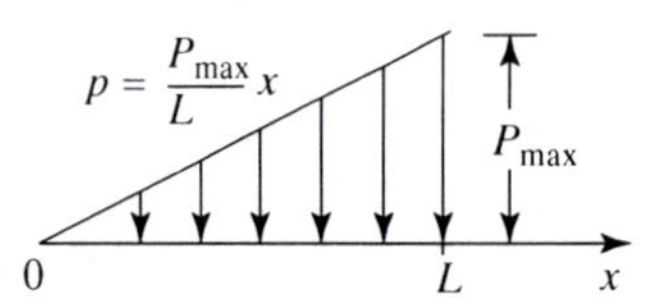

Problem 1.3.

1.4. Suppose that a 1-m^3 metal container holds air at standard conditions ($P = 1$ atm and $T = 300$ K).

(a) Calculate the pressure inside the container if it is heated up to 400 K.
(b) Calculate the density ρ inside the container. $[R = 286.6\,\mathrm{m}^2/(\mathrm{s}^2\,\mathrm{K})]$.

1.5. A 2D velocity field is given by the following formulation:

$$u = \frac{x}{x^2 + z^2}, \quad w = \frac{z}{x^2 + z^2}.$$

Calculate the value of the velocity vector q at a point (1,3).

1.6. On a warm day the thermometer reads 30 °C. Calculate the absolute temperature in degrees Kelvin and also the temperature in degrees Fahrenheit.

1.7. 1 m^3 of air at 1 atm and 300 K is sealed in a container. Calculate the pressure inside the container

(a) if the volume is reduced to 0.5 m^3 but the temperature is cooled off to 300 K,
(b) when the temperature is 350 K, and the volume is still 0.5 m^3.

1.8. A 1-m^3 balloon is filled with helium at an ambient temperature of 30 °C. The pressure inside the balloon is 1.1 atm and outside it is 1.0 atm. The molecular weight of helium is about 4 and that of the surrounding air is about 29. Calculate the weight of the helium inside the balloon. What is the weight of the outside air that has the same volume as the balloon? What is the meaning of this weight difference?

1.9. Usually we check the tire pressure in our car early in the morning when the temperature is cold. Suppose the temperature is 288 K (about 15 °C), the volume of the air inside is 0.025 m^3, the outside air density (at standard conditions) is 1.22 kg/m^3, and the tire pressure gauge indicates a pressure of 2 atm (2 times 1.1013×10^5 N/m^2) above ambient pressure.

(a) What is the tire pressure when the car is left in the summer sunshine and the tire temperature reaches 333 K?
(b) Suppose the tire is inflated with helium ($M \sim 4$) instead of air ($M \sim 29$); then how much weight is saved?

1.10. A 200-cm^3 container is filled with air at standard conditions. Estimate the number of air molecules in the container.

1.11. The temperature inside the container in the previous question was raised to 350 K. Calculate the pressure, density, and the number of air molecules inside the container.

1.12. A 3-m^3 tank is filled with hydrogen at standard conditions. If the molecular weight of helium is 4.0, calculate the mass of the gas inside the tank.

1.13. The tire pressure in a car was measured in the morning, at 280 K, and was found to be 2.5 atm. After a long trip on a warm afternoon the pressure rose to 3.1 atm. Assuming no change in the tire volume, calculate the air temperature inside the tire.

1.14. A flat plate is floating above a 0.05-cm-thick film of oil and is being pulled to the right at a speed of 1m/s (see sketch for Problem 1.15). If the fluid viscosity is 0.4 N s/m^2, calculate the shear force τ on the lower and upper interfaces (e.g., on the floor and below the plate) and at the center of the liquid film.

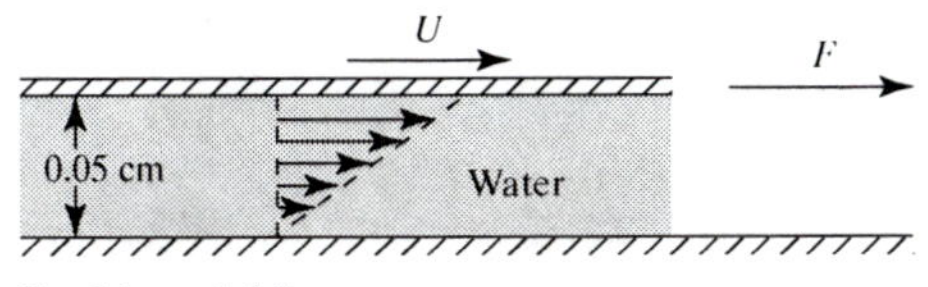

Problem 1.15.

1.15. A flat plate floating on a 0.05-cm-thick water film is pulled by the force F. Calculate F for an area of 1 m^2 and for $U = 1$ m/s [note that for water $\mu = 0.001$ kg/(m s)].

1.16. A flat plate is pulled to the right above a 0.1-cm-thick layer of viscous liquid (see sketch of Problem 1.15) at a speed of 1 m/s. If the force required for pulling the plate is 200 N/1 m^2, calculate the viscosity of the liquid.

1.17. Consider a stationary vertical line in the figure of Problem 1.15 (fixed to the lower surface). Calculate how much water per 1 m width is flowing during 1 s to the right across that line ($U = 1$ m/s).

1.18. The 2D velocity distribution above a solid surface placed on the x coordinate is

$$u = 3z - 3z^3,$$

$$w = 0.$$

Calculate the shear stress on the surface at $z = 0$ and at $z = 0.5$.

1.19. A thin oil film is covering the surface of an inclined plane, as shown in the figure. Develop an expression for the terminal velocity of a block of mass m sliding down the slope. Assume that the oil film thickness and viscosity are known, as well as the incline angle θ and the contact surface area.

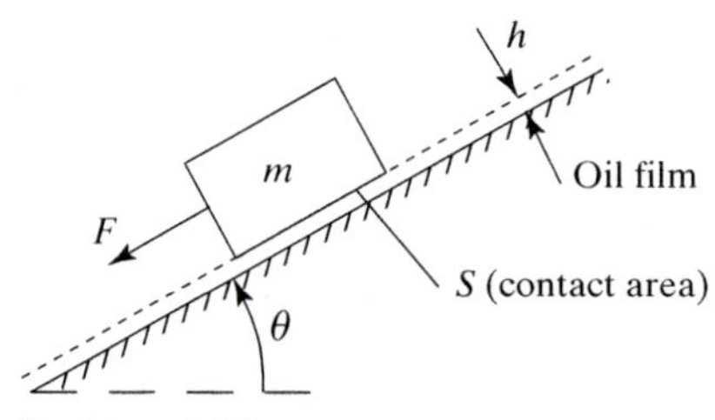

Problem 1.19.

1.20. Calculate the terminal velocity of a 0.2-m wide, 0.3-m long, 5-kg block sliding down an incline of 30°, as shown in the figure for Problem 1.19. Assume the oil film thickness is 1 mm and the oil viscosity (from Table 1.1) is 0.29 N s/m^2.

1.21. The block in the figure of Problem 1.19 slides at a velocity of 2 m/s because of the force F. In this case, however, the slope $\theta = 0$. Calculate the magnitude of the force if the oil film thickness is 1 mm and the oil viscosity (from Table 1.1) is 0.29 N s/m^2.

1.22. A thin plate is pulled to the right between two parallel plates at a velocity U, as shown in the figure. It is separated by two viscous fluids with viscosities μ_1 and μ_2 and the spacing is h_1 and h_2, accordingly. Assuming that the plates are very large, calculate the force per unit area required for pulling the central plate.

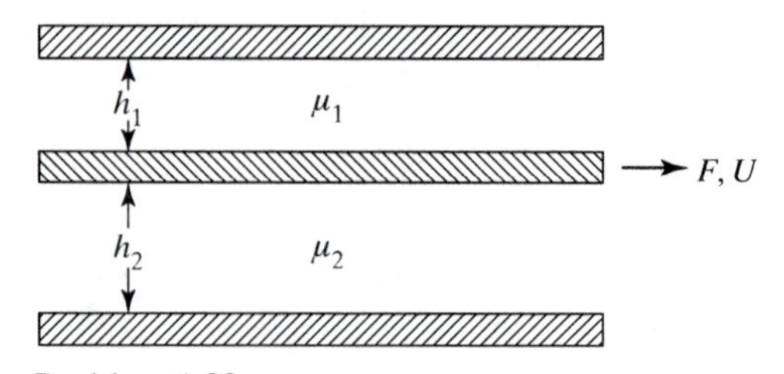

Problem 1.22.

1.23. A small bird with a characteristic length of $L = 0.2$ m flies near the ocean at a speed of 14 m/s. The mean free path of air molecules at sea level is about $\lambda = 6.8 \times 10^{-8}$ m, and the average molecular speed can be estimated as $c = 468$ m/s. Calculate the Knudsen number. Can you assume that the fluid is continuous?

1.24. At standard atmospheric condition the average speed of air molecules is estimated at $c = 468$ m/s (see Fig. 1.16). Calculate the speed of sound for this condition (at $T = 300$ K). Can you explain the large difference?

1.25. An important parameter for grouping different flow regimes (called the Reynolds number) represents the ratio between the actual and the molecular scaling of length times velocity. It can be approximated by the following formulation: $\mathrm{Re} = 2\frac{V}{c}\frac{L}{\lambda}$ (see Section 6.2). Calculate this ratio for the small bird of Problem 1.23 flying at a speed of 14 m/s (recall that $c = 468$ m/s, and $\lambda = 6.8 \times 10^{-8}$ m).

1.26. A 0.3-m wide, 0.5-m long, 10-kg block (m_1) is sliding on a 1 mm thin oil film, pulled by the mass m_2, as shown in the figure. Calculate the terminal velocity by using the oil viscosity from Table 1.1 (0.29 N s/m^2).

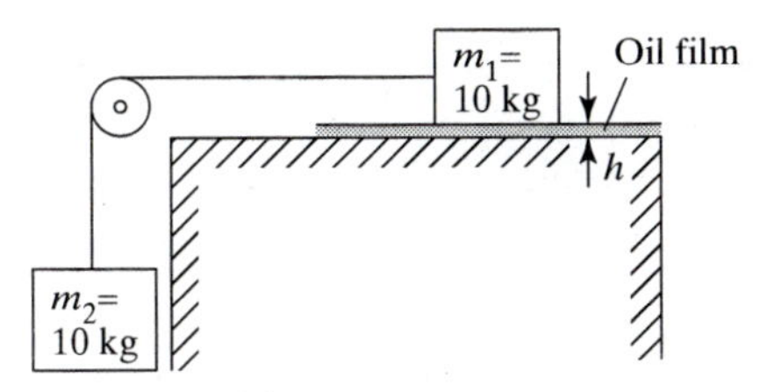

Problem 1.26.

1.27. Calculate the terminal velocity in the previous problem, but now $m_1 = 20$ kg.

1.28. Calculate the terminal velocity in Problem 1.26, but now $m_1 = 20$ kg and $m_2 = 5$ kg.

1.29. The disk shown in the figure rotates at a speed of 50 RPM above a stationary plane separated by an oil film. If the oil viscosity is $\mu = 0.01$ kg/(m s), the spacing between the disk and the stationary surface is 2 mm, and $R = 5$ cm; calculate the torque required for rotating the disk (assume a linear velocity distribution in the gap).

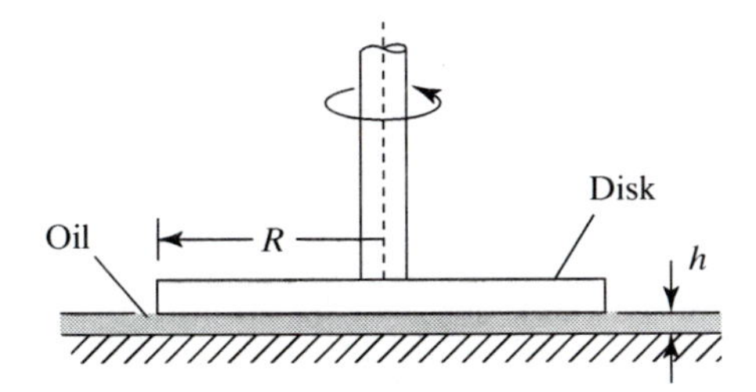

Problem 1.29.

1.30. A rotary damper consists of a disk immersed in a container, as shown in the figure. Assuming that the gap h is the same on both sides and the viscosity μ and the disk radius R are known, calculate the torque required for rotating the disk at a particular RPM.

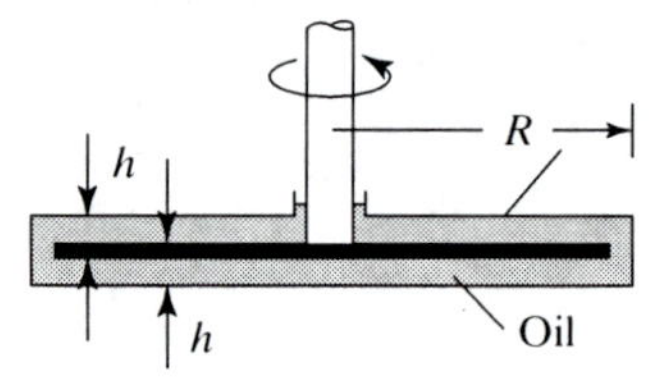

Problem 1.30.

1.31. The diameter of the rotary damper shown in the figure for Problem 1.30 is $2R = 20$ cm. The oil viscosity is $\mu = 0.29$ kg/(m s) and the gap is $h = 1$ mm. Calculate the torque on the shaft at 1000 RPM.

1.32. Suppose the gap is increased to $h = 2$ mm on both sides. How much would the torque change?

1.33. Two concentric cylinders with radiis R_1 and R_2 are separated by an oil film with viscosity μ, as shown in the figure. Next the inner cylinder is rotated and a linear velocity distribution is assumed in the gap between the cylinders (the lower surface is not active). Develop a formula for the torque on the shaft as a function of rotation speed.

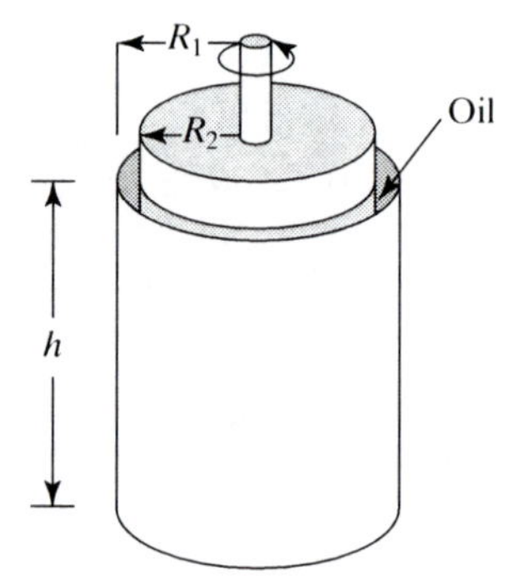

Problem 1.33.

1.34. The two concentric cylinders shown in the figure for Porblem 1.33 are separated by an oil film with viscosity $\mu = 0.023$ kg/(m s). If the shaft rotates at 200 RPM, calculate the torque on the shaft ($R_1 = 15.12$ cm, $R_2 = 15$ cm, and $h =$ 70 cm).

1.35. The device, based on the two concentric cylinders (shown in the figure for Problem 1.33) can be used to measure the viscosity of a fluid. Assuming that the shaft rotates at 200 RPM and the torque measured is 6 N m, calculate the viscosity of the fluid. (Use the dimensions from the previous problem ($R_1 = 15.12$ cm, $R_2 =$ 15 cm, and $h = 70$ cm).

1.36. Some desalination processes are based on evaporating seawater. Energy can be saved by reducing the boiling temperature of the water. From Fig. 1.13, determine the boiling temperature of the water if the pressure is lowered to 0.5 atm.

1.37. Suppose we approximate the tiny tubes carrying water up to the branches of a tall tree by a simple tube of diameter 0.005 mm. Taking the values for $\sigma = 0.073$ N/m and $\theta = 15°$, calculate how high the water can rise in such a tube.

1.38. The disk shown in the figure rotates at a speed of $\omega = 50$ rad/s above a stationary plane separated by an oil film. If the oil viscosity is $\mu = 0.01$ kg/(m s), the spacing between the disk and the stationary surface is 2 mm, and $D = 10$ cm, calculate the torque required for rotating the disk.

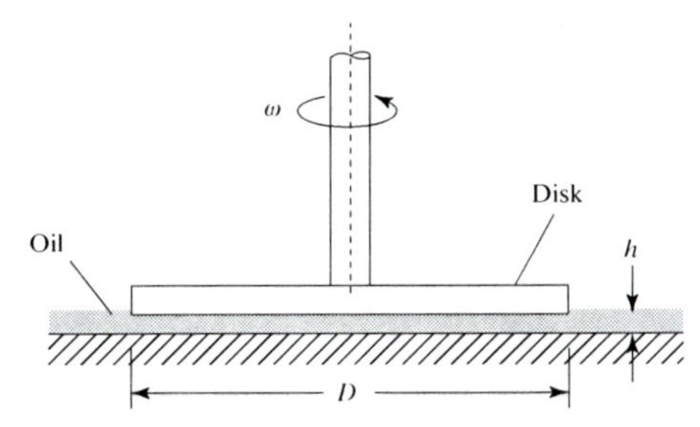

Problem 1.38.

1.39. Oil with a viscosity of μ is flowing between two parallel plates, as shown in the figure. Suppose the velocity distribution is given as

$$u(z) = -k\left(\frac{z}{h} - \frac{z^2}{h^2}\right);$$

plot and calculate the shear stress as a function of z. Where (in terms of z) is the highest and where is the lowest shear stress?

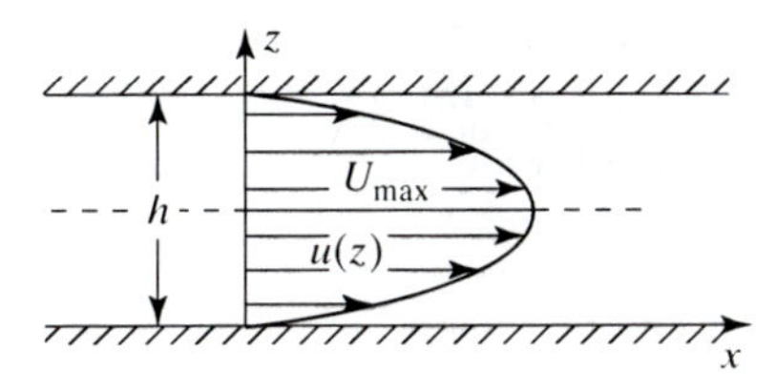

Problem 1.39.

1.40. A 1-mm inner-diameter tube is inserted into a container with mercury, as shown in the figure. Assume that for this condition the surface tension is $\sigma =$ 0.514 N/m, $\theta = 140°$, and $\rho = 13{,}600$ kg/m^3. Calculate how far the meniscus is depressed.

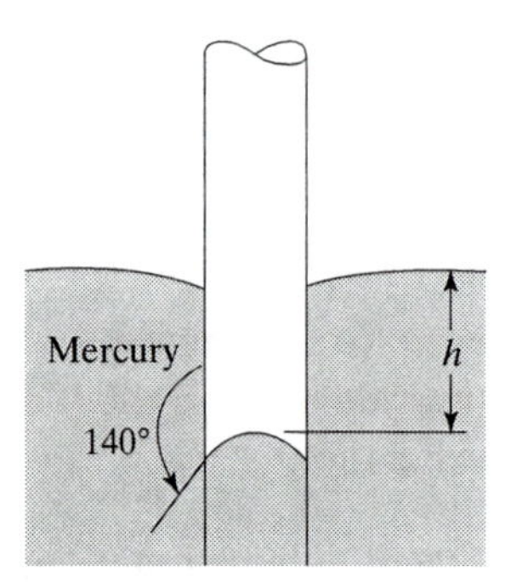

Problem 1.41.

1.41. Two circular tubes are inserted vertically into a large container filled with a liquid that has a surface tension of $\sigma = 0.07$ N/m and a contact angle of $\theta = 10°$. The density of the liquid is 960 kg/m^3, the diameter of the first tube is 1.5 mm, and the diameter of the second tube is 2.5 mm.

(a) Calculate how high the liquid columns will rise in each tube.
(b) Is this in conflict with Pascal's law? Explain.

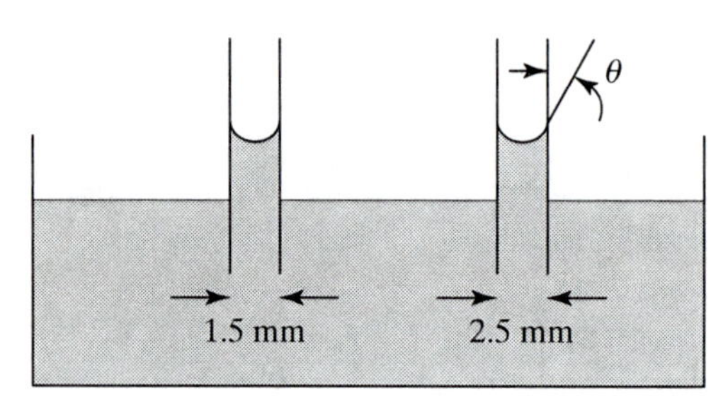

Problem 1.42.

1.42. Calculate the pressure difference that is due to surface tension for a spherical water droplet with a diameter of 0.5 cm (use the surface-tension values from Table 1.1). Is the pressure higher inside or outside the droplet?

1.43. Calculate the pressure difference for a 5-cm-diameter and a 15-cm-diameter spherical water and soap bubble. The surface tension of the water and soap mixture is about $\sigma = 2.5 \times 10^{-2}$ N/m.

1.44. A circular tube is inserted into a large container filled with a liquid that has a surface tension of $\sigma = 0.07$ N/m and a contact angle of $\theta = 10°$. If the density of the liquid is 960 kg/m^3 and the tube diameter is 2 mm, calculate how high the liquid column will rise.

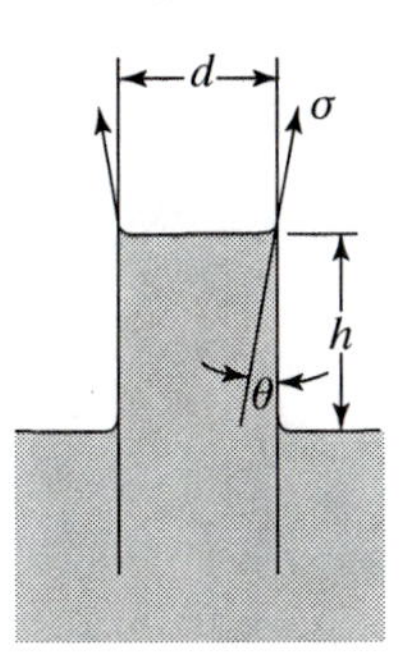

Problem 1.45.

1.45. A circular tube (similar to the one described in Problem 1.45) is inserted into a large container filled with a liquid that has a surface tension of $\sigma = 0.07$ N/m and a contact angle of $\theta = 10°$. If the density of the liquid is 1100 kg/m^3 and the tube diameter is 3 mm, calculate how high the liquid column will rise.

1.46. Two layers of fluid are dragged along by the motion of an upper plate as shown in the figure (without mixing). The bottom plate is stationary. The top fluid generates a shear stress on the upper plate, and the lower fluid generates a shear stress on the bottom plate. Calculate the velocity of the boundary between the two fluids.

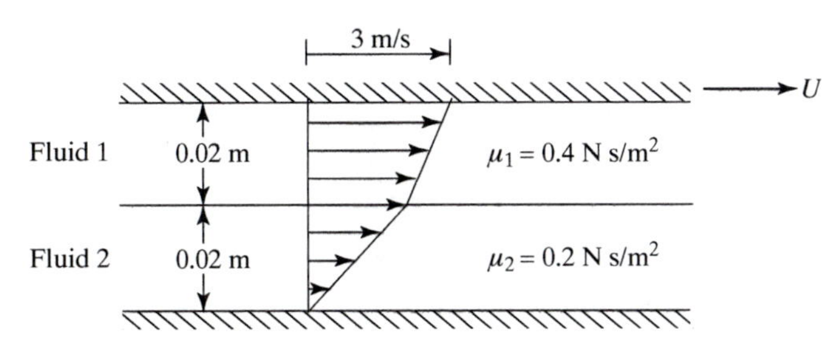

Problem 1.47.

1.47. A small steel ball is placed gently on the surface in a large water tank. Estimate the largest diameter of the ball that will float on the surface because of the surface tension (assume $\sigma = 0.073$ N/m, $\theta = 15°$, and $\rho_{\text{steel}} = 7800$ kg/m^3). Assume the surface tension is acting at the maximum perimeter (as shown).

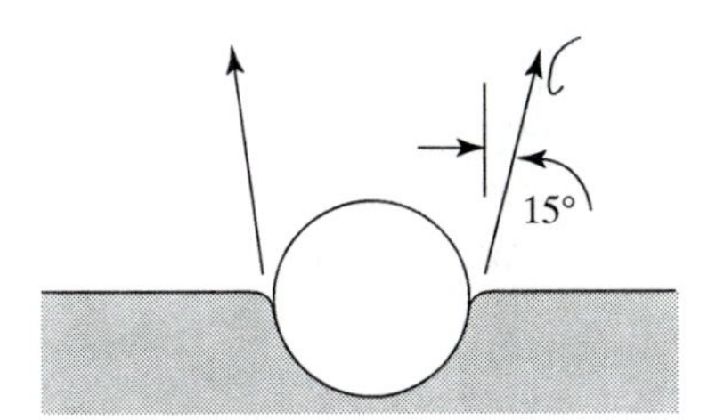

Problem 1.48.

1.48. The mosquito shown in the sketch is supported above the water by the surface tension. Determine the minimum length of this interface needed to support the bug. Assume the mosquito weighs 10^{-4} N and that the surface-tension force acts vertically upward on both sides of the contact line ($\sigma = 0.073$ N/m).

Problem 1.49.

1.49. A noise created by small earthquake at a depth of 1200 m in the ocean propagates upward and eventually reaches a bird flying above at an altitude of 400 m. Calculate how long it takes for the noise to reach the bird. For seawater use $E = 2.34 \times 10^9$ N/m^2, $\rho = 1030$ kg/m^3, and for air, $\gamma = 1.4$ and $T = 270$ K.

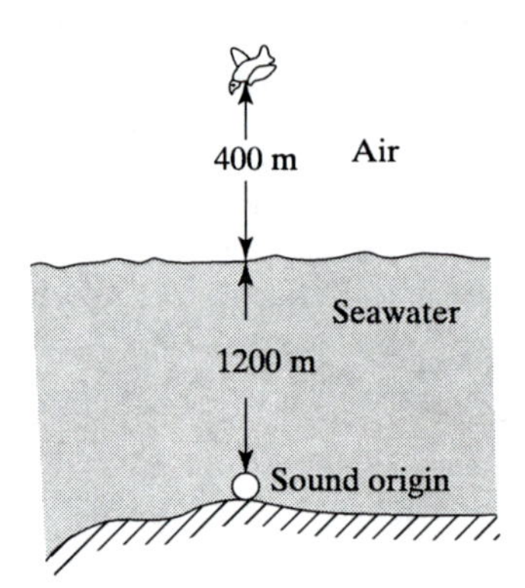

Problem 1.50.

1.50. A small explosion in the ocean is 3000 m from two swimmers (horizontal distance). The first has his ears under the water and the second swimmer's head is above the water. How soon will each hear the noise of the explosion? For seawater use $E = 2.34 \times 10^9$ N/m^2, $\rho = 1030$ kg/m^3 and for air $\gamma = 1.4$ and $T = 300$ K).

1.51. The speed of an airplane is frequently stated in terms of the Mach number, which is simply the ratio between the actual speed and the speed of sound: $M = U/a$. Suppose an airplane flies at $M = 0.8$ at sea level, where the temperature is 27 °C; calculate the actual speed of the airplane. Next, calculate the speed at the same Mach number but at an altitude of 13 km, where the temperature is −57 °C (for air, $\gamma = 1.4$).

1.52. A piston is floating over a 1-m-high column of water enclosed in a 2-cm-diameter pressure-tight cylinder. Calculate how deep the 100-kg weight will push the cylinder down. Assume the water modulus of elasticity is $E = 2.34 \times 10^9$ N/m^2.

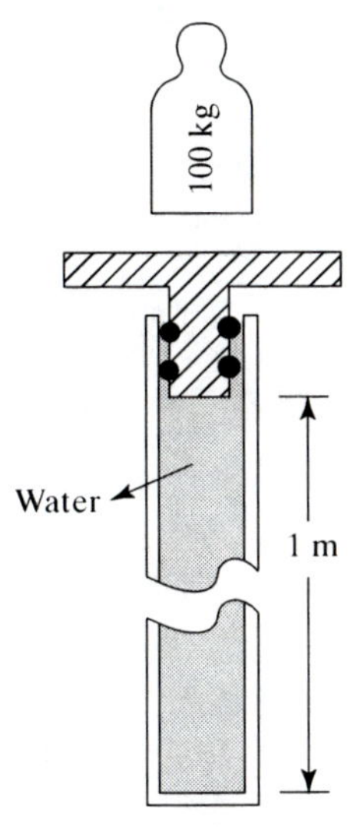

Problem 1.53.

2 The Fluid Dynamic Equation

2.1 Introduction

One approach in teaching introductory fluid mechanics is to avoid the presentation of complex fluid dynamic equations. This is done for very good reasons, including the lack of preparation in partial differential equations and the overall complexity of the problem. Although avoiding the introduction of complex equations is welcome by the average student, the negative and long-term outcome of this approach is that there is no clear rationale and a connecting string among the various chapters that follow. Therefore students are asked to be patient and "suffer quietly" at the beginning and the benefits of a clear roadmap will surface with the systematic approach that follows.

The mechanisms controlling fluid motion may include elements of basic mechanics, heat transfer, phase change, chemical reactions, and even molecular mechanics. Limiting the discussion to the simple mechanics of fluids, usually leads to principles such as the conservation of mass momentum and energy. In this text, however, we concentrate on the conservation of mass (continuity equation) and the conservation of momentum and assume a simple Newtonian fluid without heat transfer. Hence our objective in this chapter is to derive these two conservation laws, the conservation of mass and momentum.

The conservation of mass simply states that there is no change in the total mass (usually referring to a control volume):

$$\frac{d(m)}{dt} = 0. \tag{2.1}$$

The momentum equation simply applies Newton's second law of motion, which states that the change in linear momentum is equal to the sum of the forces acting on the mass m (e.g., in a control volume):

$$\frac{d(m\vec{q})}{dt} = \sum \vec{F}. \tag{2.2}$$

As will be demonstrated for the case of fluids, these equations will be quite complicated and their exact solutions are very limited. Therefore the discipline of fluid mechanics is based on simplified models, in which only portions of the full equations

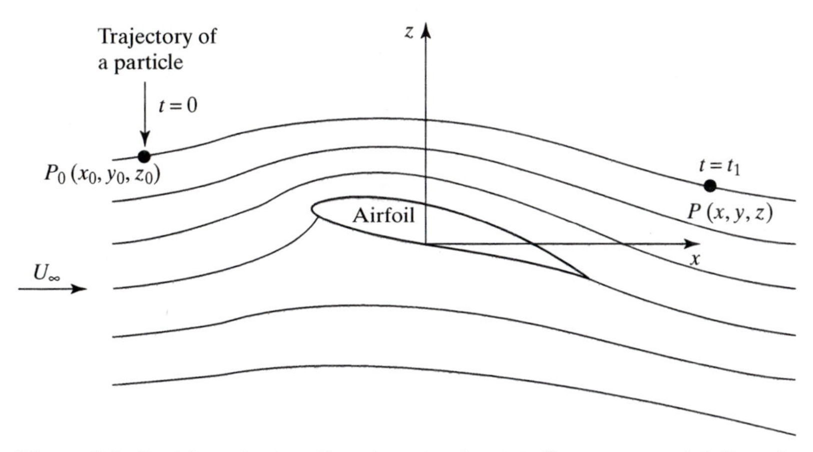

Figure 2.1. Paricle trajectory lines in a steady-state flow over an airfoil as viewed from a body-fixed coordinate system.

are used. We can accomplish this simplification by identifying the dominant terms in the formulation and neglecting terms of much lower orders of magnitude. This approach provides the practicing engineer with the ability to appreciate both the power and the limitations of the techniques that are presented in this text. Furthermore, it makes the learning process easier, as the links among the various chapters are understood.

NOTE TO THE INSTRUCTOR. When teaching an introductory course (and dedicating only one or two lectures to this chapter), it is easier to focus in class on the integral approach (Section 2.6). The rest of the chapter can be assigned as a reading assignment at home followed by a brief discussion in class. The overall objective is for the student to be familiar with the continuity and momentum equations and with the origins of the terms appearing in these equations.

2.2 Description of Fluid Motion

The fluid studied here is modeled as a continuum, and infinitesimally small regions of the fluid (with a fixed mass) are called fluid elements or fluid particles. The motion of the fluid can be described by two different methods. One adopts the particle point of view and follows the motion of the individual particles. The other adopts the field point of view and provides the flow variables as functions of position in space and time. The particle point of view, which uses the approach of classical mechanics, is called the *Lagrangian* method (after the Italian scientist Joseph-Louis Lagrange, 1736–1813). To trace the motion of each fluid particle, it is convenient to introduce a Cartesian coordinate system with the coordinates x, y, and z. The position of any fluid particle P (see Fig. 2.1) is then given by

$$\begin{aligned} x &= x_P(x_0, y_0, z_0, t), \\ y &= y_P(x_0, y_0, z_0, t), \\ z &= z_P(x_0, y_0, z_0, t), \end{aligned} \tag{2.3}$$

where (x_0, y_0, z_0) is the position of P at some initial time $t = 0$. [Note that the quantity (x_0, y_0, z_0) represents the vector with components x_0, y_0, and z_0.] The components of the velocity of this particle are then given by

$$\begin{aligned} u &= \partial x/\partial t, \\ v &= \partial y/\partial t, \\ w &= \partial z/\partial t, \end{aligned} \tag{2.4}$$

and the acceleration by

$$\begin{aligned} a_x &= \partial^2 x/\partial t^2, \\ a_y &= \partial^2 y/\partial t^2, \\ a_z &= \partial^2 z/\partial t^2. \end{aligned} \tag{2.5}$$

The Lagrangian formulation requires the evaluation of the motion of each fluid particle. For most practical applications, this abundance of information is neither necessary nor useful and the analysis is cumbersome. The field point of view, called the Eulerian method, provides the spatial distribution of flow variables at each instant during the motion. For example, if a Cartesian coordinate system is used, the components of the fluid velocity are given by

$$\begin{aligned} u &= u(x, y, z, t), \\ v &= v(x, y, z, t), \\ w &= w(x, y, z, t). \end{aligned} \tag{2.6}$$

The Eulerian approach provides information about the fluid variables that is consistent with the information supplied by most experimental techniques and is in a form that is appropriate for most practical applications. For these reasons the Eulerian description of fluid motion is the most widely used (and is developed later in this chapter). Also note that we use the notation for the velocity vector:

$$\vec{q} = (u, v, w); \tag{2.7}$$

a constant velocity of U_∞ in the x direction is therefore

$$\vec{q} = (U_\infty, 0, 0). \tag{2.7a}$$

2.3 Choice of Coordinate System

For the following chapters, when possible, primarily a Cartesian coordinate system will be used. Other coordinate systems such as curvilinear, cylindrical, or spherical are introduced and used if necessary, mainly to simplify the treatment of certain problems. Also, from the kinematic point of view, a careful choice of a coordinate system can considerably simplify the solution of a problem. As an example, consider the forward motion of an airfoil, with a constant speed U_∞, in a fluid that is otherwise at rest, as shown in Fig. 2.1. Here the origin of the coordinate system is attached to the moving airfoil and the trajectory of a fluid particle inserted at point P_0 at $t = 0$ is shown in the figure. By following the trajectories of several particles, we obtain a more complete description of the flow field in the figure. It is important

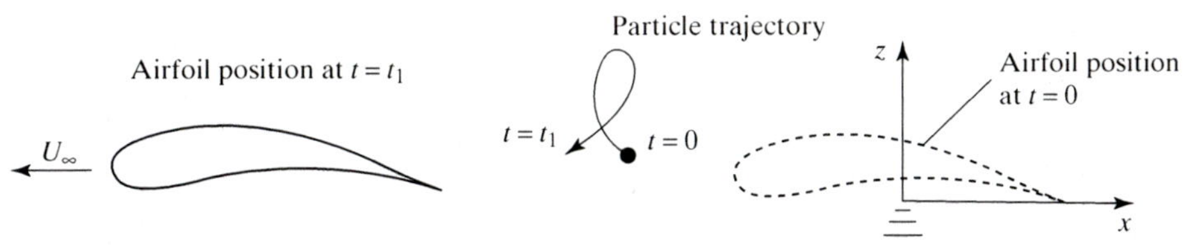

Figure 2.2. A single-particle trajectory line for an airfoil moving left at a constant speed U_∞ (as in Fig. 2.1) as viewed from a stationary frame of reference.

to observe that, for a constant-velocity (unsteady) forward motion of the airfoil, in this frame of reference, these trajectory lines become independent of time. That is, if various particles are introduced at the same point in space, they will follow the same trajectory. Now let's examine the same flow, but from a coordinate system that is fixed relative to the undisturbed fluid. At $t = 0$, the airfoil was on the right-hand side of Fig. 2.2 (dashed curve), and as a result of its constant-velocity forward motion (with a speed U_∞ toward the left-hand side of the page), later at $t = t_1$ it moves to the new position indicated in the figure. A typical particle's trajectory line between $t = 0$ and $t = t_1$ for this case is also shown. The particle's motion (as viewed from the stationary frame of reference) now depends on time, and a new trajectory has to be established for each particle. This simple example depicts the importance of a "good" coordinate system selection. For many problems in which a constant velocity and a fixed geometry (with time) are present, the use of a body-fixed frame of reference will result in a steady or time-independent flow.

Figure 2.2 also demonstrates the principle of the Lagrangian formulation. In this case it is quite common to use a nonmoving, inertial frame of reference and to trace the dynamics of the particles (or group of particles) as shown in the figure. It is also clear that each group of particles must be followed individually, thereby significantly complicating the process. In contrast, the Eulerian approach observes a fixed control volume attached to the airfoil (as shown in Fig. 2.3), and as noted the streamlines and the whole problem appear to be independent of time. In most cases this results in a major simplification, and therefore the Eulerian approach is widely used in classical fluid mechanics (and throughout this book).

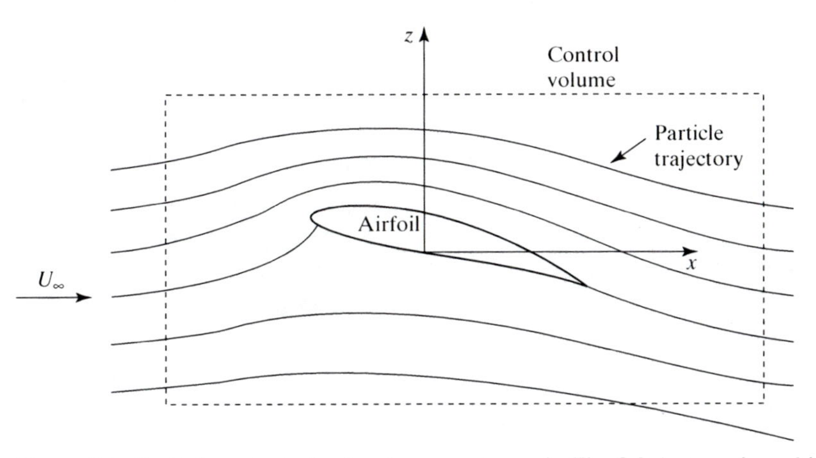

Figure 2.3. Particle trajectories for the case shown in Fig. 2.2, but as viewed in a control volume attached to the airfoil.

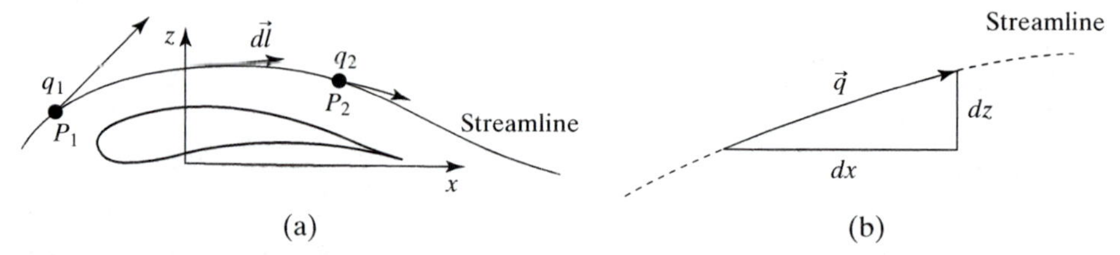

Figure 2.4. Schematic description of streamlines in a steady-state flow: The velocity vector $\vec{q}$ is parallel to the streamline $\vec{l}$.

2.4 Pathlines, Streak Lines, and Streamlines

Three sets of curves are normally associated with providing a pictorial description of a fluid motion; pathlines, streak lines, and streamlines.

Pathlines: A curve describing the trajectory of a fluid element is called a pathline or a particle path. Pathlines are obtained in the Lagrangian approach by an integration of the equations of dynamics for each fluid particle. If the velocity field of a fluid motion is given in the Eulerian framework by Eqs. (2.6) in a body-fixed frame, the pathline for a particle at P_0 in Fig. 2.1 can be obtained by an integration of the velocity. For steady flows the pathlines in the body-fixed frame become independent of time and can be drawn as in the case of flow over the airfoil shown in Fig. 2.1 or Fig. 2.3.

Streak lines: In many cases of experimental flow visualization, particles (e.g., dye or smoke) are introduced into the flow at a fixed point in space. The line connecting all of these particles is called a streak line. To construct streak lines by use of the Lagrangian approach, draw a series of pathlines for particles passing through a given point in space and at a particular instant in time, and then connect the ends of these pathlines.

Streamlines: Another set of curves can be obtained (at a given time) by lines that are parallel to the local velocity vector. To express analytically the equation of a streamline at a certain instant of time, at any point P in the fluid, the velocity $\vec{q}$ must be parallel to the streamline element $d\vec{l}$ [Fig. 2.4(a)]. Therefore, on a streamline,

$$\vec{q} \times d\vec{l} = 0. \tag{2.8}$$

If the velocity vector is $\vec{q} = (u, v, w)$, then vector equation (2.8) reduces to the following scalar equations:

$$\begin{aligned} w dy - v dz &= 0, \\ u dz - w dx &= 0, \\ v dx - u dy &= 0, \end{aligned} \tag{2.9}$$

or, in a differential equation form,

$$\frac{dx}{u} = \frac{dy}{v} = \frac{dz}{w}. \tag{2.9a}$$

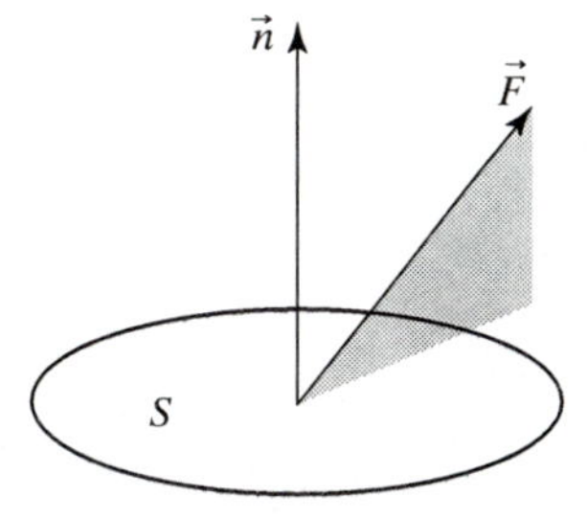

Figure 2.5. A force $\vec{F}$ is acting on a surface S.

This equation is described schematically in Fig. 2.4(b). For example, the velocity components in the x–z plane are $\vec{q} = (u, w)$ and the slope of the streamline (at a point) is dz/dx, which is equal to the slope of the velocity vector,

$$\frac{dz}{dx} = \frac{w}{u},$$

and this is identical to Eq. 2.9(a). Also, in Eq. (2.9a), the velocity (u, v, w) is a function of the coordinates and of time. However, for steady flows the streamlines are independent of time, and streamlines, pathlines, and streak lines become identical, as shown in Fig. 2.1.

2.5 Forces in a Fluid

Prior to discussing the dynamics of fluid motion, we should identify the types of forces that act on a fluid element. Here, forces such as body forces per unit mass $\vec{f}$ and surface forces that are a result of the stress vector $\vec{t}$ are considered. The body forces are independent of any contact with the fluid, as in the case of gravitational or magnetic forces, and their magnitude $\Delta\vec{F}$ is proportional to the local mass,

$$\Delta\vec{F} = \rho\vec{f}\Delta V, \tag{2.10}$$

where ΔV is the volume increment. To define the stress vector $\vec{t}$ at a point, consider the force $\vec{F}$ acting on a planar area S (shown in Fig. 2.5) with $\vec{n}$ being an outward normal to S. Then

$$\vec{t} = \lim_{S\to 0}\left(\frac{\vec{F}}{S}\right). \tag{2.11}$$

Note that in Eq. (1.6) we used a similar formulation to define the pressure. As we shall see soon, the pressure is the normal component of the stress vector. To obtain the components of the stress vector, consider the force equilibrium on an infinitesimal cubical fluid element, shown in Fig. 2.6. To simplify the discussion let us use momentarily an indicial notation. Note that τ_{ij} acts in the x_i direction on a surface whose outward normal points in the x_j direction. This indicial notation allows a simpler presentation of the equations, and the subscripts 1, 2 and 3 denote the coordinate directions x, y, and z, respectively. For example,

$$x_1 = x, \quad x_2 = y, \quad x_3 = z,$$
$$q_1 = u, \quad q_2 = v, \quad q_3 = w.$$

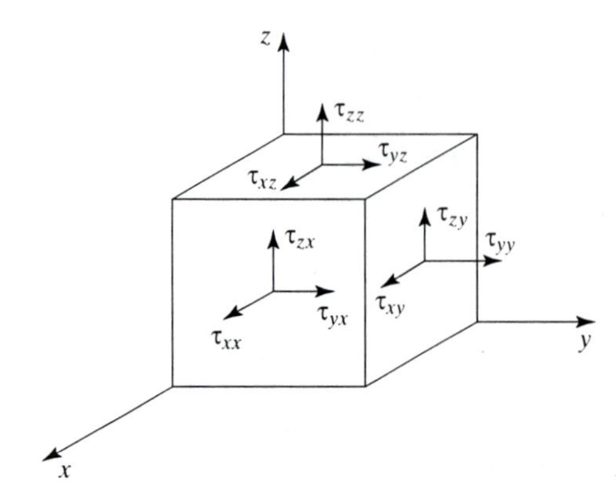

Figure 2.6. Stress components on an infinitesimal cubical fluid element.

The stress components shown on an infinitesimal cubical fluid element of Fig. 2.6. can be summarized in matrix form or in an indicial form as follows:

$$\begin{bmatrix} \tau_{xx} & \tau_{xy} & \tau_{xz} \\ \tau_{yx} & \tau_{yy} & \tau_{yz} \\ \tau_{zx} & \tau_{zy} & \tau_{zz} \end{bmatrix} = \begin{bmatrix} \tau_{11} & \tau_{12} & \tau_{13} \\ \tau_{21} & \tau_{22} & \tau_{23} \\ \tau_{31} & \tau_{32} & \tau_{33} \end{bmatrix} = \tau_{ij} \tag{2.12}$$

A treatment of the moment equilibrium results in the symmetry of the stress vector components so that $\tau_{ij} = \tau_{ji}$. Also, it is customary to sum over any index that is repeated such that

$$\sum_{j=1}^{3} \tau_{ij} n_j \equiv \tau_{ij} n_j \quad \text{for} \quad i = 1, 2, 3, \tag{2.13}$$

and to interpret an equation with a free index [as in in Eq. (2.13)] as being valid for all values of that index. For a Newtonian fluid (in which the stress components τ_{ij} are linear in the derivatives $\frac{\partial q_i}{\partial x_j}$), the stress components are related to the velocity field by (see, e.g., [1, p. 147])

$$\tau_{ij} = \left(-p - \frac{2}{3}\mu\frac{\partial q_k}{\partial x_k}\right)\delta_{ij} + \mu\left(\frac{\partial q_i}{\partial x_j} + \frac{\partial q_j}{\partial x_i}\right), \tag{2.14}$$

where μ is the viscosity coefficient, p is the pressure, the dummy variable k is summed from 1 to 3, and δ_{ij} is the Kronecker delta function defined by

$$\delta_{ij} \equiv \begin{cases} 1 & i = j \\ 0 & i \neq j \end{cases}. \tag{2.15}$$

Equation (2.14) is quite complex, and only simple cases are used in this book. For example, when the fluid is at rest, the tangential stresses vanish and the normal stress component becomes simply the pressure. Thus the stress components become

$$\tau_{ij} = \begin{bmatrix} -p & 0 & 0 \\ 0 & -p & 0 \\ 0 & 0 & -p \end{bmatrix} \tag{2.16}$$

Another interesting case of Eq. (2.14) is the one-degree-of-freedom shear flow between a stationary plate and a moving infinite plate with a velocity U_∞, which

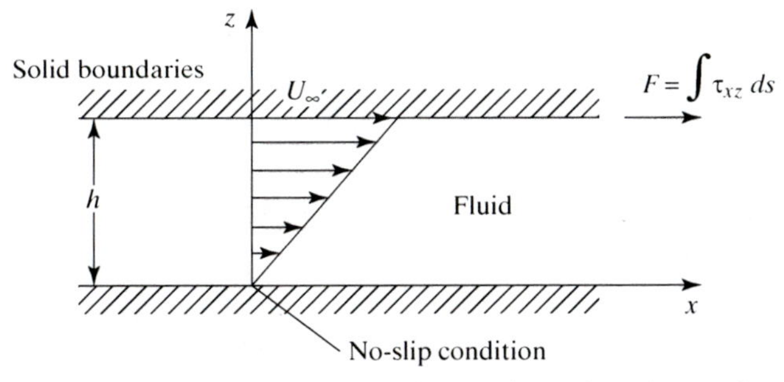

Figure 2.7. Flow between a stationary (lower) and a moving (upper) plate.

was discussed in Chapter 1 (see Fig. 1.5). Let us repeat this case in Fig. 2.7 and assume that there is no pressure gradient and the fluid motion is due to the motion of the upper plate. This flow is called *Couette flow* and was used in Chapter 1 to demonstrate the shear stress in a moving fluid.

Because this is a steady-state problem, with only the x component of the velocity, $\vec{q} = (u, 0, 0)$, and no pressure gradient, Eq. (2.14) the shear stress reduces to

$$\tau_{xz} = \mu \frac{\partial u}{\partial z} = \frac{\mu U_\infty}{h}. \tag{2.17}$$

This example also clarifies the use of the notation xz. The shear stress τ_{xz} points in the x direction and acts on the surface, which normaly is pointing in the z direction. Because there is no pressure gradient in the flow, the fluid motion in the x direction is entirely due to the action of the viscous forces. The force $\vec{F}$ on the plate can be found by integration of τ_{xz} on the upper moving surface.

2.6 Integral Form of the Fluid Dynamic Equations

To develop the governing integral and differential equations describing the fluid motion, the various properties of the fluid are investigated in an arbitrary control volume that is stationary and submerged in the fluid (Fig. 2.8). These properties can be density, momentum, energy, etc., and of course may change with time. A typical accounting for the change in one of those properties inside the control volume must be the sum of the accumulation of the property inside the control volume and the transfer of this property through the control-volume boundaries. Selection of the control volume usually includes the area of interest, as will be demonstrated later. As an example, we can analyze the conservation of mass by observing the changes

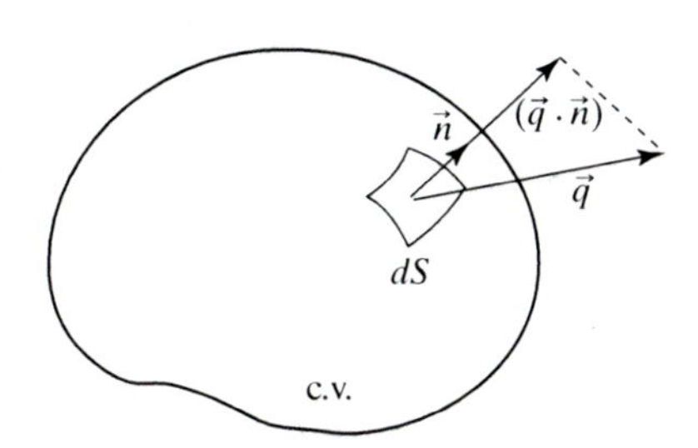

Figure 2.8. Control volume in the fluid. Note that to calculate the flux of the fluid crossing a surface element ds we use only the normal (to the surface) component.

in fluid density ρ for the control volume (c.v.) in Fig. 2.8. The mass $m_{\text{c.v.}}$ within the control volume is then

$$m_{\text{c.v.}} = \int_{\text{c.v.}} \rho dV, \tag{2.18}$$

where dV is the volume element. The accumulation of mass within the control volume is simply the time derivative of Eq. (2.18):

$$\frac{\partial m_{\text{c.v.}}}{\partial t} = \frac{\partial}{\partial t}\int_{\text{c.v.}} \rho dV. \tag{2.18a}$$

The change in the mass within the control volume, because of the mass leaving (m_{out}) and the mass entering (m_{in}) through the control-volume boundaries (c.s.) is

$$m_{\text{out}} - m_{\text{in}} = \int_{\text{c.s.}} \rho(\vec{q}\cdot\vec{n})dS, \tag{2.19}$$

where $\vec{q}$ is the velocity vector (u, v, w) and $\rho(\vec{q}\cdot\vec{n})dS$ is the rate of mass leaving normal to the surface element dS ($\vec{n}$ is the outward normal unit vector), as shown in Fig. 2.8. Because mass is conserved and no new material is being produced, it is clear that

$$\frac{\partial m_{\text{c.v.}}}{\partial t} = m_{\text{in}} - m_{\text{out}}.$$

Replacing the left- and right-hand terms with Eq. (2.18a) and Eq. (2.19), respectively, yields

$$\frac{\partial}{\partial t}\int_{\text{c.v.}} \rho dV + \int_{\text{c.s.}} \rho(\vec{q}\cdot\vec{n})dS = 0, \tag{2.20}$$

or we can state that $m_{\text{c.v.}}$ is conserved (conservation of mass):

$$\frac{dm_{\text{c.v.}}}{dt} = \frac{\partial}{\partial t}\int_{\text{c.v.}} \rho dV + \int_{\text{c.s.}} \rho(\vec{q}\cdot\vec{n})dS = 0. \tag{2.20a}$$

As noted, Eq. (2.20) is the integral representation of the conservation of mass (often called the continuity equation). It is the equivalent of Eq. (2.1), and it simply states that any change in the mass of the fluid in the control volume is equal to the rate of mass being transported across the control-surface (c.s.) boundaries. In a similar manner, the rate of change in the momentum of the fluid flowing through the control volume at any instant $d(m\vec{q})_{\text{c.v.}}/dt$ is the sum of the accumulation of the momentum $\rho\vec{q}$ per unit volume within the control volume and of the change of the momentum across the control-surface boundaries [so all we need to do is to multiply Eq. (2.20) by $\vec{q}$]:

$$\frac{d(m\vec{q})_{\text{c.v.}}}{dt} = \frac{\partial}{\partial t}\int_{\text{c.v.}} \rho\vec{q}dV + \int_{\text{c.s.}} \rho\vec{q}(\vec{q}\cdot\vec{n})dS. \tag{2.21}$$

This change in the momentum, as given in Eq. (2.2), according to Newton's second law, must be equal to the forces $\sum\vec{F}$ applied to the fluid inside the control volume:

$$\frac{d(m\vec{q})_{\text{c.v.}}}{dt} = \sum\vec{F}. \tag{2.22}$$

The forces acting on the fluid in the control volume in the x_i direction are either body forces ρf_i per unit volume or surface forces $n_j \tau_{ij}$ per unit area, as discussed in Section 2.4:

$$\left(\sum \vec{F}\right)_i = \int_{\text{c.v.}} \rho f_i dV + \int_{\text{c.s.}} n_j \tau_{ij} dS, \tag{2.23}$$

where $\vec{n}$ is the unit normal vector that points outward from the control volume. By substituting Eqs. (2.21) and (2.23) into Eq. (2.22), we obtain the integral form of the momentum equation in the i direction:

$$\frac{\partial}{\partial t}\int_{\text{c.v.}} \rho q_i dV + \int_{\text{c.s.}} \rho q_i(\vec{q}\cdot\vec{n})dS = \int_{\text{c.v.}} \rho f_i dV + \int_{\text{c.s.}} n_j \tau_{ij} dS. \tag{2.24}$$

Sometimes the pressure and the shear terms are separated, and Eq. (2.24) will have the form

$$\frac{\partial}{\partial t}\int_{\text{c.v.}} \rho q_i dV + \int_{\text{c.s.}} \rho q_i(\vec{q}\cdot\vec{n})dS = \int_{\text{c.v.}} \rho f_i dV + \int_{\text{c.s.}} n_j \tau_{ij} dS - \int_{\text{c.s.}} p n_i dS, \tag{2.24a}$$

where the minus sign in front of the pressure term is a result of the outward normal unit vector pointing opposite to the pressure force [see Eq. (1.7)]. Also, τ_{ij} now represents the shear stress only.

The preceding approach can be used to develop additional governing equations, such as the energy equation. However, for the fluid dynamic cases that are considered here, the mass and the momentum equations are sufficient to describe the fluid motion.

EXAMPLE 2.1. FLOW THROUGH A STREAM TUBE. Upto this point we have developed the equivalents of the two basic equations [Eqs. (2.1) and (2.2)] in integral form and applied them to a control volume. To demonstrate their applicability to fluid motion let us consider 1D flow through a stream tube. Here the fluid is flowing in the x direction (Fig. 2.9) and it enters the stream tube through inlet A_1 and exits through A_2.

We may further assume an incompressible (ρ = constant) flow, and the 1D flow assumption suggests that the velocity entering the stream tube at section 1 is uniform $\vec{q}_1 = (u_1, 0, 0)$ as well as the exiting velocity $\vec{q}_2 = (u_2, 0, 0)$. The control surface contains the entire stream tube in Fig. 2.9 and the outside pressure is p_a. Continuity equation (2.20) in steady state becomes

$$\int_{\text{c.s.}} \rho(\vec{q}\cdot\vec{n})dS = 0. \tag{2.25}$$

Next we need to consider all the areas where fluid is crossing the control volume (namely areas A_1 and A_2) in the figure. The velocity vector in section 1

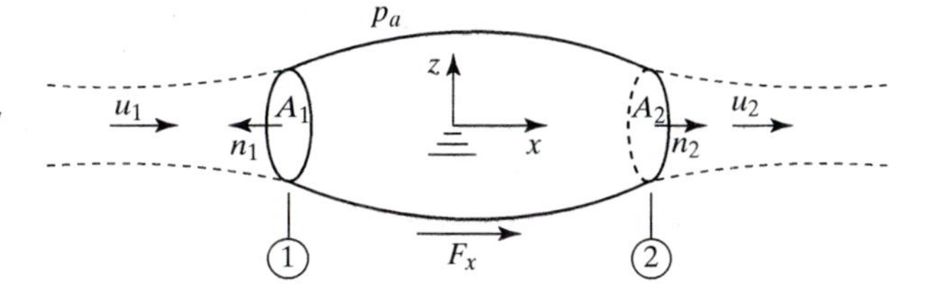

Figure 2.9. Example: 1D flow through a stream tube.

is $(u_1, 0, 0)$ and in section 2 it is $(u_2, 0, 0)$. Also the unit normal vector is defined as pointing outside the control volume and therefore $\vec{n}_1 = (-1, 0, 0)$ and $\vec{n}_2 = (1, 0, 0)$, as shown in the figure. The integration on all other areas is zero because the velocity is not penetrating the surface. Consequently Eq. (2.25) for this case becomes

$$\rho u_1(-A_1) + \rho u_2(+A_2) = 0,$$

where the minus sign in the first term is due to $\vec{n}_1 = (-1, 0, 0)$. Or we can write

$$\dot{m} = \rho u_1 A_1 = \rho u_2 A_2, \tag{2.26}$$

where $\dot{m}$ is the mass flow rate. So basically this states that all the flow entering the control volume at station 1 must leave at the other end. Let us now continue with the momentum equation for the steady-state incompressible case without body forces:

$$\int_{\text{c.s.}} \rho q_i(\vec{q} \cdot \vec{n}) dS = \int_{\text{c.s.}} n_j \tau_{ij} dS. \tag{2.27}$$

By looking at the forces (right-hand side of the equation), we can separate the pressure [as in Eq. (2.24a)] acting on surfaces A_1 and A_2 and call the shear forces acting on the axisymmetric stream-tube shell F_x. Consequently a simpler form of the force term is now

$$\int_{\text{c.s.}} n_j \tau_{ij} dS = F_x - \int_{\text{c.s.}} p \cdot \vec{n} dS.$$

Steady-state momentum equation (2.24) with this modified force term has the following form:

$$\int_{\text{c.s.}} \rho q_x(\vec{q} \cdot \vec{n}) dS = F_x - \int_{\text{c.s.}} p \cdot \vec{n} dS. \tag{2.28}$$

Also, instead of using p as the pressure we can use the pressure difference $(p - p_a)$. This has no effect on the result because the integral of constant pressure p_a over a closed body is zero. Consequently momentum equation (2.28), when applied to the stream tube, has the following form

$$\rho u_1(-u_1 A_1) + \rho u_2(+u_2 A_2) = F_x - (p_1 - p_a)(-A_1) - (p_2 - p_a)(+A_2),$$

and, after rearranging,

$$F_x = \rho u_2^2 A_2 - \rho u_1^2 A_1 + (p_2 - p_a)A_2 - (p_1 - p_a)A_1. \tag{2.29}$$

So the force F_x on the stream tubes is a result of the change in the linear momentum plus the change in pressure on the two open ends. Also the force F_x is positive in the positive x direction, so for the control volume to stay stationary the momentum and pressure forces are balanced by the external force F_x. This equation is useful for calculating forces that are due to jets, such as the thrust created by jet engines, and more examples are provided in Chapter 4.

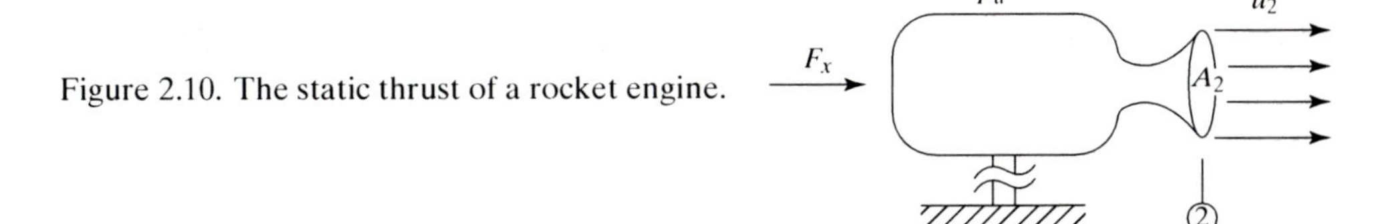

Figure 2.10. The static thrust of a rocket engine.

EXAMPLE 2.2. STATIC THRUST OF A ROCKET. As a second example let us use this method to derive the static thrust of a rocket engine, as shown in the Fig. 2.10.

In this case as well, the control volume encloses the whole engine and the fluid exchange takes place only through exit area A_2. Consequently, by using Eqs. (2.26) and (2.29), we get

$$F_x = \rho u_2^2 A_2 + (p_2 - p_a)A_2,$$

and by using the continuity equation

$$\dot{m} = \rho u_2 A_2,$$

we can write

$$F_x = \dot{m}u_2 + (p_2 - p_a)A_2.$$

This indicates that the thrust of the rocket engine consists of the exhaust momentum and the pressure term at the exit.

EXAMPLE 2.3. CONSERVATION OF MASS. As a numerical example for the principle of continuity [Eq. (2.26)], consider the flow into a pipe (see Fig. 2.11), but instead of one exit as in Fig. 2.9, let us have two! Assume that water enters at station 1, where $A_1 = 30\ \text{cm}^2$ ($0.003\ \text{m}^2$) at an average velocity of $u_1 = 1$ m/s and leaves at station 3, where $A_3 = 20\ \text{cm}^2$ and $u_3 = 1.2$ m/s. The rest of the flow leaves through section 2, where the cross-section area is $A_2 = 20\ \text{cm}^2$. Calculate the flow rates entering at station 1 and leaving at station 2, and the average exit velocity at station 2.

Solution: We can imagine a control volume surrounding the pipes, which have one entrance and two exits. We can calculate the flow rate entering at station 1 by using Eq. (2.26),

$$\dot{m}_1 = \rho u_1 A_1 = 1000\frac{\text{kg}}{\text{m}^3}1\frac{\text{m}}{\text{s}}0.003\,\text{m}^2 = 3\frac{\text{kg}}{\text{s}},$$

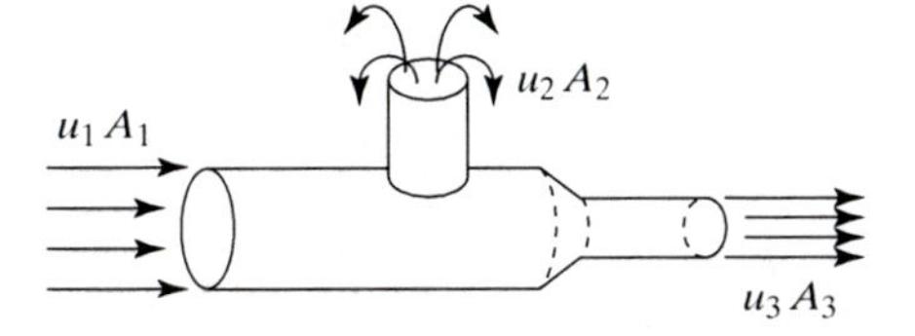

Figure 2.11. Simple example for the conservation of mass.

and here we used the water density from Table 1.1. Also note that the volume flow rate is 3 L/s because 1 L water has a mass of 1 kg. The flow leaving at station 3 is calculated with the same method:

$$\dot{m}_3 = \rho u_3 A_3 = 1000\frac{\text{kg}}{\text{m}^3}1.2\frac{\text{m}}{\text{s}}0.002\,\text{m}^2 = 2.4\frac{\text{kg}}{\text{s}}.$$

The conservation of mass implies that

$$\dot{m}_2 = \dot{m}_1 - \dot{m}_3 = 0.6\frac{\text{kg}}{\text{s}}.$$

The average velocity leaving at station 2 is then

$$u_2 = \frac{\dot{m}_2}{\rho A_2} = \frac{0.6}{1000 \times 0.002} = 0.3\frac{\text{m}}{\text{s}}.$$

EXAMPLE 2.4. THE MOMENTUM OF A JET. To demonstrate the application of Eq. (2.29), let us calculate the force on the water pipe that is due to the jet, shown in Fig. 2.12. The average exit velocity is 5 m/s, the exit area is 20 cm^2, and the pressure at the exit is equal to the pressure outside [so there is no pressure term in Eq. (2.29)].

Solution: The pipe and the tap assembly are surrounded by a control volume, and our interest is to find the horizontal force only. In this case the momentum in the horizontal direction is leaving only through the exit, as in the case of the rocket in Example 2.2. To calculate the force that is due to the jet, we can use the same equation:

$$F_x = \dot{m}u_e = \rho u_e^2 A_e = 1000 \times 5^2 \times 0.002 = 50\,\text{N}.$$

This force acts on the control volume (to the right in Fig. 2.12), and the reaction force (acting on the pipe) is in the opposite direction.

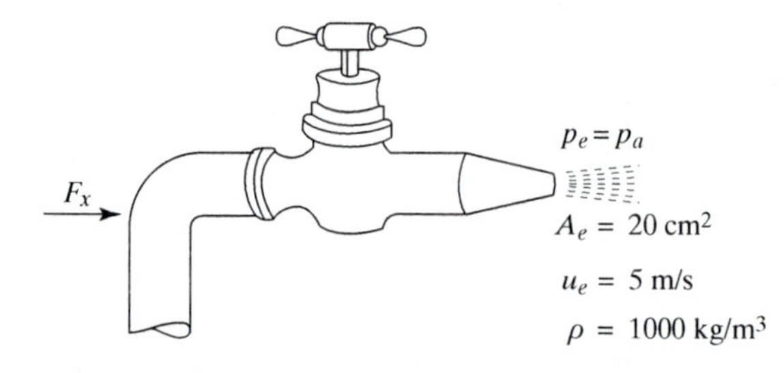

Figure 2.12. Force that is due to a water jet.

2.7 Differential Form of the Fluid Dynamic Equations

Equations (2.20) and (2.24) are the integral forms of the conservation of mass and momentum equations. In many cases, though, the differential representation is more useful. To derive the differential form of the conservation of mass equation, both integrals of Eq. (2.20) should be volume integrals. This can be accomplished by the use of the divergence theorem (see [1, p. 39]), which states that for a vector $\vec{q}$,

$$\int_{\text{c.s.}} \vec{n} \cdot \vec{q}\, dS = \int_{\text{c.v.}} \nabla \cdot \vec{q}\, dV. \tag{2.30}$$

If $\vec{q}$ is the velocity vector then this equation states that the fluid flux through the boundary of the control surface (left-hand side) is equal to the rate of expansion of the fluid (right-hand side) inside the control volume. In Eq. (2.30), ∇ is the gradient operator and, in Cartesian coordinates, is

$$\nabla = \vec{i}\frac{\partial}{\partial x} + \vec{j}\frac{\partial}{\partial y} + \vec{k}\frac{\partial}{\partial z}$$

or in indicial form

$$\nabla = \vec{e}_j \frac{\partial}{\partial x_j},$$

where $\vec{e}_j$ is the unit vector ($\vec{i}, \vec{j}, \vec{k}$, for j = 1, 2, 3). Thus the indicial form of the divergence theorem becomes

$$\int_{\text{c.s.}} n_j q_j dS = \int_{\text{c.v.}} \frac{\partial q_j}{\partial x_j} dV. \tag{2.30a}$$

An application of Eq. (2.30) to the surface integral term in Eq. (2.20) transforms it to a volume integral:

$$\int_{\text{c.s.}} \rho(\vec{q} \cdot \vec{n}) dS = \int_{\text{c.v.}} (\nabla \cdot \rho\vec{q}) dV.$$

This allows the two terms in Eq. (2.20) to be combined as one volume integral:

$$\int_{\text{c.v.}} \left(\frac{\partial \rho}{\partial t} + \nabla \cdot \rho\vec{q} \right) dV = 0,$$

where the time derivative is taken inside the integral as the control volume is stationary. Because the equation must hold for an arbitrary control volume anywhere in the fluid, the integrand is also equal to zero. Thus the following differential form of conservation of mass or the continuity equation is obtained:

$$\frac{\partial \rho}{\partial t} + \nabla \cdot \rho\vec{q} = 0. \tag{2.31}$$

Expansion of the second term of Eq. (2.31) yields

$$\frac{\partial \rho}{\partial t} + \vec{q} \cdot \nabla\rho + \rho\nabla\vec{q} = 0, \tag{2.31a}$$

and in Cartesian coordinates

$$\frac{\partial \rho}{\partial t} + u\frac{\partial \rho}{\partial x} + v\frac{\partial \rho}{\partial y} + w\frac{\partial \rho}{\partial z} + \rho\left(\frac{\partial u}{\partial x} + \frac{\partial v}{\partial y} + \frac{\partial w}{\partial z} \right) = 0. \tag{2.31b}$$

We can use the material derivative, which is defined as

$$\frac{D}{Dt} \equiv \frac{\partial}{\partial t} + \vec{q} \cdot \nabla = \frac{\partial}{\partial t} + u\frac{\partial}{\partial x} + v\frac{\partial}{\partial y} + w\frac{\partial}{\partial z}.$$

Because this is an important operator it is discussed in more detail in Section 2.8. When this operator is used, Eq. (2.31) becomes

$$\frac{D\rho}{Dt} + \rho\nabla \cdot \vec{q} = 0. \tag{2.31c}$$

The material derivative $\frac{D}{Dt}$ represents the rate of change following a fluid particle. For example, the acceleration of a fluid particle is given by

$$\vec{a} = \frac{D\vec{q}}{Dt} = \frac{\partial \vec{q}}{\partial t} + \vec{q} \cdot \nabla \vec{q}. \tag{2.32}$$

An incompressible fluid is a fluid whose elements cannot experience volume change (think of water). Because, by definition, the mass of a fluid element is constant, the fluid elements of an incompressible fluid must have constant density. (A homogeneous incompressible fluid is therefore a constant-density fluid.) Continuity equation (2.31) for an incompressible fluid reduces to

$$\nabla \cdot \vec{q} = \frac{\partial u}{\partial x} + \frac{\partial v}{\partial y} + \frac{\partial w}{\partial z} = 0. \tag{2.33}$$

Note that the incompressible continuity equation does not have time derivatives (but time dependency can be introduced by means of time-dependent boundary conditions). To obtain the differential form of the momentum equation, we follow the same approach, and the divergence theorem [Eq. (2.30a)] is applied to the surface integral terms of Eq. (2.24):

$$\int_{\text{c.s.}} \rho q_i (\vec{q} \cdot \vec{n}) dS = \int_{\text{c.v.}} \nabla \cdot \rho q_i \vec{q} dV,$$

$$\int_{\text{c.s.}} n_j \tau_{ij} dS = \int_{\text{c.v.}} \frac{\partial \tau_{ij}}{\partial x_j} dV.$$

Substituting these results into Eq. (2.24) yields

$$\int_{\text{c.v.}} \left[\frac{\partial}{\partial t}(\rho q_i) + \nabla \cdot \rho q_i \cdot \vec{q} - \rho f_i - \frac{\partial \tau_{ij}}{\partial x_j} \right] dV = 0. \tag{2.34}$$

Because this integral holds for an arbitrary control volume, the integrand must be zero, and therefore

$$\frac{\partial}{\partial t}(\rho q_i) + \nabla \cdot \rho q_i \vec{q} = \rho f_i + \frac{\partial \tau_{ij}}{\partial x_j} \quad (i = 1, 2, 3). \tag{2.35}$$

Expanding the left-hand side of Eq. (2.35) first and then using the continuity equation will reduce the left-hand side to

$$\frac{\partial}{\partial t}(\rho q_i) + \nabla \cdot (\rho q_i \vec{q}) = q_i \left[\frac{\partial \rho}{\partial t} + \nabla \cdot \rho \vec{q} \right] + \rho \left[\frac{\partial q_i}{\partial t} + \vec{q} \cdot \nabla q_i \right] = \rho \frac{Dq_i}{Dt}.$$

because the terms in the first square brackets are equal to zero [see Eq. (2.31)]. Note that the fluid acceleration [the material derivative, as noted in Eq. (2.32)] is

$$a_i = \frac{Dq_i}{Dt} = \frac{\partial q_i}{\partial t} + \vec{q} \cdot \nabla q_i,$$

which, according to Newton's second law, when multiplied by the mass per unit volume must be equal to $\sum F_i$. So, after substituting this form of the acceleration term into Eq. (2.35), we find that the differential form of the momentum equation becomes $\rho a_i = \sum F_i$ or

$$\rho a_i = \rho f_i + \frac{\partial \tau_{ij}}{\partial x_j} \quad (i = 1, 2, 3), \tag{2.36}$$

or, by using the material derivative, we obtain

$$\rho \frac{Dq_i}{Dt} = \rho f_i + \frac{\partial \tau_{ij}}{\partial x_j} \quad (i = 1, 2, 3), \tag{2.36a}$$

and, in Cartesian coordinates,

$$\rho \left(\frac{\partial u}{\partial t} + u\frac{\partial u}{\partial x} + v\frac{\partial u}{\partial y} + w\frac{\partial u}{\partial z} \right) = \sum F_x = \rho f_x + \frac{\partial \tau_{xx}}{\partial x} + \frac{\partial \tau_{xy}}{\partial y} + \frac{\partial \tau_{xz}}{\partial z} \tag{2.36b}$$

$$\rho \left(\frac{\partial v}{\partial t} + u\frac{\partial v}{\partial x} + v\frac{\partial v}{\partial y} + w\frac{\partial v}{\partial z} \right) = \sum F_y = \rho f_y + \frac{\partial \tau_{yx}}{\partial x} + \frac{\partial \tau_{yy}}{\partial y} + \frac{\partial \tau_{yz}}{\partial z}, \tag{2.36c}$$

$$\rho \left(\frac{\partial w}{\partial t} + u\frac{\partial w}{\partial x} + v\frac{\partial w}{\partial y} + w\frac{\partial w}{\partial z} \right) = \sum F_z = \rho f_z + \frac{\partial \tau_{zx}}{\partial x} + \frac{\partial \tau_{zy}}{\partial y} + \frac{\partial \tau_{zz}}{\partial z}. \tag{2.36d}$$

The stress components on the fluid element were described in Fig. 2.6. Also note that in Eqs. (2.36b)–(2.36d) the symmetry of the stress vector has been enforced. For a Newtonian fluid the stress components τ_{ij} are given by Eq. (2.14), and by substituting them into Eqs. (2.36b)–(2.36d), we obtain the Navier–Stokes equations:

$$\rho \left(\frac{\partial q_i}{\partial t} + \vec{q} \cdot \nabla q_i \right) = \rho f_i - \frac{\partial}{\partial x_i} \left(p + \frac{2}{3}\mu \nabla \cdot \vec{q} \right) + \frac{\partial}{\partial x_j} \mu \left(\frac{\partial q_i}{\partial x_j} + \frac{\partial q_j}{\partial x_i} \right) \quad (i = 1, 2, 3), \tag{2.37}$$

and in Cartesian coordinates we have

$$\rho \left(\frac{\partial u}{\partial t} + \vec{q} \cdot \nabla u \right) = \rho f_x - \frac{\partial p}{\partial x} + \frac{\partial}{\partial x} \left\{ \mu \left[2\frac{\partial u}{\partial x} - \frac{2}{3}(\nabla \cdot \vec{q}) \right] \right\}$$
$$+ \frac{\partial}{\partial y} \left[\mu \left(\frac{\partial u}{\partial y} + \frac{\partial v}{\partial x} \right) \right] + \frac{\partial}{\partial z} \left[\mu \left(\frac{\partial w}{\partial x} + \frac{\partial u}{\partial z} \right) \right], \tag{2.37a}$$

$$\rho \left(\frac{\partial v}{\partial t} + \vec{q} \cdot \nabla v \right) = \rho f_y - \frac{\partial p}{\partial y} + \frac{\partial}{\partial y} \left\{ \mu \left[2\frac{\partial v}{\partial y} - \frac{2}{3}(\nabla \cdot \vec{q}) \right] \right\}$$
$$+ \frac{\partial}{\partial z} \left[\mu \left(\frac{\partial v}{\partial z} + \frac{\partial w}{\partial y} \right) \right] + \frac{\partial}{\partial x} \left[\mu \left(\frac{\partial u}{\partial y} + \frac{\partial v}{\partial x} \right) \right], \tag{2.37b}$$

$$\rho \left(\frac{\partial w}{\partial t} + \vec{q} \cdot \nabla w \right) = \rho f_z - \frac{\partial p}{\partial z} + \frac{\partial}{\partial z} \left\{ \mu \left[2\frac{\partial w}{\partial z} - \frac{2}{3}(\nabla \cdot \vec{q}) \right] \right\}$$
$$+ \frac{\partial}{\partial x} \left[\mu \left(\frac{\partial w}{\partial x} + \frac{\partial u}{\partial z} \right) \right] + \frac{\partial}{\partial y} \left[\mu \left(\frac{\partial v}{\partial z} + \frac{\partial w}{\partial y} \right) \right]. \tag{2.37c}$$

Typical boundary conditions for this problem require that, on stationary solid boundaries (Fig. 2.13), both the normal and tangential velocity components reduce to zero (e.g., the no-slip boundary conditions):

$$q_n = 0 \quad \text{(on solid surface)}, \tag{2.38a}$$

$$q_t = 0 \quad \text{(on solid surface)}. \tag{2.38b}$$

The number of exact solutions to the Navier–Stokes equations is small because of the nonlinearity of the differential equations. However, in many situations some

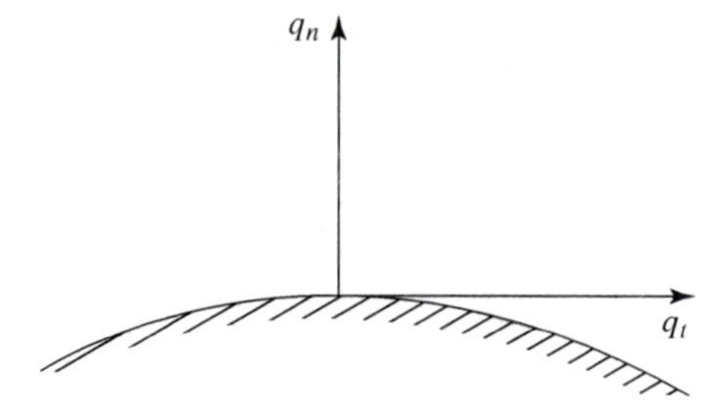

Figure 2.13. Definition of tangential and normal velocity components near a solid boundary.

terms can be neglected so that simpler equations can be obtained. For example, by assuming a constant-viscosity coefficient μ, we find that Eq. (2.37) becomes

$$\rho\left(\frac{\partial \vec{q}}{\partial t}+\vec{q}\cdot\nabla\vec{q}\right)=\rho\vec{f}-\nabla p+\mu\nabla^2\vec{q}+\frac{\mu}{3}\nabla(\nabla\cdot\vec{q}). \tag{2.39}$$

Furthermore, by assuming an incompressible fluid [for which continuity equation (2.33) becomes $\nabla\cdot\vec{q}=0$], we find that Eq. (2.37) reduces to

$$\rho\left(\frac{\partial \vec{q}}{\partial t}+\vec{q}\cdot\nabla\vec{q}\right)=\rho\vec{f}-\nabla p+\mu\nabla^2\vec{q}. \tag{2.40}$$

For an inviscid compressible fluid,

$$\frac{\partial \vec{q}}{\partial t}+\vec{q}\cdot\nabla\vec{q}=\vec{f}-\frac{\nabla p}{\rho}. \tag{2.41}$$

This equation is called the Euler equation. The general approach for simplifying (neglecting certain termes) the Navier–Stokes equations is discussed in Chapter 6. In situations in which the problem has cylindrical or spherical symmetry, the use of appropriate coordinates can simplify the solution. As an example, the fundamental equations for an incompressible fluid with constant viscosity are presented. The cylindrical coordinate system is described in Fig. 2.14, and for this example the r, θ coordinates are in a plane normal to the x coordinate. The operators ∇, ∇^2, and $\frac{D}{Dt}$ in the r, θ, x system are (see [3, p. 38] or [4, p. 132])

$$\nabla=\left(\vec{e}_r\frac{\partial}{\partial r},\vec{e}_\theta\frac{1}{r}\frac{\partial}{\partial\theta},\vec{e}_x\frac{\partial}{\partial x}\right), \tag{2.42}$$

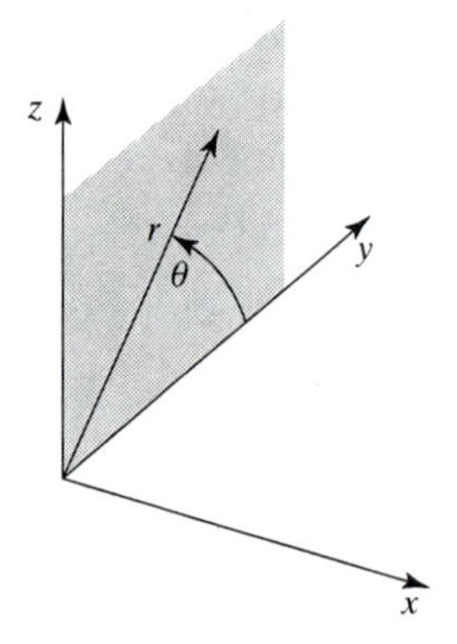

Figure 2.14. A cylindrical coordinate system.

$$\nabla^2 = \frac{\partial^2}{\partial r^2} + \frac{1}{r}\frac{\partial}{\partial r} + \frac{1}{r^2}\frac{\partial^2}{\partial \theta^2} + \frac{\partial^2}{\partial x^2}, \tag{2.43}$$

$$\frac{D}{Dt} = \frac{\partial}{\partial t} + q_r\frac{\partial}{\partial r} + \frac{q_\theta}{r}\frac{\partial}{\partial \theta} + q_x\frac{\partial}{\partial x}. \tag{2.44}$$

The continuity equation in cylindrical coordinates for an incompressible fluid then becomes

$$\frac{\partial q_r}{\partial r} + \frac{1}{r}\frac{\partial q_\theta}{\partial \theta} + \frac{\partial q_x}{\partial x} + \frac{q_r}{r} = 0. \tag{2.45}$$

The momentum equation for an incompressible fluid in the r direction is

$$\rho\left(\frac{Dq_r}{Dt} - \frac{q_\theta^2}{r}\right) = \rho f_r - \frac{\partial p}{\partial r} + \mu\left(\nabla^2 q_r - \frac{q_r}{r^2} - \frac{2}{r^2}\frac{\partial q_\theta}{\partial \theta}\right), \tag{2.46}$$

in the θ direction is

$$\rho\left(\frac{Dq_\theta}{Dt} + \frac{q_r q_\theta}{r}\right) = \rho f_\theta - \frac{1}{r}\frac{\partial p}{\partial \theta} + \mu\left(\nabla^2 q_\theta + \frac{2}{r^2}\frac{\partial q_r}{\partial \theta} - \frac{q_\theta}{r^2}\right), \tag{2.47}$$

and in the x direction is

$$\rho\frac{Dq_x}{Dt} = \rho f_x - \frac{\partial p}{\partial x} + \mu\nabla^2 q_x. \tag{2.48}$$

A spherical coordinate system with the coordinates r, θ, φ is described in Fig. 2.15. The operators ∇, ∇^2, and $\frac{D}{Dt}$ in the r, θ, φ system are ([5, Chapter 2] or [4, p. 132])

$$\nabla = \left(\vec{e}_r\frac{\partial}{\partial r}, \vec{e}_\theta\frac{1}{r}\frac{\partial}{\partial \theta}, \vec{e}_\varphi\frac{1}{r\sin\theta}\frac{\partial}{\partial \varphi}\right), \tag{2.49}$$

$$\nabla^2 = \frac{1}{r^2}\frac{\partial}{\partial r}\left(r^2\frac{\partial}{\partial r}\right) + \frac{1}{r^2\sin\theta}\frac{\partial}{\partial \theta}\left(\sin\theta\frac{\partial}{\partial \theta}\right) + \frac{1}{r^2\sin^2\theta}\frac{\partial^2}{\partial \varphi^2}, \tag{2.50}$$

$$\frac{D}{Dt} = \frac{\partial}{\partial t} + q_r\frac{\partial}{\partial r} + \frac{q_\theta}{r}\frac{\partial}{\partial \theta} + \frac{q_\varphi}{r\sin\theta}\frac{\partial}{\partial \varphi}. \tag{2.51}$$

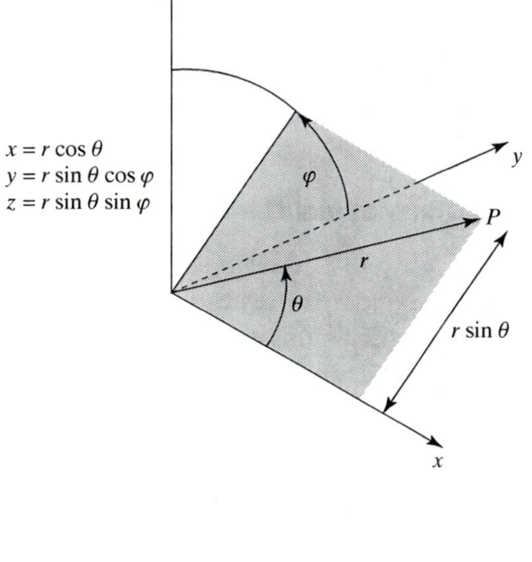

Figure 2.15. The spherical coordinate system.

The continuity equation in spherical coordinates for an incompressible fluid becomes ([3, p. 40])

$$\frac{1}{r}\frac{\partial(r^2 q_r)}{\partial r} + \frac{1}{\sin\theta}\frac{\partial(q_\theta \sin\theta)}{\partial\theta} + \frac{1}{\sin\theta}\frac{\partial q_\varphi}{\partial\varphi} = 0. \tag{2.52}$$

The momentum equation for an incompressible fluid is ([3, p. 40]), in the r direction is

$$\rho\left(\frac{Dq_r}{Dt} - \frac{q_\varphi^2 + q_\theta^2}{r}\right)$$
$$= \rho f_r - \frac{\partial p}{\partial r} + \mu\left(\nabla^2 q_r - \frac{2q_r}{r^2} - \frac{2}{r^2}\frac{\partial q_\theta}{\partial\theta} - \frac{2q_\theta \cot\theta}{r^2} - \frac{2}{r^2\sin\theta}\frac{\partial q_\varphi}{\partial\varphi}\right), \tag{2.53}$$

in the θ direction is

$$\rho\left(\frac{Dq_\theta}{Dt} + \frac{q_r q_\theta}{r} - \frac{q_\varphi^2\cot\theta}{r}\right)$$
$$= \rho f_\theta - \frac{1}{r}\frac{\partial p}{\partial\theta} + \mu\left(\nabla^2 q_\theta + \frac{2}{r^2}\frac{\partial q_r}{\partial\theta} - \frac{q_\theta}{r^2\sin^2\theta} - \frac{2\cos\theta}{r^2\sin^2\theta}\frac{\partial q_\varphi}{\partial\varphi}\right), \tag{2.54}$$

and in the φ direction is

$$\rho\left(\frac{Dq_\varphi}{Dt} + \frac{q_\varphi q_r}{r} + \frac{q_\theta q_\varphi\cot\theta}{r}\right)$$
$$= \rho f_\varphi - \frac{1}{r\sin\theta}\frac{\partial p}{\partial\varphi} + \mu\left(\nabla^2 q_\varphi - \frac{q_\varphi}{r^2\sin^2\theta} + \frac{2}{r^2\sin\theta}\frac{\partial q_r}{\partial\varphi} + \frac{2\cos\theta}{r^2\sin^2\theta}\frac{\partial q_\theta}{\partial\varphi}\right). \tag{2.55}$$

When a 2D flow field is treated in this book, it is described in either a Cartesian coordinate system with coordinates x and z or in a corresponding polar coordinate system with coordinates r and θ (see Fig. 2.16). In this polar coordinate system, we obtain the continuity equation for an incompressible fluid from Eq. (2.45) by eliminating $\frac{\partial q_x}{\partial x}$ and the r- and θ-momentum equations for an incompressible fluid are identical to Eqs. (2.46) and (2.47), respectively.

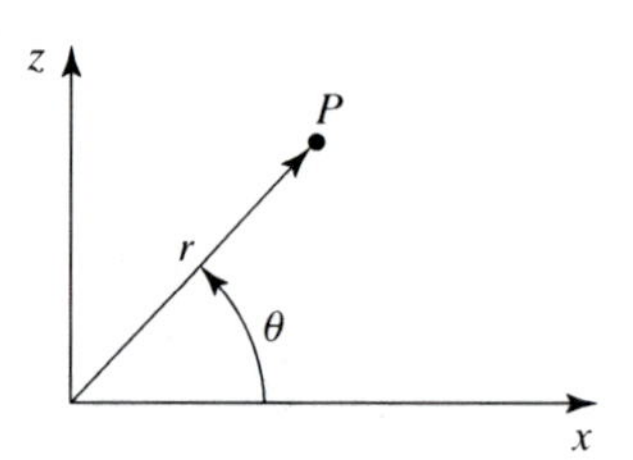

Figure 2.16. 2D polar coordinate system.

2.8 The Material Derivative

In the process of developing the differential form of the momentum equation, a sequence of derivatives called material derivatives was obtained. In this section we return to discuss this operator and to demonstrate that it represents fluid acceleration. Consider an Eulerian description of the fluid motion in which the velocity of a particle is

$$\vec{q} = (u, v, w).$$

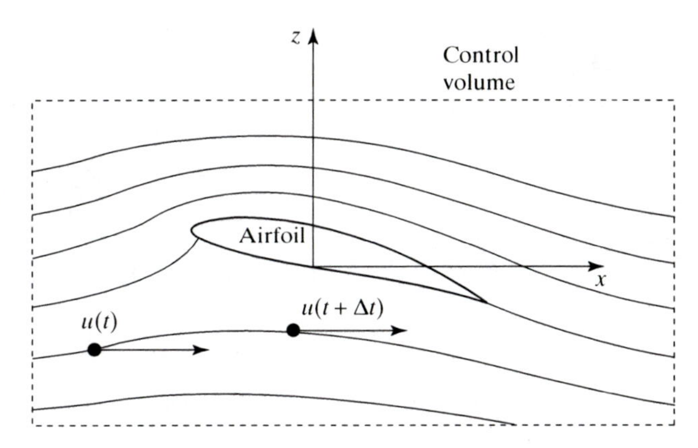

Figure 2.17. The motion of a particle in a fixed control volume.

The Eulerian control volume of Fig. 2.3 is redrawn in Fig. 2.17, and the velocity components (e.g., u, v, or w) are a function of location and time. For example, the velocity in the x direction of a selected particle [shown for simplicity as $u(t)$ in the figure] is described by the function f:

$$u = f(x, y, z, t). \tag{2.56}$$

After a short time interval Δt, the particle has moved to a new location, as shown, and its velocity is now $u + \Delta u$ [shown as $u(t + \Delta t)$ in the figure]. To estimate the change in the velocity we can use Eq. (2.56) as

$$u + \Delta u = f(x + \Delta x, y + \Delta y, z + \Delta z, t + \Delta t). \tag{2.57}$$

Assuming all quantities appearing with Δ are small, we can expand Eq. (2.57) into a Taylor series:

$$\begin{aligned} u + \Delta u = f(x, y, z, t) + \frac{\partial f}{\partial x}\Delta x + \frac{\partial f}{\partial y}\Delta y + \frac{\partial f}{\partial z}\Delta z + \frac{\partial f}{\partial t}\Delta t \\ + \frac{\partial^2 f}{\partial x^2}\frac{(\Delta x)^2}{2} + \frac{\partial^2 f}{\partial y^2}\frac{(\Delta y)^2}{2} + \cdots . \end{aligned} \tag{2.58}$$

The distance the particle traveled during the time interval Δt is

$$\Delta x = u\Delta t, \quad \Delta y = v\Delta t, \quad \Delta z = w\Delta t. \tag{2.59}$$

Substituting these into Eq. (2.58), recalling that $u = f(x, y, z, t)$, and neglecting the higher-order terms, we obtain Eq. (2.58) as

$$\Delta u = \frac{\partial f}{\partial x}u\Delta t + \frac{\partial f}{\partial y}v\Delta t + \frac{\partial f}{\partial z}w\Delta t + \frac{\partial f}{\partial t}\Delta t.$$

Dividing by Δt, recalling that f is actually u, and taking the limit for very small increments, we have

$$a_x = \lim \frac{\Delta u}{\Delta t_{\to 0}} = u\frac{\partial u}{\partial x} + v\frac{\partial u}{\partial y} + w\frac{\partial u}{\partial z} + \frac{\partial u}{\partial t}. \tag{2.60}$$

Thus the derivative of the x component of the velocity is the particle acceleration a_x in the x direction.* As noted, this operator is called the *material derivative*, and it can be applied to any other property:

$$\frac{D}{Dt} = \frac{\partial}{\partial t} + u\frac{\partial}{\partial x} + v\frac{\partial}{\partial y} + w\frac{\partial}{\partial z}. \tag{2.61}$$

Of course, accelerations in the y and z directions will have a similar form; for the 3D particle acceleration we can recall Eq. (2.32):

$$\vec{a} = \frac{D\vec{q}}{Dt} = \frac{\partial \vec{q}}{\partial t} + \vec{q} \cdot \nabla \vec{q}. \tag{2.32}$$

To rewrite the momentum equation in the x direction, recall the acceleration term from Eq. (2.60),

$$a_x = \frac{\partial u}{\partial t} + u\frac{\partial u}{\partial x} + v\frac{\partial u}{\partial y} + w\frac{\partial u}{\partial z}, \tag{2.60a}$$

and the acceleration times the density must be equal to the sum of the forces acting in the x direction:

$$\rho a_x = \rho\left(\frac{\partial u}{\partial t} + u\frac{\partial u}{\partial x} + v\frac{\partial u}{\partial y} + w\frac{\partial u}{\partial z}\right) = \sum F_x. \tag{2.62}$$

Once the force term is found, the Navier–Stokes equation will result.

The general approach for simplifying (neglecting certain terms) in the Navier–Stokes equation is discussed in Chapter 6.

2.9 Alternative Derivation of the Fluid Dynamic Equations

The differential form of the continuity and momentum equations so far is based (for sake of simplicity) on the integral formulation. However, these equations can be derived by use of elementary principles. To demonstrate the method, let us consider an infinitesimal (very small) cubical control volume, as shown in Fig. 2.18. For example, the flow in the y direction enters the control volume across an area of $ds = dxdz$ at a speed of v. The mass flow rate across this plane is therefore $(\rho v)dxdz$. On the other side of the cube (at a distance of dy) both velocity and density may have

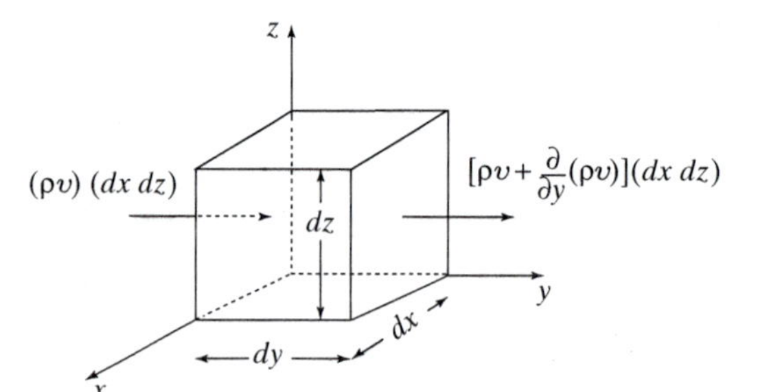

Figure 2.18. Infinitesimal control volume (Cartesian coordinates).

* The approach used here was aimed at depicting the meaning of the material derivative. The acceleration into the x direction expressed by Eq. (2.60) can be obtained by use of the chain rule for the derivative of $f(x, y, z, t)$.

$$\frac{du}{dt} = \frac{\partial f}{\partial t} + \frac{\partial f}{\partial x}\frac{\partial x}{\partial t} + \frac{\partial f}{\partial y}\frac{\partial y}{\partial t} + \frac{\partial f}{\partial z}\frac{\partial z}{\partial t} = \frac{\partial u}{\partial t} + u\frac{\partial u}{\partial x} + v\frac{\partial u}{\partial y} + w\frac{\partial y}{\partial z}.$$

changed and the mass flow rate across the exit plane is $\left[\rho v + \frac{\partial}{\partial y}(\rho v)\,dy\right] dxdz$. The net change in the y direction is therefore

$$(\rho v)dxdz - \left[\rho v + \frac{\partial}{\partial y}(\rho v)\,dy\right] dxdz = -\left[\frac{\partial}{\partial y}(\rho v)\right] dxdydz,$$

and similar expressions can be formulated in the x and z directions, as well.

The total change of mass in the cubical control volume is the sum of the changes in the three orthogonal directions:

$$-\left[\frac{\partial}{\partial x}(\rho u)\right] dxdydz - \left[\frac{\partial}{\partial y}(\rho v)\right] dxdydz - \left[\frac{\partial}{\partial z}(\rho w)\right] dxdydz.$$

Now let us examine the conservation of mass principle, as stated in Eq. (2.20). According to this principle, the accumulation of mass inside the cubical element is:

$$\frac{\partial m}{\partial t} = \frac{\partial \rho}{\partial t}(dV) = \frac{\partial \rho}{\partial t} dxdydz,$$

which must be equal to the change that is due to the flux in and out of the control volume:

$$\frac{\partial \rho}{\partial t} dxdydz = -\left[\frac{\partial}{\partial x}(\rho u)\right] dxdydz - \left[\frac{\partial}{\partial y}(\rho v)\right] dxdydz - \left[\frac{\partial}{\partial z}(\rho w)\right] dxdydz.$$

Canceling the volume of the cubical element ($dx\,dy\,dz$) results in the continuity equation in Cartesian coordinates:

$$\frac{\partial \rho}{\partial t} + \frac{\partial}{\partial x}(\rho u) + \frac{\partial}{\partial y}(\rho v) + \frac{\partial}{\partial z}(\rho w) = 0, \tag{2.63}$$

and this is exactly the same as Eq. (2.31b):

$$\frac{\partial \rho}{\partial t} + u\frac{\partial \rho}{\partial x} + v\frac{\partial \rho}{\partial y} + w\frac{\partial \rho}{\partial z} + \rho\left(\frac{\partial u}{\partial x} + \frac{\partial v}{\partial y} + \frac{\partial w}{\partial z}\right) = 0. \tag{2.31b}$$

The momentum equation, too, can be developed with the cubical element in Fig. 2.18, but now the momentum balance must be made in the three directions (of the Cartesian coordinates selected). For example, we obtain the momentum change in the x direction by replacing ρ with ρu in Eq. (2.63). However, the result will include more terms, which can be eliminated by use of the continuity equation [Eq. (2.63)]. The result, of course, is the acceleration, as discussed in the previous section (about the material derivative). Because we already developed this term in Eq. (2.60), it can be readily applied. Let us demonstrate this in the x direction:

$$a_x = \frac{\partial u}{\partial t} + u\frac{\partial u}{\partial x} + v\frac{\partial u}{\partial y} + w\frac{\partial u}{\partial z}, \tag{2.64}$$

and the acceleration times the density must be equal to the sum of the forces acting in the x direction;

$$\rho a_x = \rho\left(\frac{\partial u}{\partial t} + u\frac{\partial u}{\partial x} + v\frac{\partial u}{\partial y} + w\frac{\partial u}{\partial z}\right) = \sum F_x. \tag{2.65}$$

The forces on the infinitesimal cubical element consist of body force, shear stress, and pressure, as discussed in Section 2.5. The body force acts on the mass inside the

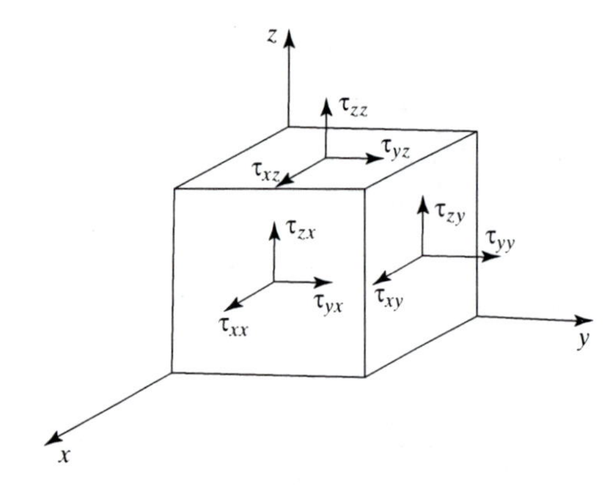

Figure 2.19. Stress components on a cubical fluid element.

cube and its magnitude in the x direction (per unit volume) is ρf_x. The stress on the cubical element was discussed in reference to Fig. 2.6, which is reintroduced here in Fig. 2.19.

The force in the x direction that is due to the pressure on the panels facing the plus and minus x directions is

$$\frac{\partial \tau_{xx}}{\partial x} = -\frac{\partial p}{\partial x}.$$

The force in the x directions that is due to the panels facing the plus and minus y directions is a result of the change in the shear on those panels:

$$\frac{\partial \tau_{xy}}{\partial y}.$$

Finally, a similar term that is due to the change in shear on the panels facing the plus and minus z directions is

$$\frac{\partial \tau_{xz}}{\partial z}.$$

The total force on the cubical element in the x direction is therefore the sum of these components:

$$\sum F_x = \rho f_x + \frac{\partial \tau_{xx}}{\partial x} + \frac{\partial \tau_{xy}}{\partial y} + \frac{\partial \tau_{xz}}{\partial z}.$$

The momentum balance in the x direction, as noted earlier, is

$$\rho a_x = \sum F_x,$$

and by substituting the acceleration and the force terms (for the x direction only), we get the same result as that presented earlier in Eq. (2.36b):

$$\rho\left(\frac{\partial u}{\partial t} + u\frac{\partial u}{\partial x} + v\frac{\partial u}{\partial y} + w\frac{\partial u}{alz}\right) = \sum F_x = \rho f_x + \frac{\partial \tau_{xx}}{\partial x} + \frac{\partial \tau_{xy}}{\partial y} + \frac{\partial \tau_{xz}}{\partial z}. \quad (2.66)$$

Of course, the same procedure must be repeated for the y and z directions. Finally, replacing the stress terms with their definitions in Eq. (2.14) results in the Navier–Stokes equations as given by Eqs. (2.37).

2.10 Summary and Concluding Remarks

The equations developed in this chapter provide the foundations for fluid dynamics and are used in almost every chapter that follows. Therefore it is useful to regroup them at the end of the chapter where it is easier to locate them. Let us start with the integral fom of the equations. The first, the continuity equation, states that no fluid is lost:

$$\frac{\partial}{\partial t}\int_{\text{c.v.}} \rho dV + \int_{\text{c.s.}} \rho(\vec{q}\cdot\vec{n})dS = 0. \tag{2.20}$$

The second equation, the conservation of momentum, is

$$\frac{\partial}{\partial t}\int_{\text{c.v.}} \rho q_i dV + \int_{\text{c.s.}} \rho q_i(\vec{q}\cdot\vec{n})dS = \int_{\text{c.v.}} \rho f_i dV + \int_{\text{c.s.}} n_j \tau_{ij} dS. \tag{2.24}$$

The same equations derived in a differential form, which is more frequently used for computations, are (starting with the continuity equation)

$$\frac{\partial\rho}{\partial t} + \nabla\cdot\rho\vec{q} = 0, \tag{2.31}$$

and for an incompressible fluid where $\rho = \text{const.}$,

$$\nabla\cdot\vec{q} = \frac{\partial u}{\partial x} + \frac{\partial v}{\partial y} + \frac{\partial w}{\partial z} = 0. \tag{2.33}$$

Here also the Cartesian form was added. The momentum equation still states the same thing, that the acceleration is a result of the forces acting on the fluid:

$$\rho a_i = \sum F_i. \tag{2.36}$$

Writing this in differential form for a Newtonian fluid, we obtain

$$\rho\left(\frac{\partial q_i}{\partial t} + \vec{q}\cdot\nabla q_i\right) = \rho f_i - \frac{\partial}{\partial x_i}\left(p + \frac{2}{3}\mu\nabla\cdot\vec{q}\right) + \frac{\partial}{\partial x_j}\mu\left(\frac{\partial q_i}{\partial x_j} + \frac{\partial q_j}{\partial x_i}\right) \quad (i = 1, 2, 3), \tag{2.37}$$

and by assuming a constant-viscosity coefficient μ, we have

$$\rho\left(\frac{\partial\vec{q}}{\partial t} + \vec{q}\cdot\nabla\vec{q}\right) = \rho\vec{f} - \nabla p + \mu\nabla^2\vec{q} + \frac{\mu}{3}\nabla(\nabla\cdot\vec{q}). \tag{2.39}$$

For an incompressible fluid we have

$$\rho\left(\frac{\partial\vec{q}}{\partial t} + \vec{q}\cdot\nabla\vec{q}\right) = \rho\vec{f} - \nabla p + \mu\nabla^2\vec{q}. \tag{2.40}$$

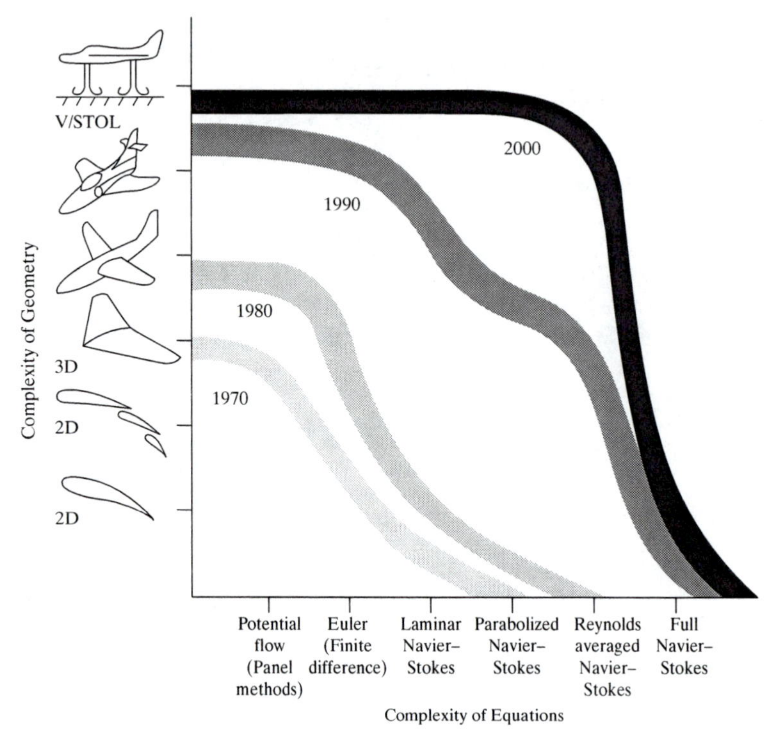

Figure 2.20. Progress in solving the fluid dynamic equations.

For an inviscid compressible fluid we end up with the Euler equation:

$$\frac{\partial \vec{q}}{\partial t} + \vec{q} \cdot \nabla \vec{q} = \vec{f} - \frac{\nabla p}{\rho}. \tag{2.41}$$

These equations are complex, and analytical solutions exist only for simplified forms, as will be demonstrated later. However, tremendous progress has been made in numerical solutions (see Chapter 9) in recent years, and this is sumarized in Fig. 2.20.

In Fig. 2.20 the complexity of the flow field is shown schematically on the ordinate and the complexity of the applicable equations is on the abscissa. For example, in the 1970s, the flow over a 3D wing could be solved by use of potential flow methods. Some 20 years later, we could solve (numerically) the same problem by solving the Navier–Stokes equation and modeling turbulent effects. After the year 2000, the complexity of the geometry that can be modeled increased gradually, as shown in the figure.

REFERENCES

[1] Batchelor, G. K., *An Introduction to Fluid Mechanics,* Cambridge University Press, New York, 1967.
[2] Kellogg, O. D., *Foundation of Potential Theory*, Dover, New York, 1953.
[3] Pai, Shih-I, *Viscous Flow Theory*, Van Nostrand, New York, 1956.
[4] Yuan, S. W., *Foundations of Fluid Mechanics*, Prentice-Hall, Englewood Cliffs, NJ, 1969.
[5] Karamcheti, K., *Principles of Ideal-Fluid Aerodynamics*, Krieger, New York, 1980.

PROBLEMS

2.1. Calculate the net force that is due to the shear forces acting on two opposite sides of the cube shown in the figure. What is the direction of the resultant (net) force? (Cube dimensions are 2 m by 2 m by 2 m).

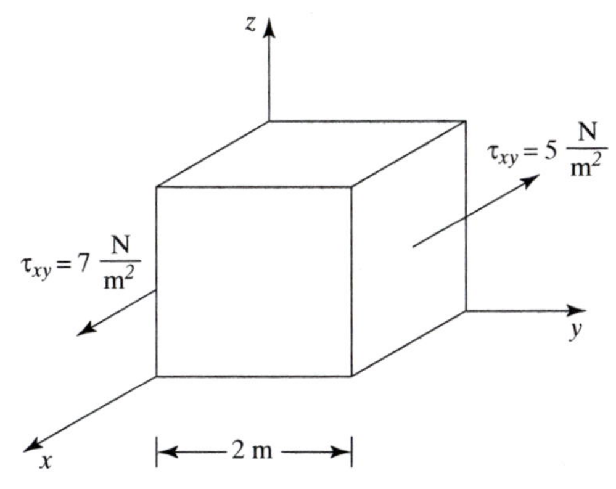

Problem 2.1.

2.2. The velocity field in the x–z plane is given by the the following expressions:

$$u = x,$$
$$w = -z.$$

Determine the equation describing the streamline passing through point (2,1) and sketch the flow field described by this formula.

2.3. The velocity field in the x–z plane is given by the following expressions:

$$u = -3z^2,$$
$$w = -6x.$$

Determine the equation describing the streamline passing through point (1,1) and sketch the flow field described by this formula.

2.4. A 2D velocity field is given by the following equations:

$$u = \frac{x}{x^2 + z^2},$$
$$w = \frac{z}{x^2 + z^2}.$$

Check if these satisfy the incompressible continuity equation.

2.5. The velocity distribution between two parallel horizontal plates is given by the following expression:

$$u(z) = -k\left[\frac{z}{h} - \frac{z^2}{h^2}\right],$$

where k is a constant and h (in the z direction) is the clearance between the two plates. Calculate the shear along a vertical line (between $z = 0$ and $z = h$). Where does the shear force reaches its maximum value?

2.6. Based on Eq. (2.37) write the 2D Euler equation in Cartesian coordinates.

2.7. Based on Eqs. (2.46) and (2.47) write the 2D Euler equation in cylindrical coordinates (r–θ directions).

2.8. Based on Eq. (2.37) write the 3D incompressible Navier–Stokes equation in Cartesian coordinates.

2.9. A 2D steady-state flow between two parallel plates is given by the following functions:

$$u = a\left[\frac{z}{h} - \left(\frac{z}{h}\right)^2\right], \quad 0 < z < h,$$
$$w = 0.$$

Note that the velocity is in the x direction only, but the flow is still 2D.

(a) Find the maximum velocity.

(b) Calculate the flow rate Q at a station $x =$ const., which is given by the integral from the continuity equation,

$$Q = \int_{\text{c.s.}} \rho(\vec{q} \cdot \vec{n})dS.$$

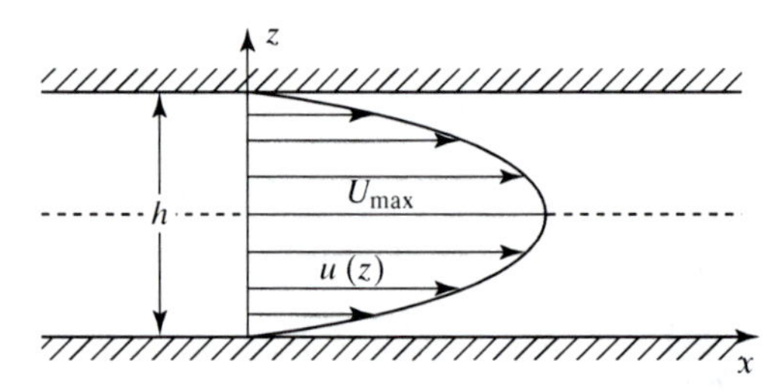

Problem 2.9.

2.10. Calculate the force per unit area (in the previous problem) acting on the surface $z = 0$ and on the surface $z = h$. Assume fluid viscosity, μ, is known.

2.11. Fluid is flowing through a circular pipe into a cubical control volume, as shown in the figure. If the cube dimensions are $a = 30$ cm and $S = 3$ cm^2, the velocity vector is (0. – 1,0) m/s, and the outside density is $\rho_a = 1.0$ kg/m^3, determine the value of the integral

$$\int_{\text{c.s.}} \rho(\vec{q} \cdot \vec{n})dS$$

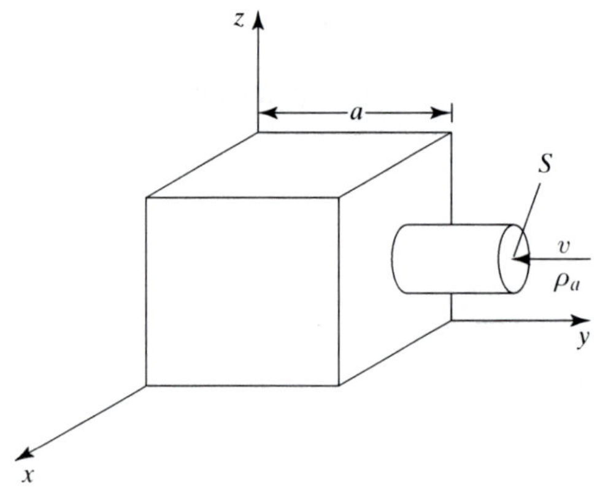

Problem 2.11.

2.12. Evaluate the value of the integral $\frac{\partial}{\partial t}\int_{\text{c.v.}} \rho dV$ for the control volume discussed in the previous problem. Assume initial conditions at $t = 0$, $\rho = 0$.

2.13. Calculate the force and direction of the force acting on the cube. Assume the pressure at the pipe inlet is the same as the pressure surrounding the cube.

2.14. A 2D steady-state velocity field is described by the following velocity components: $u = 5x$, $w = -5z$. Calculate the corresponding acceleration field (using the material derivative). What is the magnitude and direction of the acceleration at point $x = 1$, $z = 1$?

2.15. A 2D steady-state velocity field is described by the following velocity components: $u = \frac{x}{x^2+z^2}$, $w = \frac{z}{x^2+z^2}$. Provide a graphical representation of the velocity distribution along the following lines: (a) $z = 0$ and (b) $x = 0$.

2.16. A 2D steady-state velocity field is described by the following velocity components: $u = \frac{x}{x^2+z^2}$, $w = \frac{z}{x^2+z^2}$. Calculate the corresponding acceleration field (using the material derivative). What is the magnitude and direction of the acceleration at point $x = 0$, $z = 1$?

2.17. Try to sketch the streamlines from the previous problem, starting at the origin. What is the shape of this flow?

2.18. A 2D steady-state velocity field is described by the following velocity components: $u = \frac{z}{x^2+z^2}$, $w = \frac{x}{x^2+z^2}$. Calculate the corresponding acceleration field (using the material derivative). What is the magnitude and direction of the acceleration at point $x = 0$, $z = 1$?

2.19. Try to sketch the streamlines from the previous problem, starting at the origin. What is the shape of this flow?

2.20. A 2D steady-state velocity field is described by the following velocity components: $u = 1 + 2x + z$, $w = 1 - 2x + 3z$. Calculate the corresponding acceleration field (using the material derivative). What is the magnitude and direction of the acceleration at point $x = 1$, $z = 1$?

2.21. An incompressible (1D as in Fig. 2.9) steady-state flow in a circular diffuser is flowing from station 1 at $x = 0$ (where the velocity is u) to station 2 at $x = x_2$. The inlet radius at station 1 is r_1 and at station 2 the radius is r_2. Assuming $r_2 > r_1$, develop an expression for the fluid acceleration as a function of x.

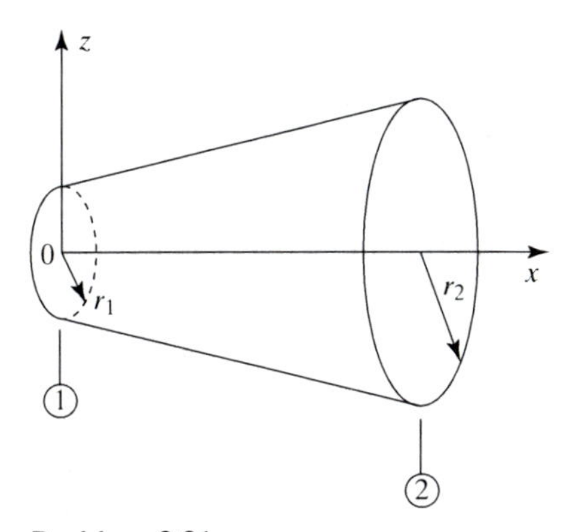

Problem 2.21.

2.22. An incompressible (1D) steady-state flow in a circular diffuser is flowing from station 1 at $x_1 = 0$ (where the velocity is $u = 3$ m/s) to station 2 at $x_2 = 10$ cm. The inlet radius at station 1 is $r_1 = 5$ cm and at station 2 the radius is $r_2 = 10$ cm. Calculate the fluid acceleration as a function of x.

2.23. An incompressible (1D) steady-state flow in a nozzle with a rectangular cross section is flowing from station 1 at $x = 0$ (where the velocity is u) to station 2 at $x = x_2$. The inlet height at station 1 is b_1 and at station 2 is b_2. Assuming that $b_2 < b_1$, develop an expression for the fluid acceleration as a function of x.

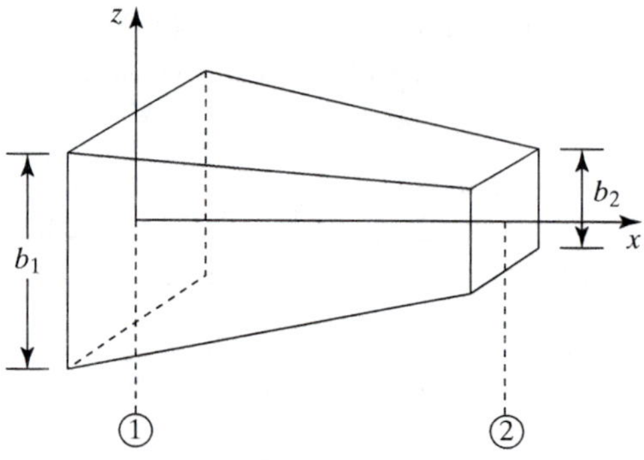

Problem 2.23.

2.24. An incompressible (1D) steady-state flow in a nozzle with a rectangular cross section is flowing from station 1 at $x = 0$ (where the velocity is $u = 3$ m/s) to station 2 at $x = 10$ cm. If the inlet height at station 1 is $b_1 = 10$ cm and at station 2 is $b_2 = 5$ cm, calculate the fluid acceleration as a function of x.

2.25. A 2D steady-state flow is given by the following functions:

$$u = a_1 z + a_2 z^2, \quad 0 < z < h,$$
$$w = 0.$$

Note that the velocity is in the x direction only, but the flow is still 2D.

(a) Sketch the velocity distribution along a vertical line (between $z = 0$ and $z = h$).

(b) Calculate the shear stress τ_{xz} and its value at $z = 0$.

2.26. The velocity profile above a horizontal flat plate (within the range of $0 < z < \delta$) is described by the following function:

$$\frac{u}{U_e} = \sin\left(\frac{\pi z}{2\delta}\right), \quad z \le \delta.$$

Calculate the force per unit area acting on the surface $z = 0$. Assume fluid viscosity of μ.

2.27. The velocity profile above a horizontal flat plate (within the range of $0 < z < \delta$) is described by the following function:

$$\frac{u}{U_e} = \left(\frac{z}{\delta}\right)^{\frac{1}{9}}, \quad z \le \delta.$$

(a) Calculate the force per unit area acting on the surface $z = 0$. Assume fluid viscosity of μ.

(b) Compare with the force calculated in the previous problem. Which one is higher? (Note that these two formulas relate to the shape of a boundary layer above a flat plate; the latter is for turbulent flows).

2.28. Wine in poured from a barrel into a liter bottle, as shown in the figure. At a certain point, the fluid velocity is 1 m/s, and the cross-section area is 0.5 cm^2. How long does it take to fill the 1-L bottle?

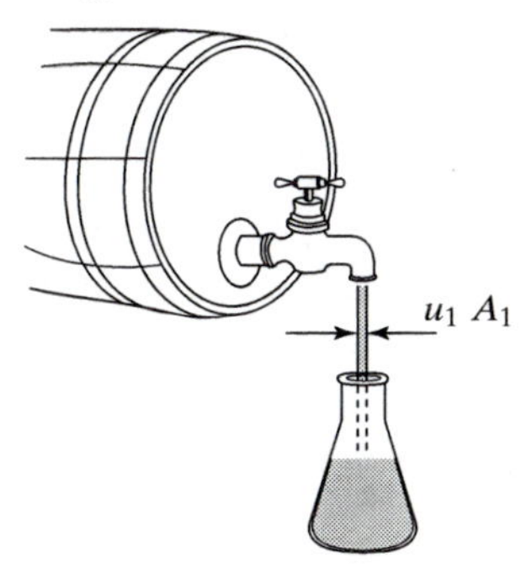

Problem 2.28.

2.29. A firefighter holds a water hose, as shown in the figure. The water leaves the nozzle at a velocity of 10 m/s, and the diameter of the circular jet is 0.03 m. Assuming the pressure at the exit of the nozzle is atmospheric, calculate the force pushing the firefighter backward ($\rho_{water} = 1000$ kg/m^3).

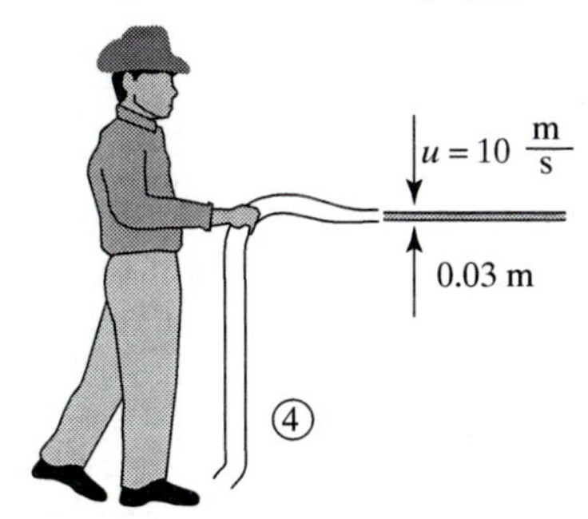

Problem 2.29.

2.30. A small fan engine (as shown in the figure) is tested and the incoming speed u_1 is 100 m/s. The inlet area is $A_1 = 0.1$ m^2 and the exhaust speed is $u_2 = 200$ m/s. Assuming that the flow is accelerated by the fan and that the pressure and density are the same at the inlet and exit (and $\rho = 1.2$ kg/m^3), calculate the exit area A_2. Also calculate the thrust generated by this unit.

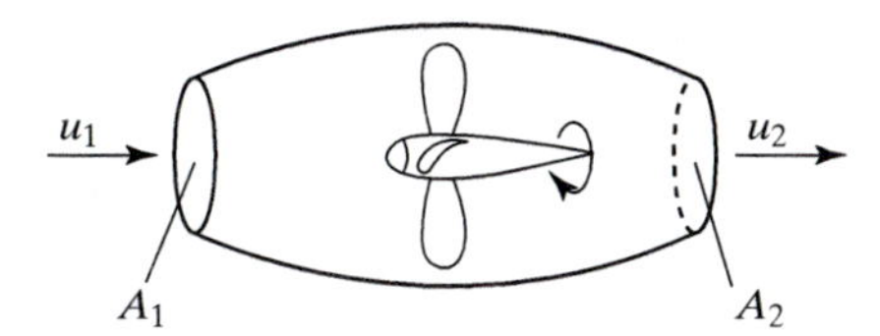

Problem 2.30.

2.31. Water enters a 0.05-m-diameter tube at section 1 at an average speed of 0.5 m/s and exits through stations 2 and 3, as shown. The diameter of the exit at station 2 is 0.03 m and the average velocity there is 0.5 m/s. as well. Calculate the average exit velocity at station 3 if the tube diameter there is 0.02 m.

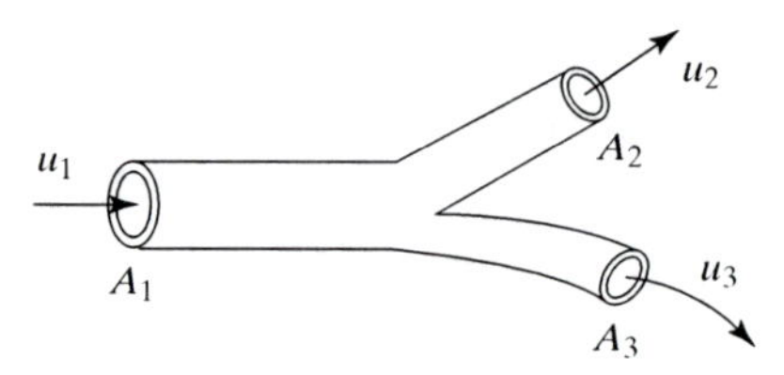

Problem 2.31.

2.32. At $t = 0$, the pipe above an empty cylindrical container is opened, pouring water at a rate of $Q_1 = 1$ L/s. At the bottom of the container, an open pipe can drain the container at a rate of $Q_2 = 0.3$ L/s. How long it will take to fill up the container (with the dimensions shown in the figure)? Assume that Q_2 is independent of the water level in the container.

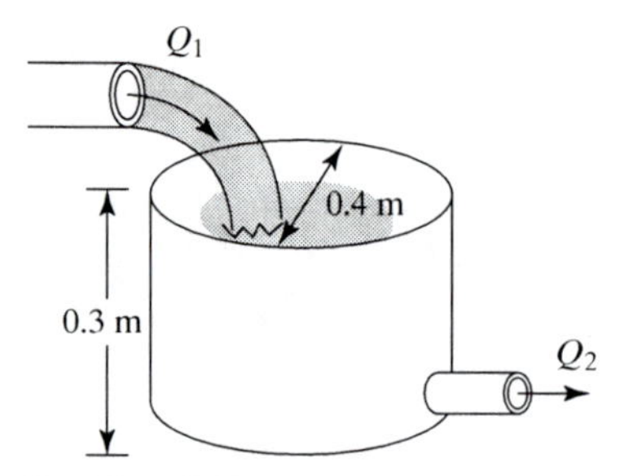

Problem 2.32.

2.33. A water stream is flowing from a pipe at a rate of $Q = 1$ L/s into a container placed on a scale, as shown in the figure. Assuming that the effective height of the water column is 1 m (e.g., the vertical velocity is zero at that height), calculate the force measured by the scale.

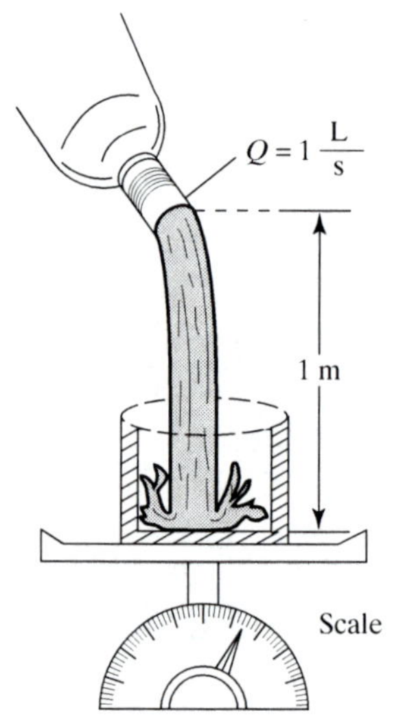

Problem 2.33.

2.34. Oil is poured at a rate of $Q = 1$ L/s into a conical funnel of $r = 0.3$ m, as shown in the figure. Assuming it exits at the bottom of the funnel at a constant average speed of $u = 0.8$ m/s through a 0.03-m-diameter tube, calculate how long it will take to fill up the conical section (the volume of a cone = $1/3 \times$ base area $\times$ height).

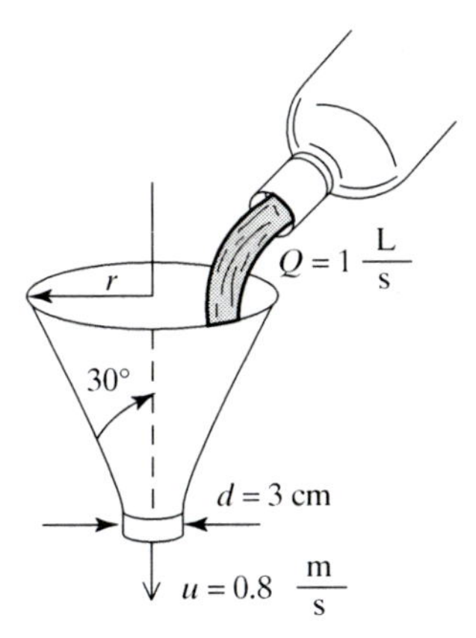

Problem 2.34.

2.35. A water jet hits horizontally a 50-kg block. The friction coefficient between the block and the ground is 0.9. What is the minimum diameter d of the water jet in order for the block to slide to the left? Assume that the jet speed, as it hits the block, is 20 m/s.

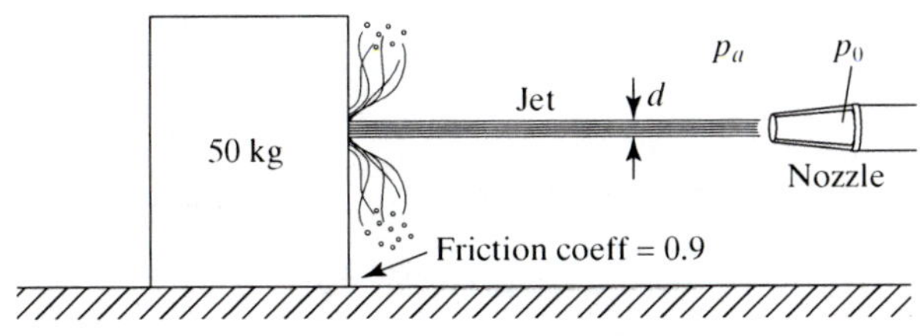

Problem 2.35.

2.36. Consider the flow in a circular pipe of radius R. Assume that the axis symmetric velocity distribution is given by the expression $q_x = A[1 - (\frac{r}{R})^2]$, where A is a constant. Calculate the volumetric flow rate and the average velocity.

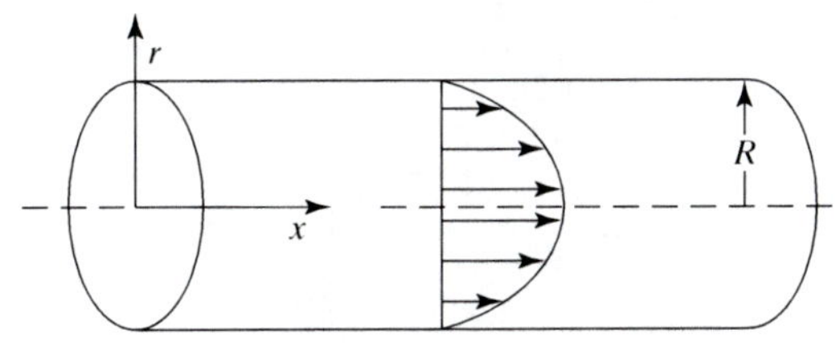

Problem 2.36.

2.37. Calculate the shear stress on the inner wall of the pipe shown in the figure for Problem 2.36 and estimate the shear force per unit length of the pipe. Find where in the fluid the shear stress is minimum.

2.38. A fluid jet exits a large container, and the 2D velocity distribution can be described as $u(z) = U(z/h)$, where U is the maximum velocity at the top.

Assuming the exit pressure is the same as the ambient ($p_e = p_a$), calculate the flow rate per unit width, the average velocity, and the force on the container.

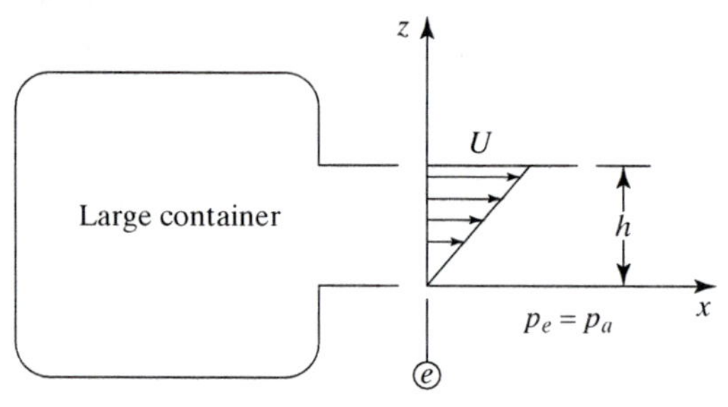

Problem 2.38.

2.39. The 2D velocity profile between a plate moving at a velocity U and a stationary plate is given by

$$u(z) = U \sin\left(\frac{\pi z}{2h}\right), \quad 0 \le z \le h.$$

Calculate the 2D flow rate (per unit width) crossing the $x = 0$ plane.

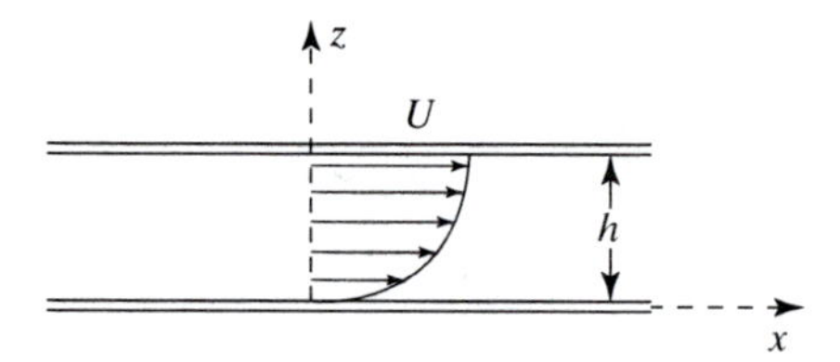

Problem 2.39.

2.40. Calculate the shear stress on the upper and lower walls in the previous problem.

3 Fluid Statics

3.1 Introduction

This chapter deals with the simplest form of the fluid dynamic equations, namely, when there is no motion at all.

Because there is no fluid motion, there is no shear strain, and the shear stress (and force) disappears. This provides the opportunity to focus on the effects and the forces resulting from the pressure. Important principles such as the center of pressure are also discussed.

3.2 Fluid Statics: The Governing Equations

In this case the fluid is at rest and the velocity vector is $\vec{q} = 0$. Consequently all terms in the continuity equations becomes zero because they contain the velocity vector. The momentum equation, however, contains other terms and will provide information on the forces that may exist in a resting fluid. As noted, because there is no velocity, there is no strain and no shear stress. Consequently the general stress formula (Eq. 2.12) becomes

$$\tau_{ij} = \begin{bmatrix} -p & 0 & 0 \\ 0 & -p & 0 \\ 0 & 0 & -p \end{bmatrix}. \tag{3.1}$$

This means that the only force acting on the fluid element described in Fig. 2.5 is the pressure, and its action is equal to all directions (at a point). This observation was documented by Blaise Pascal (French scientist, 1623–1662), who postulated that the pressure at a point in a fluid is independent of direction (sometimes called isetropic). Suppose we'd like to state this in a Cartesian system; then clearly

$$p_x = p_y = p_z = p. \tag{3.2}$$

With these assumptions, momentum equation (Eq. 2.36) in Cartesian coordinates reduces to

$$\rho a_x = \rho f_x - \frac{\partial p}{\partial x}, \tag{3.3a}$$

$$\rho a_y = \rho f_y - \frac{\partial p}{\partial y}, \tag{3.3b}$$

$$\rho a_z = \rho f_z - \frac{\partial p}{\partial z}, \tag{3.3c}$$

or in vector form

$$\rho \vec{a} = \rho \vec{f} - \nabla p \tag{3.4}$$

or

$$\rho(\vec{f} - \vec{a}) = \nabla p. \tag{3.4a}$$

Note that we have left the acceleration term $\vec{a}$ and at the same time we assumed that the velocity is zero. This is done in order to discuss cases in which we wish to include body forces that are due to solid-body acceleration (see Section 3.5). Therefore, in these cases, there is no "fluid motion" because the fluid particles do not move relative to each other.

In the following section we discuss the pressure field that is due to gravitational force and due to solid body acceleration. Once the pressure is known, the resulting forces on submerged surfaces including buoyancy can be calculated.

3.3 Pressure Due to Gravity

Let us consider a case in which the body forces $\vec{f}$ consist of only the gravitational force g, which acts in the vertical direction. The body-force vector is then

$$\vec{f} = (0, 0, -g), \tag{3.5}$$

and the "solid-body acceleration" is zero:

$$\vec{a} = 0.$$

Equations (3.3) then reduce to

$$0 = -\frac{\partial p}{\partial x}, \tag{3.6a}$$

$$0 = -\frac{\partial p}{\partial y}, \tag{3.6b}$$

$$0 = -\rho g - \frac{\partial p}{\partial z}. \tag{3.6c}$$

Assuming incompressible fluid and integrating the first two equations, we find that the pressure is constant in the x and y directions:

$$p(x) = p(y) = \text{const.} \tag{3.7}$$

Next, by integrating the third equation, we get

$$p(z) = -\int \rho g dz = -\rho g z + c, \tag{3.8}$$

where c is the constant of integration. Now let us select an x–z coordinate system attached to the top surface of a liquid, as shown in Fig. 3.1:

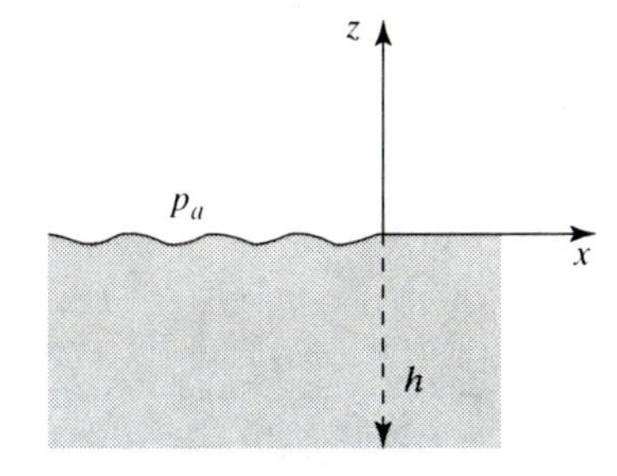

Figure 3.1. Coordinate system used to describe the hydrostatic pressure variation.

In this coordinate system, above the liquid and at $z = 0$, the pressure is $p(0) = p_a$, and by substituting this into Eq. (3.8) we get

$$p(z) = p_a - \rho g z. \tag{3.9a}$$

Or we can measure the depth of the liquid from the surface as h, and then

$$p(h) = p_a + \rho g h. \tag{3.9b}$$

Because the fluid is stationary we can compare this condition with that of solids. For example, let us compare it with a pile of bricks, as shown in Fig. 3.2. Assume that each brick weights 1 kg and its base area is 0.02 m^2. In the case of the left-hand column we have six blocks, and the pressure under them is

$$p_1 = \frac{6\ \text{kg}_f}{0.02\ \text{m}^2} = 300\frac{\text{kg}_f}{\text{m}^2}.$$

For the smaller pile, where $h_2 = h_1/2$, the pressure is reduced by the same ratio and therefore

$$p_2 = \frac{3\ \text{kg}_f}{0.02\ \text{m}^2} = 150\frac{\text{kg}_f}{\text{m}^2}.$$

In conclusion, the taller the column, the larger the pressure at the bottom. However, in case of the solid block pile, there is no pressure to the sides. In the case of a liquid column, there will be a pressure of magnitude $\rho g h$ acting to the sides, as well.

Now let us return to the observations of Blaise Pascal (1623–1662), who documented these relations. One of the consequences of Eqs. 3.9 is that the pressure at the bottom of the container shown in Fig. 3.3 is the same under all branches and it is $\rho g h$ (in addition to the ambient pressure p_a). Therefore the height of the liquid column in each of the branches is independent of the shape and will have the same height h.

We can continue with the example by sealing the top of the container and adding a longer tube onto one of the branches. Now we fill it with a small amount

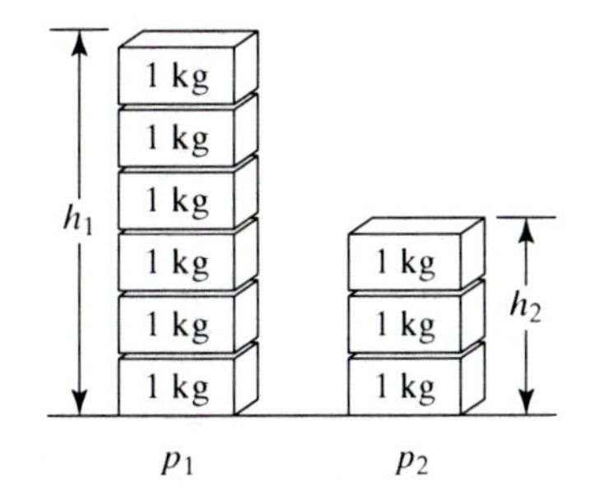

Figure 3.2. The pressure under a pile of bricks depends on the height h of the pile.

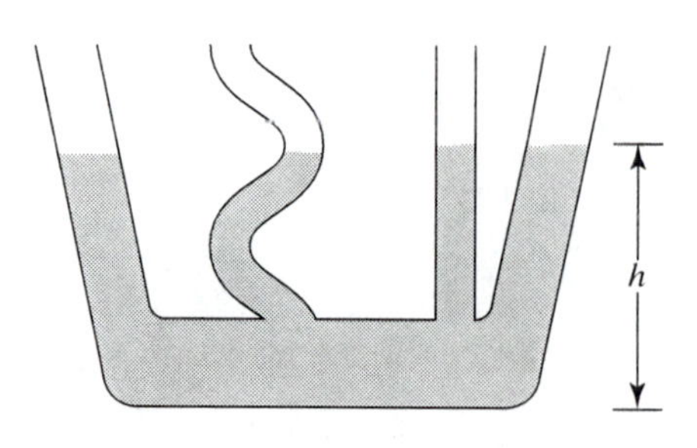

Figure 3.3. The fluid level h in a container is independent of the shape of the container.

of liquid to the height h_1 and the pressure inside will increase throughout the container. For example the pressure above the liquid level p_x (see Fig. 3.4) will increase according to Eqs. (3.9)

$$p_x = p_a + \rho g h_1,$$

and the pressure at the bottom of the container now is

$$p = p_a + \rho g(h + h_1);$$

this example shows how easy it is to increase the pressure in a pipe system. This principle is used to ensure sufficient high pressure in urban areas by the positioning of water towers on higher grounds.

From the same principle we can develop a pressure measurement device called the U-tube manometer (see Fig. 3.5). In this case a U-shaped tube is filled with a liquid with known density. When both sides of the U-tube are open the liquid column will have the same level. However, if one side is connected to the unknown pressure p_x, then the liquid column will move accordingly. In this case the measured pressure is higher than the ambient pressure p_a by

$$p_x = p_b = p_a + \rho g h.$$

Another important application for which we can credit Blaise Pascal's principle about the pressure's being isetropic is in hydraulic jacks. In this case the objective is to generate mechanical advantage by using the principle that the pressure is the same throughout the whole system (in this case we assume that the hydraulic pressure inside the system is very high and small changes in $\rho g h$ that are due to differences in the inner fluid level are negligible).

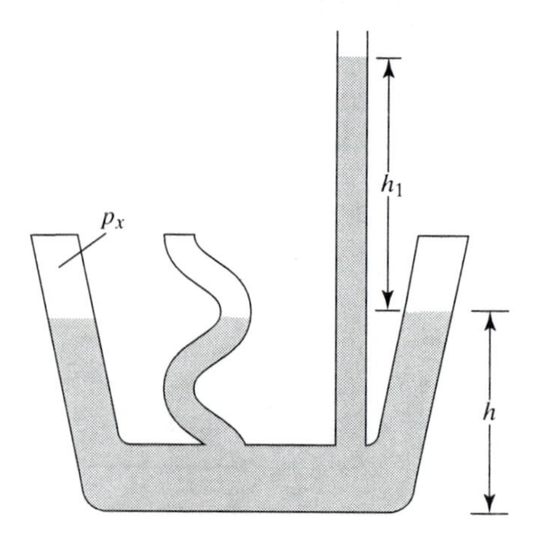

Figure 3.4. One method of increasing the pressure in a closed container by simply adding a liquid column h_1, as shown.

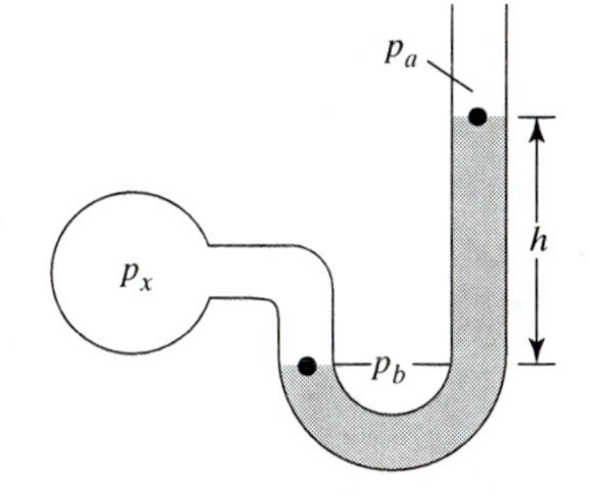

Figure 3.5. Principle of U-tube manometer. The liquid column height h allows us to measure the pressure p_x.

Now let us refer to Fig. 3.6. The pressure inside the hydraulic lift system is increased by application of a force F_1 on the smaller piston having an area of A_1. The resulting pressure is

$$p = \frac{F_1}{A_1}.$$

Because the pressure is the same in the hydraulic fluid (as noted, we neglect small differences between the fluid levels because the pressure inside is much larger than ρgh) we can write

$$p = \frac{F_1}{A_1} = \frac{F_2}{A_2}.$$

Consequently, the mechanical advantage for lifting the heavier car is

$$F_2 = F_1 \frac{A_2}{A_1}. \tag{3.10}$$

Although we gained in magnifying the lifting force, the work invested pV is unchanged because the force F_1 is pushed in deeper to displace the same volume of fluid V. So the total work invested w, is simply the pressure multiplied by the volume displaced:

$$w = F_1 h_1 = pA_1 h_1 = pV$$

and h_1 is the stroke in pump 1 in Fig. 3.6.

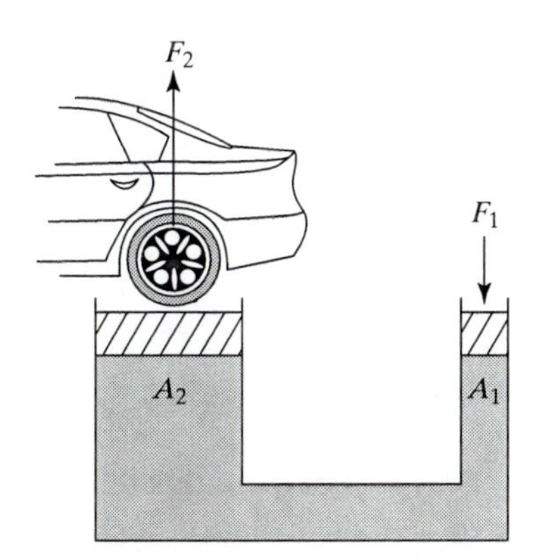

Figure 3.6. The principle of a hydraulic lift: using a smaller force F_1 to create a mechanical advantage in order to lift the vehicle.

3.4 Hydrostatic Pressure in a Compressible Fluid

Let us repeat the previous exercise, but now the stationary fluid is compressible. Therefore it is only logical that the lower layers (facing the higher pressure) will be compressed more. As in the previous example, let us assume that the only force acting is gravitation:

$$\vec{f} = (0, 0, -g). \tag{3.5}$$

Furthermore, we assume that the compressible fluid behaves like an ideal gas and fulfills the state equation,

$$\frac{p}{\rho} = RT, \tag{1.8}$$

Eqs. (3.3) reduce to the same forms as in the previous case:

$$0 = -\frac{\partial p}{\partial x}, \tag{3.6a}$$

$$0 = -\frac{\partial p}{\partial y}, \tag{3.6b}$$

$$0 = -\rho g - \frac{\partial p}{\partial z}. \tag{3.6c}$$

Again, by integrating the first two equations we find that the pressure is constant in the x and y directions,

$$p(x) = p(y) = \text{const.},$$

and by substituting Eq. (1.8) into (3.6c) we get

$$\frac{dp}{dz} = -\rho g = -\frac{p}{RT} g.$$

After separating the variables we get

$$\frac{dp}{p} = -\frac{g}{RT} dz,$$

which allows the integration of both sides:

$$\ln p \Big|_{p_2}^{p_1} = -\frac{g}{RT} z \Big|_{z_1}^{z_2}.$$

For a constant temperature T we can write

$$\ln \frac{p_2}{p_1} = -\frac{g}{RT}(z_2 - z_1) \tag{3.11}$$

or

$$p_2 = p_1 e^{-\frac{g}{RT}(z_2 - z_1)}. \tag{3.11a}$$

Now, compare this relation with the linear one in Eq. (3.9). You'll note that the exponential variation in Eq. (3.11a) compares well with the experimental data in Fig 3.7.

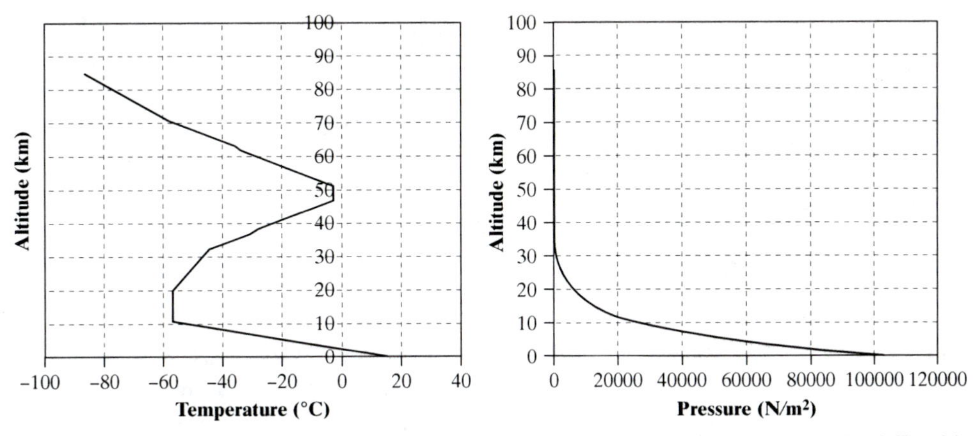

Figure 3.7. Variations of temperature and pressure versus altitude in the standard Earth's atmosphere.

EXAMPLE 3.1. HYDROSTATIC PRESSURE VARIATION IN A COMPRESSIBLE FLUID. To demonstrate the use of this equation let us calculate the pressure changes we face when climbing the elevator in one of the world's tallest buildings. If we approximate the height of the Empire State Building as 380 m, then the expected change in pressure, based on Eq. (3.11), is

$$\frac{p_2}{p_1} = e^{-\frac{9.81}{286.6\times 300}(380)} = 0.9575;$$

here we assume a constant temperature of $T = 300\,\text{K}$ and a gas constant for air of $R = 286.6\ \text{m}^2/(\text{s}^2\,\text{K})$. At the base of the building we have an atmospheric pressure of, say 101,325 N/m²; then the difference in pressure is

$$\Delta p = (1 - 0.9575)101{,}325 = 4299\ \text{N/m}^2.$$

Now if we were to use the incompressible formula, the pressure difference would be

$$\Delta p = \rho g h = 1.22 \times 9.81 \times 380 = 4547\ \text{N/m}^2,$$

and this inaccuracy is significant.

The compressibility effect is best observed in the Earth's atmosphere. However, some of our assumptions are not holding there, mainly because of the nonlinear temperature variations. The temperature changes with altitude are shown in Fig. 3.7 along with the changes in pressure (and density). It appears that the density too, is following in general the exponential behavior of Eq. (3.11); however, the large temperature changes create significant inaccuracies.

3.5 "Solid-Body" Acceleration of Liquids

In this case the fluid inside a container is accelerated such that the fluid remains static relative to the container. Initial effects (such as the shape of the fluid body before the acceleration began) are not considered, and during the constant acceleration there is no relative motion between fluid particles. In principle we can regard

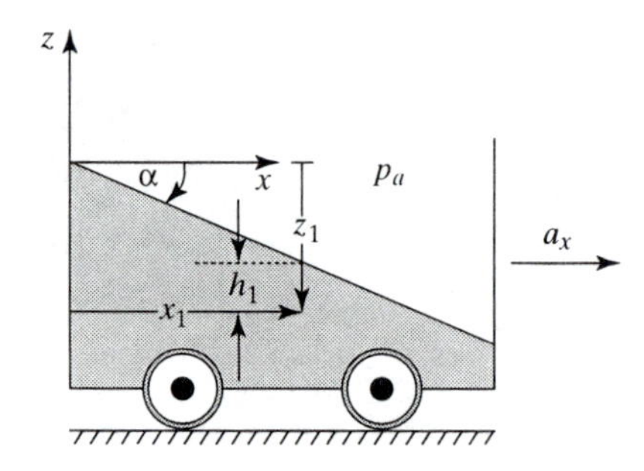

Figure 3.8. Linear acceleration of a fluid.

the resulting body forces in the same manner in which we treated the gravitational acceleration. Therefore this case is similar to the earlier cases with the stagnant fluid, but now we can add another force that is due to solid-body-type acceleration $\vec{a}$, and Eq. (3.4a) becomes

$$\rho(\vec{f} - \vec{a}) = \nabla p. \tag{3.12}$$

Let us demonstrate this principle for two cases: (a) linear acceleration and (b) steady rotation.

3.5.1 Linear Acceleration

Assume a container is filled with a liquid as shown in Fig. 3.8. The container is subject to a constant forward acceleration, and after a while the fluid inside the container appears as depicted in the figure.

We can speculate that the fluid will be pushed backward by a body force, as shown in the figure. Equation (3.12) in Cartesian coordinates now becomes

$$-\rho a_x = \frac{\partial p}{\partial x}, \tag{3.13a}$$

$$0 = \frac{\partial p}{\partial y} \tag{3.13b}$$

$$-\rho g = \frac{\partial p}{\partial z}. \tag{3.13c}$$

Note that from the "fluid point of view" both a_x and g are the same and the only difference is their direction. This is described schematically in Fig. 3.9, and we can visualize this case as a static fluid subject to a body force of $(-a_x, 0, -g)$.

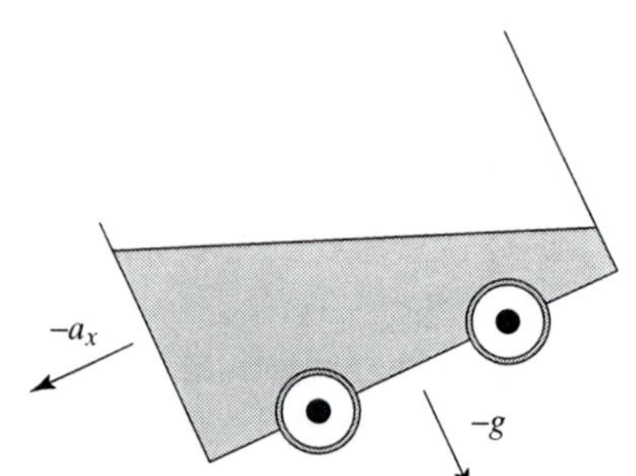

Figure 3.9. This case is the same as a stationary fluid subject to a body force $\vec{f} = (-a_x, 0, -g)$.

To proceed and solve Eqs. (3.13), let us assume an incompressible fluid, and by integrating Eq. (3.13b) we find that the pressure is constant in the y direction:

$$p(y) = \text{const.}$$

Next we can integrate Eqs. (3.13a) and (3.13c) to get

$$p(x) = -\int \rho a_x dx = -\rho a_x x + c_1,$$

$$p(z) = -\int \rho g dz = -\rho g z + c_2,$$

where c_1 and c_2 are the constants of integration. By combining the two independent solutions we get

$$p(x, z) = -\rho(a_x x + gz) + c, \tag{3.14}$$

where c replaces the two previous constants of integration. Now we select a coordinate system with the origin at the upper left-hand corner, as shown in Fig. 3.8. Consequently, at the origin, $x = z = 0$ above the liquid surface, the ambient pressure is $p = p_a$. Substituting this boundary condition into the previous result determines the constant c:

$$c = p_a.$$

Substituting this result into Eq. (3.14) shows the pressure distribution inside the liquid:

$$p(x, z) - p_a = -\rho(a_x x + gz). \tag{3.15}$$

We can find the shape of the liquid surface by letting $p(x, z) = p_a$. Substituting this into Eq (3.15) we get

$$0 = -\rho(a_x x + gz),$$

and by rearranging this relation we get the curve describing the surface,

$$z = \frac{-a_x}{g} x, \tag{3.16}$$

and the slope of the surface

$$\tan\alpha = \frac{dz}{dx} = \frac{-a_x}{g}. \tag{3.17}$$

This angle α is shown at the top of Fig. 3.8. Note that the basic formula

$$p(x, z) - p_a = \rho g h$$

works here as well when h represents the liquid height at any point (inside the liquid). Let us examine this observation by checking the pressure at an arbitrary point $x = x_1$ inside the liquid in the container, where the liquid depth is h_1. The z_1 coordinate at a depth of h_1 is then the sum of the liquid's surface position combined with the depth h_1.

$$z_1 = \frac{-a_x}{g} x_1 - h_1.$$

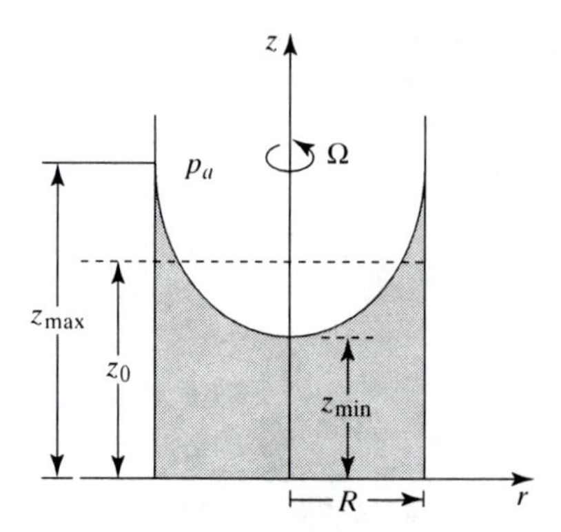

Figure 3.10. Solid-body rotation of a fluid.

Now we can use Eq. (3.15) to calculate the pressure at this point (x_1, z_1):

$$p(x,z) - p_a = -\rho(a_x x_1 + g z_1) = -\rho\left[a_x x_1 + g\left(\frac{-a_x}{g}x_1 - h_1\right)\right] = \rho g h_1.$$

As expected, this shows that we can calculate the hydrostatic pressure at any point by using the local liquid column height.

3.5.2 Solid-Body Rotation of a Fluid

Assume a cylindrical container of radius R filled with a liquid, as shown in Fig 3.10. The dashed horizontal line shows the fluid level (z_0) when the container is not rotating. Next the container is rotated about the z axis at a rate of Ω. We assume that after a while the fluid inside the container will rotate as a solid body and its upper surface will assume a new shape, as shown in Fig. 3.10. Also, there is no relative motion between the container and the fluid (on the boundary) or inside the fluid (between the fluid particles); hence we can look at this as a fluid statics problem. As in the previous example, our objective is to find the pressure distribution inside the liquid and the shape of the upper surface.

We can start with the fluid statics equation, Eq. (3.12):

$$\rho(\vec{f} - \vec{a}) = \nabla p. \tag{3.12}$$

Because this is an axisymmetric case we use the r–z cylindrical coordinate system, as shown in the figure. Next, we need to identify the forces (or accelerations). The gravitational acceleration remains as before,

$$\vec{f} = (0, 0, -g), \tag{3.5}$$

whereas for the radial acceleration a_r we can write

$$a_r = -r\Omega^2. \tag{3.18}$$

Equation (3.12) in cylindrical coordinates (r, θ, z) now becomes

$$\rho r\Omega^2 = \frac{\partial p}{\partial r}, \tag{3.19a}$$

$$0 = \frac{\partial p}{\partial \theta}, \tag{3.19b}$$

$$-\rho g = \frac{\partial p}{\partial z}. \tag{3.19c}$$

Again, assuming an incompressible liquid, we can integrate Eqs. (3.19), and the result is

$$p(r) = \int \rho r \Omega^2 dr = \rho \frac{(r\Omega)^2}{2} + c_1,$$

$$p(\theta) = \text{const.} = c_2,$$

$$p(z) = -\int \rho g dz = -\rho g z + c_3,$$

where c_1, c_2, and c_3 are the constants of integration. Also, the second equation suggests that there are no tangential variations (with θ) inside the liquid. By combining these independent solutions we get

$$p(r, z) = \rho \left[\frac{(r\Omega)^2}{2} - gz \right] + c, \tag{3.20}$$

where c replaces the previous constants of integration. To calculate this constant we need to specify the boundary conditions. For example, we observe that at the center of the rotating liquid where $z = z_{min}$, just above the liquid, the pressure is equal to the ambient pressure p_a. Substituting this into Eq. (3.20) results in

$$p(0, z_{min}) = p_a = \rho\,(0 - g z_{min}) + c$$

or

$$c = p_a + \rho g z_{min}.$$

Substituting this result into Eq. (3.20) provides the expression for the pressure distribution inside the liquid:

$$p(r, z) - p_a = \rho \left[\frac{(r\Omega)^2}{2} - g(z - z_{min}) \right]. \tag{3.21}$$

We can find the shape of the liquid surface by letting $p(r, z) = p_a$. Substituting this into Eq (3.21) we get

$$0 = \rho \left[\frac{(r\Omega)^2}{2} - g(z - z_{min}) \right],$$

and after rearranging we get the shape of the liquid's upper surface, which is a parabola:

$$z = z_{min} + \frac{(r\Omega)^2}{2g}, \tag{3.22}$$

and we can calculate z_{max} by simply substituting $r = R$ into Eq. (3.22). At this point, it is important to establish the relation between liquid height z_0 before the rotation and the two heights z_{min} and z_{max} (appearing in Fig. 3.10) once the rotation is established. We can use the conservation of mass principle stating that the liquid volume before and after the rotation is the same:

$$V_{before} = V_{after}.$$

Of course, the initial volume is

$$V_{before} = \pi R^2 z_0$$

and the volume of the rotating fluid is

$$V_{\text{after}} = \int_0^R z(2\pi r)dr = \int_0^R \left[z_{\min} + \frac{(r\Omega)^2}{2g}\right] 2\pi r(dr)$$
$$= \pi R^2 z_{\min} + \frac{\pi}{g}\Omega^2 \frac{R^4}{4} = \pi R^2 \left[z_{\min} + \frac{(R\Omega)^2}{4g}\right].$$

Comparing the two volumes

$$\pi R^2 z_0 = \pi R^2 \left[z_{\min} + \frac{(R\Omega)^2}{4g}\right].$$

and solving for z_0, we get

$$z_0 = z_{\min} + \frac{(R\Omega)^2}{4g}, \tag{3.23}$$

and by substituting $r = R$ in Eq. (3.22):

$$z_{\max} = z_{\min} + \frac{(R\Omega)^2}{2g}. \tag{3.24}$$

We can rearrange these equations based on the initial liquid height z_0:

$$z_{\min} = z_0 - \frac{(R\Omega)^2}{4g}, \tag{3.25}$$

$$z_{\max} = z_0 + \frac{(R\Omega)^2}{4g}. \tag{3.26}$$

This shows that the initial height of the liquid is at the centerline between $z_{\min}$ and $z_{\max}$ once the liquid is rotated. Also note that the basic formula,

$$p(r, z) - p_a = \rho g h,$$

works well when h represents the liquid height at any point. Let us examine this observation by checking the pressure at the bottom of the container at $r = R$. Based on this simple assumption, the pressure there must be

$$p(R, 0) - p_a = \rho g z_{\max}.$$

According to Eq. (3.21) the pressure there is

$$p(R, 0) - p_a = \rho \left[\frac{(R\Omega)^2}{2} + g z_{\min}\right]. \tag{3.27}$$

However, based on Eqs. (3.25) and (3.26),

$$z_{\max} - z_{\min} = \frac{(R\Omega)^2}{2g}.$$

Substituting this into Eq. (3.27) results in

$$p(R, 0) - p_a = \rho\left[g(z_{\max} - z_{\min}) + g z_{\min}\right] = \rho g z_{\max}, \tag{3.28}$$

which is exactly as expected.

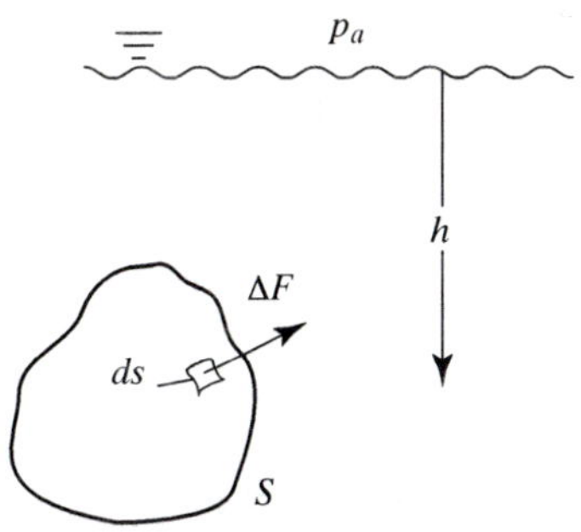

Figure 3.11. The pressure acts normal to the submerged surface S.

3.6 Hydrostatic Forces on Submerged Surfaces and Bodies

In the previous sections we demonstrated that the pressure may vary in a stationary fluid because of body forces (gravitational or even solid-body acceleration). In this section we try to calculate the forces that are due to pressure in a stationary fluid (called hydrostatic) acting on a submerged surface or on a closed body. Consider a body with surface S, as shown in Fig. 3.11, submerged in an incompressible fluid. The force acting at a point on this body was defined by Eq. (1.4) such that

$$\Delta F = -p\vec{n}dS, \tag{1.4}$$

and dS is an infinitesimal surface element, as shown.

So if we need to find the total force F acting on a specific area S_1 because of a variable-pressure field, then we need to integrate this equation over a desirable surface S_1:

$$F = -\int_{S_1} p\vec{n}dS. \tag{3.29}$$

Assuming that only gravitational forces act on the fluid, then the pressure will vary with the depth h, as given by Eq. (3.9b)

$$p(h) = p_a + \rho gh. \tag{3.9b}$$

The meaning of this relation is explained visually in Fig 3.12, where the pressure (and resulting force) will increase on the sides of a container with the depth h. At the bottom of the container, at a depth of H, the pressure is constant, as shown in the figure. So basically all the weight of the liquid is supported by the bottom surface, but in addition, the sides will experience horizontally acting hydrostatic force (because pressure at a point acts equally to all directions). In comparison, if the

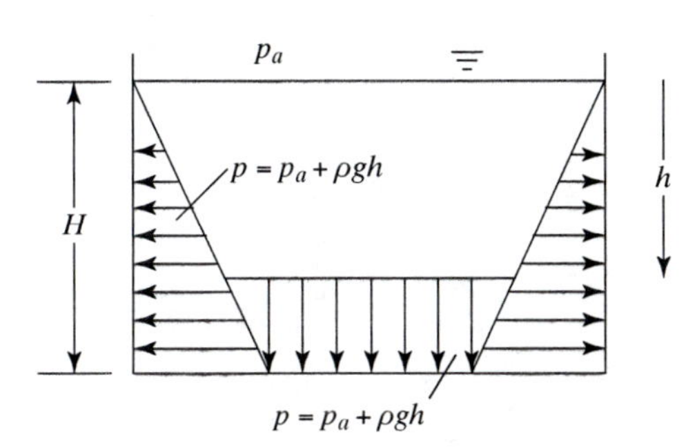

Figure 3.12. Variation of the hydrostatic pressure in a stationary container.

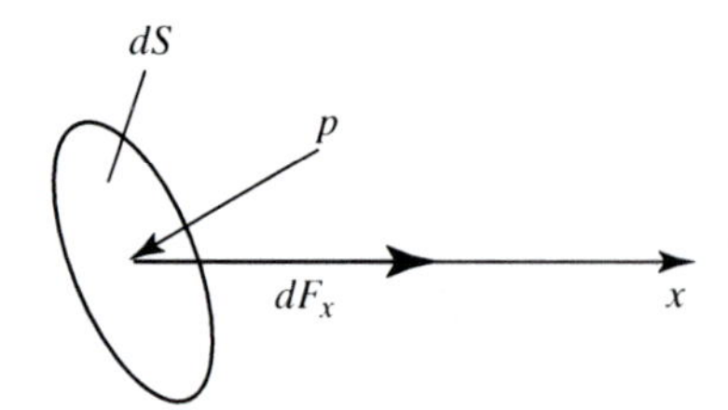

Figure 3.13. The force components acting on a surface are split by the unit normal vector $\vec{n} = (n_x, n_y, n_z)$.

same weight of bricks were piled in the container, then only the loads on the bottom surface will be present.

To continue the determination of the force acting on the submerged surface we substitute the pressure from Eq. (3.9b) into force equation (3.29) to get

$$F = -\int_{S_1} (p_a + \rho g h)\vec{n} dS. \tag{3.30}$$

Note that Eq. (3.30) is a vector expression and the force components must be evaluated in a well-defined coordinate system. For example, Fig. 3.13 shows a surface element dS, and clearly the pressure acts normal to it. The force in the x direction, however, accounts for only the projection of the area in this direction:

$$F_x = -\int_{S_1} (p_a + \rho g h) n_x dS, \tag{3.31}$$

where the unit normal vector is $\vec{n} = (n_x, n_y, n_z)$. Because the integration is now in the y–z plane we can also write

$$F_x = -\int_{S_1} (p_a + \rho g h) dy dz. \tag{3.31a}$$

In the following subsections we use this equation to determine the forces acting on submerged surfaces or submerged bodies.

3.6.1 Hydrostatic Forces on Submerged Planar Surfaces

The discussion on hydrodynamic forces can be divided into two parts. The first part focuses on calculating the magnitude of the hydrostatic force; the second part is aimed at determining the center of pressure.

3.6.1.1 Hydrostatic Force on a Submerged Surface

Let us consider the liquid reservoir shown in Fig. 3.14 and assume that the right wall is planar and that it is inclined at an angle α, as shown in the figure. The origin of the Cartesian coordinate system is attached to the upper right-hand corner (at the

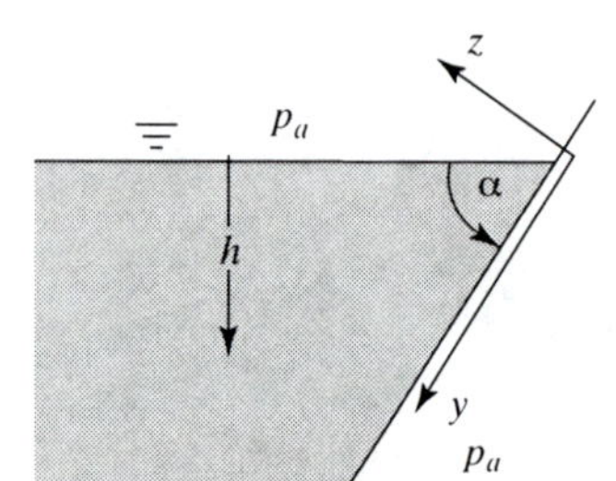

Figure 3.14. Nomenclature for calculating the hydraulic force on an inclined planar surface.

fluid level) and the x coordinate points into the page. The y and z coordinates are positioned, as shown in the figure. In addition we assume that the ambient pressure p_a is constant above the fluid level and behind the wall.

Based on Eq. (3.30), the force that is due to pressure on the wall is

$$F_z = -\int_S (p_a + \rho g h) n_z dS,$$

and here the normal points in the z-coordinate direction: $\vec{n} = (0, 0, 1)$. Because the constant pressure p_a acts on both sides of the wall, we cancel it in this equation. Also, we can replace h with $h = y \sin\alpha$ to get

$$F_z = -\int_S (\rho g y \sin\alpha) dS = -\rho g \sin\alpha \int_S y dS. \tag{3.32}$$

However, the remaining integral is similar to the process of finding the average coordinate $\overline{y}$ for the area S (or the centroid):

$$\overline{y} = \frac{1}{S}\int y dS. \tag{3.33}$$

Consequently, with the aid of Eq. (3.32), we can calculate the force as

$$F_z = -\rho g \sin\alpha \overline{y} S. \tag{3.34}$$

This result can be interpreted such that the force is simply the average pressure $\overline{p} = p(y = \overline{y})$, multiplied by the area S:

$$F_z = -\overline{p} S. \tag{3.35}$$

Location of the average $\overline{y}$ is at the centroid of the area, and it is shown for several common geometrical shapes in Fig. 3.15.

3.6.1.2 The Center of Pressure

Next we'll try to find the center of pressure. This point is very important in engineering because it represents the location where the resultant force acts. This is demonstrated in Fig. 3.16(a), where bricks are piled up on a planar board. To find the action line of the resultant force, we place a small triangle underneath the board and try to balance the board. Once this point is found, then the force F shown is the reaction to the resultant force and acts in the opposite direction.

We can repeat a similar exercise, but now we use a linear variation of the pressure along the plate, as shown in Fig. 3.16(b). Let us assume that the pressure on the plate varies linearly such that

$$p(x) = p_{\max}\frac{x}{l}.$$

Based on Eq. 3.29 the total force per unit width is then

$$F_z = -\int_0^l p dx = -\frac{p_{\max}}{l}\int_0^l x dx = -\frac{l p_{\max}}{2}. \tag{3.36}$$

This is the resultant force, and it confirms Eq. (3.35) because $\overline{p} = \frac{p_{\max}}{2}$. Next we must answer the question about the location of the equivalent force. For example, we can select the force F in Fig. 3.16(b) such that the total sum of moments about the origin is zero (e.g., the system is in equilibrium). Consequently the moment that is due to

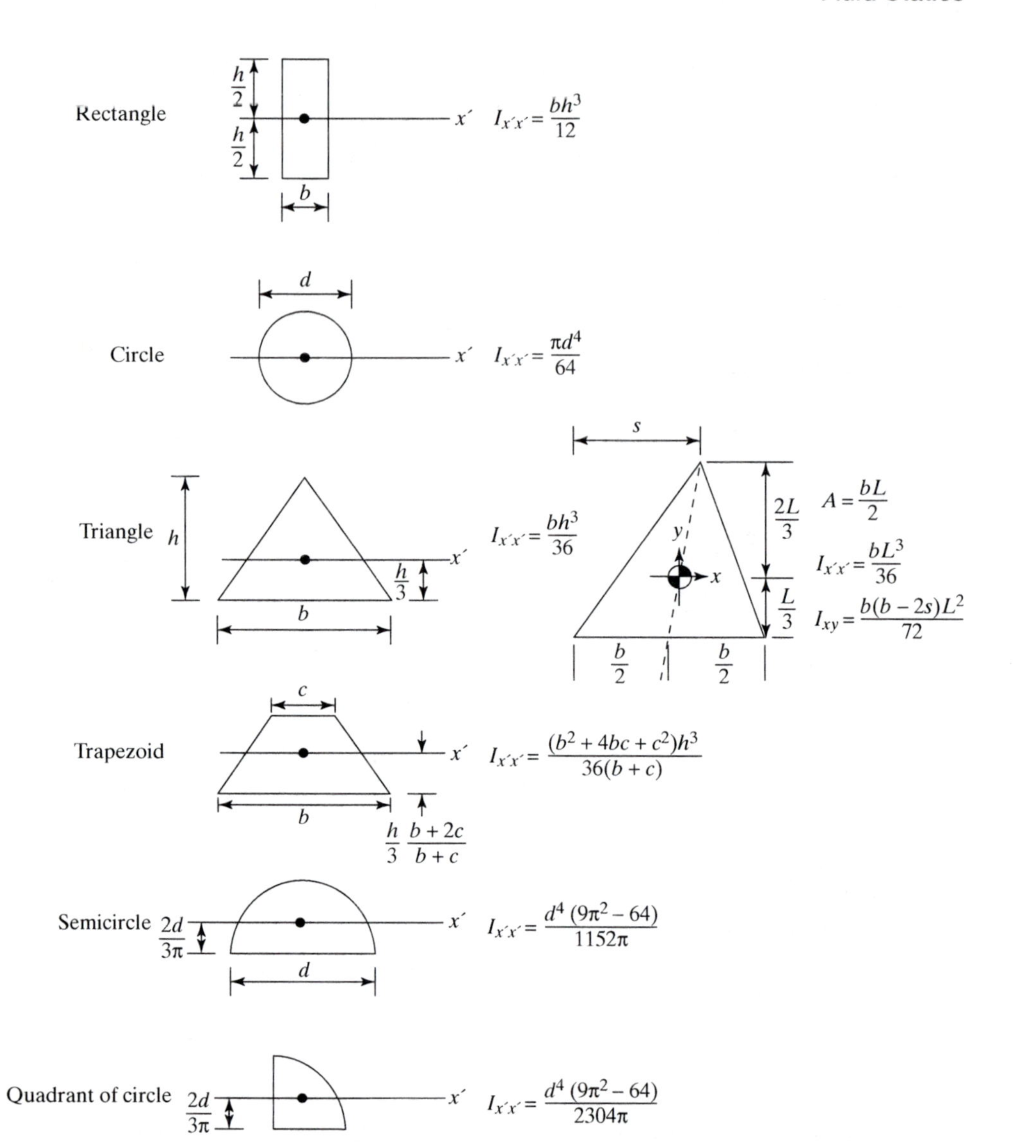

Figure 3.15. Moments of inertia for several common geometries (calculated about the centroid, as shown.

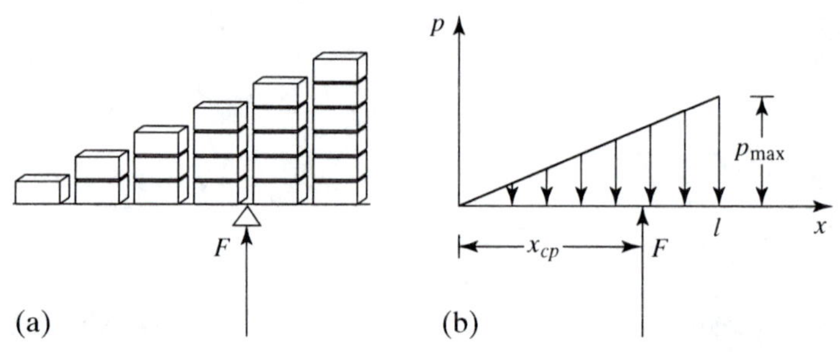

Figure 3.16. The principle of the center of pressure, representing the point where the resultant force acts.

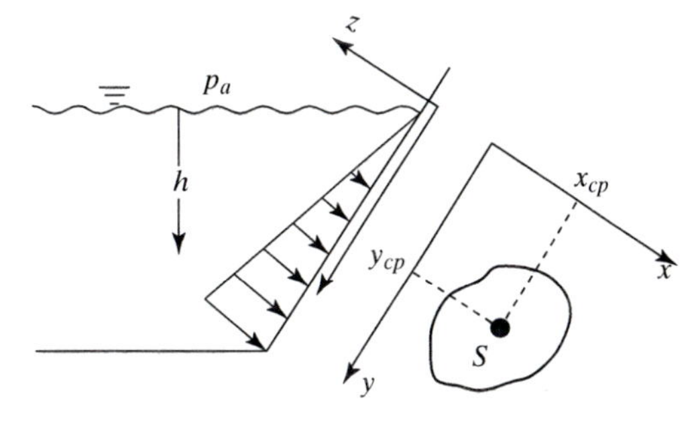

Figure 3.17. Calculation of the center of pressure for a segment S lying on the inclined planar surface.

the resultant force (but opposite in direction) multiplied by x_{cp} must balance the moment due to the pressure from above the plate:

$$F_z x_{\text{cp}} = \int_0^l xpdx = \frac{p_{\max}}{l} \int_0^l x^2 dx = -\frac{l^2 p_{\max}}{3}. \tag{3.37}$$

Using the expression for F_z from Eq. (3.36) and solving for the center of pressure, we get

$$x_{\text{cp}} = \frac{2}{3} l. \tag{3.38}$$

This location is closer to the right-hand side where the pressure is higher.

Now we can continue to calculate the center of pressure for the inclined wall in Fig. 3.14. The projection of the area of interest S is shown in Fig. 3.17 and the center-of-pressure coordinates are at $(x_{\text{cp}}, y_{\text{cp}})$, as shown in the figure. Using the same principle, we require that, at this point, the moments will be balanced by the resultant force (with the minus sign). Let us start with the y coordinate:

$$F_z y_{\text{cp}} = \int_S ypdS = \int_S y(\rho g y \sin\alpha) dS = \rho g \sin\alpha \int_S y^2 dS. \tag{3.38}$$

By substituting F_x from Eq. (3.34), we can solve for the y coordinate of the center of pressure:

$$y_{\text{cp}} = \frac{\int_S y^2 dS}{\int_S ydS}. \tag{3.39}$$

But $\int_S y^2 dS = I_{xx}$ is the (area) moment of inertia and $\int_S ydS = \overline{y}S$ [as in Eq. (3.33)]; therefore

$$y_{\text{cp}} = \frac{I_{xx}}{\overline{y}S}, \tag{3.40}$$

and, as noted, I_{xx} is the area moment of inertia about the horizontal x axis. In a similar manner we can derive the center-of-pressure x coordinate x_{cp}. To balance the moments about the y axis we write:

$$F_z x_{\text{cp}} = \int_S xpdS = \int_S x(\rho g y \sin\alpha) dS = \rho g \sin\alpha \int_S xydS. \tag{3.41}$$

Substituting F_x from Eq. (3.34) we can solve for the y coordinate of the center of pressure:

$$x_{\text{cp}} = \frac{\int_S xydS}{\int_S ydS}. \tag{3.42}$$

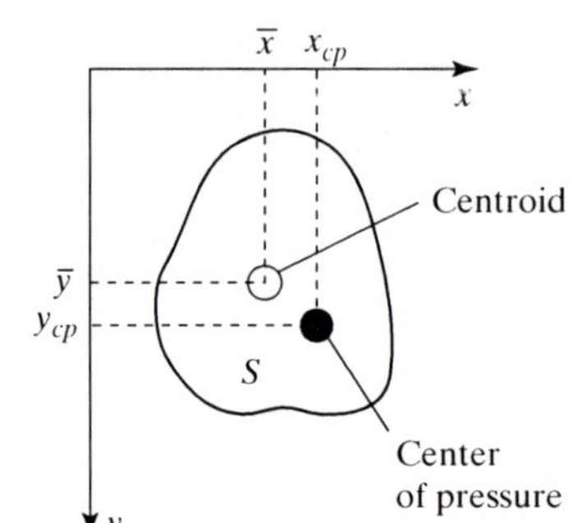

Figure 3.18. Application of the parallel-axes theorem for shifting the moment of inertia between parallel axes.

But $\int_S xydS = I_{xy}$ and $\int_S ydS = \overline{y}S$ and therefore we can summarize:

$$x_{\mathrm{cp}} = \frac{I_{xy}}{\overline{y}S},$$

$$y_{\mathrm{cp}} = \frac{I_{xx}}{\overline{y}S}, \tag{3.43}$$

and the moments of inertia are calculated relative to the coordinate axes shown in Fig. 3.17.

Next, let us redraw in Fig. 3.18 the x–y plane and the area of interest S from Fig. 3.17. The centroid for the area S is at $(\overline{x}, \overline{y})$, as shown. The moments of inertia in Fig. 3.15, and in most engineering reference books, are given relative to the centroid of the area and *not* relative to the x–y frame in Fig. 3.18. To use those formulas we need to use the parallel-axes theorem that allows us to shift the calculated moment of inertia between parallel axes:

$$I_{xx} = I_{\overline{xx}} + \overline{y}^2 S,$$

$$I_{xy} = I_{\overline{xy}} + \overline{x}\,\overline{y}S. \tag{3.44}$$

Substituting these relations into Eqs. (3.43) allows us to use the area centroid-based moments of inertia ($I_{\overline{xx}}$ and $I_{\overline{xy}}$), as given in Fig. 3.15:

$$x_{\mathrm{cp}} = \frac{I_{\overline{xy}}}{\overline{y}S} + \overline{x},$$

$$y_{\mathrm{cp}} = \frac{I_{\overline{xx}}}{\overline{y}S} + \overline{y}. \tag{3.45}$$

This can be rearranged to show the displacement between the centroid and the center of pressure:

$$x_{\mathrm{cp}} - \overline{x} = \frac{I_{\overline{xy}}}{\overline{y}S},$$

$$y_{\mathrm{cp}} - \overline{y} = \frac{I_{\overline{xx}}}{\overline{y}S}, \tag{3.45a}$$

and $\overline{y}$ is measured from the water level, as shown in Fig. 3.14. Note that if the area is symmetric about a line (crossing the centroid) and parallel to the y axis, then $I_{\overline{xy}}$ is zero (as shown in Fig. 3.15).

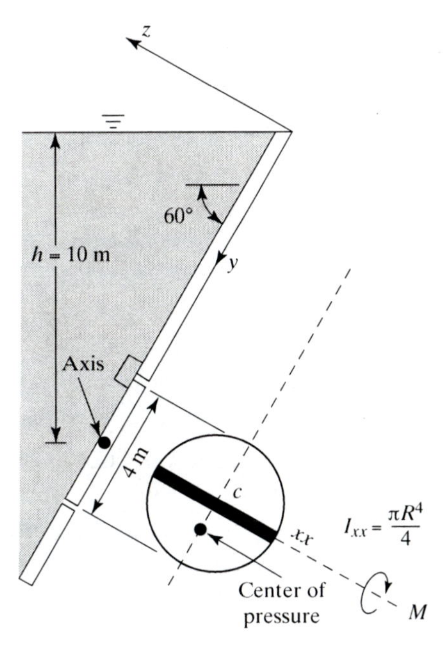

Figure 3.19. Force on a submerged circular valve in a pool.

EXAMPLE 3.2. HYDROSTATIC FORCE ON A CIRCULAR VALVE. As an example, consider the 4-m-diameter circular valve submerged in a pool of water, as shown in Fig. 3.19. The horizontal rotation axis crosses the center of the circle, and this point is located at a depth of 10 m. The wall inclination is 60°, and we would like to calculate the hydrostatic force on the valve, the center of pressure, and the magnitude of the moment required to open it.

First let us find the force acting on the circular valve. We can use Eq. (3.35) because we know the centroid location is at $\overline{h} = 10$ m. The average pressure at the centroid is

$$\overline{p} = \rho g \overline{h} \rightarrow = 1000 \times 9.8 \times 10 = 9.8 \times 10^4 \frac{\text{N}}{\text{m}^2}$$

and the valve area is $S = \pi 2^2 = 12.56$ m^2. The force is then

$$F_z = -\overline{p}S = -9.8 \times 10^4 \times 12.56 = -1.23 \times 10^6 \text{ N},$$

and of course the minus sign means that the normal force is pushing the valve in the $-z$ direction. Because of the lateral symmetry, the center of pressure will not be shifted to the side, but it will be below the centroid. To calculate this point we use Eqs. (3.45):

$$x_{\text{cp}} = \overline{x}$$

$$y_{\text{cp}} - \overline{y} = \frac{I_{\overline{xx}}}{\overline{y}S} = \frac{\pi R^4}{4(h/\sin 60)\pi R^2} = 0.0866 \text{ m}.$$

Note that $\overline{y}$ is the distance in the coordinate system of Fig. 3.19 (with the origin at the upper liquid surface). Next we need to calculate the moment required to open the valve. The opening moment M is

$$M = F_z(y_{\text{cp}} - \overline{y}) = 1.23 \times 10^6 \times 0.0866 = 1.06 \times 10^5 \text{ Nm},$$

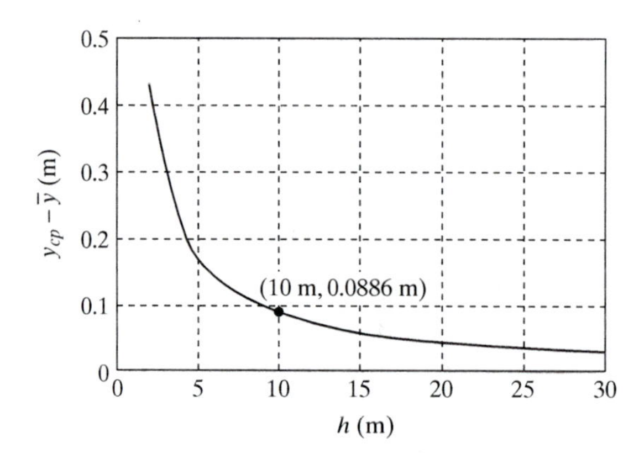

Figure 3.20. Shift in center of pressure location for the valve in Example 3.2.

and the required direction is shown in the figure. It is important to show that the distance between the centroid and the center of pressure depends on the depth h. This is logical because the deeper the valve in the pool, the smaller the relative changes in the pressure across the valve. The trend in the distance between these two points versus the valve depth is depicted in Fig. 3.20, and the distance clearly decreases with increasing depth.

EXAMPLE 3.3. HYDROSTATIC FORCE ON A TRIANGULAR SURFACE. In this second example we demonstrate a case in which the submerged surface is not symmetric and therefore the center of pressure will not be immediately below the centroid. The geometry of the submerged triangular surface at the bottom of a pool, filled with water, is shown in Fig. 3.21. The apex is located at a depth of 5 m and the triangle base is at $h = 11$ m, with a surface inclination of 30°, as shown. Again, we would like to calculate the hydrostatic force on the triangular surface and the location of the center of pressure.

Based on Fig. 3.15 the area centroid for the triangle is at 1/3 of its height and 1/3 of its base, as shown. The depth $\bar{h}$ at this point is

$$\bar{h} = 5 + 8 \sin 30^\circ = 9 \text{ m},$$

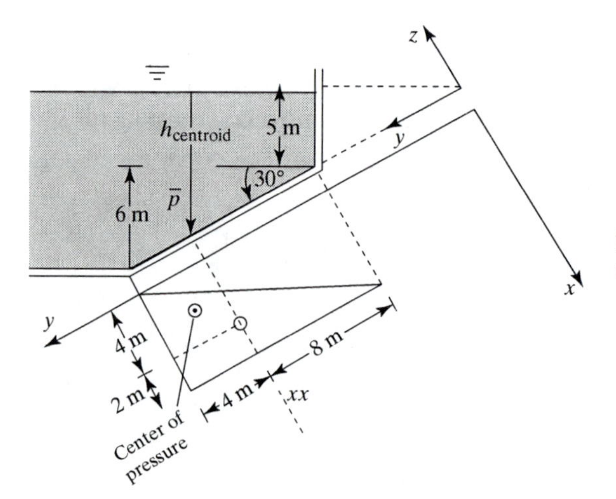

Figure 3.21. Force on a submerged triangular valve in a pool.

and the hydrostatic pressure at this point is

$$\overline{p} = \rho g\overline{h} = 1000 \times 9.8 \times 9 = 88{,}200 \frac{\text{N}}{\text{m}^2}.$$

The area S of the triangle is

$$S = \frac{6 \times 12}{2} = 36 \text{ m}^2,$$

the force on the triangular surface, using Eq. (3.35), is

$$F_z = -\overline{p}S = -88{,}200 \times 36 = -3.175 \times 10^6 \text{ N},$$

and the minus sign means that the force is pointing in the $-z$ direction. Next we need to calculate the position of the center of pressure. The formulas for the triangle's moment of inertia are taken from Fig. 3.15; first we calculate those about the area centroid;

$$I_{\overline{xx}} = \frac{bL^3}{36} = \frac{6 \times 12^3}{36} = 288 \text{ m}^4,$$

$$I_{\overline{xy}} = \frac{b(b-2s)L^2}{72} = \frac{6(6 - 2 \times 6)12^2}{72} = -72 \text{ m}^4.$$

We now find the center of pressure by using Eqs. (3.45). First, however, we must find $\overline{y}$:

$$\overline{y} = \frac{\overline{h}}{\sin 30^\circ} = \frac{9}{0.5} = 18 \text{ m}.$$

Substituting this into Eqs. (3.45), we obtain

$$x_{\text{cp}} - \overline{x} = \frac{I_{\overline{xy}}}{\overline{y}S} = \frac{-72}{18 \times 36} = -0.111 \text{ m},$$

$$y_{\text{cp}} - \overline{y} = \frac{I_{\overline{xx}}}{\overline{y}S} = \frac{288}{18 \times 36} = 0.444 \text{ m}.$$

The resultant force F_z acts at this point and its location is shown in the figure. Note that it is below the centroid and the minus sign indicates that it is shifted in the $-x$ direction (closer to the y coordinate) relative to the centroid.

3.6.2 Hydrostatic Forces on Submerged Curved Surfaces

The calculation of the hydrostatic forces on nonplanar surfaces is quite similar and is based on Eq. (3.29). For complex submerged surface shapes, however, the integration process may be complicated (because of the complex math). Let us demonstrate the method by calculating the hydrostatic forces on a dam having a simple parabolic shape, as shown in Fig. 3.22.

Let us use an x–z coordinate system placed at the bottom floor of the reservoir, as shown in the figure, and calculate the forces on this dam, per unit width. The shape of the surface is given by $z = x^2/a$, where a is a constant and H is the depth

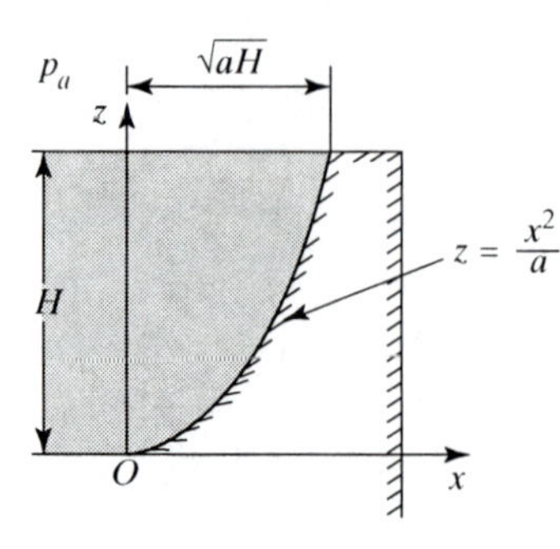

Figure 3.22. Forces on a parabolic dam.

of the water. The x coordinate at the upper-right corner of the water surface is $x = \sqrt{aH}$, as shown. As noted, we use Eq. 3.29 to calculate the forces:

$$F_x = -\int_S p dS_x = -\int_S \rho g(H - z) dydz.$$

And here the projection of the area element pointing in the $-x$ direction is $dS_x = dydz$, and calculating per unit width ($dy = 1$) we get

$$F_x = \int_S \rho g(H - z)dz = \rho g \left(Hz - \frac{z^2}{2} \right)\Bigg|_0^H = \rho g \frac{H^2}{2}.$$

This, as expected, means that the average pressure at $H/2$ (which is $\rho g \frac{H}{2}$) is multiplied by the projected area, $H \times 1$. Now let us proceed to calculate the vertical component of the hydrostatic force F_z:

$$F_z = -\int_S p dS_z = -\int_S \rho g(H - z) dxdy.$$

And here the projection of the area element pointing in the $+z$ direction is $dS_z = dxdy$, and calculating per unit width ($dy = 1$) we get

$$F_z = -\int_S \rho g(H - z)dx = -\int_S \rho g \left(H - \frac{x^2}{a} \right) dx = -\rho g \left(Hx - \frac{x^3}{3a} \right)\Bigg|_0^{\sqrt{aH}}$$
$$= -\rho g \frac{2}{3} H\sqrt{aH}.$$

Note that, because the integration is on the x–y plane, z is replaced with the curve shape $\frac{x^2}{a}$. The negative sign means that the force acts downward, opposite to the $+z$ coordinate. The mass m per unit width of the liquid above the surface is

$$m = \rho \int_0^{\sqrt{aH}} (H - z)dx = \rho \left(Hx - \frac{x^3}{3a} \right)\Bigg|_0^{\sqrt{aH}} = \rho \frac{2}{3} H\sqrt{aH}.$$

So it is clear that the vertical component of the force is equal to the mass of fluid above it. We can generalize these conclusions as follows:

> The horizontal component of the hydrostatic force is the product of the projected area into the horizontal direction multiplied by the pressure acting at the centroid of this area.
>
> The vertical component of the hydrostatic force acting on a curved surface is equal to the weight of the liquid column above it.

Note that in this section we did not include the force that is due to the ambient pressure p_a, which must be included to calculate the absolute pressure values.

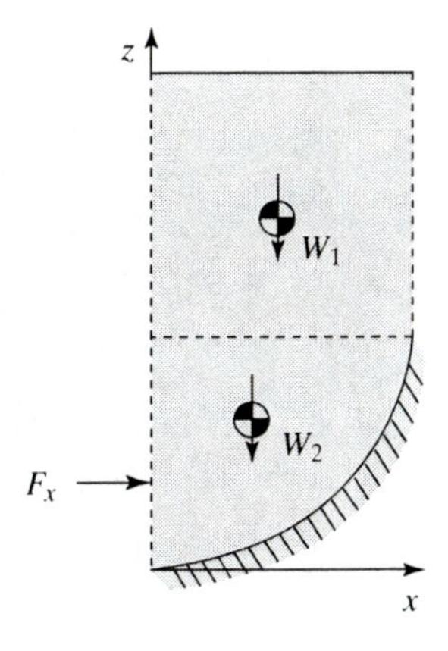

Figure 3.23. The principle of calculating the forces on a submerged curved surface.

The preceding two principles are demonstrated schematically in Fig. 3.23, which indicates that the force on the bottom surface is equal to the weight of the fluid above it whereas we can calculate the side force by using the preceding calculations using the average pressure at the centroid of the projected area. For simplicity, the volume in Fig. 3.23 is divided into upper and lower sections with weights of W_1 and W_2, respectively. The vertical force on the curved lower surface is then $W_1 + W_2$. The lateral force F_x is the same as the force acting on the projected area at the left, as shown.

Now let us continue with the example of Fig. 3.22 and calculate the center of pressure for the horizontal and vertical forces. To calculate z_{cp} let us consider the moment balance about origin of the x–z coordinate system:

$$z_{cp} F_x = -\int_S pzdS_x = -\int_S \rho g(H-z)zdz = \rho g\left(H\frac{z^2}{2} - \frac{z^3}{3}\right)\Big|_0^H = \rho g\frac{H^3}{6}$$

Substituting the results for F_x we get

$$z_{cp} = \frac{H}{3}$$

Repeating the calculation for the x coordinate of the center of pressure we get

$$x_{cp} F_z = -\int_S pxdS_z = -\int_S \rho g(H-z)xdx = -\rho g\left(H\frac{x^2}{2} - \frac{x4}{4a}\right)\Big|_0^{\sqrt{aH}} = -\rho g\frac{aH^2}{4}$$

Again by substituting the vertical force component we get

$$x_{cp} = \frac{3}{8}\sqrt{aH}$$

In conclusion, to find the force acting on complex surfaces, we can still use Eq. (3.30), and only the integration process becomes more complicated.

3.7 Buoyancy

Calculation of the hydrostatic forces acting on bodies immersed or floating in a liquid is important to numerous disciplines and to the design of ships, balloons, etc. Some of these problems are depicted schematically in Fig. 3.24. A ship, for example, that is floating on the water surface is lifted by the buoyancy force that is equal

Figure 3.24. Cases of objects in a liquid.

to its weight. A submarine or a fish planning to sink or rise in the water must account for buoyancy forces. The third case, that of a structure on the bottom floor of the sea, is similar to the cases discussed in the previous section, and the normal force on it is equal to the weight of the fluid column above it.

Let us start with a simple analysis of a completely submerged object. Assume that a cube is placed inside the fluid such that its upper and lower surfaces are horizontal, as shown in Fig. 3.25. It is clear that the left-to-right and fore-and-aft forces on the cube are the same because of symmetry, whereas the forces on the upper and lower surfaces are not identical.

The downward-pointing force on the upper surface is $F_1 = \rho g h_1 S$ and the upward-pointing force on the lower panel is $F_2 = \rho g h_2 S$. The net force L (lift) is therefore

$$L = F_2 - F_1 = \rho g(h_2 - h_1)S.$$

But the cube volume $V = (h_2 - h_1)S$, and therefore we can conclude that the buoyancy force is

$$L = \rho g V, \tag{3.46}$$

where V is the volume of the displaced fluid. This was discovered many years ago by the Greek physicist Archimedes (287–212 B.C.E.) who made these statements:

1. A body immersed in a fluid experiences a vertical buoyancy force equal to the weight of the displaced fluid.
2. A floating body experiences a vertical buoyancy force equal to its weight (which is also equal to the weight of the displaced fluid).
3. The buoyancy force acts through the centroid of the displaced fluid volume.

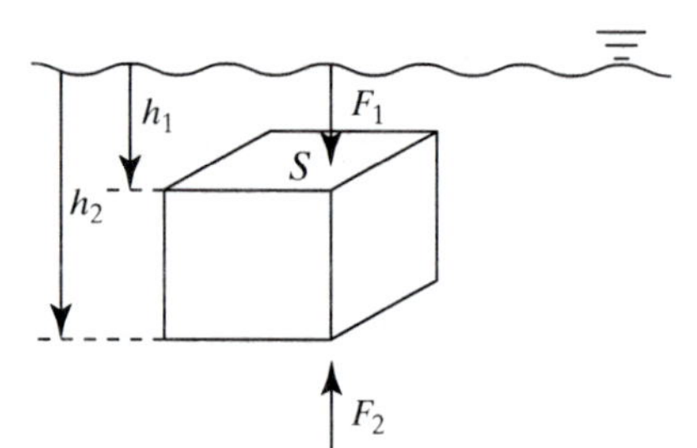

Figure 3.25. Simple approach to estimating the forces on a submerged object.

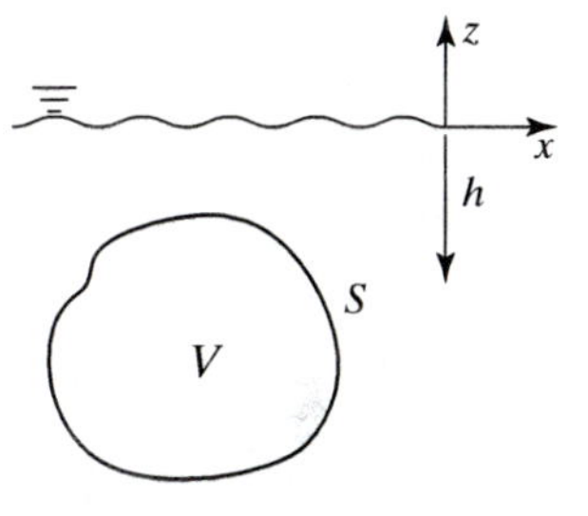

Figure 3.26. The use of the Gauss divergence theorem for calculating the buoyancy force.

To prove those observations let us apply the Gauss divergence theorem [see Eq. (2.30)] to Eq. (3.29). The benefit of this is that, instead of integrating the pressure on the surface of the submerged body, we get a much simpler volume integral:

$$F = -\int_{S_1} p\vec{n}dS = -\int_V \nabla p dV. \tag{3.47}$$

Let us use this result for the buoyancy-force calculation. We can use the same Cartesian coordinate system as before and the z coordinate points up (opposite to h in Fig. 3.26). The hydrostatic pressure in the fluid is $p = \rho g(-z)$ and $\frac{\partial p}{\partial z} = -\rho g$. Therefore

$$\nabla p = (0, 0, -\rho g).$$

Substituting this into Eq. (3.47) gives

$$L = -\int_V (0, 0, -\rho g) dV = \rho g \int_V dV = \rho g V, \tag{3.48}$$

and the force is in the positive z direction. This result is exactly the same as the one presented in Eq. (3.46)! At this point we have verified Archimede's first and second observations. However, the buoyancy-force line of action requires more attention. We can apply the same principles used in the previous sections. Referring to Fig. 3.27, assume that the buoyancy force acts at a point shown and therefore the moment about the origin is: $\overline{x}L = \int x dL$. However, the buoyancy force is given in Eq. (3.48) and therefore

$$\overline{x}\rho g V = \int x dL = \int x \rho g dV.$$

By canceling ρg on both sides we simply end up with the volume centroid location,

$$\overline{x} = \frac{1}{V}\int x dV, \tag{3.49}$$

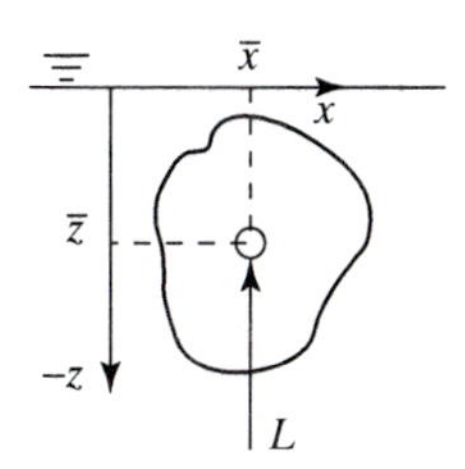

Figure 3.27. Nomenclature used to prove that the buoyancy force acts at the centroid of the displaced volume.

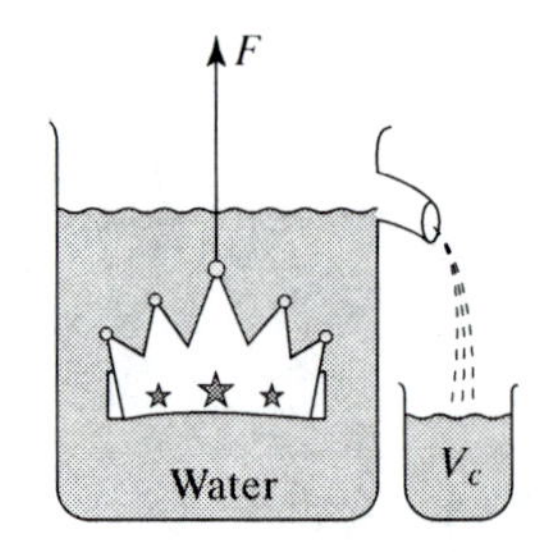

Figure 3.28. Archimedes' experiment to determine the volume of complex shapes.

and by simply repeating the same procedure we can write

$$\bar{z} = \frac{1}{V}\int z dV, \tag{3.50}$$

which proves that the buoyancy force acts at the centroid of the displaced volume.

EXAMPLE 3.4. THE KING'S CROWN. One of the most famous anecdotes about Archimedes involves the newly crafted crown made for Syracuse's king Hiero II. The king suspected that the goldsmith skimmed some of the gold and asked the renowned scientist to conduct one of the early undestructive tests. While taking a bath, Archimedes discovered how to measure the volume of a complex-shaped object (the crown), as depicted in Fig. 3.28. He was so excited that he ran to the street loosing his towel and crying "Eureka" (got it – in Greek). Next (one would hope that he had put his towel back on by then.) he went to the king and held the crown in a container full of water, as shown.

The force F to hold the crown in position was 20.9 N, and the volume of water displaced by submerging the crown was $V_C = 310$ cm^3. The weight of the crown is the sum of F and the weight of displaced water (ρ_w is the water density):

$$W_c = F + \rho_w g V_c = 20.9\ \mathrm{N} + 1000 \times 9.8 \times 310 \times 10^{-6} = 23.94\ \mathrm{N}.$$

The crown density ρ_c therefore is

$$\rho_c = \frac{W_c}{V_c g} = \frac{23.94}{310 \times 10^{-6} \times 9.8} = 7880 \frac{\mathrm{kg}}{\mathrm{m}^3}.$$

Of course, pure gold is 19.3 times heavier than water (whereas iron is only about 7.8). So the fate of the goldsmith was doomed by this clever measurement.

EXAMPLE 3.5. HOT-AIR BALLOON. Let us assume that a hot-air balloon's structural weight, including the passengers, is 275 kg, the ambient temperature is 290 K, and the air density is $\rho = 1.22$ kg/m^3. The balloon is then inflated with hot air and its volume can be approximated by a sphere of diameter $D = 15$ m (see Fig. 3.29). The question we ask is at which internal temperature will the liftoff begin?

To solve this problem we must find out at which temperature the buoyancy forces are equal to the structural weight. Based on Eq. (3.48), the lift is $\rho g V$, but the weight of the hot air inside must be taken into account as well. Consequently, the lift that is due to buoyancy is

$$L = \rho_{\mathrm{cold}} g V - \rho_{\mathrm{hot}} g V = (\rho_{\mathrm{cold}} - \rho_{\mathrm{hot}}) g \frac{\pi D^3}{6}.$$

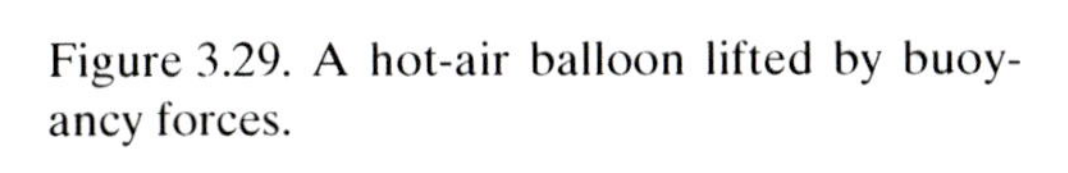

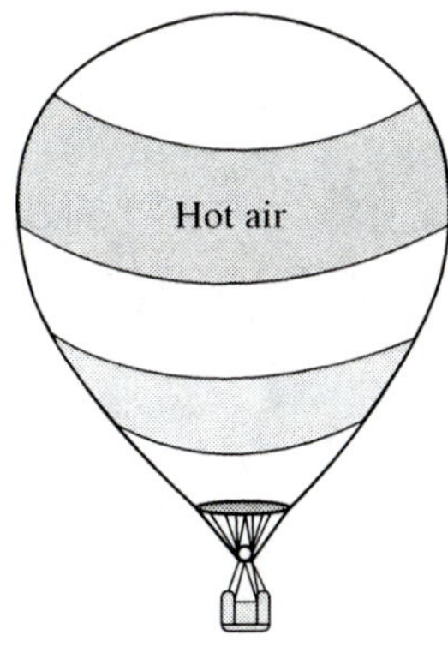

Figure 3.29. A hot-air balloon lifted by buoyancy forces.

Because we know the "cold-air" temperature we can use the ideal-gas relation to calculate the density of the hot air. Using Eg. (1.8) we can write

$$\frac{p}{\rho_{\text{cold}} T_{\text{cold}}} = R = \frac{p}{\rho_{\text{hot}} T_{\text{hot}}},$$

but because the pressure inside the balloon is the same as that outside we get

$$\rho_{\text{cold}} T_{\text{cold}} = \rho_{\text{hot}} T_{\text{hot}}.$$

Substituting this into the lift equation and equating to the weight of $275 \times (9.8 = g)$ we get

$$L = \rho_{\text{cold}} \left(1 - \frac{T_{\text{cold}}}{T_{\text{hot}}}\right) g \frac{\pi D^3}{6} = 275 \times 9.8 \text{ N}.$$

Solving for the average inside temperature T_{hot} we get

$$\frac{T_{\text{cold}}}{T_{\text{hot}}} = 1 - \frac{6L}{\rho_{\text{cold}}\, g \pi D^3} = 0.8724,$$

and after substituting the ambient temperature we get

$$T_{\text{hot}} = 332.4 \text{ K} = 59.2\,^{\circ}\text{C},$$

which is quite hot but not too hot to burn holes in the fabric of the balloon.

3.8 Stability of Floating Objects

Buoyancy effects play an important role in the design of naval vessels or even airborne balloons. Let us discuss briefly the lateral-stability problem of floating objects, as shown in Fig. 3.30.

Three important cases are shown schematically in Fig. 3.30. For example, case (a) on the left-hand side shows a floating vessel and to its right a condition in which it was rolled slightly by a perturbation. Under normal conditions the vessel is horizontal but its center of gravity is above the centroid of the area representing the displaced fluid volume. At first sight it appears that this situation is not stable. To check this we introduce a disturbance and roll the vessel by an angle ϕ. However, as a result of this change, the centroid of the displaced fluid is now moved to the right, creating a restoring moment M. So if we define the positive direction for the

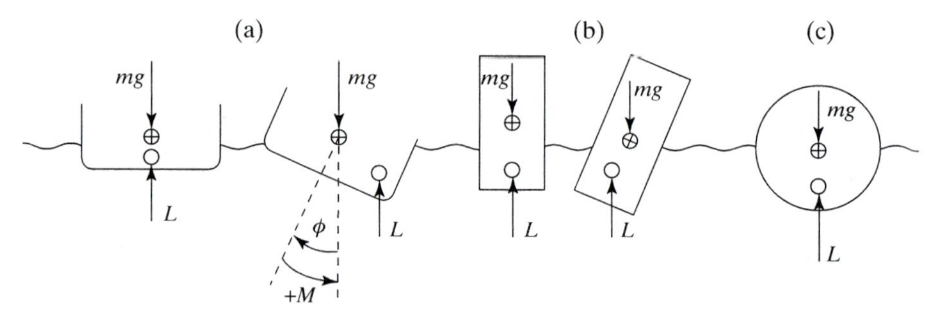

Figure 3.30. Lateral stability of floating objects.

roll angle and the moment in the same direction then clearly the moment acts in the opposite direction. Consequently we can write:

$$\frac{\partial M}{\partial \phi} < 0, \quad \text{stable,} \tag{3.51}$$

and this is the condition for a "statically stable" vessel. An unstable condition is depicted in case (b) in Fig. 3.30. In this case a roll in the positive direction results in a nonrestoring moment, which increases with the roll angle ϕ. Therefore we conclude that this case is unstable

$$\frac{\partial M}{\partial \phi} > 0, \quad \text{unstable,} \tag{3.51a}$$

and of course the vessel will roll over. These simple observations have significant meaning, when various ground, sea, or airborne vehicles are being designed. To conclude the discussion, we may look at case (c) in Fig. 3.30 in which a sphere or a cylinder will not be affected by a rotational perturbation, and we can call this condition neutral.

EXAMPLE 3.6. STABILITY OF A FLOATING BLOCK. Many floating objects may initially appear as neutral, but a closer examination shows that they are stable in roll (see the left-hand side of Fig. 3.30). As an example, consider a floating rectangular block, as shown on the left-hand side of Fig. 3.31. When floating, undisturbed, the block's center of gravity is exactly above the center of buoyancy. Next, let us create a roll displacement to the right, as shown in the figure. The center of gravity is still at the centroid of the block, but at the condition shown, the displaced fluid has a triangular shape (from the front view). The centroid of this area is at $2l/3$, from the left, clearly resulting in a restoring moment. We can easily estimate the magnitude of this moment by calculating the

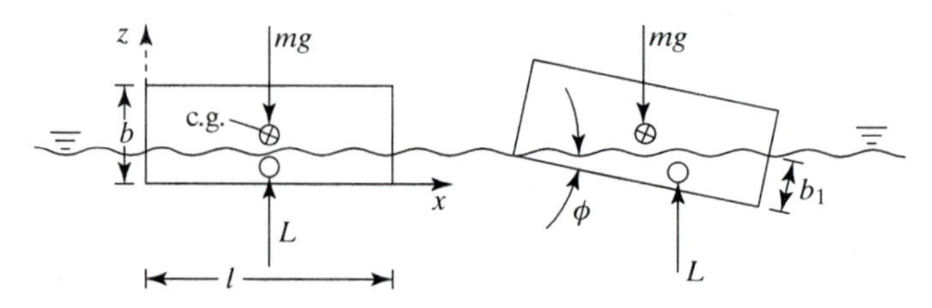

Figure 3.31. Roll stability of a floating block.

horizontal distance between the center of gravity and the center of buoyancy. Let us select an x–z coordinate system attached to the block, as shown on the left-hand side of the figure. In these coordinates, the center of gravity (cg) is located at $(l/2, b/2)$, and the buoyancy center is at $(2l/3, b_1/3)$. On the right-hand side of the figure, the block is rotated by ϕ and the horizontal locations of these points are obtained by a simple transformation. The horizontal locations of the center of gravity x_{cg} and buoyancy x_b are

$$x_{cg} = \frac{l}{2}\cos\phi + \frac{b}{2}\sin\phi,$$

$$x_b = \frac{2l}{3}\cos\phi + \frac{b_1}{3}\sin\phi,$$

the horizontal moment arm is

$$x_b - x_{cg} = \frac{l}{6}\cos\phi + \left(\frac{b_1}{3} - \frac{b}{2}\right)\sin\phi,$$

the restoring moment M is

$$M = -mg(x_b - x_{cg}) \tag{3.52}$$

and the roll angle in terms of the block geometry is

$$\tan\phi = \frac{b_1}{l}. \tag{3.53}$$

Clearly, a positive roll angle results in a negative moment (restoring), indicating a stable flotation.

3.9 Summary and Conclusions

In this chapter we discussed stationary fluids and the resulting hydrostatic pressure field. This gave us the opportunity to evaluate forces that are due to pressure only (recall that in a moving fluid there are additional forces that are due to shear). The first important observation is that the pressure (because of gravity) increases with the depth h:

$$p(h) = p_a + \rho g h. \tag{3.9b}$$

Thus the first task in engineering calculation is to evaluate the pressure field. Once the pressure field is known, the resulting forces can be calculated by summing up the local effects:

$$F = -\int_{S_1} p\vec{n}dS. \tag{3.29}$$

Recall that the pressure acts normal to a surface, as indicated by the vector $\vec{n}$. Once the force that is due to the pressure is calculated, it is desirable to establish an equivalent resultant force that acts at the center of pressure. This provides valuable information for engineering calculations, and methods to calculate the center of pressure were discussed. When the hydrostatic force is integrated over a body, a buoyancy force results for both submerged and floating objects. The buoyancy force is simply the weight of the displaced fluid:

$$L = \rho g V. \tag{3.48}$$

PROBLEMS

3.1. Two interconnected cylinders are filled with oil of density 850 kg/m^3, as shown in the figure. A force $F_1 = 200$ N is applied to the smaller piston of diameter $d_1 = 4$ cm. Assuming negligible piston weight, calculate the magnitude of the force F_2 acting on the larger piston with a diameter of $d_2 = 10$ cm and positioned at $h = 2$ m.

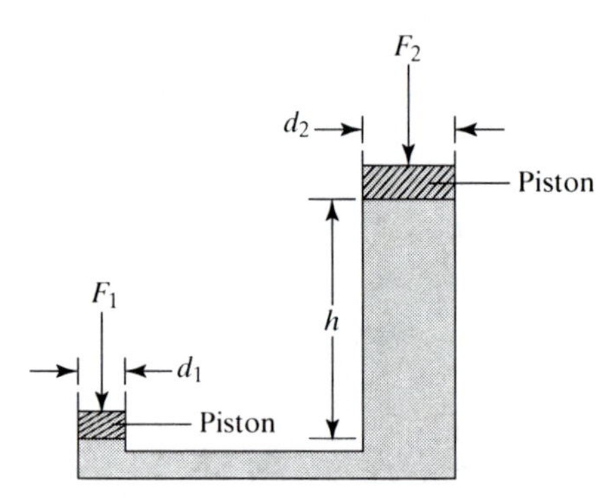

Problem 3.1.

3.2. A hydraulic jack is used to lift a 5-metric-ton weight. The diameter of the small piston is 1 cm and that of the large piston is 6 cm. Using the geometry of the handle, as shown in the figure, determine the force F_1 needed to lift the weight. How much will the 5-ton weight raise if the small piston travel is 3 cm?

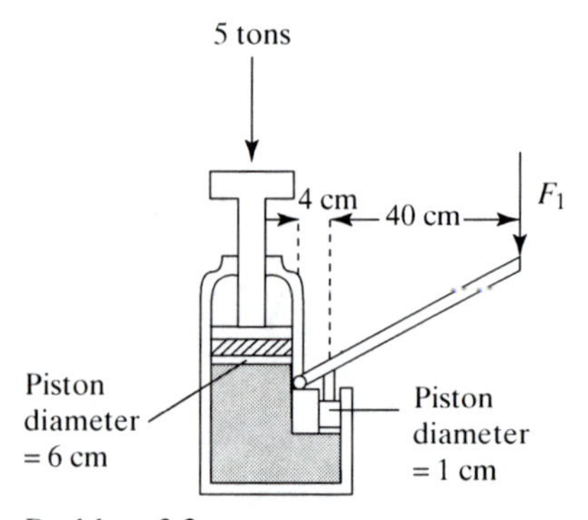

Problem 3.2.

3.3. A rectangular container (with a square base: 0.6 m by 0.6 m) is partially filled with water ($h_1 = 0.15$ m). Calculate the total force on the bottom surface of the tank. Also calculate the total weight of the water for this case. Next, the tank is filled up to the mark shown by $h_2 = 1.0$ m through a 1-cm-diameter tube. Again, calculate the total force on the bottom surface of the tank. Is this force equal to the total weight of the water?

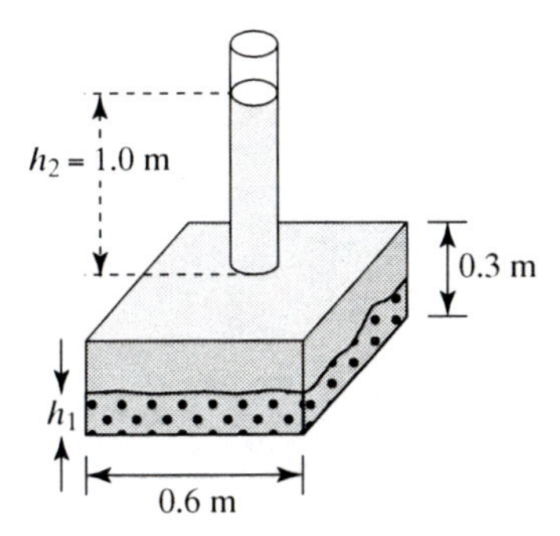

Problem 3.3.

3.4. The apparatus shown in the figure is filled with air and the unknown pressure p_x is measured by a $h = 20$ cm high mercury column. Calculate the pressure difference $p_x - p_a$ (take the density of mercury from Table 1.1).

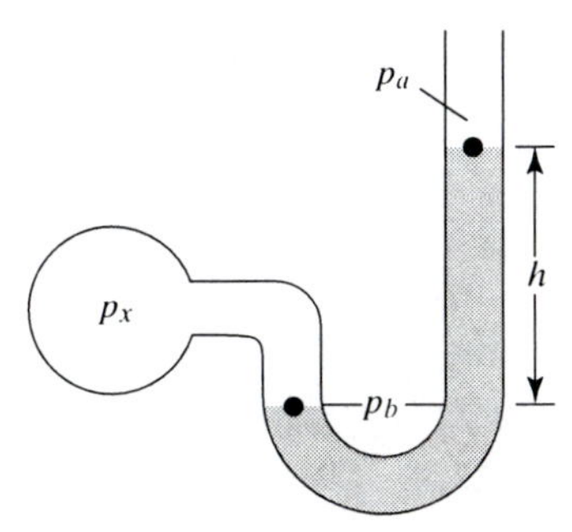

Problem 3.4.

3.5. The hull dimensions of a 100,000-ton cruise ship can be estimated by a 40 m wide and 290 m long square.

(a) Calculate how deep the hull will sink in seawater (based on a simple box-shaped hull).

(b) If 4500 passengers and crew are boarding the ship, with each adding about 100-kg weight (including luggage), calculate the additional depth the hull will sink.

3.6. A circular container of diameter $D = 0.1$ m is filled with water. A weightless piston is placed on the liquid surface and the water level stabilizes (also in the small tube) to a level of h. Next, a weight of 10 kg is placed on the piston causing the liquid in the small tube to rise. Calculate how high will the liquid rise in the small tube and how deep the piston will sink,

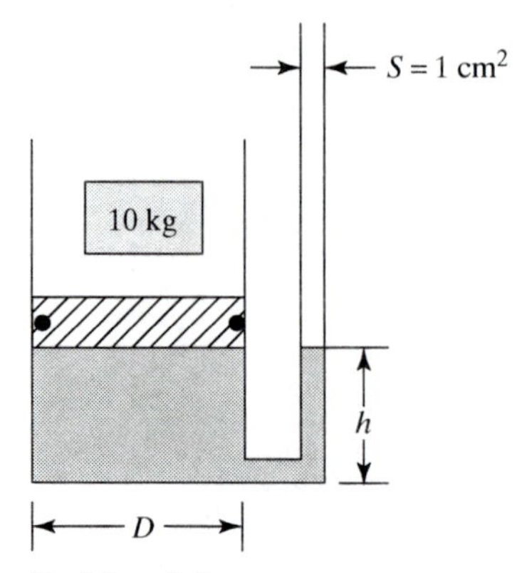

Problem 3.6.

3.7. A rectangular gate that is $b = 2$ m wide and $L = 4$ m high is located in the vertical wall of a tank containing water, as shown in the figure (case a).

(a) Calculate the center-of-pressure location.
(b) What is the magnitude of the horizontal force on the gate?

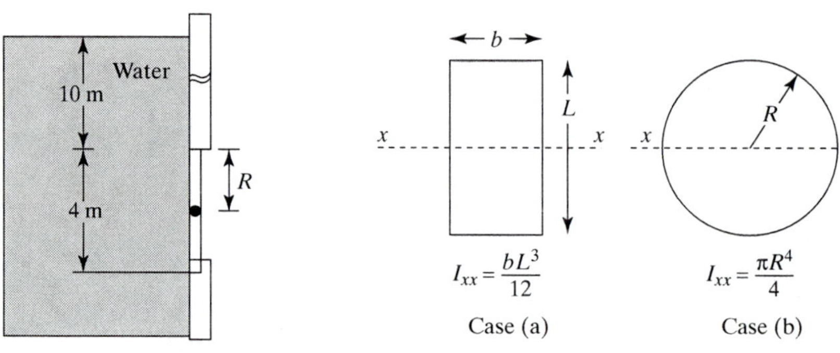

Problems 3.7 and 3.8.

3.8. A circular gate is located on the vertical wall of a tank containing water [same figure, case (b)]. Use the condition shown in the figure.

(a) Calculate the horizontal force on the gate.
(b) How deep is the center of pressure on the gate (measured from the top)?
(c) Calculate the moment required to open the gate (assuming that the horizontal axis is at $R = 2$ m,
(d) Is this moment holding the gate close or trying to push it open? ($\rho_{\text{water}} = 1000$ kg/m^3.)

3.9. Water is flowing in a V-shaped irrigation channel, as shown in the figure. Its sides are supported by poles, mounted normal to the channel walls, as shown. Calculate the axial force F_1 on the support, per 1-m length (into the page) of the channel.

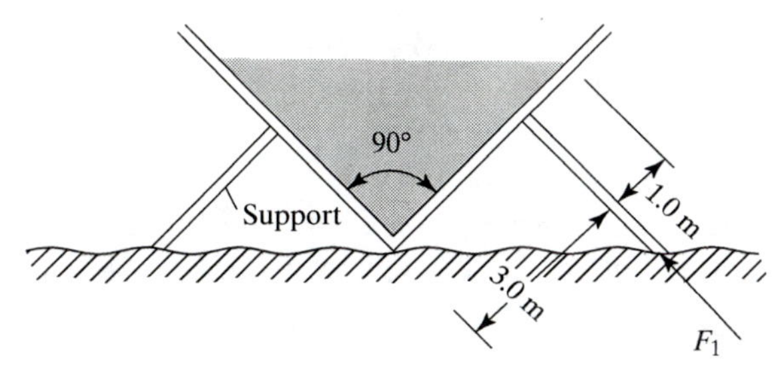

Problem 3.9.

3.10. A gate at the bottom of a pool is held close by the hydrostatic pressure. Determine the vertical force F to just start opening the 2-m-wide gate in the figure. Also, calculate the location of the center of pressure.

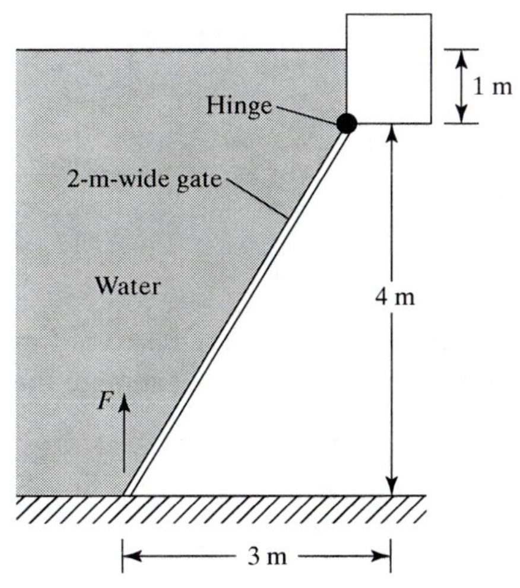

Problem 3.10.

3.11. A 3-m-wide rectangular plate is sealing the bottom of a pool filled with water, as shown in the figure. If $h = 5$ m, find the magnitude of the hydrostatic force acting on the plate. Find the location of the center of pressure along the plate (measured from the top).

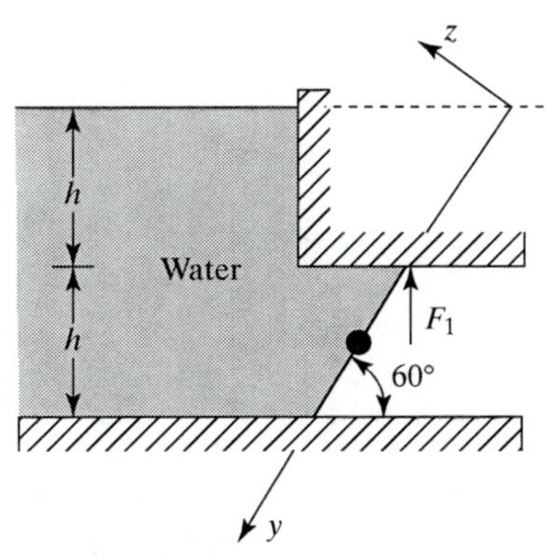

Problem 3.11.

Suppose the plate is hinged at the bottom and held in position by a vertical force F_1, as shown. What is the magnitude of this force?

3.12. The 2D pressure distribution along a flat plate (of length $= c$) is given by the following function: $p = A_1 \sin(\pi x/c)$, where A_1 is the maximum pressure. Calculate the magnitude of the resultant force and the center-of-pressure location.

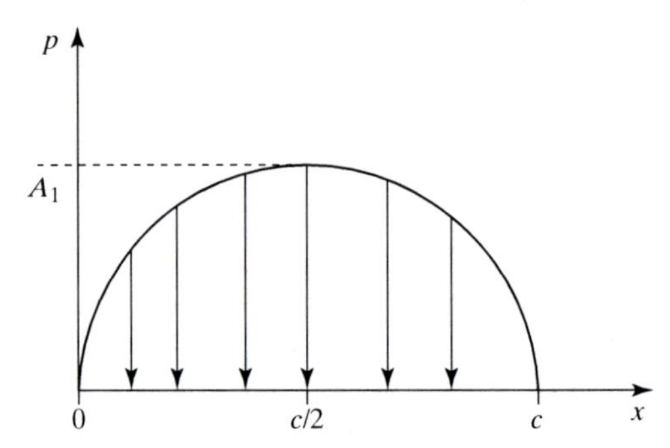

Problem 3.12.

3.13. A circular-arc-shaped gate is located at the bottom of a water reservoir, as shown in the figure. Based on the principle demonstrated in Fig. 3.23, estimate the hydrodynamic force per unit width and its direction on the gate.

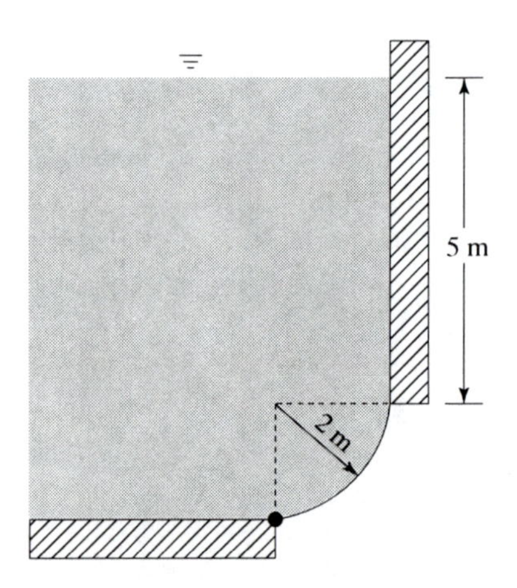

Problem 3.13.

3.14. A circular-arc-shaped gate is aligned with the water surface in the reservoir, as shown in the figure. Based on the principle demonstrated in Fig. 3.23, estimate the hydrodynamic force per unit width and its direction on the gate. Assume $R = 2$ m.

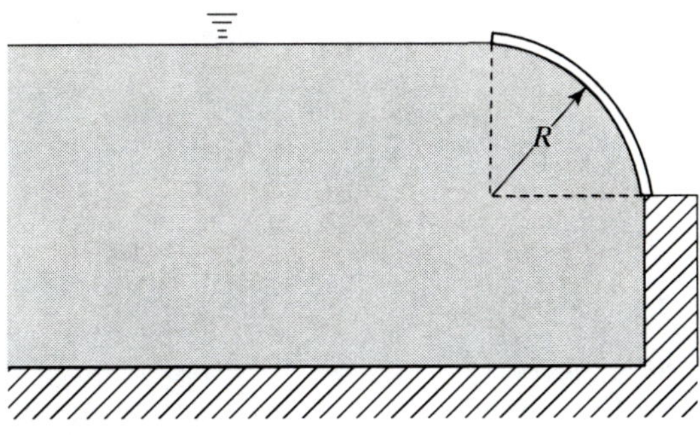

Problem 3.14.

3.15. Solve the previous problem by direct integration of the pressure on the quarter-circle surface (unit depth into the page). Use the coordinate system proposed in the figure. Consequently, the pressure as a function of θ becomes $p = \rho g h = \rho g R(1 - \cos\theta)$. The results must be the same as before!

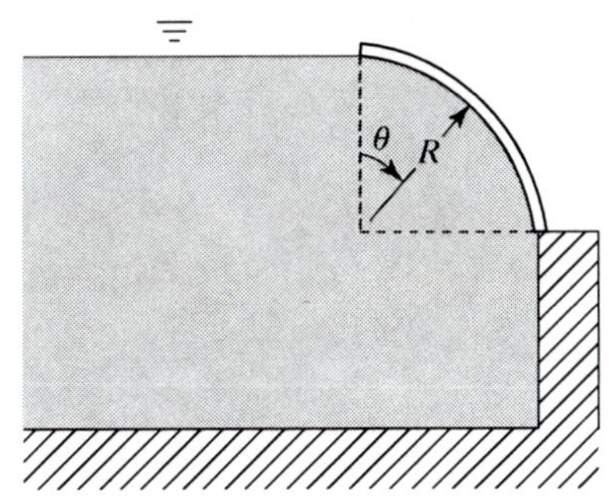

Problem 3.15.

3.16. An automatic gate is based on a heavy semi-cylindrical valve, which controls the water height in the reservoir. It is designed to open at the condition shown in the figure. Estimate the horizontal and vertical forces, per unit width, on the cylindrical valve.

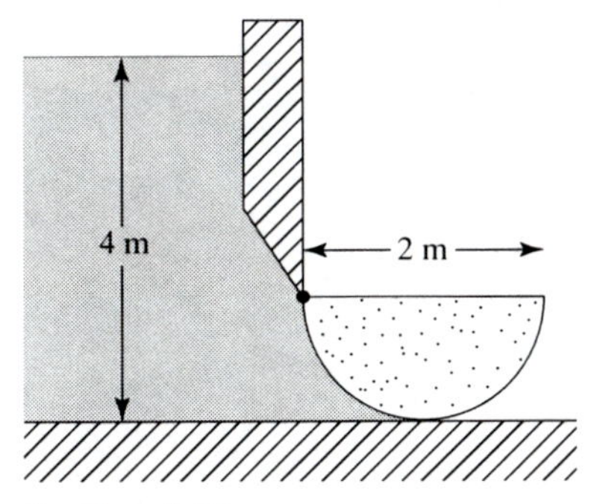

Problem 3.16.

3.17. A rectangular gate (4 m $\times$ 4 m) is hinged at the upper surface in a water reservoir. Calculate the resultant force on the gate, the center-of-pressure location, and the magnitude of the force F_1 on the lock.

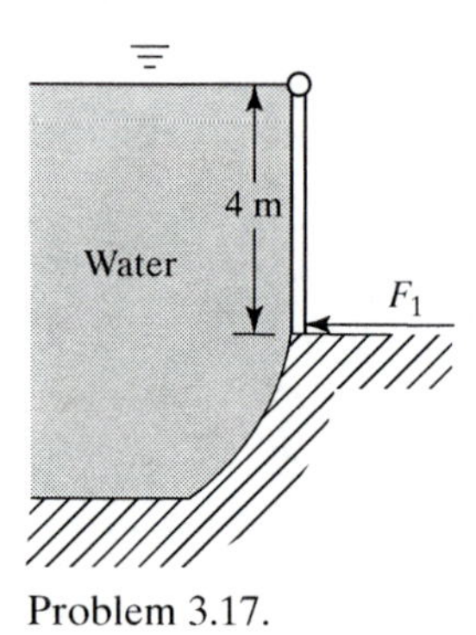

Problem 3.17.

3.18. A rectangular gate (4 m × 4 m) is held in place by the two forces F_1 and F_2. Calculate the resultant force on the gate, its direction, the center-of-pressure location, and the magnitude of the forces F_1 and F_2.

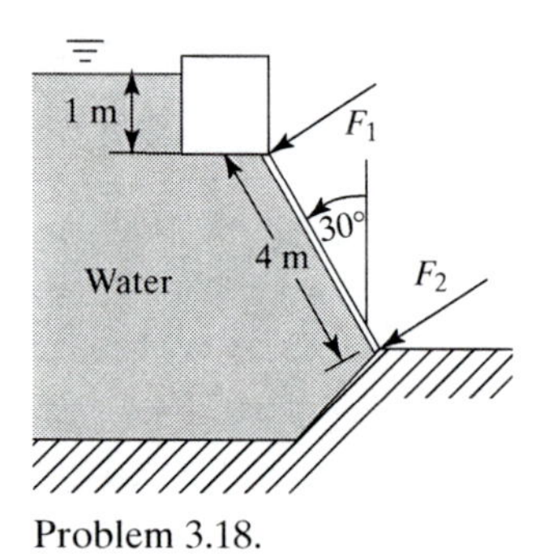

Problem 3.18.

3.19. A circular gate of 2-m radius is held in place by the two forces F_1 and F_2 (refer to the previous figure). Calculate the resultant force on the gate, its direction, the center-of-pressure location, and the magnitude of the forces F_1 and F_2.

3.20. Water is flowing into a reservoir equipped with a gate (2 m wide, into the page) that will open once the water level reaches 3 m. Calculate the weight of the mass m required for proper operation of the gate.

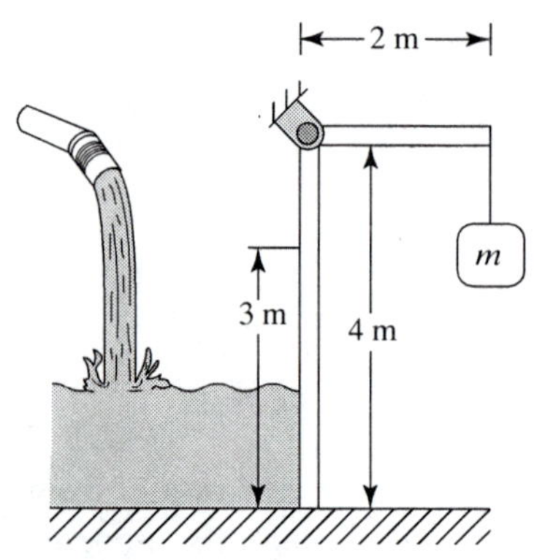

Problem 3.20.

3.21. A cylindrical container is filled with water as shown in the figure.

(a) Calculate the slope of the liquid–air interface if the container is accelerated at $0.7g$ in the x direction,

(b) What is the pressure at the bottom of the container (at point A)?

(c) At what acceleration will the water spill out?

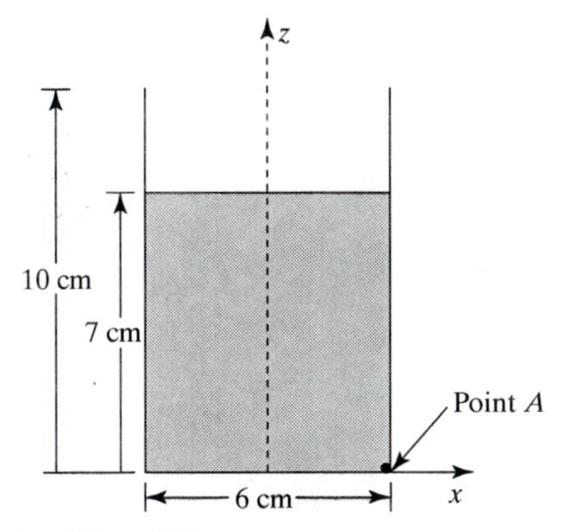

Problem 3.21.

3.22. The container of the previous problem is spinning about its vertical axis.

(a) At which rotation speed will the water spill from the container?

(b) While the container is spinning at the rate calculated in (a), calculate the pressure at point A.

3.23. A 0.3-m-wide and 0.4-m-long container is filled with water up to 0.15 m, as shown. Later, it is accelerated in the x direction at a constant rate of $a_x =$ 4 m/s^2. Calculate the percentage of fluid lost during the acceleration.

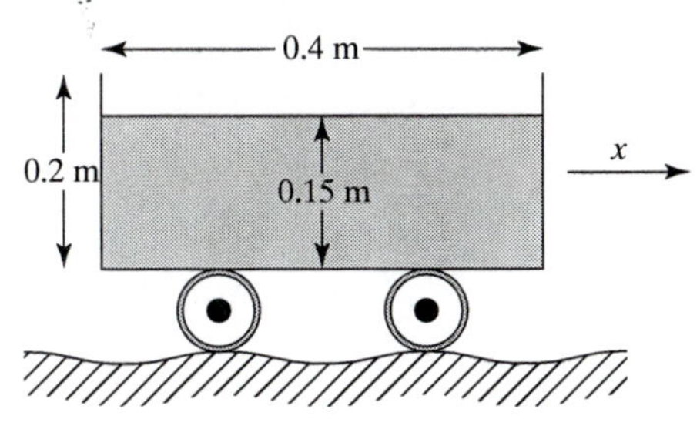

Problem 3.23.

3.24. A circular container is filled with a liquid to a height of 0.15 m, as shown. At a certain point the container is rotated about its vertical axis and after a while it

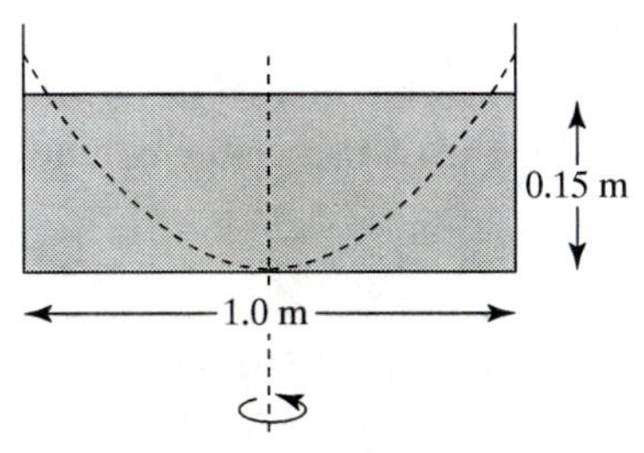

Problem 3.24.

reaches a "solid body rotation." How fast, in terms of RPM, must the container spin in order for the center of the liquid's upper surface to touch the bottom of the container?

3.25. A 0.5-m-long container is filled to 0.1-m height with water. How fast should the container accelerate in order for the upper water level to reach the bottom of the container at point A?

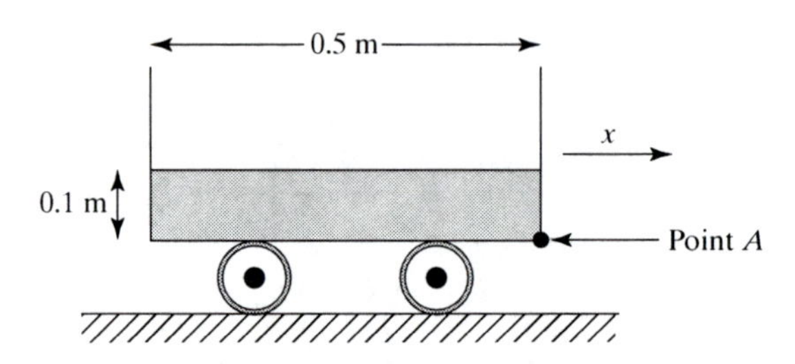

Problem 3.25.

3.26. The 0.5-m-long container shown in the figure accelerates downhill, so that the liquid surface is parallel to the road.

(a) Estimate how fast the container should accelerate in order to maintain the shown condition.

(b) Suppose the container slows down gradually and stops at the slope in the condition shown; estimate the volume of liquid spilled (assume the container width, into the page, is 0.2 m).

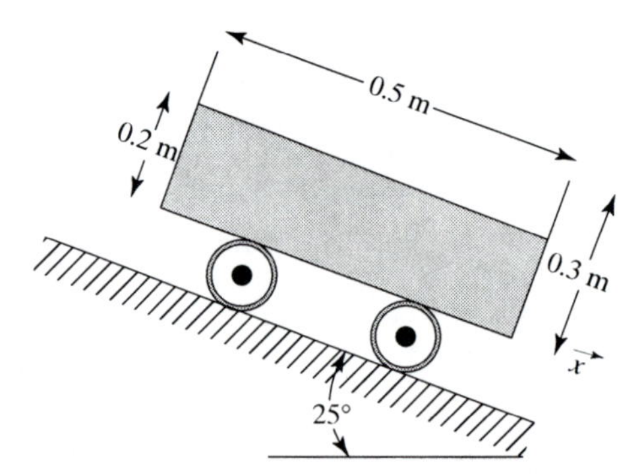

Problem 3.26.

3.27. A cylindrical container of diameter $D = 0.5$ m is filled with water up to $z_0 = 0.5$ m. Next, it begins spinning about the z axis until it reaches a "solid-body rotation." How fast should it rotate for the pressure at point A to be 7500 N/m^2 above the ambient pressure? At that condition, what is the height of the water column above point A?

3.28. A cylindrical U tube is filled with water up to $z_0 = h = 0.3$ m. Next, it is rotated about the z axis at 200 RPM. At this condition the upper surface of the liquid in the left leg of the tube is touching the center of the container (e.g., $z_{min} = 0$). Estimate the radius R of the container.

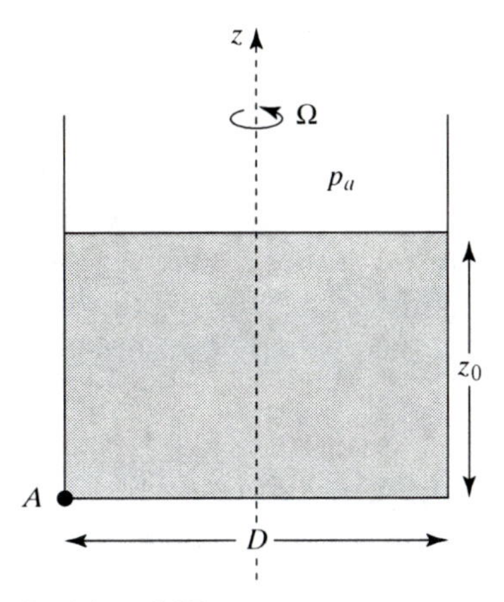

Problem 3.27.

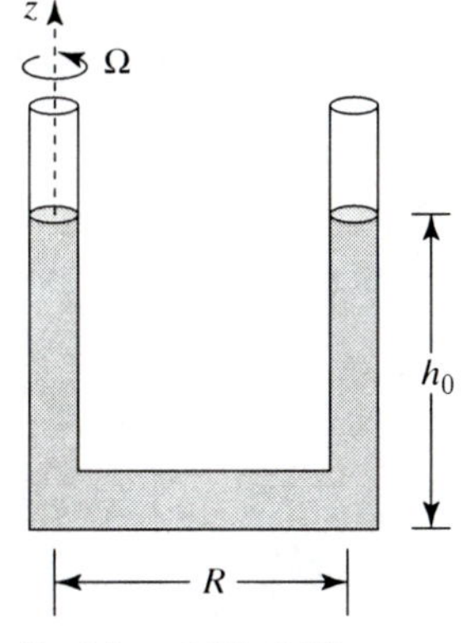

Problem 3.28, 3.29.

3.29. The U-tube shown in the figure is filled with an unknown liquid up to a height of $h_0 = 0.2$ m. Calculate the difference in the height of the two liquid columns when the U-tube is rotated about the z axis, at 40 RPM (assume R = 0.25 m). Does the liquid density has an effect on your answer?

3.30. At what RPM there will be no liquid in the left arm of the U-tube of the previous problem?

3.31. The U-tube shown in the figure of Problem 3.29 is filled with an unknown liquid up to a height of $h = 0.2$ m. At this time, however, the rotation axis is 5 cm to the right of the left tube and R is still 0.25 m. Calculate the difference in the height of the two liquid columns when the U-tube is rotated at 40 RPM.

3.32. A container filled with water is accelerated along the x axis (as shown in the figure). If the maximum liquid height h_0, the container length l, and its width b, are given, calculate the acceleration, the total force F_z on the lower (bottom) horizontal surface, and the center of the pressure.

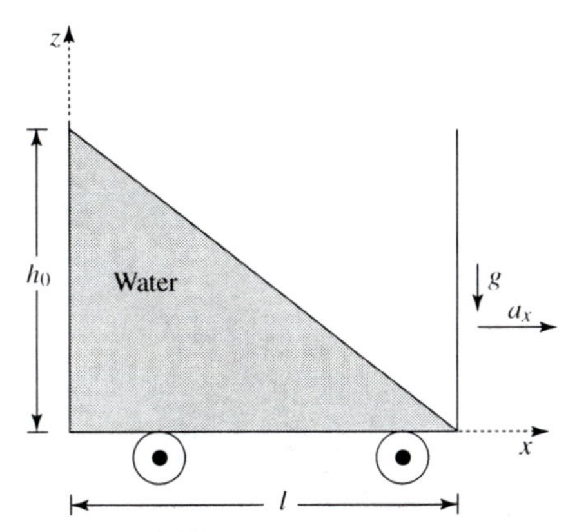

Problem 3.32.

3.33. The water height in the previous problem is $h_0 = 0.2$ m, the container length $l = 0.5$ m and its width is $b = 0.3$ m. Calculate the acceleration, the total force F_x, on the rear (left) vertical surface, and the center of the pressure (height).

3.34. The average radius of the earth is 6371 km and the average pressure at sea level is $1.013 \times 10_5$ N/m^2. If the pressure at sea level is a result of the fluid column weight above, then estimate the total mass of air around the globe (the surface of a sphere is πD^2).

3.35. A $0.1 \times 0.1 \times 0.1$ m cube is placed on a (mechanical spring) scale, registering a weight of 7.5 kg. Next it was placed on the same scale, but on the bottom of a pool filled with water ($\rho = 1000$kg/m^3). What weight will the scale show?

3.36. A 90 deg inverted cone is filled with water as shown. The volume of the water in the cone, for the 90 deg cone case, is given by $V = \pi h^3/3$. The initial depth of the water is $h = 10$ cm. Next, a block with a volume of 200 cm^3 and a specific gravity of 0.6 is floated in the water. Calculate the raise Δh in the water surface height (inside the cone).

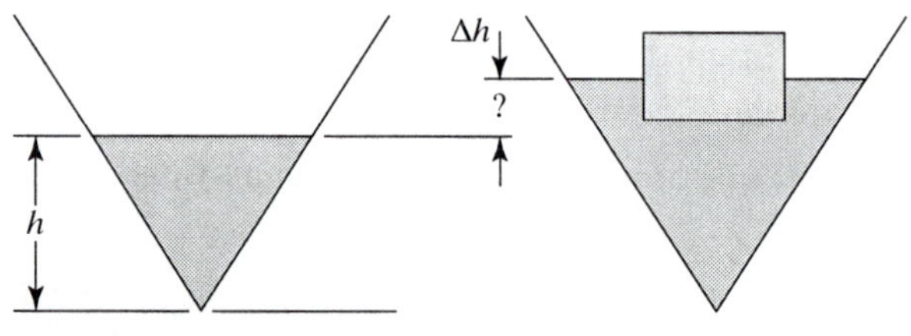

Problem 3.36.

3.37. A buoy consists of a hemispherical bottom and a conical top, as shown. The diameter of the hemisphere is 1 m and the cone angle is 60° (see figure). If the mass of the buoy is 450 kg, then find the water level when the buoy floats on seawater ($\rho = 1030$ kg/m^3).

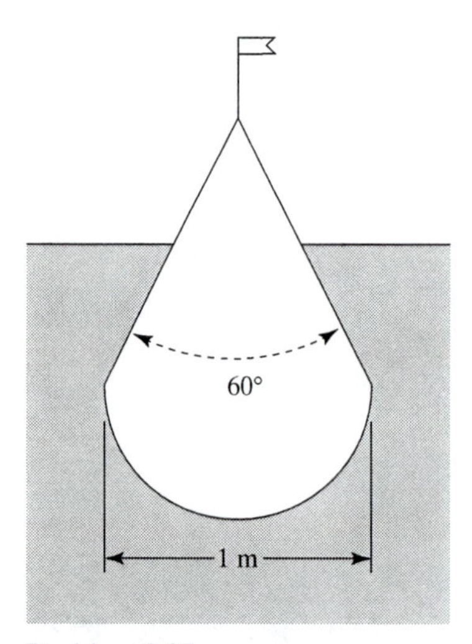

Problem 3.37.

3.38. A rectangular box with negligible weight is floating on the water (dimensions are given in the figure). Suppose a weight of 30 kg (at sea level) is placed at its center; calculate how deep the box will submerge ($h = ?$).

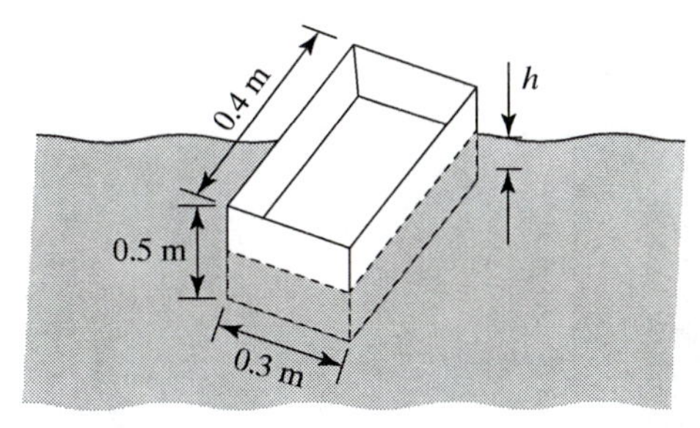

Problem 3.38.

3.39. A 0.4-m-diameter and 0.6-m-high cylinder, with negligible weight, is floating in a pool of water. To stabilize it, 45 kg of liquid with $\rho = 2000$ kg/m^3 is poured inside. In addition a 5-kg block of wood is thrown in. Determine the height h between the pool surface and the highest point on the cylinder ($\rho_{\text{water}} = 1000$ kg/m^3).

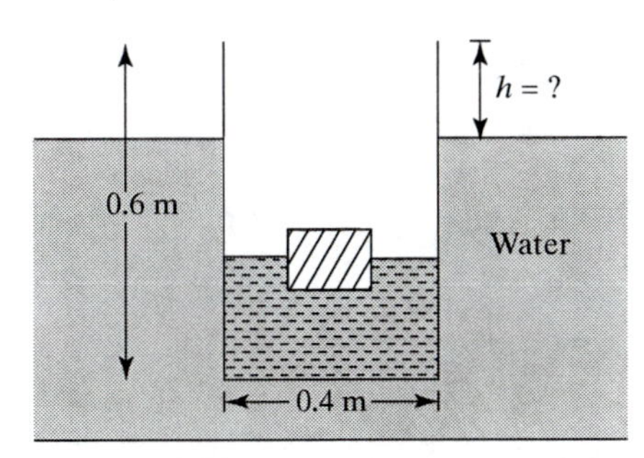

Problem 3.39.

3.40. A cylinder is floating in a pool of water as shown in the figure. Assuming the mass of the cylinder is 50 kg and most of it is concentrated at the bottom (so it floats as shown), determine the height h between the pool surface and the highest point on the cylinder ($\rho_{\text{water}} = 1000$ kg/m^3).

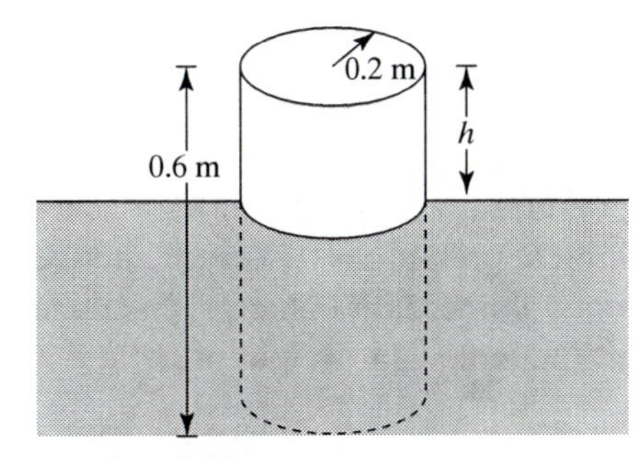

Problem 3.40.

3.41. A cylindrical container is filled with water to a level of 0.3 m as shown ($D =$ 0.3 m). Next, a block of wood measuring $0.2 \times 0.2 \times 0.2$ m (density $\rho = 800$ kg/m^3) is placed inside the container (water density is $\rho = 1000$ kg/m^3).

(a) How deep will the block sink into the water?
(b) Calculate the water level rise Δh in the container.

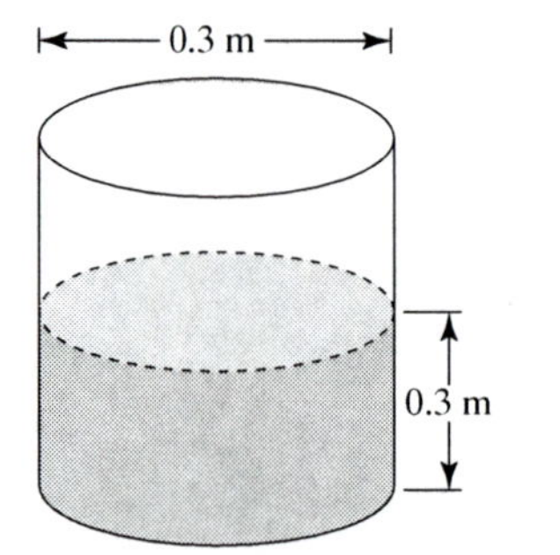

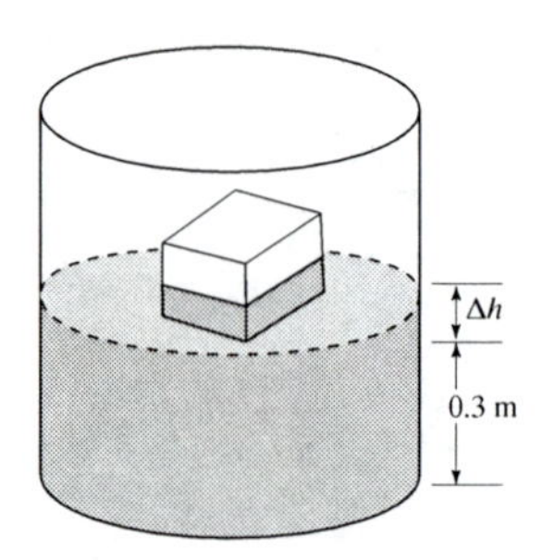

Problem 3.41.

3.42. A rectangular container is filled with water to a level of 0.3 m as shown ($L = 0.3$ m). Next, a block of cylindrical wood with a diameter of 0.2 m and height of 0.2 m (density $\rho = 800$ kg/m^3) is placed inside the container (water density is $\rho = 1000$ kg/m^3).

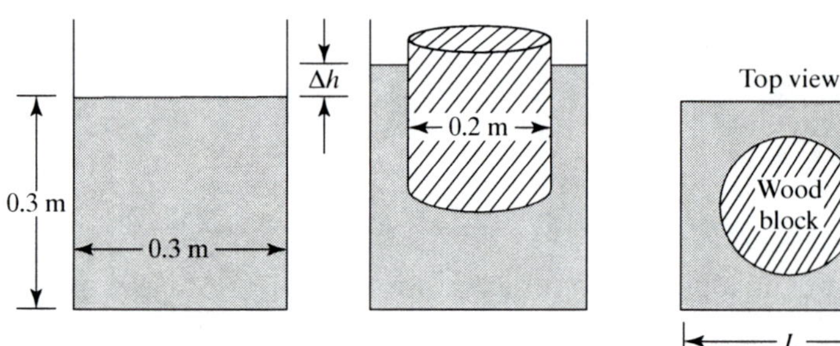

Problem 3.42.

(a) How deep will the block sink into the water?
(b) Calculate the water level rise Δh in the container.

3.43. A 10-m^3 balloon filled with helium at a pressure of 1 atm at 300 K, reaches (and stays at), an altitude of 11 km, where the ambient pressure is 0.224 atm and the temperature is 217 K. Assuming that the internal volume didn't change, calculate the structural weight of the balloon (assume that the molecular weight of helium is about 4 and of the surrounding air is about 29).

3.44. A 1.2-m-wide gate is holding oil ($\rho = 800$ kg/m^3) inside the tank, as shown. Calculate the total force and the location of the center of pressure. Also calculate the moment on the bottom hinge.

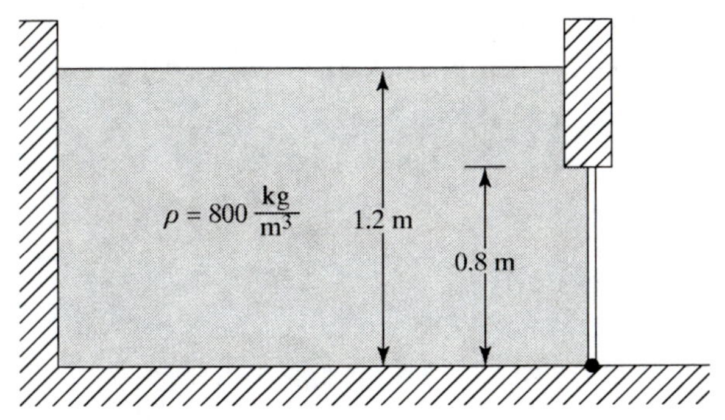

Problem 3.44.

3.45. The triangular gate shown in the figure is pivoted along the lower horizontal plane (A–C). Assuming the water level in the tank is $h = 10$ m, the height of the triangle is $H = 9$ m, and its width $W = 4$ m, calculate the total force on the gate and the location of the center of pressure. Also calculate the moment on the bottom hinge.

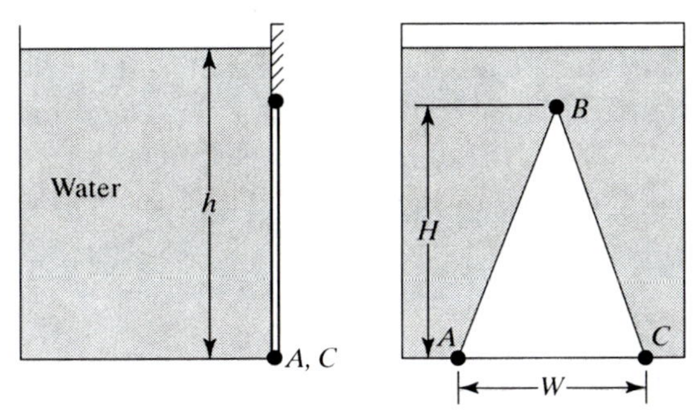

Problem 3.45.

3.46. A hot-air balloon's total weight, including the passengers, is 250 kg, and the ambient temperature is 290 K, and the air density is $\rho = 1.22$ kg/m^3. The balloon is then filled with hot air at an average temperature of 60 °C. Approximating the shape of the balloon as a sphere, what is the minimum diameter for the balloon to take off?

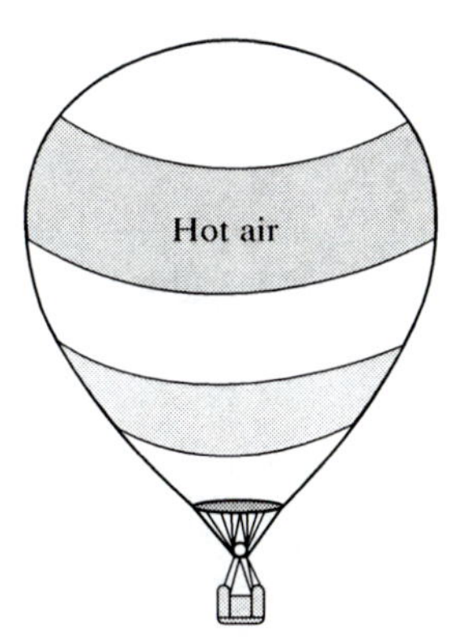

Problem 3.46.

3.47. A cylinder with negligible mass is held in a vertical position by a weight $W = 20\ \text{kg}_f$. If its dimensions are $D = 0.4$ m and $l = 0.7$ m, calculate how deep it will sink in the water.

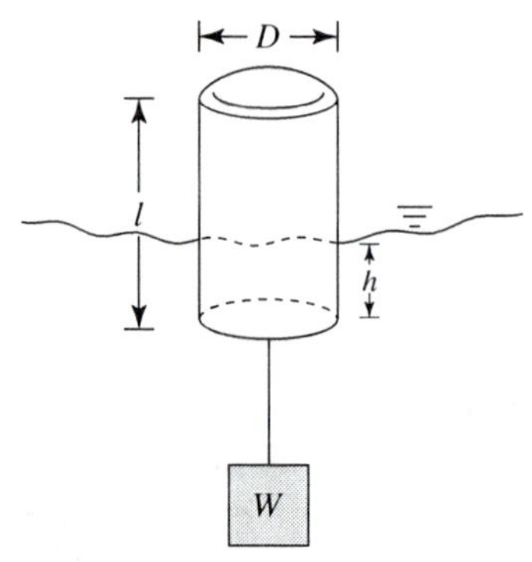

Problem 3.47.

3.48. A cylinder with negligible mass is floating in a pool filled with water. It is held in a vertical position by a weight $W = 60\ \text{kg}_f$, which is placed on a scale. The cylinder dimensions are $D = 0.4$ m and $l = 0.7$ m. If the cylinder is submerged to a depth of $l_1 = 0.4$ m, calculate the weight measured by the scale at the bottom of the pool.

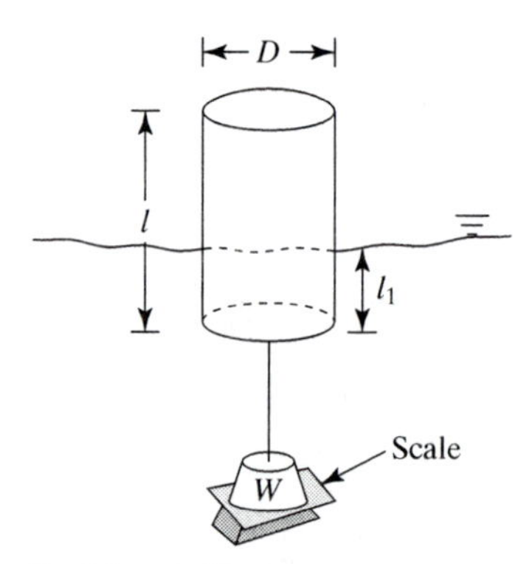

Problem 3.48.

3.49. A cylinder with a mass of 5 kg is placed at the bottom of a pool filled with water. Its length is $l_1 = 0.3$ m and its frontal area is $S = 0.015\ \text{m}^2$.

(a) Calculate the relative density of the cylinder.

(b) To demonstrate the principles used by submarines to rise to the water surface, the length of the cylinder is extended to $l = 0.7$ m. Calculate the percentage of the body volume that is submerged when the cylinder is floating at its extended mode.

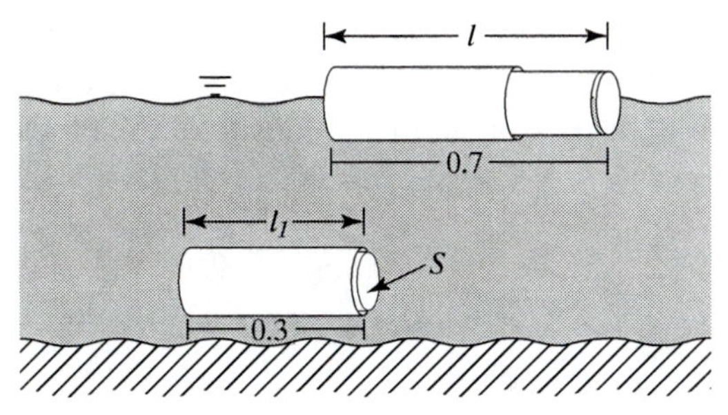

Problem 3.49.

3.50. The water level in a flush tank is controlled by the mechanism shown in the figure. The condition shown represents the "tank-full" condition and the water supply valve is closed. Assuming a negligible mass for the half-sunken ball (of radius $R = 0.08$ m), calculate the vertical force on the valve at point B (note that the arm is hinged at point A and $l_1 = 0.15$ m and $l_2 = 0.05$ m).

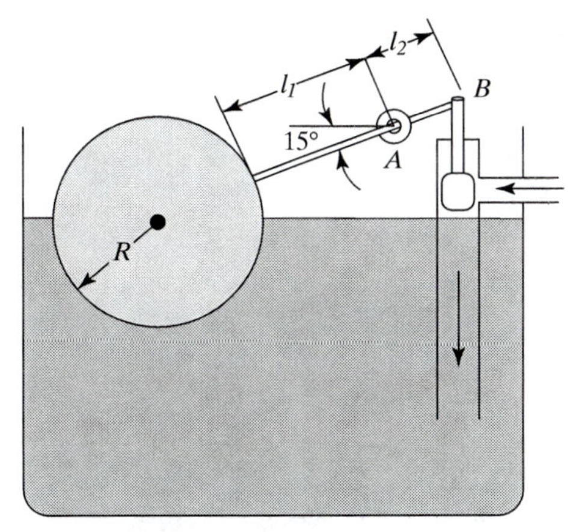

Problem 3.50.

3.51. A water container, filled to the top, is placed on a scale that is measuring 30 kg_f (the weight of the container not included). Next a metal block of volume 0.005 m^3 and weighing of 13.5 kg_f is dropped into the container, and the excess water is spilled at point A (so the water level is still full). Calculate the weight that the scale will show.

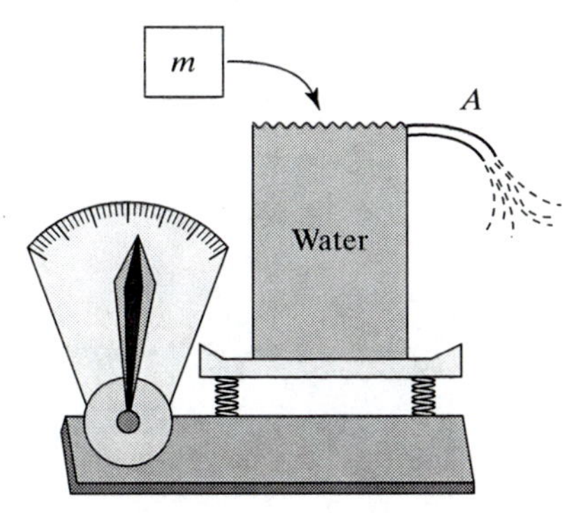

Problem 3.51.

3.52. A 0.6-m-long wooden block is floating on the water. Its height is $b = 0.1$ m and width is $l = 0.5$ m, as shown in the figure. Because of a torque the block is rotated and its right-hand side sinks down to $b_1 = 0.08$ m. Calculate the density of the block and the magnitude of the torque.

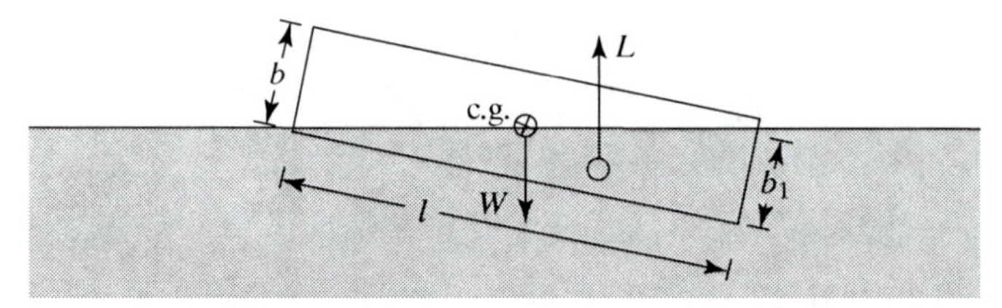

Problem 3.52.

3.53. A 0.6-m-long wooden block is floating on the water (length is the dimension vertical to the page). Its height is $b = 0.1$ m and width is $l = 0.5$ m. Because of a vertical force F, the block assumes the position shown with its lower surface submerged for $l_1 = 0.4$ m. Calculate the block density and the magnitude of the force F.

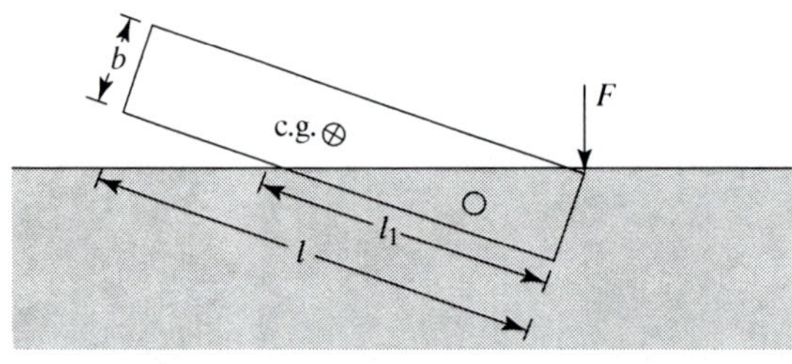

Problem 3.53.

4 Introduction to Fluid in Motion – One-Dimensional (Frictionless) Flow

4.1 Introduction

In the previous chapter, the effects of pressure in a fluid were isolated (because the fluid was not moving) and the resulting forces were investigated. If we use the Navier–Stokes equations as a roadmap for gradually increasing the complexity of the models, then the next level requires the addition of the inertia terms in the momentum equations. For simplicity, it is assumed that the effects of viscosity are negligible, and the examples focus on cases for which such a simplification is acceptable. The addition of the viscosity term (and resulting friction) is discussed in the next chapter.

This is the first chapter in which solutions for fluid in motion are studied. By starting with the simple 1D model the basic principles can be easily demonstrated. The conservation of momentum, for example, closely resembles the classical mechanical formulation and easily can be explained. At first, however, we must clarify the meaning of the 1D flow assumption. Let us start by observing the velocity distribution inside a stream of fluid leaving a pipeline (Fig. 4.1). It is likely that the exiting flow velocity will not be uniform, and the size of the arrows in the figure describes the velocity distribution. If this velocity distribution has an axisymmetric shape then we can call the flow 2D because the velocity u will have a distribution $u = u(r)$, where r is the radial direction (e.g., we must consider both the x and r variables). However, if we define an average velocity $\overline{u}$ for the exiting stream (as shown in the lower sketch in the figure) having the same mass flow, then we can call this a 1D flow, because only one quantity (namely $\overline{u}$) exists when the flow is described in the x direction. Of course, in reality the flow could have a much more complex 3D velocity distribution.

The basic equations used for such 1D flows are the integral form of the continuity and momentum equations (as developed in Section 2.6). However, an additional equation, called the Bernoulli equation, is frequently used, and therefore we must derive it first.

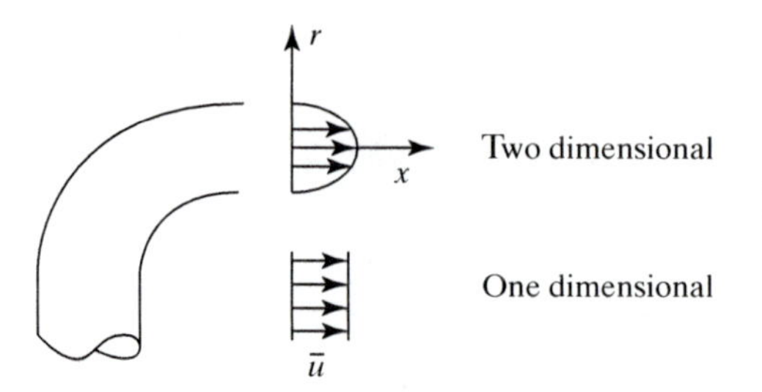

Figure 4.1. 1D and 2D velocity distributions.

4.2 The Bernoulli Equation

In the early discussion about the kinetic theory of gases (see Fig. 1.17), we speculated how Daniel Bernoulli connected (in the mid-1700s) velocity with pressure variations in a moving fluid. This equation is widely used in numerous engineering applications, but it is initially limited to flows without friction (in fact, we modify this later by including a head-loss term). The objective of this section is to introduce the Bernoulli equation for inviscid flows, and a more detailed discussion of its limitations follows in Section 8.3. To clarify the applicability of this equation, consider the flow over a moving vehicle, as shown in Fig. 4.2. We may follow a particle moving along a streamline as shown, but we assume that there are no losses such as friction along this path. This assumption suggests that we must use the inviscid form of the momentum equation [Eq. (2.41)], called the Euler equation:

$$\frac{\partial \vec{q}}{\partial t} + \vec{q} \cdot \nabla \vec{q} = \vec{f} - \frac{\nabla p}{\rho}. \tag{2.41}$$

The x component of the Euler equation may be rewritten (because we are discussing 1D flow only) as

$$\frac{\partial u}{\partial t} + u\frac{\partial u}{\partial x} + v\frac{\partial u}{\partial y} + w\frac{\partial u}{alz} = f_x - \frac{1}{\rho}\frac{\partial p}{\partial x}. \tag{4.1}$$

The only body force we consider for this case is gravity, which acts in the vertical direction [see Eq. (3.5)]:

$$\vec{f} = (0, 0, -g). \tag{3.5}$$

This force is conservative and the work along the streamline depends only on h. Consequently we can write that the force is a gradient of a conservative potential $(-gh)$:

$$\vec{f} = \nabla(0, 0, -gh). \tag{3.5a}$$

Figure 4.2. Frictionless flow along a streamline.

Note that here (see Fig. 4.2) h is positive in the $+z$ direction! Next, let us limit the discussion to steady flows ($\partial u/\partial t = 0$) and replace the x coordinate with a coordinate s along the streamline. Because this is a 1D model (along the streamline) we do not have velocity components in the other directions, and Eq. (4.1) reduces to

$$u\frac{\partial u}{\partial s} = \frac{\partial}{\partial s}\left(-gh - \frac{p}{\rho}\right).$$

Here only one term remains on the inertia side (left-hand side), and the force component that is due to gravity and pressure is on the right-hand side. This can be rearranged such that

$$\frac{\partial}{\partial s}\left(gh + \frac{p}{\rho} + \frac{u^2}{2}\right) = 0.$$

To calculate the changes along the streamline we integrate along s (e.g., between point a and b in Fig. 4.2) to get

$$gh + \frac{p}{\rho} + \frac{u^2}{2} = \text{const.} \tag{4.2}$$

This equation is named after Daniel Bernoulli (1700–1782), and a more rigorous development of this equation, clarifying its applicability, follows later in Section 8.2. As an example we can apply Eq. (4.2) to the two points in Fig. 4.2 to get

$$gh_a + \frac{p_a}{\rho} + \frac{u_a^2}{2} = gh_b + \frac{p_b}{\rho} + \frac{u_b^2}{2}, \tag{4.3}$$

and here we assume that the fluid is incompressible. This equation shows the change in pressure versus the change in velocity (taking into account the loss of energy that is due to altitude change). A much more common use of the equation is to refer to a vehicle moving at a speed of U_∞ and to apply the equation to the far undisturbed field (at ∞) and to a point where the flow stops relative to the vehicle (e.g., point c in the figure where the velocity u_c is zero). Neglecting the changes in altitude we can write

$$p_c - p_\infty = \frac{\rho}{2}U_\infty^2. \tag{4.4}$$

This equation shows the increase of pressure (ram effect) ahead of the cooling system as a function of vehicle speed. The higher pressure at point c is called *total pressure* because velocity is zero there, whereas p_∞ in this case is called the *static pressure*. The right-hand side ($\frac{\rho}{2}U_\infty^2$) is called the *dynamic pressure*. Consequently we can rewrite Eq. (4.4) as

$$\text{total pressure} - \text{static pressure} = \text{dynamic pressure}. \tag{4.4a}$$

4.3 Summary of the One-Dimensional Tools

The proposed 1D model is not necessarily limited to flows along a single straight line, as demonstrated by the following examples. Three equations are used for this 1D fluid flow model, depicted schematically in Fig. 4.3. The stream tube, as shown, is

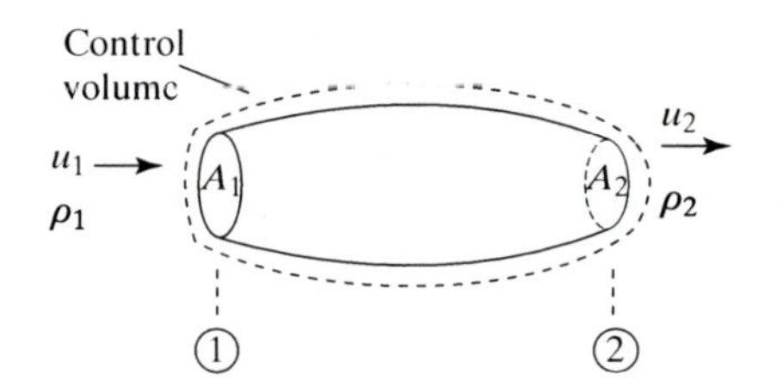

Figure 4.3. Basic model for 1D flow.

wrapped by a control volume, and the flow enters at station 1 and exits at station 2, as shown. For the continuity equation (or conservation of mass) we use Eq. (2.26),

$$\dot{m} = \rho u_1 A_1 = \rho u_2 A_2, \tag{4.5}$$

which basically states that in steady flow the fluid flow that enters the control volume is equal to the flow leaving it.

The 1D momentum equation [Eq. (2.29)] is used to calculate the forces that are due to the momentum change and the pressure acting on the two surfaces at stations 1 and 2:

$$F_x = \rho u_2^2 A_2 - \rho u_1^2 A_1 + (p_2 - p_a) A_2 - (p_1 - p_a) A_1. \tag{4.6a}$$

If there is no mass addition or subtraction inside the control volume, then by using Eq. (4.5) we can write

$$F_x = \dot{m} u_2 - \dot{m} u_1 + (p_2 - p_a) A_2 - (p_1 - p_a) A_1. \tag{4.6b}$$

Note that the momentum equation is a vector expression, and, for example, in a Cartesian coordinate system we can write three such equations in each of the directions. The continuity and momentum relate to a control volume, whereas Bernoulli is between two points, as shown in Fig. 4.2:

$$g h_a + \frac{p_a}{\rho} + \frac{u_a^2}{2} = g h_b + \frac{p_b}{\rho} + \frac{u_b^2}{2}. \tag{4.7}$$

The use of this model and the applicability of the three equations is demonstrated in the following examples.

4.4 Applications of the One-Dimensional Flow Model

In cases in which friction effects are negligible the preceding model can provide fast and simple estimates on the forces generated by a moving fluid. Because this model closely resembles classical particle dynamics it is easily understood and cases such as the forces that are due to free jets can be simply estimated.

4.4.1 Free Jets

Let us consider a liquid jet leaving a nozzle and being turned by a solid frictionless surface, as shown in Fig. 4.4. To calculate the forces acting on the structure we define the control volume (as shown) with inflow station 1 and exit station 2.

If we assume an incompressible flow then the continuity equation remains the same as Eq. (4.5):

$$\dot{m} = \rho u_1 A_1 = \rho u_2 A_2.$$

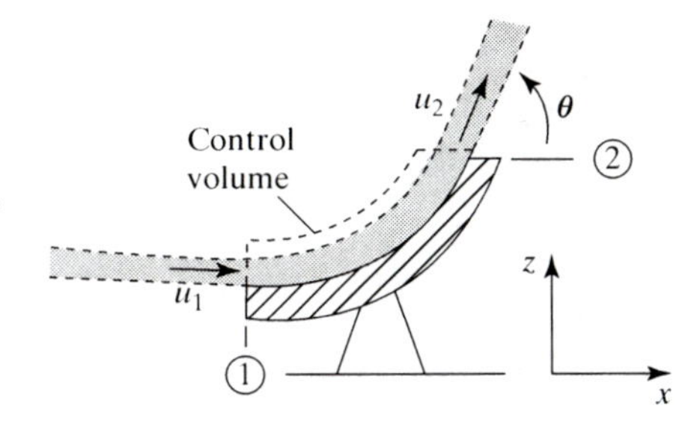

Figure 4.4. Free jet turned by a solid surface.

Because this is a free jet, the pressure is the same at both stations, $p_1 = p_2 = p_a$, and because there are no losses (friction) the velocity is also the same ([see Eq. (4.7)]:

$$u_2 = u_1.$$

At this point the pressure and the velocity (at the inlet and exit) are fixed and the Bernoulli equation is not needed anymore (assuming negligible gravitational effects). The momentum equation can be split into two components (in the x and z directions) as follows:

$$F_x = \dot{m}u_2 \cos\theta - \dot{m}u_1 = \dot{m}u(\cos\theta - 1), \tag{4.8a}$$

$$F_z = \dot{m}w_2 - \dot{m}w_1 = \dot{m}u \sin\theta \qquad (w_1 = 0). \tag{4.8b}$$

Note that for $\theta < 90°$ the x-direction force is negative and the normal force is positive. These are the forces acting on the control volume. Of course, the reaction force on the curved stand is pushing it to the right and down.

EXAMPLE 4.1. FORCE ON A TURBINE BLADE. As a numerical example let us consider a simple model of a turbine blade. The water jet in Fig. 4.5 is leaving the nozzle with an area of 1 cm^2 at a velocity of $u_n = 15$ m/s while the blade moves in the positive x direction at a velocity of $u_b = 10$ m/s. We attach the control volume to the (constant-speed) blade, and therefore the relative velocity at station 1 is

$$u_1 = 5 \text{ m/s} = u_2,$$

which is the same as the velocity at station 2

The forces can be found by use of Eq. (4.8):

$$F_x = \rho u^2 A(\cos\theta - 1) = 1000 \times 5^2 \times 0.0001(\cos 120 - 1) = -3.75 \text{ N},$$
$$F_z = \rho u^2 A \sin\theta = 1000 \times 5^2 \times 0.0001 \cdot \sin 120 = 2.16 \text{ N}.$$

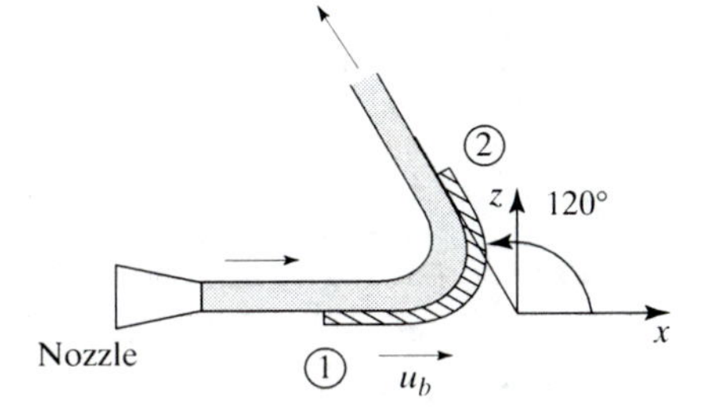

Figure 4.5. Free jet hitting a turbine blade.

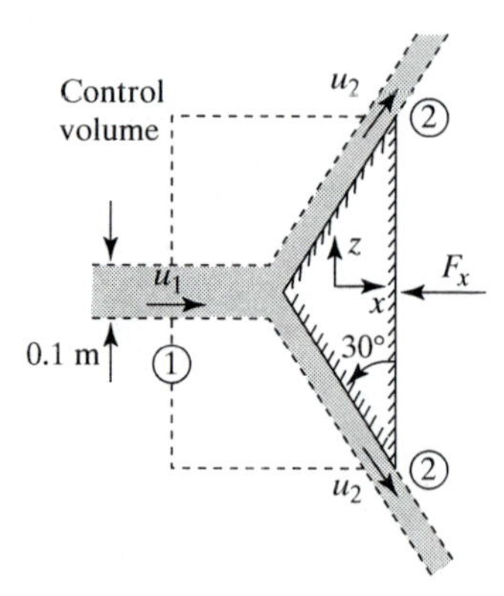

Figure 4.6. Free jet hitting a symmetric wedge.

Again the force in the x direction is negative, indicating the direction of the external force acting on the control volume. This means that the fluid (reaction) is pushing the turbine blade to the right and the z component of the force is pushing the blade down. The power W generated by this turbine blade is simply the force times its translation velocity u_b:

$$W = F_x u_b = 37.5 \text{ W}.$$

EXAMPLE 4.2. FORCE ON A WEDGE. A circular water jet of velocity $u_1 = 15$ m/s and diameter 0.1 m is impinging on a wedge, as shown in Fig. 4.6. The jet is split into two equal jets, and the force on the wedge must be calculated.

We start by defining the control volume as shown in the figure, assuming no changes in the pressure (free jet). Also, as in Example 4.1, we assume that there are no losses and the velocity at the exit remains the same (as in the previous example):

$$u_1 = u_2 = 15 \text{ m/s}.$$

Because the flow is incompressible, the total jet cross-section area is not changing ($A_1 = 2A_2 = A$). Now we can write the momentum equation (4.6) in the horizontal direction, assuming symmetry between the upper and lower exiting jets:

$$F_x = 2\rho u_2^2 \frac{1}{2} A_2 \cos\theta - \rho u_1^2 A_1 = \rho u^2 A(\cos\theta - 1)$$

$$= 1000 \times 15^2 \left(\frac{\pi 0.1^2}{4}\right)(\cos 60 - 1) = -883.6 \text{ N}.$$

The vertical components of the momentum equation cancel each other and there is no force in that direction:

$$F_z = \frac{1}{2}\rho u^2 A \sin\theta - \frac{1}{2}\rho u^2 A \sin\theta = 0.$$

EXAMPLE 4.3. THRUST OF A JET ENGINE. A jet engine flies at a speed of U_∞, as depicted in Fig. 4.7, and the surrounding pressure is p_a. To calculate the thrust of the engine we need to use both the continuity; and the momentum

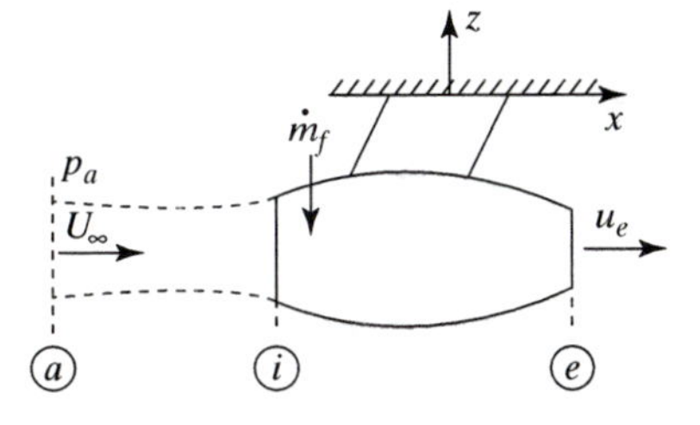

Figure 4.7. Calculating the thrust of a jet engine.

equations. Let us wrap the nacelle with a control volume; the air enters through inlet section i and leaves at exhaust section e. Based on Eq. (4.6), the thrust T is

$$T = F_x = \rho u_e^2 A_e - \rho u_i^2 A_i + (p_e - p_a)A_e - (p_i - p_a)A_i. \tag{4.9}$$

We can use the continuity equation to quantify the incoming and exiting mass flows rates:

$$\dot{m}_a = \rho_i u_i A_i,$$
$$\dot{m}_e = \rho_e u_e A_e.$$

However, the fuel flow $\dot{m}_f$ must be taken into account:

$$\dot{m}_e = \dot{m}_a + \dot{m}_f.$$

Substituting these results into Eq. (4.9) yields

$$T = (\dot{m}_a + \dot{m}_f)u_e - \dot{m}_a u_i + (p_e - p_a)A_e - (p_i - p_a)A_i. \tag{4.10}$$

Usually the exhaust conditions and flight speed are known whereas the inlet conditions at station i are not. To further simplify this relation (and eliminate the unknowns at the inlet) we can introduce a new control volume in front of the nacelle, between a far-field station a and inlet i. By applying the momentum equation to this control volume and noting that there is no thrust generated by it, we get

$$0 = \dot{m}_a u_i - \dot{m}_a U_\infty + (p_i - p_a)A_i - (p_a - p_a)A_a,$$

and after rearranging, we get

$$\dot{m}_a U_\infty = \dot{m}_a u_i + (p_i - p_a)A_i.$$

Substituting this into Eq. (4.10) yields

$$T = (\dot{m}_a + \dot{m}_f)u_e - \dot{m}_a U_\infty + (p_e - p_a)A_e. \tag{4.11}$$

In modern engines the nozzle converts most of the pressure to exhaust velocity and the last term is small relative to the other two terms. Consequently we can approximate the thrust as

$$T = (\dot{m}_a + \dot{m}_f)u_e - \dot{m}_a U_\infty. \tag{4.11a}$$

Compare this with the rocket (at the end of Section 2.6) and you'll see that the rocket has a higher thrust because there is no incoming momentum.

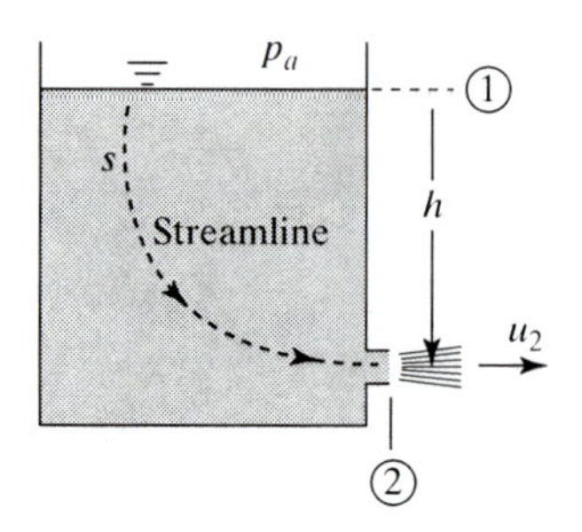

Figure 4.8. Flow leaving a large reservoir.

4.4.2 Examples for Using the Bernoulli Equation

In these cases we must identify the two points for applying the Bernoulli equation and make sure that there are no losses as the fluid moves from one point to the other.

EXAMPLE 4.4. THE VELOCITY OF A JET LEAVING A LARGE RESERVOIR. To demonstrate the applicability of the Bernoulli equation, let us analyze the flow leaving a large reservoir, as shown in Fig. 4.8. Exit section 2 is very small, and the fluid loss that is due to the flow through this exit has negligible effect. The control-volume exit section is placed slightly outside the container, and therefore the outside pressure (in the free jet) is the same as on the top (e.g., $p_1 = p_2 = p_a$). Now we assume that there are no losses such as friction along the path s, and we apply Eq. (4.7) at the top of the reservoir (station 1) and at the exit (station 2), but it is clear that there is no velocity at station 1:

$$gh_1 + \frac{p_a}{\rho} + \frac{0^2}{2} = gh_2 + \frac{p_a}{\rho} + \frac{u_2^2}{2}.$$

Note that here h is positive in the $+z$ direction, as shown in Fig. 4.2. After rearranging the terms we get

$$g(h_1 - h_2) = gh = \frac{u_2^2}{2},$$

and the velocity is exactly the same as calculated by assuming a free fall (recall we assumed no friction losses):

$$u_2 = \sqrt{2gh}. \tag{4.12}$$

EXAMPLE 4.5. THE FORCES ON A WATER HOSE. This example is similar to the free-jet examples, but now we'll try to use the Bernoulli equation as well. Assume that the pressure inside the pipeline, shown in Fig. 4.9, is p_0 and is much larger

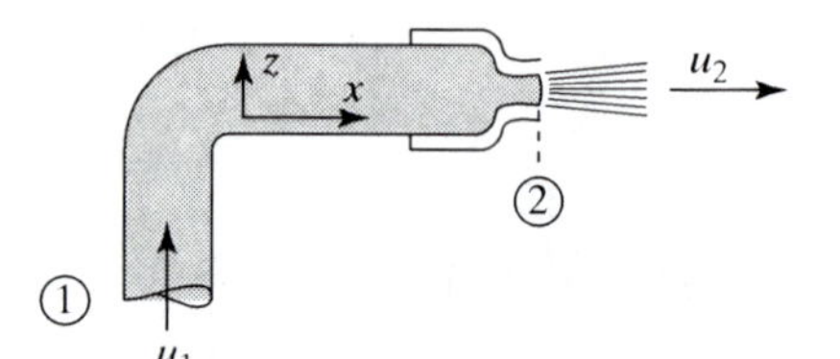

Figure 4.9. Forces on a bent fire hose.

than the ambient pressure p_a. Also, the inflow u_1 and cross-section area A_1 are known. It appears, though, that both the continuity equation and the Bernoulli equation control the exit velocity. Of course, we all know that for a smooth jet we need to adjust the jet exit area. For example, if it is too small then the excess pressure will brake the jet, and then we cannot assume that there are no losses in this flow. So in our example let us first find the ideal exit velocity, by using the Bernoulli equation, assuming no losses through the nozzle for this incompressible flow:

$$\frac{u_2^2}{2} = \frac{u_1^2}{2} + \frac{p_0 - p_a}{\rho}. \tag{4.13}$$

From the continuity equation we can calculate the required exit area for this matched flow:

$$A_2 = A_1 \frac{u_1}{u_2}. \tag{4.14}$$

We now proceed with the force calculation by using the momentum equation (4.6) (recall that the exit area was selected such that $p_2 = p_a$) for both directions:

$$\begin{aligned} F_x &= \rho u_2^2 A_2, \\ F_z &= -\rho u_1^2 A_1 - (p_0 - p_a) A_1. \end{aligned} \tag{4.15}$$

Therefore the external force on the control volume acts down and to the right. To a person holding the hose, the reaction force is felt pushing back (to the left) and up. If we want to rewrite this in terms of the mass flow rate $\dot{m}$ and the incoming velocity u_1, then with Eq. (4.13) (recall – no losses), this becomes

$$\begin{aligned} F_x &= \dot{m}\sqrt{u_1^2 + \frac{2(p_0 - p_a)}{\rho}}, \\ F_z &= -\dot{m} u_1 - (p_0 - p_a) A_1. \end{aligned} \tag{4.16}$$

4.4.3 Simple Models for Time-Dependent Changes in a Control Volume

Certain time-dependent problems can be approximated by use of the 1D flow model. As an example, the flow from a container that is due to gravity or the flow leaving a pressurized container are presented here. In these examples an incompressible fluid is assumed; a similar model but for compressible flow is presented later in Section 10.3.

EXAMPLE 4.6. FLOW (DUE TO GRAVITY) THROUGH A SMALL HOLE IN A CONTAINER. For the first case, consider a container with a (fixed) cross section A_1, as shown in Fig. 4.10. The momentary height of the liquid is h and the outside pressure is p_a. The flow exits the container at the bottom through a small pipe of cross-section area A_2. It is clear that the exit velocity depends on the liquid height h, and therefore a solution for this variable (with time) is sought.

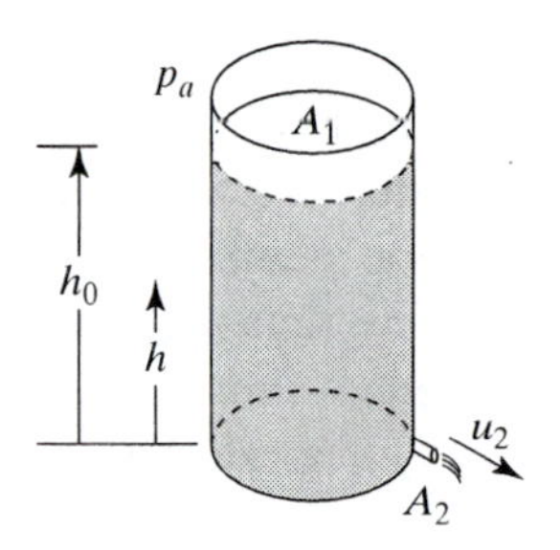

Figure 4.10. Time-dependent flow that is due to gravity.

The exit velocity (at the bottom) can be calculated with Eq. (4.12),

$$u_2 = \sqrt{2gh},$$

and the mass flow rate leaving the container is

$$\dot{m} = \rho u_2 A_2 = \rho \sqrt{2gh} A_2.$$

Next, consider a control volume containing all the liquid inside the container. The conservation of mass principle requires that no mass be lost:

$$\frac{d}{dt}(m_{\text{tank}}) = \frac{d}{dt}(\rho A_1 h) = -\rho A_2 \sqrt{2gh}.$$

This means that the mass reduction in the tank is equal to the fluid leaving at the bottom (hence the minus sign). This equation can be rearranged, with a separation of the variables, as follows:

$$\frac{dh}{\sqrt{h}} = -\frac{A_2}{A_1}\sqrt{2g}dt.$$

After integrating both sides of the equation, we get

$$2\sqrt{h} = -\frac{A_2}{A_1}\sqrt{2g}t + C,$$

and C is the constant of integration. Solving for h, we get

$$h = \left(C_1 - \frac{A_2}{A_1}\sqrt{\frac{g}{2}}t\right)^2,$$

and C_1 is a modified constant of integration to be calculated by use of the initial conditions. Assuming that at $t = 0$, $h = h_0$, we get $c_1 = \sqrt{h_0}$, and therefore the time-dependent height of the liquid in the container is

$$h = \left(\sqrt{h_0} - \frac{A_2}{A_1}\sqrt{\frac{g}{2}}t\right)^2. \tag{4.17}$$

The corresponding exit velocity at the bottom is

$$u_2 = \left(\sqrt{2gh_0} - \frac{A_2}{A_1}gt\right), \tag{4.18}$$

and the time to drain the container [e.g., $t(h = 0)$] can be calculated by setting $u_2 = 0$ in this equation

$$t_{h=0} = \frac{A_1}{A_2}\sqrt{\frac{2h_0}{g}}. \tag{4.19}$$

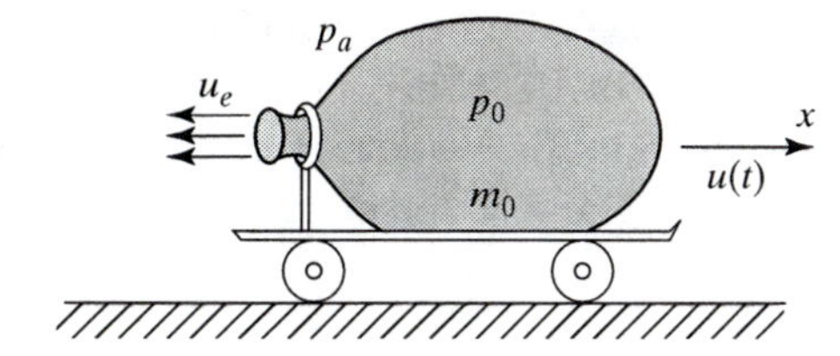

Figure 4.11. A water balloon placed on frictionless wheels.

EXAMPLE 4.7. THE VELOCITY OF A WATER BALLOON. This example is similar to the example in Fig 4.8, but now a 1D equation of motion (for the system) is added. Consider a popular children's toy (Fig. 4.11) consisting of a water-inflated balloon, placed on a small rolling platform. The objective is to calculate the momentary velocity of the apparatus.

This problem can be solved with a few simplifying assumptions. First, assume that the exit velocity and internal pressure p_0 do not change as the balloon deflates. Second, assume the nozzle losses are negligible, and we can use the Bernoulli equation (Eq. 4.4) to calculate the water jet exit velocity u_e:

$$u_e = \sqrt{\frac{2(p_0 - p_a)}{\rho}} \approx \text{const.}$$

With these assumptions the equation of motion in the x direction becomes

$$m(t)\frac{du(t)}{dt} = \dot{m}u_e = \frac{-dm(t)}{dt}u_e,$$

where $u(t)$ is the apparatus velocity and the minus sign on the right-hand side is a result of the mass $m(t)$ shrinking with time. After separating the variables and neglecting friction forces in the wheels, we get

$$du(t) = \frac{-dm(t)}{m(t)}u_e.$$

After integration we get

$$u(t)\big|_0^{t_1} = u_e \ln m(t)\big|_0^{t_1}.$$

Assuming initial conditions at $t = 0$,

$$u(0) = 0 \quad \text{and} \quad m(0) = m_0,$$

and replacing the variable t_1 with t, we get

$$u(t) = u_e \ln \frac{m_0}{m(t)} = \sqrt{\frac{2(p_0 - p_a)}{\rho}} \ln \frac{m_0}{m(t)}. \tag{4.20}$$

These results indicate that both acceleration and momentary speed will increase with time. If the mass of the system without the water is m_s, then the final velocity u_f becomes

$$u_f = \sqrt{\frac{2(p_0 - p_a)}{\rho}} \ln \frac{m_0}{m_s}. \tag{4.21}$$

Of course in most cases the pressure inside the balloon may change, and then the equations are more complex.

As a numeric example, consider a balloon filled with 1-kg water at a pressure of 10^5 N/m^2. The nozzle area is 1 cm^2 and the mass of the whole system, without the water, is 0.5 kg.

The solution starts with calculating the exit velocity as

$$u_e = \sqrt{\frac{2 \times 10^5}{1000}} = 14.14 \frac{\text{m}}{\text{s}}.$$

The final velocity of the system, based on Eq. (4.21), is

$$u_f = 14.14 \ln \frac{1.5}{0.5} = 15.53 \frac{\text{m}}{\text{s}}.$$

Note that the final velocity is higher than the jet exit speed!

4.5 Flow Measurements (Based on Bernoulli's Equation)

The preceding 1D models (mainly based on Bernoulli's equation) led to the developments of several important flow-measuring devices. The principle of their operation is discussed next.

4.5.1 The Pitot Tube

The principle introduced by Bernoulli's equation allows us to measure speed at a point in the flow by simply measuring a pressure difference. The device utilizing this principle is called the Pitot tube and was named after Henry Pitot (1695–1771), a French hydraulic engineer who invented this device to measure river flows. The basic apparatus is described schematically in Fig. 4.12 and consists of two concentric tubes. It is assumed that at the tip the velocity comes to a halt. Therefore the inner tube measures at its tip the higher, total pressure, which increases as flow speed increases (as at point c in Fig. 4.2). The holes surrounding the outer tube are exposed to a speed of U_∞ and measure the static pressure, and this should be equal to the undisturbed pressure (at ∞), which is not affected by the vehicle's speed. The difference between the pressure in the two concentric tubes (the dynamic pressure) can be measured and connected to a display that shows the speed of the fluid stream.

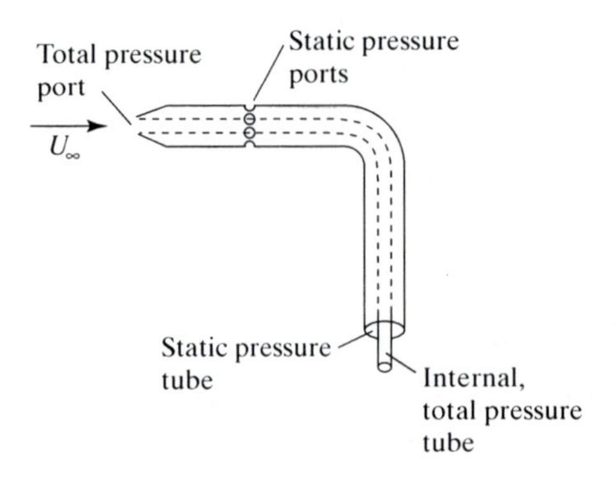

Figure 4.12. The Pitot tube.

Now if we attach the Pitot tube to a moving vehicle and make sure that it points straight into the flow, then, from Eq. (4.4), we can measure the vehicle's speed as

$$U_\infty = \sqrt{\frac{2(p_{\text{tot}} - p_\infty)}{\rho}}. \tag{4.22}$$

Pitot tubes are widely used on airplanes, ships, and in wind tunnels to measure the flow speed. For best accuracy, the tube must be aligned with the flow and placed away from disturbances created by the moving vehicle. Therefore Pitot tubes are frequently mounted on long rods extending ahead of an airplane or any other test vehicle.

EXAMPLE 4.8. MEASURING AIRSPEED IN A WIND TUNNEL. A Pitot tube was inserted into the flow in a wind tunnel, and the pressure difference between the tubes was measured by a water column of 20-cm H_2O. If air density is about $\rho_a = 1.2\ \text{kg/m}^3$ then calculate the airspeed in the wind tunnel.

The solution is given by Eq. (4.22); however, the pressure difference is given as

$$\Delta p = \rho_w g h = 1000 \times 9.8 \times 0.2 = 1960\ \text{N/m}^2.$$

Substituting this into Eq. (4.22) gives us

$$U_\infty = \sqrt{\frac{2 \times 1960}{1.2}} = 57.15 \frac{\text{m}}{\text{s}} = 205.75 \frac{\text{km}}{\text{h}}.$$

4.5.2 The Venturi Tube

Another device that can be used to measure fluid flow (or average velocity) is the Venturi tube (or meter), named after the Italian physicist G. B. Venturi (1746–1822), who was the first to investigate its operating principles in 1791. The Venturi meter consists of a tube with a narrowed center section, as shown in Fig. 4.13, and when in operation, the air flow (or water flow or any other fluid flow) moves faster through the narrow section, as expected. Usually, the Venturi meter is used to measure the flow inside a pipe whereas the Pitot tubes used mainly for external flows.

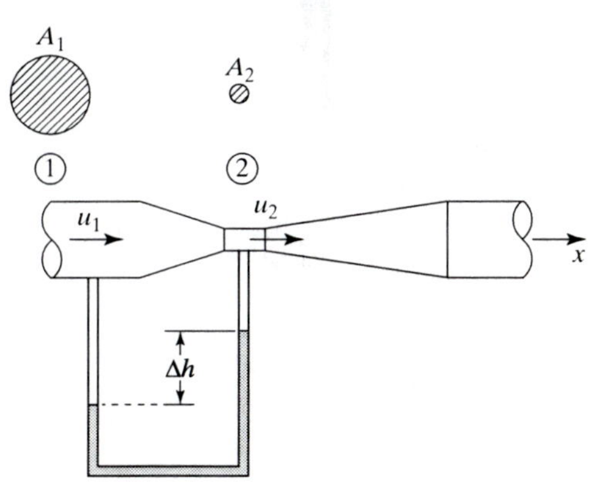

Figure 4.13. Nomenclature for the Venturi tube.

To measure the pressure difference, a thinner tube is connected to the large and narrow sections (as shown in the figure) and filled partially with (sometimes heavier) liquid. The pressure difference between the wide and narrow sections of the Venturi tube causes the fluid to rise in the lower-pressure side of the tube. The level of the fluid in the thin tube can be directly related (and calibrated) to the velocity inside the Venturi tube. In general, the pressure difference (and signal) created by a Venturi meter is smaller than the signal from a Pitot tube. This makes the Venturi meter less desirable for measuring external velocity (e.g., airspeeds), and in actual practice Venturi meters are used primarily to measure liquid flow rates in pipes. To demonstrate the principle of operation let us assume that the fluid inside the tube is incompressible and the relation between the two velocities and cross section areas (based on the continuity equation) is

$$u_2 = u_1 \frac{A_1}{A_2}. \qquad (*)$$

Now we can apply the Bernoulli equation between these two points:

$$p_1 - p_2 = \rho \left(\frac{u_2^2}{2} - \frac{u_1^2}{2} \right).$$

Next, we substitute u_2 from $(*)$ into the Bernoulli equation to get

$$p_1 - p_2 = \rho \frac{u_1^2}{2} \left[\left(\frac{A_1^2}{A_2^2} \right) - 1 \right].$$

Solving for u_1, we get

$$u_1 = \sqrt{\frac{2(p_1 - p_2)}{\rho \left[\left(\frac{A_1^2}{A_2^2} \right) - 1 \right]}}. \qquad (4.23)$$

Usually the flow rate is more important, and therefore

$$\dot{m} = \rho u_1 A_1 = \sqrt{\frac{2\rho(p_1 - p_2)}{\frac{1}{A_2^2} - \frac{1}{A_1^2}}}. \qquad (4.24)$$

Note that when the pressure difference is measured with a liquid column, as shown in Fig. 4.13, then

$$p_1 - p_2 = \rho_m g \Delta h,$$

and here the measuring fluid density ρ_m is used.

To incorporate the losses in the pipeline that are due to the installation of this measuring device, a discharge or loss coefficient C_D is used. Consequently the frictionless Venturi formula is modified by simply adding the discharge coefficient C_D:

$$\dot{m} = C_D \sqrt{\frac{2\rho(p_1 - p_2)}{\frac{1}{A_2^2} - \frac{1}{A_1^2}}}. \qquad (4.25)$$

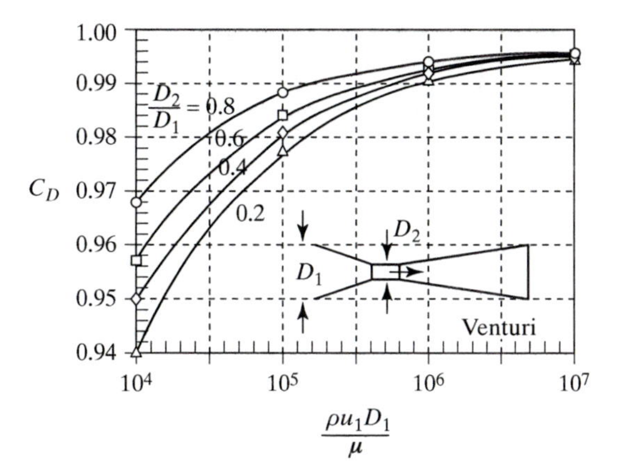

Figure 4.14. Experimental results for the discharge coefficient of a Venturi tube.

In reality, the performance of the Venturi tube depends on the fluid properties and speed and on the tube geometry. Typical experimental results for the discharge coefficient are given in Fig. 4.14 for a range of diameter ratios.

The Venturi tube is very efficient and the losses are quite small, as seen in Fig. 4.14 where typical values for C_D are between 0.94 to 0.98! Also note that the discharge coefficient varies with the internal flow speed. However, instead of the speed u_1, a nondimensional number $\frac{\rho u_1 D_1}{\mu}$ was used. This number is called the Reynolds number and is discussed in the next chapter.

4.5.3 The Orifice

The orifice shown in Fig. 4.15 is a popular flow-measuring device, and its advantage is in its simplicity. Basically a simple flange is added to the pipeline with a narrower hole, as shown in the figure. Because of this narrower passage, the flow accelerates near the smaller hole and the pressure drops accordingly. Contrary to the smoother flow in a Venturi tube, the flow in the orifice is separated, causing more losses.

The flow rate in the orifice is measured with the same formula developed for the Venturi tube [Eq. (4.25)] but the loss coefficient C_D is much larger. The discharge coefficient depends much more on the ratio between the two areas and typically varies between 0.55 for smaller holes to 0.65 for the relatively larger openings. Experimental data showing the discharge coefficient versus the nondimensional flow speed are shown in Fig. 4.16.

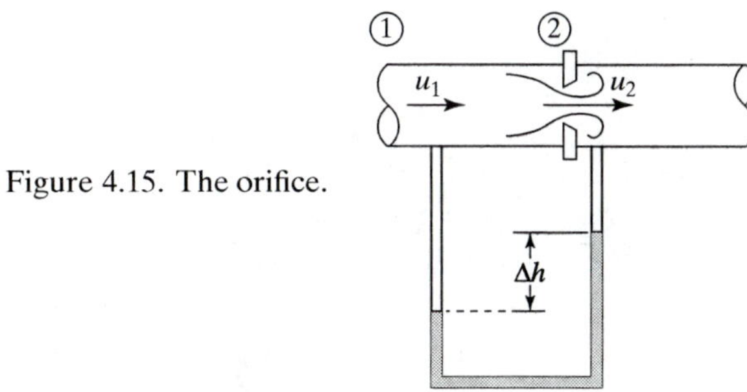

Figure 4.15. The orifice.

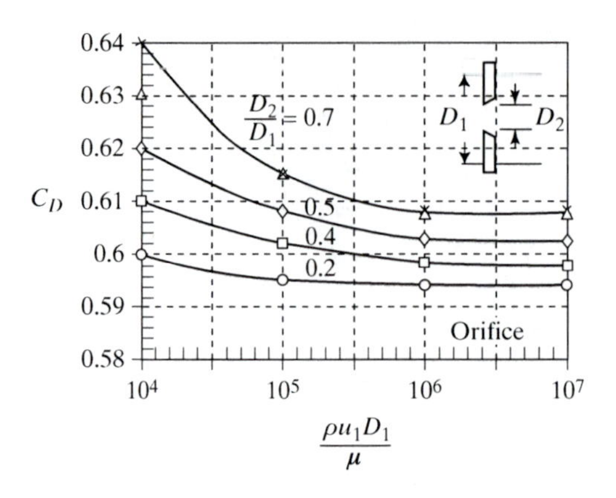

Figure 4.16. Discharge coefficients for an orifice.

4.5.4 The Sluice Gate

A quite simple flow-measuring device for open channel flows is the sluice gate, shown schematically in Fig. 4.17. It is operated by raising the gate and allowing the liquid (often water) to flow under it. By applying the Bernoulli equation between point 0 and point 2, we can develop a quite good approximation for the flow rate:

$$gh_0 + \frac{p_a}{\rho} + \frac{0^2}{2} = g\frac{h_2}{2} + \frac{p_a}{\rho} + \frac{u_2^2}{2}.$$

Here we assume that the reservoir on the left is large and the velocity at point 0 is negligible. Furthermore, assuming no friction losses and that the ambient pressure p_a is the same at the two points, we find that the velocity at station 2 becomes

$$u_2 = \sqrt{2g\left(h_0 - \frac{h_2}{2}\right)}.$$

Figure 4.17 suggests that the water stream exiting the gate contracts from a height of h_1 to h_2. Usually, when $h_0/h_1 > 5$, taking $h_2 = 0.6h_1$ is considered a reasonable approximation.

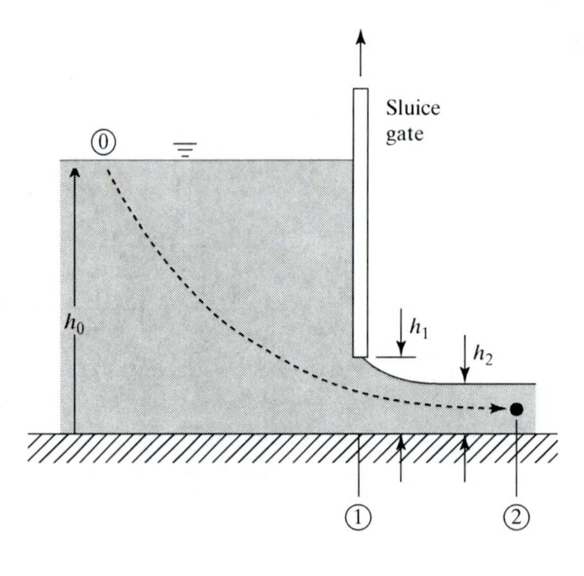

Figure 4.17. Schematic description of a sluice gate. It is operated by simply raising the gate and allowing the water to flow under it.

With the preceding considerations, the flow rate under the sluice gate is

$$Q = \rho u_2 A_2 = \rho h_2 b \sqrt{2g\left(h_0 - \frac{h_2}{2}\right)} \tag{4.26}$$

and here b is the width of the gate.

4.5.5 Nozzles and Injectors

The discussion in this section was aimed at flow-measuring devices. Nozzles and injectors can also be viewed as flow-control devices because they deliver a premeasured flow rate. Furthermore, the flow rate of liquids through the narrow orifice (of area A_i) is controlled by the same formulas used in this section. Three generic injector shapes are shown in Fig. 4.18, all having the same exit area, A_i. If the stagnation pressure inside the nozzle is p_0 and the pressure outside is p_a, then the 1D exit velocity u can be estimated by use of Eq. (4.22):

$$u = C_D \sqrt{\frac{2(p_0 - p_a)}{\rho}}. \tag{4.27}$$

Here C_D is the discharge coefficient, as defined earlier. Typical values for the discharge coefficient are also shown in Fig. 4.18.

Frequently the flow rate of the injector is sought, and then, by using the continuity equation, we get

$$\dot{m} = \rho u A_i = C_D A_i \sqrt{2\rho(p_0 - p_a)}. \tag{4.28}$$

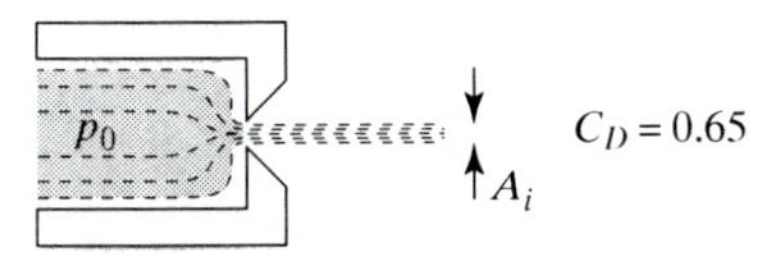

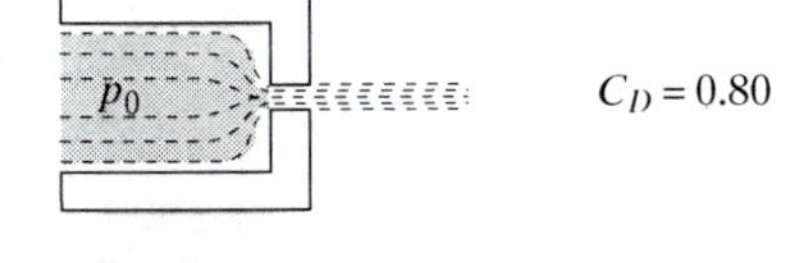

Figure 4.18. Simple injector–nozzle geometries (for liquids) and typical discharge coefficients.

4.6 Summary and Conclusions

In this chapter a simple 1D model was introduced that could be very useful when the average velocity can be estimated. Calculation of the average velocity in pipe flows is discussed in the next chapter. Because viscosity effects were not discussed directly, the forces that are due to momentum changes were easily evaluated.

The Bernoulli equation introduced the relation between velocity and pressure in a frictionless environment, and several flow-measuring devices using this principle were introduced.

Sometimes the simplicity of this model creates the impression that fluid mechanics is an easy subject (and some students stop paying attention to the following chapters). Of course, this is great misconception.

PROBLEMS

4.1. Water flows through the circular nozzle shown in the figure at a rate of 0.3 m^3/s (or 300 kg/s). The diameter at section 1 is 0.3 m and at section 2 is 0.1 m. Using the simple stream-tube model, calculate the velocities at stations 1 and 2 and the force acting on the flange. Assume that the pressure at station 1 inside the nozzle is 700,000 N/m^2 and is zero at station 2.

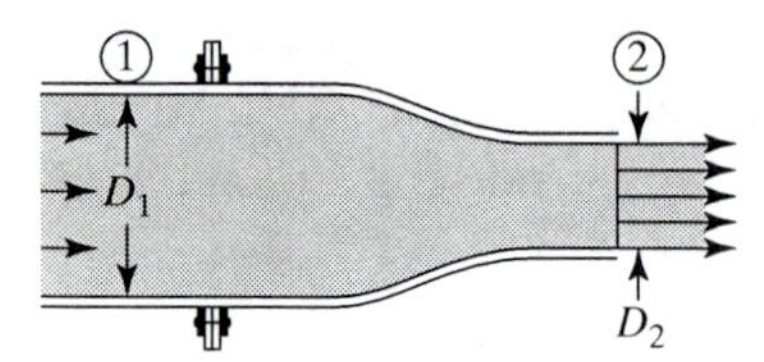

Problem 4.1.

4.2. The rocket motor shown in the figure is tested under static conditions. Suppose the exit velocity is 500 m/s, exit area $A_2 = 0.1$ m^2, and the density of the jet at the exit is $\rho_2 = 0.6$ kg/m^3; then calculate the static thrust (e.g., F_x). Assume the exit pressure is $p_2 = 1.5$ atm and the ambient pressure is $p_a = 1$ atm.

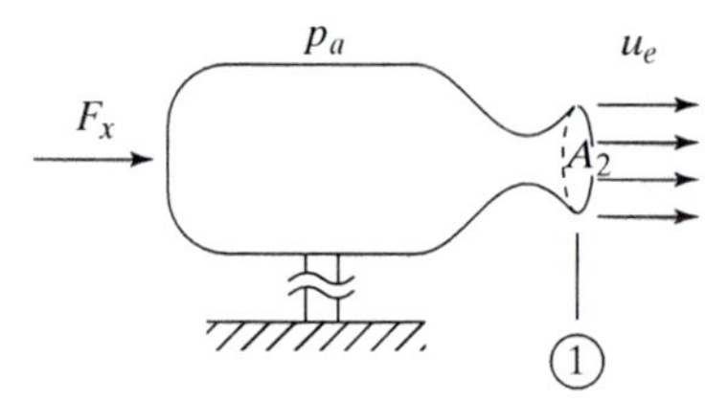

Problem 4.2.

4.3. The rocket motor of the previous problem is fired in space. Calculate its thrust. Is it different from the value obtained during static tests at sea level?

4.4. The container shown in the figure is filled with water. The water discharges through a hole of area A_2 at the lower right.

(a) Develop an expression for the discharge velocity u_2.
(b) Calculate the rate at which the water level drops at point 1?

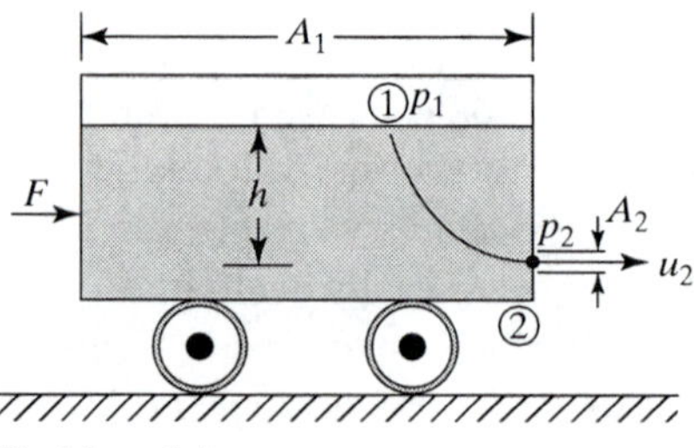

Problem 4.4.

(c) Develop an expression for the force required for holding the container in place.

4.5. The Pitot tube shown in the figure is mounted in front of a moving vehicle. The pressure in the stagnation tube is higher than the one measured in the static pressure tube by 10 cm of water. Calculate the speed of the vehicle in terms of km/h and mph. Assume the air density is $\rho = 1.2$ kg/m^3.

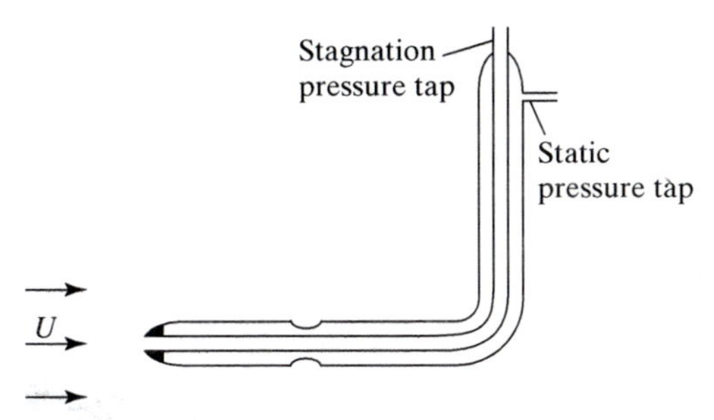

Problem 4.5.

4.6. A conical weight of 4 kg$_f$ is "floating" on the jet, as shown (ignore the wire, which is there to stabilize the weight). The jet leaving the orifice at the bottom has an upward velocity of 15 m/s and a diameter of 3 cm. Note that the jet speed is reduced with height! Find the height h at which the weight will remain stationary.

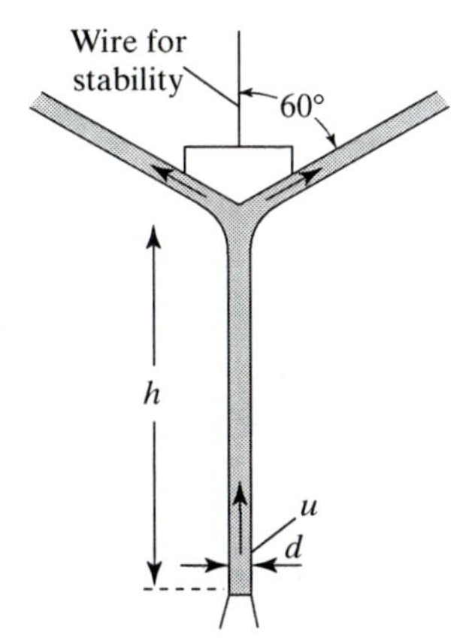

Problem 4.6.

4.7. According to Bernoulli's principle the pressure ahead of an automobile radiator will increase with increasing U_∞, thereby pushing the cooling air across the dense cooling fins. Assuming that the air ahead of the radiator (point A) stops completely, estimate the pressure rise for speeds of 60, 100, and 140 km/h (take air density as 1.22 kg/m^3).

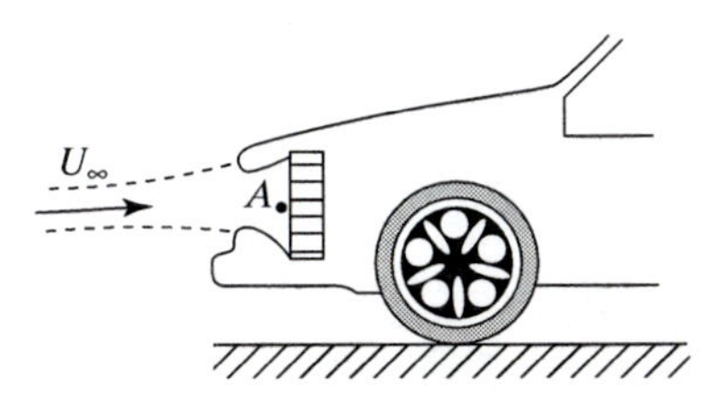

Problem 4.7.

4.8. A 70-km/h wind is blowing on a concave dish antenna, with a radius of $R = 0.4$ m. Estimate the pressure acting on the frontal surface and the resulting forces on the frontal surface of the antenna (compare your results with the results of Example 8.9).

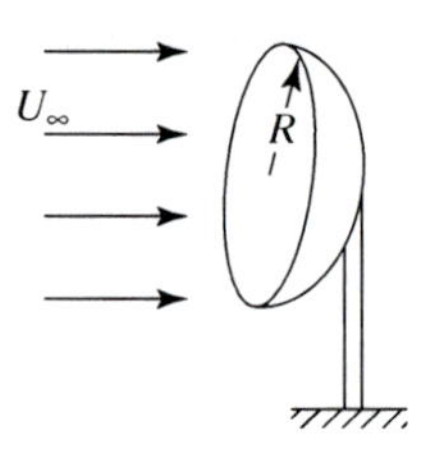

Problem 4.8.

4.9. A stationary water jet impinges on a turning vane mounted on a container with frictionless wheels, as shown on the left-hand side of the figure. If the jet cross-section area is 5 cm^2 and its horizontal velocity is 10 m/s, calculate the force F_x in the x direction while the container is not moving. Calculate its initial acceleration if its initial mass is 5 kg. How will the force change when the container is moving to the right at a speed of 3 m/s (assume mass is the same).

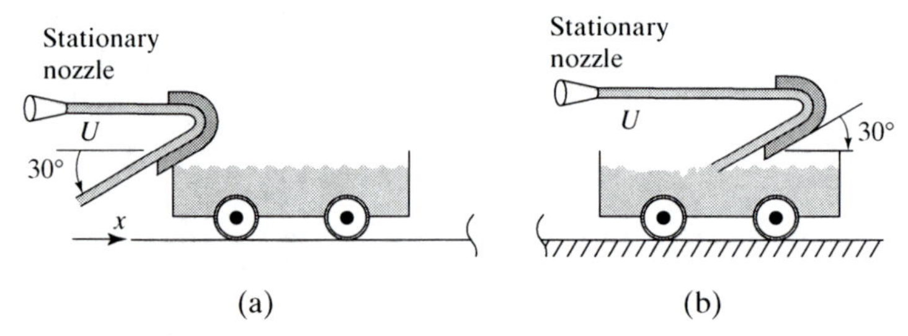

Problem 4.9.

4.10. A stationary water jet impinges on a turning vane mounted on a container with frictionless wheels (right-hand side of the figure in Problem 4.9). In this case, however, the vane is mounted forward and the returning jet hits the water in the container. If the jet cross-section area is 5 cm^2 and its horizontal velocity is 10 m/s, calculate the force F_x in the x direction while the container is not moving. Calculate its initial acceleration if its initial mass is 5 kg. How will the force change when the container is moving to the right at a speed of 3 m/s (assume mass is the same).

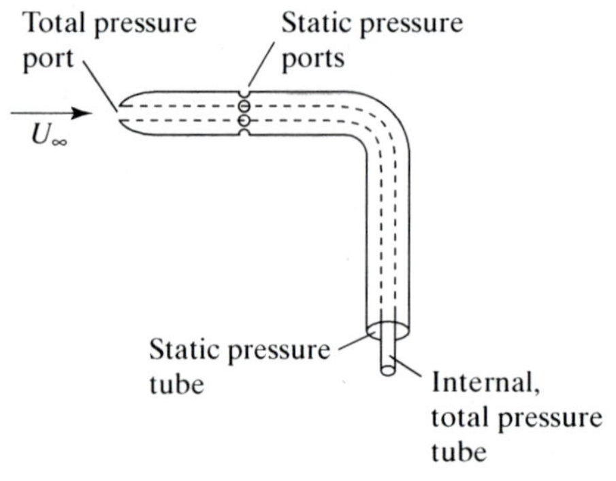

Problem 4.11.

4.11. An airplane is flying at a speed of 500 km/h. It is using a simple Pitot tube to measure its velocity. Calculate the pressure difference the probe will register where air density is 1.1 kg/m^2. Provide your answer in atmospheres (atm).

4.12. A strong wind is blowing at a velocity of U_∞ past a probe pointing directly into the wind. The lower end of the tube is inserted into a small water tank, and the inside water level reaches $\Delta h = 4$ cm below the water level in the tank. Calculate the wind speed (assume air density is 1.22 kg/m^3).

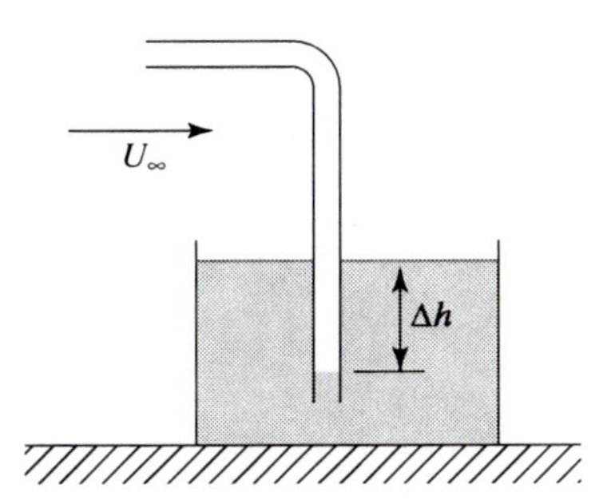

Problem 4.12.

4.13. A 6-cm-diameter circular jet with a velocity of 20 m/s is hitting the turning vane, as shown. Assuming the vane is moving away from the nozzle at a speed of 7 m/s, then (assuming no friction) calculate the horizontal and vertical forces on the vane.

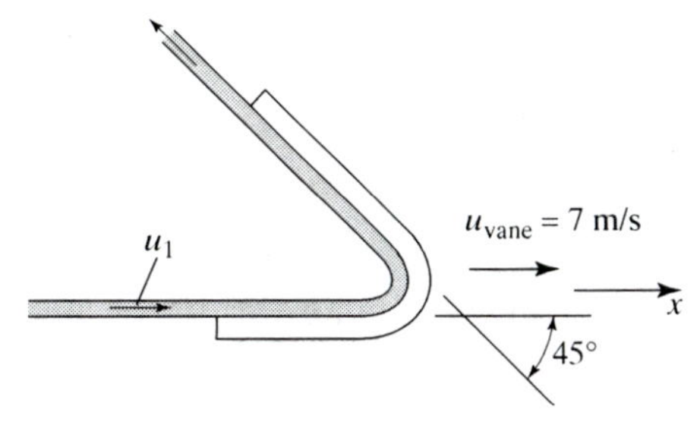

Problem 4.13.

4.14. A simple model of a turbofan engine is shown and, at the exit, two concentric jets can be seen. Assume a flight speed of 300 m/s, and that the total incoming air flow rate is 300 kg/s. At the exit, however, the inner jet has an exit velocity of 1000 m/s and mass flow rate of 100 kg/s, whereas the outer flow speed is 600 m/s and the mass flow rate is 200 kg/s (fuel mass is negligible). Calculate the thrust of this unit.

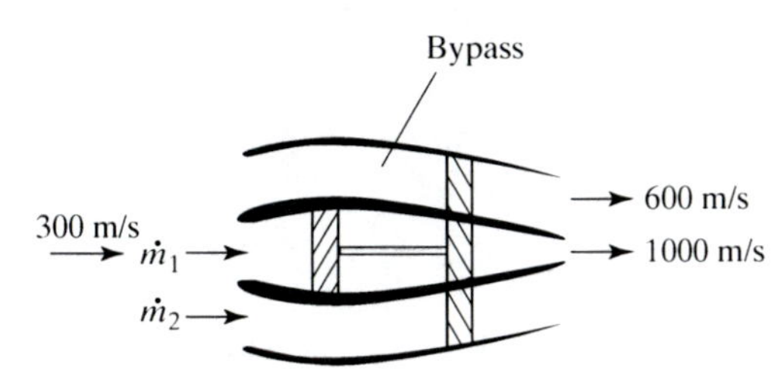

Problem 4.14.

4.15. A turbofan engine is modeled by two concentric jets; the outer fan flow is leaving the engine at a speed of 260 m/s and the inner hot flow leaves at 800 m/s. Assuming a core mass flow rate of 20 kg/s and a bypass ratio of 4 (the outer cold flow is 4 times larger than the hot core flow) calculate the thrust of this engine at a flight speed of 200 m/s.

4.16. Consider a water jet being turned by a solid frictionless surface, as shown in the figure. Calculate the forces acting on the structure when $u_1 = 10$ m/s, the circular jet cross section is 5 cm^2, and the angle $\theta = 20°$.

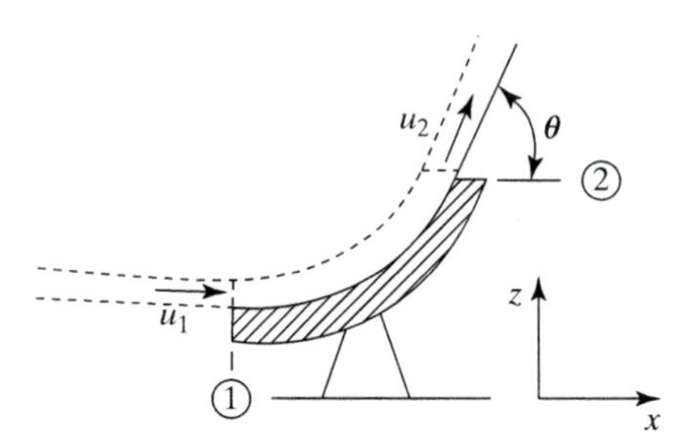

Problem 4.16.

4.17. A circular water jet (as in the previous problem) is turned by a solid frictionless surface, and a downward force of 20 N is measured on the structure. The incoming jet velocity is $u_1 = 15$ m/s and its cross section is 5 cm^2.

(a) calculate the turning angle θ.
(b) Calculate the horizontal force on the solid surface.

4.18. An airplane flies at a speed of $U_\infty = 250$ m/s, the air mass flow rate into the engine is $\dot{m}_a = 50$ kg/s, and the average exhaust speed is $u_e = 800$ m/s.

(a) If the air to fuel ratio is about 2%, calculate the engine thrust (assume $p_e = p_a$).
(b) Suppose the airplane lift-to-drag ratio is 10 and it has two engines; calculate the airplane weight.

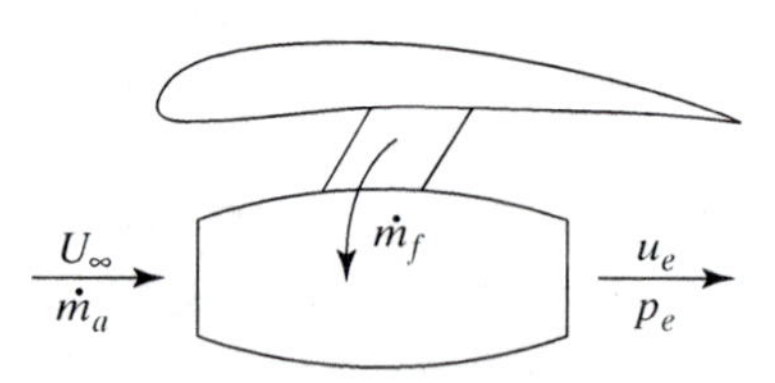

Problem 4.18.

4.19. Juliet empties a bucket of water on Romeo from her window, which is 5 m above his head. Assume the jet diameter is 5 cm at section 2 (just above Romeo's head).

(a) Calculate the jet velocity (at point 2) as it hits Romeo's head.
(b) What is the force acting on Romeo's head if the water is splashed sideways (no vertical speed)?

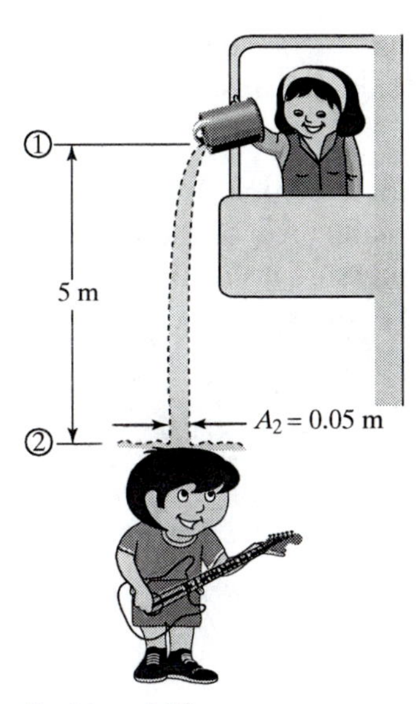

Problem 4.19.

4.20. A jet of fluid hits a flat plate and leaves without any axial velocity component, as shown in the figure.

(a) If the flow rate is 50 L/s and the velocity of the jet hitting the plate is 20 m/s, calculate the force on the plate.

(b) Calculate the force in the case in which the plate moves away from the jet at a speed of 50 km/h.

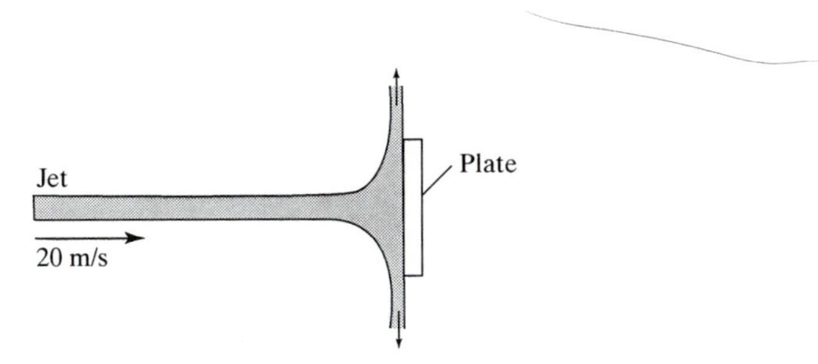

Problem 4.20.

4.21. A small boat is propelled by a water jet and is traveling at a speed of 5 m/s. The force required for propelling the boat is 100 N, and the water flow rate is 5 kg/s.

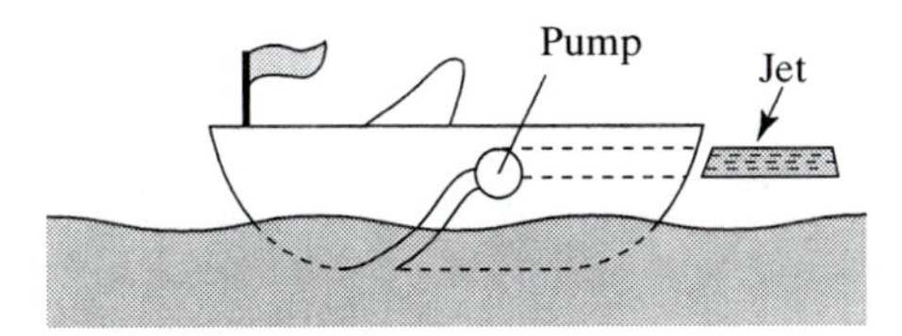

Problem 4.21.

(a) Calculate the jet exit velocity (assume exit pressure = ambient pressure).
(b) Would it matter if the exiting jet were slightly submerged? (Still, the exit pressure = ambient pressure.)

4.22. A water jet hits horizontally a 50-kg block. The friction coefficient between the block and the ground is 0.9. What is the minimum diameter d of the water jet for the block to slide to the left? Assume the stagnation pressure inside the nozzle is $p_0 = 4$ atm the pressure outside the nozzle is $p_a = 1$ atm ($1 \text{ atm} = 1.013 \times 10^5 \text{ N/m}^2$), and the flow is ideal through the nozzle.

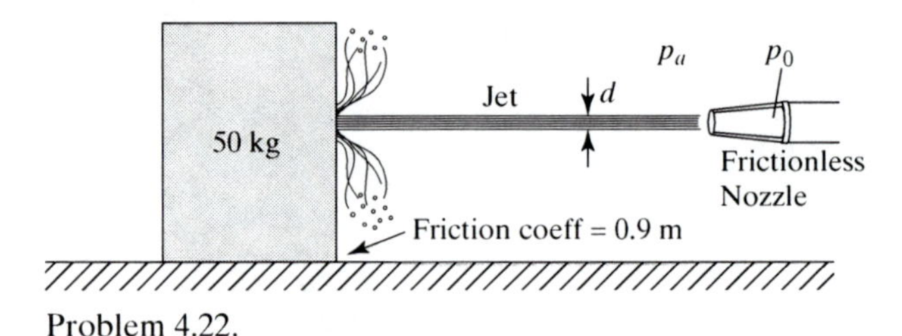

Problem 4.22.

4.23. The neighbor kid decided to water the garden while standing on his skateboard. As he opened the valve, the circular jet speed was 7 m/s and its diameter was $d_2 = 0.02$ m. If his total mass is 40 kg, then calculate the force of the jet and his initial acceleration (assuming no friction).

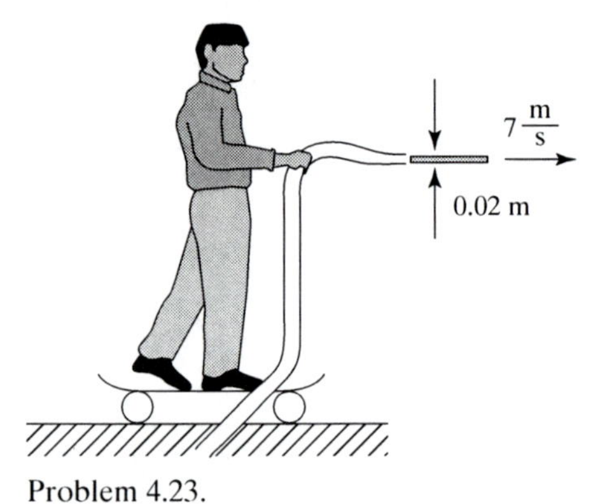

Problem 4.23.

4.24. A free stream of $U = 50$ m/s is blowing on a 0.6-m-diameter cylinder, as shown in the figure. A simple approximation of the flow behind it indicates that the velocity was reduced to 30 m/s for a strip of 0.6 m (e.g., the same as the cylinder's diameter). Suppose we define a force coefficient as $C_F = \frac{F}{0.5\rho U^2 D}$, where F is the force per unit

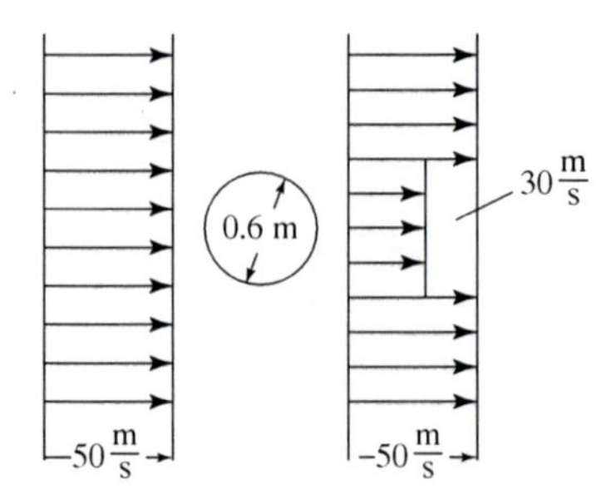

Problem 4.23.

width on the cylinder; then calculate the value of C_F (you may use Table 1.1 for the air density).

4.25. The orifice shown in the figure is used to measure the water flow in the pipe. The pressure difference across the orifice registers as a column of 10-cm water. Assuming that the water density at $\rho = 1000$ kg/m^3, $R_1 = 20$ cm and $R_2 = 10$ cm, and the discharge coefficient is $C_D = 0.6$, calculate the mass flow rate in the pipe.

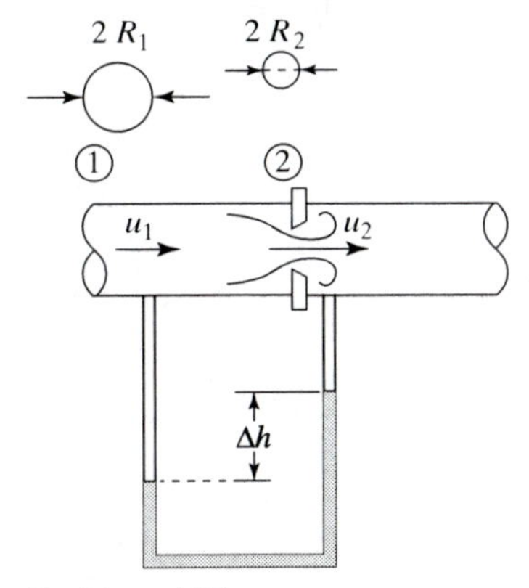

Problem 4.25.

4.26. The orifice from Problem 4.25 is used to measure the water flow in a pipe. Assuming that the water density at $\rho = 1000$ kg/m^3, $R_1 = 20$ cm and $R_2 = 10$ cm, and the discharge coefficient is $C_D = 0.6$, calculate the average speed in the pipe when the mass flow rate is 100 kg/s. What is the pressure difference (in terms of centimeters of water) for the preceding flow rate?

4.27. The flow rate across the orifice of Problem 4.25 is 100 L/s (1 L water = 1 kg). Calculate the water-column height Δh registered in the U-tube monometer.

4.28. A Venturi meter is used to measure the water flow in a 3-cm-diameter pipe. The diameter of the narrow section is 1.5 cm. The pressure difference between the large- and small-diameter pipes registers as a 5-cm column of water. Assuming a discharge coefficient of $C_D = 0.95$, calculate the mass flow rate in the pipe ($\rho =$ 1000 kg/m^3).

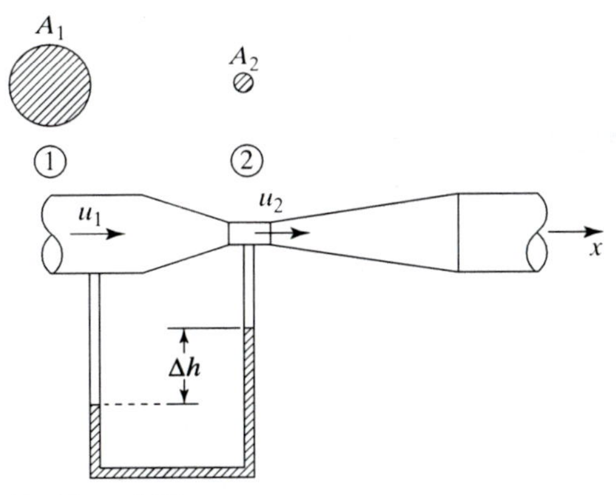

Problem 4.28.

4.29. The Venturi tube shown in the figure of Problem 4.28 is used to measure water flow in a pipe. The incoming pipe diameter is $D_1 = 20$ cm and the throat diameter is $D_2 = 10$ cm. Assuming a discharge coefficient of $C_D = 1.0$, $u_1 = 0.5$ m/s, and the density of water $\rho = 1000$ kg/m^3 calculate the following values:

(a) The water flow rate.

(b) The pressure difference measured as Δh in meters (water column)

4.30. Air is drawn into a long tube, as shown in the figure. The air velocity inside the tube is 50m/s and the pressure outside, in the still air, is 1 atm. Assuming the air density is 1.22 kg/m^3 and constant, calculate the pressure inside the tube.

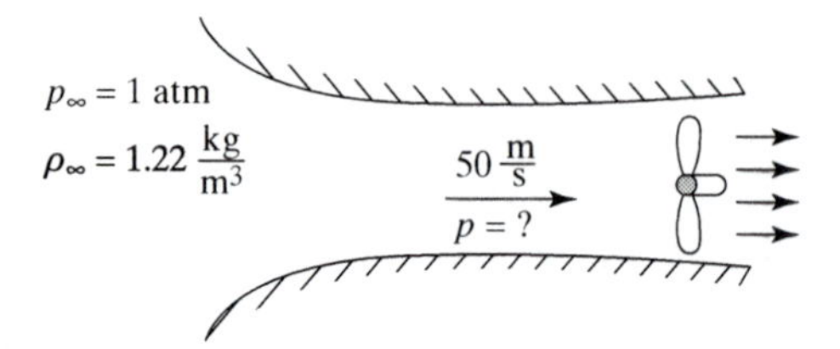

Problem 4.30.

4.31. Air is drawn into a wind tunnel during automobile testing, as shown in the figure. A simple manometer measures the velocity in terms of water-column height h. If the velocity at the test section is 120 km/h, calculate the height of the water column in centimeters ($\rho_{water} = 1000$ kg/m^3, $\rho_{air} = 1.22$ kg/m^3).

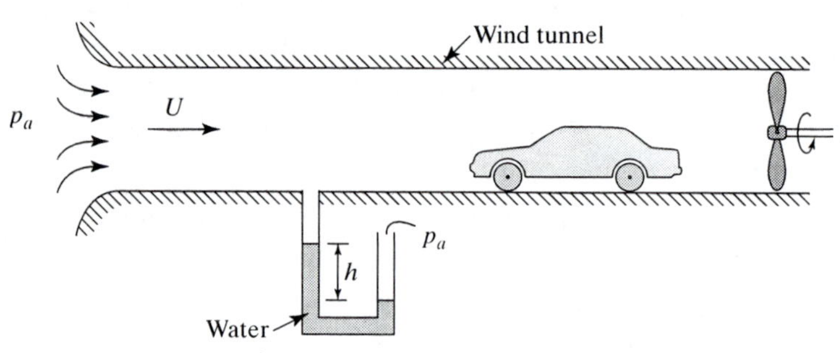

Problem 4.31.

4.32. We repeat the previous experiment, but now we use a tube to measure the total pressure, as shown in the figure. If the water-column height in the manometer $h = 8$ cm, calculate the velocity at the test section ($\rho_{water} = 1000$ kg/m^3, $\rho_{air} =$ 1.22 kg/m^3).

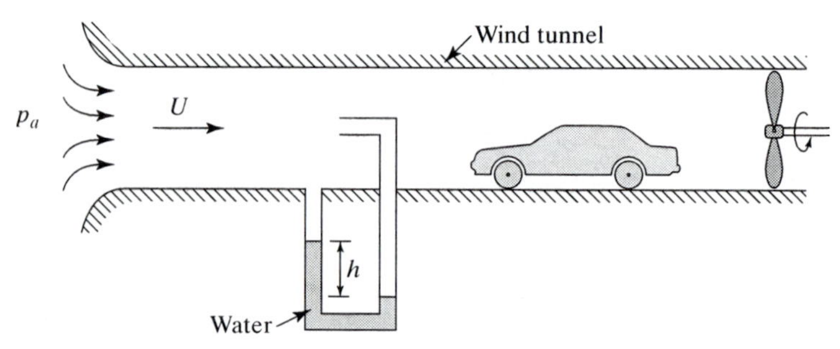

Problem 4.32.

4.33. Air flows at a rate of 1.7 kg/s through a smooth contraction in a pipe with negligible friction losses. The pressure difference between the two sections is measured by a water manometer. If the pipe diameters are $D_1 = 0.3$ m and $D_2 = 0.15$ m, calculate the height difference Δh measured between the two vertical tubes (assume air density is 1.22 kg/m^3).

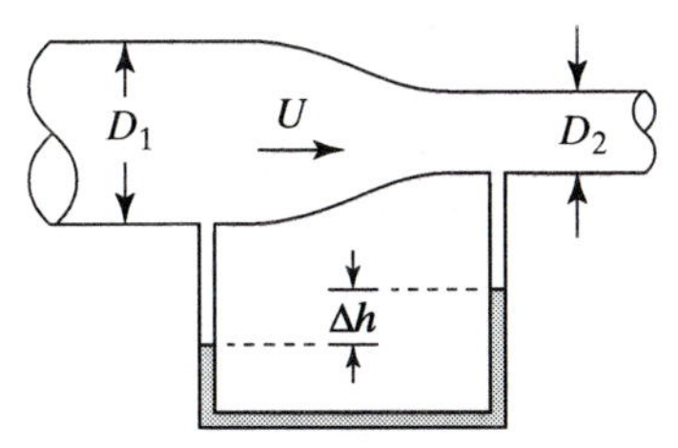

Problem 4.33.

4.34. Air flows through a smooth contraction in a pipe with negligible friction losses (the pipe diameters are $D_1 = 0.3$ m and $D_2 = 0.15$ m). The pressure difference between the two sections is measured by a water manometer. If the height difference Δh measured between the two vertical tubes is 0.2 m, calculate the mass flow rate (assume the air density is 1.22 kg/m^3).

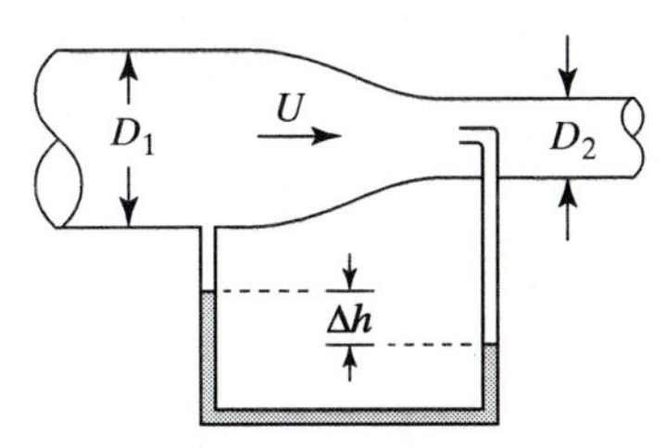

Problem 4.34.

4.35. Air flows through a smooth contraction in a pipe with negligible friction losses (the pipe diameters are $D_1 = 0.3$ m and $D_2 = 0.15$ m). The pressure difference between the two sections is measured by a water manometer. If the height difference Δh measured between the two vertical tubes is 0.15 m, calculate the mass flow rate (assume the air density is 1.22 kg/m^3).

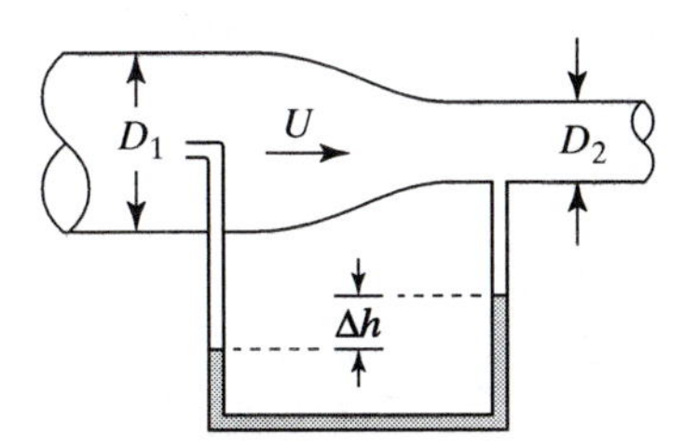

Problem 4.35.

4.36. Air flows at a rate of 3 kg/s through a smooth contraction in a pipe with negligible friction losses (the pipe diameters are $D_1 = 0.4$ m and $D_2 = 0.2$ m, as in problem 4.35). The pressure difference between the two sections is measured by two tubes connected to a water manometer. Calculate the height Δh measured between the two vertical tubes (assume the air density is 1.22 kg/m^3).

4.37. Air flows through a smooth contraction in a pipe with negligible friction losses, and the pressure difference between the two vertical tubes of problem 4.35 is measured at $\Delta h = 0.3$ m. If the pipe diameters are $D_1 = 0.3$ m and $D_2 = 0.2$ m, calculate the flow rate (assume the air density is 1.22 kg/m^3).

4.38. A small circular jet having a cross section of 5 cm^2 is flowing out at the bottom of a large container and is aimed 30° upward. At a horizontal distance of 0.5 m, the jet impinges on a vertical plate. Calculate the horizontal force required for holding the plate in place if the water level in the container is $h = 5$m. Also estimate the height from the bottom where the jet hits the plate.

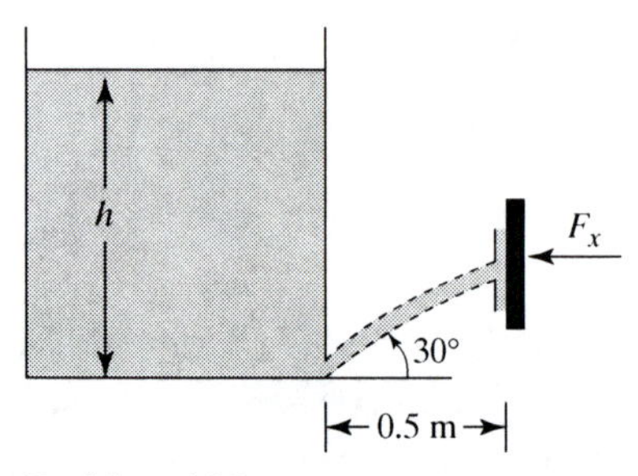

Problem 4.38.

4.39. Water is flowing through two holes from a container with frictionless wheels. The left hole is located at a height of $h_1 = 0.3$ m and its area is 3 cm^2, and the right hole area is 1 cm^2. If the water level in the container is $h_2 = 0.7$ m, calculate the resultant force and its direction on the container.

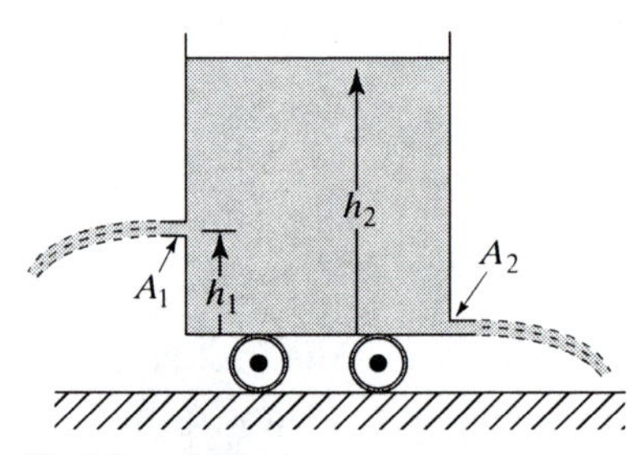

Problem 4.39.

4.40. A ball weighing 1 kg is "floating" on a circular jet, as shown in the figure (ignore the wire, which is there to stabilize the weight). The jet leaving the orifice at

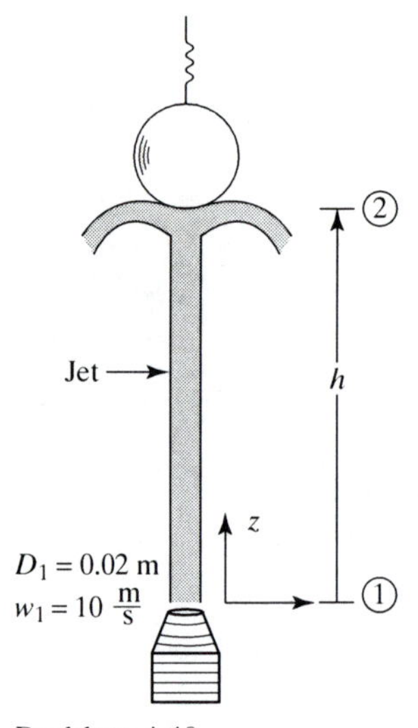

Problem 4.40.

the bottom has an upward velocity of 10 m/s, and the jet diameter at the exit is 2 cm. Assuming the jet spills sideways at 90°, find the height h at which the ball will float.

4.41. A jet flowing from the bottom of a water tank is aimed upward at 30°, as shown in the figure. Assuming no friction losses at the nozzle, calculate the maximum height H and distance of the jet (again, neglecting friction with the outside air).

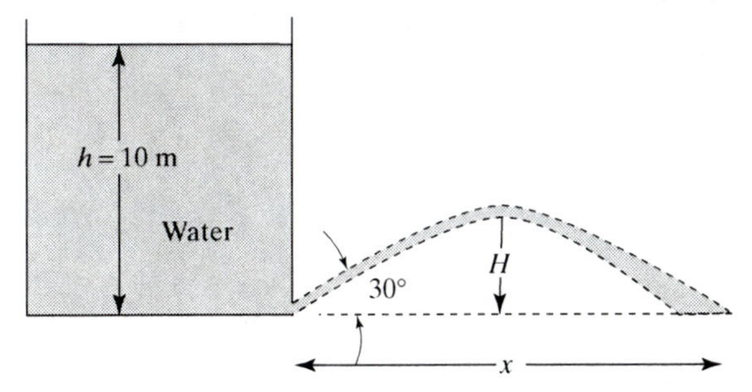

Problem 4.41.

4.42. A rectangular container is filled with water, as shown in the figure. At a time $t = 0$, a small opening with an area A_e is opened at the bottom. Develop the formulation for the exit velocity u_e and the water column height h as a function of time.

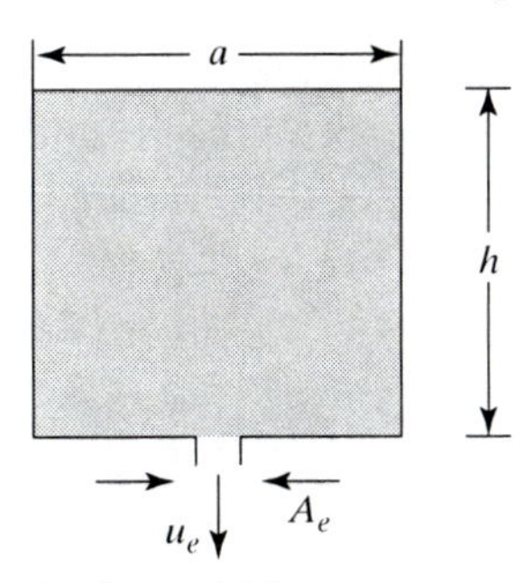

Problem 4.42.

4.43. If the width of the container in the previous problem is $a = 20$ cm, $h = 80$ cm, and the opening area A_e is 3 cm^2, calculate how long it will take to empty the container.

4.44. A 1-m-diameter spherical tank is half full with water. A circular drain of 1-cm-diameter opened at the bottom, as shown in the figure. How long does it take

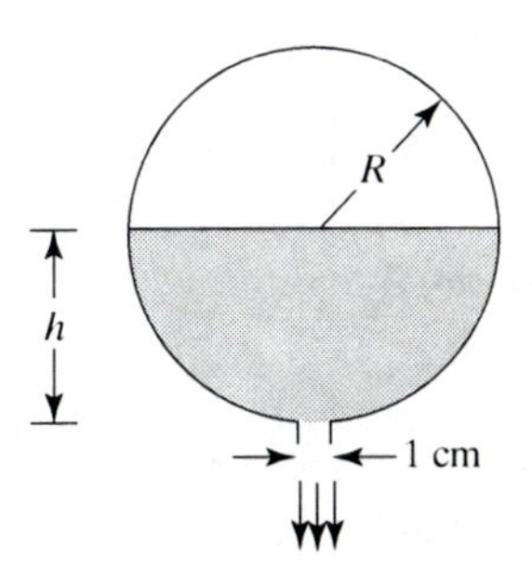

Problem 4.44.

to empty the tank? Note that the radius r of the liquid surface area inside the tank can be expressed as $\pi r^2 = \pi[R^2 - (R-h)^2]$.

4.45. A cylindrical container having a diameter of $D = 0.3$ m is filled with water to a level of $h_0 = 0.5$ m. Suddenly the drain plug with an area of 3 cm^2 is opened, allowing the water to flow out. Calculate the horizontal force versus time until the tank is completely drained. Also develop a formula for the water height versus time. How long it will take to drain the container?

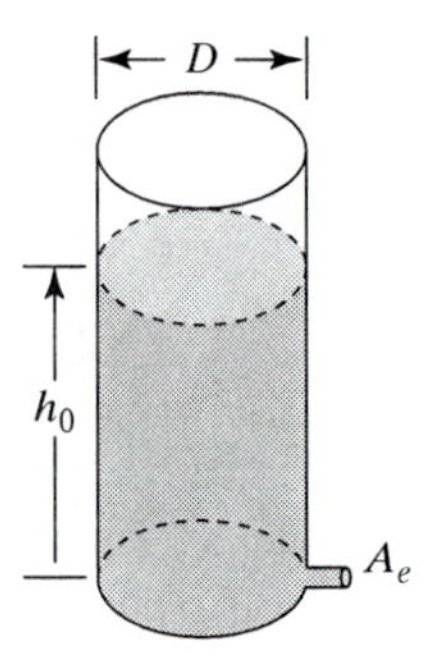

Problem 4.45.

4.46. A cylindrical container having a diameter of D_1 is filled with water to a level of h_1. A drain plug with a diameter of D_2, located at a height of h_2 above the ground is opened, allowing the water to flow out. Develop an expression for the distance x, versus time.

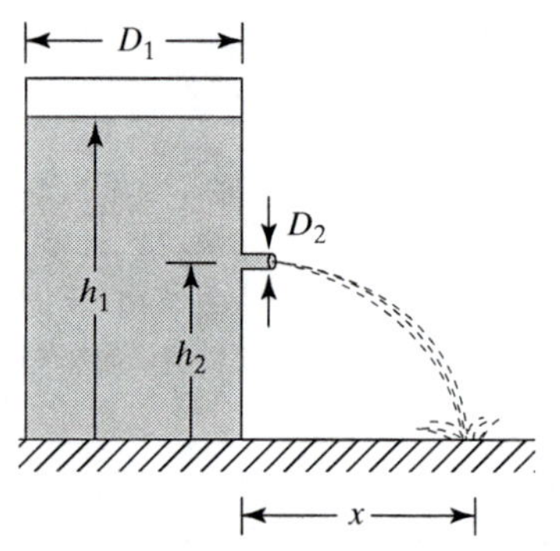

Problem 4.46.

4.47. The balloon shown in Fig. 4.11 is filled with 2 kg water at a pressure of 0.8×10^5 N/m^2, and the mass of the whole system, without the water, is 0.5 kg. Assuming constant discharge velocity (without friction losses), calculate the maximum velocity of the device. Can it move faster than the exit jet velocity?

4.48. A water jet of area $A_i = 5$ mm^2 flows out of an injector where the inside pressure p_0 is 2 atm (and the outside pressure is 1 atm). Calculate the water flow rate if the discharge coefficient is $C_D = 0.65$.

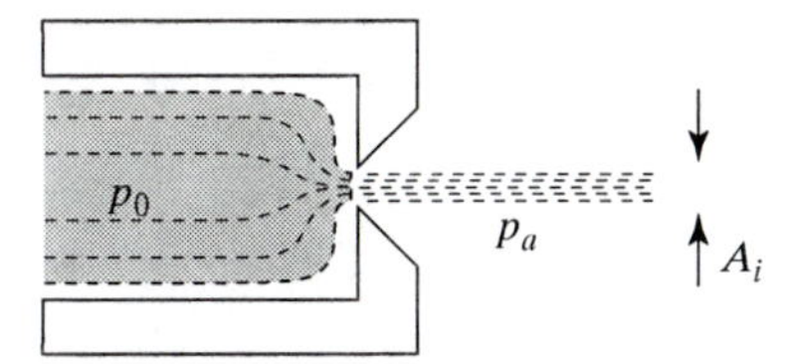

Problem 4.48.

4.49. The jet in a sprinkler system shown in the figure, is designed to deliver 2 L of water per minute. The pressure difference across the nozzle is 1.5 atm and its discharge coefficient is $C_D = 0.85$. Calculate the orifice diameter D_i.

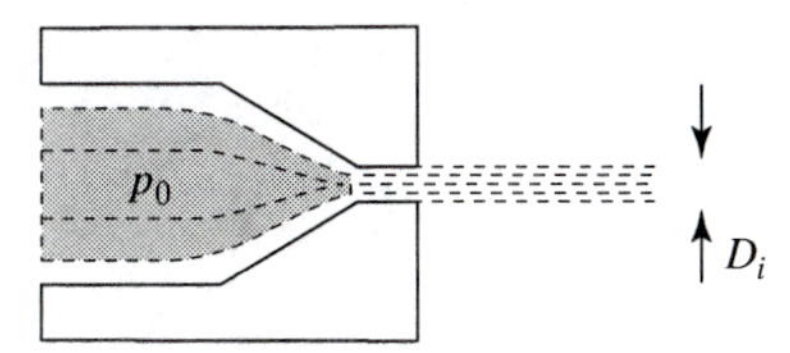

Problem 4.49.

4.50. The water height in a large reservoir is $h_0 = 5$ m. When the water is needed for irrigation, a 1-m-wide sluice gate is opened, leaving a gap of $h_1 = 0.6$ m under it. Estimate the velocity at point 2 and the water flow rate when the gate is opened.

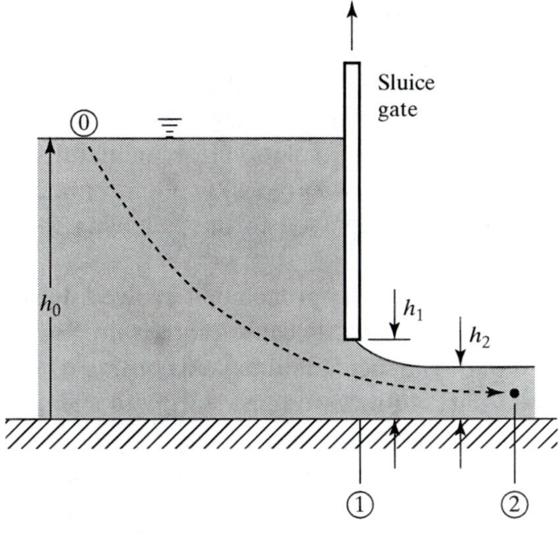

Problem 4.50.

4.51. The water height in a large reservoir is $h_0 = 6$ m. If a flow rate of 3000 L/s is needed for irrigation, how high (h_1) a 1-m-wide gate must be raised?

5 Viscous Incompressible Flow: Exact Solutions

(Leading to Some Practical Engineering Solutions)

5.1 Introduction

In Chapter 3 the effects of pressure in a fluid were isolated (because the fluid was not moving), and in Chapter 4 the inertia terms were added. The inclusion of viscosity, its effects, and the resulting velocity distribution are discussed here. For example, the velocity distribution for the laminar flow inside a pipe is formulated and the average velocity is calculated. This provides the relation between the simple 1D average velocity model (of Chapter 4) and the more complex (and realistic) 2D or 3D flows.

The solutions presented early in this chapter are often called *exact solutions*. This means that, for a few limited cases, a set of logical assumptions leads to simplification of the fluid dynamic equations, which allows their solution (for laminar flow)! Also, the cases presented in this chapter (e.g., the flow in pipes) is often termed as *internal flows*. The discussion on *external flows* is delayed to the following three chapters.

The second part of this chapter demonstrates the approach that evolved during the past 200 years for solving fluid dynamic problems (because there is no closed-form analytic solution to the complete fluid dynamic equations). According to this approach, to develop a practical engineering solution, we must start with a simple but exact solution that determines the major parameters and the basic trends of the problem (e.g., the pressure drop in a circular pipe versus the Reynolds number). Based on these parameters, an empirical database can be developed for treating a wider range of engineering problems. As an example, the viscous laminar flow model in circular pipes is extended into the high-Reynolds-number range and the effects of turbulent flow are discussed.

5.2 The Viscous Incompressible Flow Equations (Steady State)

When developing the fluid dynamic equations in Chapter 2 we noted that their general analytic solution is next to impossible. There are several simple cases, however, for which we can actually derive a solution that contains important physical information. One of the most successful examples is the viscous flow in circular pipes, which provides the rationale for developing an experimental database to help in

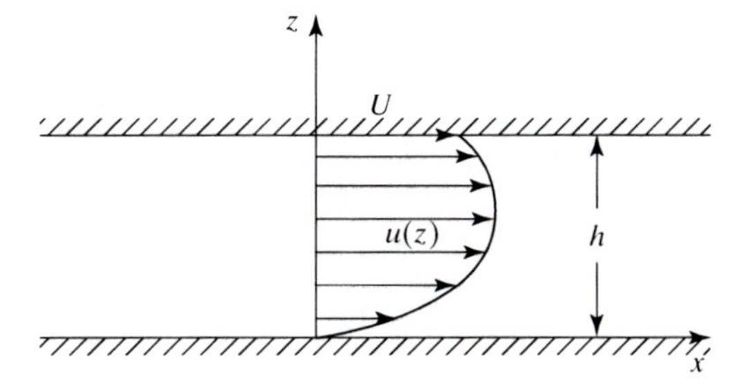

Figure 5.1. Flow between two infinite parallel plates.

more complex (but similar) engineering problems. Prior to starting, let us rewrite the governing equations for steady-state, viscous, incompressible flows. The continuity equation is [Eq. (2.33)]

$$\frac{\partial u}{\partial x} + \frac{\partial v}{\partial y} + \frac{\partial w}{\partial z} = 0, \tag{5.1}$$

and the viscous incompressible momentum equations [from Eq. (2.40)] in Cartesian coordinates are

$$u\frac{\partial u}{\partial x} + v\frac{\partial u}{\partial y} + w\frac{\partial u}{\partial z} = \frac{-1}{\rho}\frac{\partial p}{\partial x} + \frac{\mu}{\rho}\left(\frac{\partial^2 u}{\partial x^2} + \frac{\partial^2 u}{\partial y^2} + \frac{\partial^2 u}{\partial z^2}\right), \tag{5.2a}$$

$$u\frac{\partial v}{\partial x} + v\frac{\partial v}{\partial y} + w\frac{\partial v}{\partial z} = \frac{-1}{\rho}\frac{\partial p}{\partial y} + \frac{\mu}{\rho}\left(\frac{\partial^2 v}{\partial x^2} + \frac{\partial^2 v}{\partial y^2} + \frac{\partial^2 v}{\partial z^2}\right), \tag{5.2b}$$

$$u\frac{\partial w}{\partial x} + v\frac{\partial w}{\partial y} + w\frac{\partial w}{alz} = \frac{-1}{\rho}\frac{\partial p}{\partial z} + \frac{\mu}{\rho}\left(\frac{\partial^2 w}{\partial x^2} + \frac{\partial^2 w}{\partial y^2} + \frac{\partial^2 w}{\partial z^2}\right). \tag{5.2c}$$

Here we neglected the body forces. However, if gravitational force is present, its effect can be reintroduced in the pressure-gradient term.

5.3 Laminar Flow between Two Infinite Parallel Plates – The Couette Flow

Let us start with the simplest example, the flow between two parallel (infinite) plates, as shown in Fig. 5.1.

The fluid is considered viscous and incompressible (such as water or oil), and for this case we neglect the time derivatives and the body forces. We also assume that the flow is laminar, and this statement will be clarified toward the end of the chapter. Let us use a Cartesian coordinate system attached to the lower plate, as shown in Fig. 5.1, and the governing equations for this case are summarized by Eqs. (5.1) and (5.2). Observing Fig. 5.1, we can assume that the lower plate is stationary and therefore the velocity near the lower wall is zero (at $z = 0$). We may speculate about the shape of the velocity distribution (as shown); however, it is clear that the velocity at $z = h$ is equal to the velocity of the upper plate, U. This flow is called the Couette flow after Maurice Marie Alfred Couette (1858–1943), a well-known French physicist. The model is 2D and there are no changes in the y direction. Also there is no velocity in the z direction, and we summarize this as

$$v = w = 0. \tag{5.3}$$

Now, substituting this into continuity equation (5.1) results in

$$\frac{\partial u}{\partial x} = 0. \tag{5.4}$$

The conclusion therefore is that the velocity profile is a function of z and it is the same at any x station:

$$u = u(z). \tag{5.5}$$

Recall that the plates extend to infinity ($-\infty < x < \infty$) and therefore it is obvious that the velocity profile $u(z)$ is the same at any x station. Next we apply all the previous assumptions to momentum equations (5.2), which now reduce to

$$0 = \frac{-1}{\rho}\frac{\partial p}{\partial x} + \frac{\mu}{\rho}\left(\frac{\partial^2 u}{\partial z^2}\right), \tag{5.6a}$$

$$0 = \frac{-1}{\rho}\frac{\partial p}{\partial y}, \tag{5.6b}$$

$$0 = \frac{-1}{\rho}\frac{\partial p}{\partial z}. \tag{5.6c}$$

The two last equations indicate that there is no lateral or vertical change in the pressure. Hence the pressure can change only with x and is constant vertically:

$$p = p(x). \tag{5.7}$$

Also, this is a steady-state problem, and Eq. (5.4) states that there are no changes in u with x (and the velocity is changing only with z). So if the second term in Eq. (5.6a) is constant along the x axis, then we conclude that the pressure gradient must be constant too:

$$\frac{dp}{dx} = \text{const.} \tag{5.8}$$

At this point, only Eq. (5.6a) remains, and after rearranging it becomes

$$\frac{\partial^2 u}{\partial z^2} = \frac{1}{\mu}\frac{dp}{dx}. \tag{5.9}$$

The only parameter now is x, and instead of the partial derivatives, an ordinary differential equation remains. After integrating twice we get

$$u(z) = \frac{1}{\mu}\frac{dp}{dx}\frac{z^2}{2} + Az + B, \tag{5.10}$$

where A and B are the constants of integration.

5.3.1 Flow with a Moving Upper Surface

This first case is the original Couette flow, and the boundary conditions for Eq. (5.10) are

$$\begin{aligned} &\text{at } z = 0, \quad u = 0; \\ &\text{at } z = h, \quad u = U. \end{aligned} \tag{5.11}$$

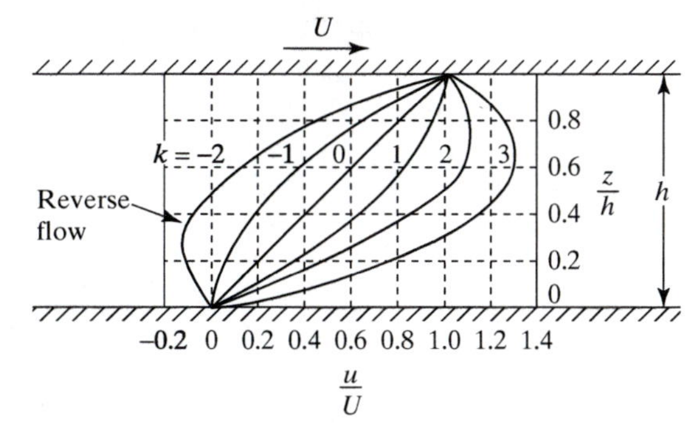

Figure 5.2. Effect of the pressure gradient on the velocity profile of the flow between two relatively moving plates.

Substituting the $z = 0$ condition into Eq. (5.10) yields that $B = 0$, and the second boundary condition results in

$$U = \frac{1}{\mu}\frac{dp}{dx}\frac{h^2}{2} + Ah.$$

After rearranging we get

$$A = \frac{U}{h} - \frac{h}{2\mu}\frac{dp}{dx}.$$

Substituting this result into Eq. (5.10) provides the velocity distribution at any x station:

$$u(z) = U\frac{z}{h} - \frac{h^2}{2\mu}\frac{dp}{dx}\left(\frac{z}{h}\right)\left(1 - \frac{z}{h}\right). \tag{5.12}$$

The first term is clearly a linear variation and reduces to the shear flow profile discussed in reference to Fig. 1.5. The second parabolic (shape) term contains the effect of the pressure gradient, and this effect is summarized in Fig. 5.2. To visualize this, let us define a constant k such that

$$k = -\frac{h^2}{2\mu}\frac{dp}{dx}. \tag{5.13}$$

When $k = 0$, then we get the basic shear flow, and when the pressure gradient is favorable (negative dp/dx or positive k) then we get an additional parabolic forward velocity distribution, as shown. When the pressure gradient is opposing the flow then this parabolic shape is reversed. At higher negative values of k, even reverse flows are possible (as shown in the figure).

5.3.2 Flow between Two Infinite Parallel Plates – The Results

With the assumptions posed for the viscous flow between parallel plates we actually arrived at the "exact solution." It is important to connect those results with engineering quantities such as the average velocity, which was used in the 1D flow model of Chapter 4. Thus the average velocity U_{av} is the velocity that, when multiplied by the inflow area, will have the same volumetric flow rate Q (per unit width):

$$Q = U_{\text{av}}h = \int_0^h u dz. \tag{5.14}$$

Consequently we define the average velocity as

$$U_{\text{av}} = \frac{1}{h}\int_0^h u dz. \tag{5.15}$$

Next, the velocity distribution from Eq. (5.12) is substituted and the integration is performed:

$$\begin{aligned} U_{\text{av}} &= \frac{1}{h}\int_0^h \left[U\frac{z}{h} - \frac{h^2}{2\mu}\frac{dp}{dx}\left(\frac{z}{h}\right)\left(1-\frac{z}{h}\right)\right]dz \\ &= \frac{U}{2}\left(\frac{z}{h}\right)^2 - \frac{h^2}{2\mu}\frac{dp}{dx}\left[\frac{1}{2}\left(\frac{z}{h}\right)^2 - \frac{1}{3}\left(\frac{z}{h}\right)^3\right]_0^h \\ &= \frac{U}{2} - \frac{h^2}{12\mu}\frac{dp}{dx}. \end{aligned} \tag{5.16}$$

Now we can also solve for the volumetric flow rate (per unit width:

$$Q \equiv hU_{\text{av}} = \frac{Uh}{2} - \frac{h^3}{12\mu}\frac{dp}{dx}. \tag{5.17}$$

Figure 5.2 shows the various velocity distributions as functions of the pressure gradient. To calculate the maximum velocity we must derive the velocity distribution,

$$\frac{du(z)}{dz} = \frac{U}{h} - \frac{h^2}{2\mu}\frac{dp}{dx}\left(\frac{1}{h} - \frac{2z}{h^2}\right) = 0,$$

and after solving for z, we get

$$\frac{z}{h} = \frac{1}{2} - \frac{\mu}{h^2}\frac{U}{\frac{dp}{dx}}. \tag{5.18}$$

Note that this equation is not valid when dp/dx approaches zero. The next important observation is that in this viscous flow there is a shear stress. This force acts between the parallel fluid layers as they slide on each other (creating friction). To calculate the shear stress (per unit width) we use the definitions introduced by Eq. (1.14) and the velocity distribution from Eq. (5.12):

$$\tau_{xz} = \mu\frac{du(z)}{dz} = \mu\frac{U}{h} - \frac{h^2}{2}\frac{dp}{dx}\left(\frac{1}{h} - \frac{2z}{h^2}\right). \tag{5.19}$$

From the engineering point of view, the determination of the shear force (or friction) on the solid surface is important. Therefore a friction coefficient C_f can be defined as the ratio between the shear stress and the average dynamic pressure:

$$C_f \equiv \frac{\tau_{xz}}{\frac{1}{2}\rho U_{\text{av}}^2}. \tag{5.20}$$

Calculation of this coefficient is important at the upper and lower walls, which can be found from Eq. (5.19)

$$\tau_{xz}\Big|_0 = \mu\frac{U}{h} - \frac{h}{2}\frac{dp}{dx}, \tag{5.21a}$$

$$\tau_{xz}\Big|_h = \mu\frac{U}{h} + \frac{h}{2}\frac{dp}{dx}. \tag{5.21b}$$

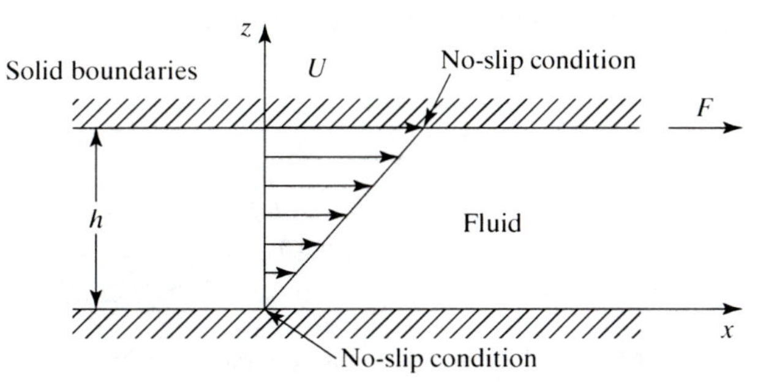

Figure 5.3. The flow between two parallel plates. The upper is moving at a velocity of U and the lower is stationary.

EXAMPLE 5.1. SHEAR FLOW (WITH NO PRESSURE GRADIENT). For this first example let us revisit our introductory model for explaining the effects of viscosity. Figure 1.5 is redrawn here in Fig. 5.3, and the problem at hand is the case in which the upper plate moves at a velocity U but there is no pressure gradient:

$$\frac{dp}{dx} = 0.$$

The average velocity is obtained from Eq. (5.16),

$$U_{\text{av}} = \frac{U}{2}, \tag{5.22}$$

and, as expected, it is the velocity at the centerline. Similarly the volumetric flow rate per unit width [from Eq. (5.17)] is

$$Q = \frac{Uh}{2}, \tag{5.23}$$

and the maximum velocity is at the contact with the upper plate ($z = h$):

$$U_{\text{max}} = U. \tag{5.24}$$

The shear force at the wall, based on Eq. (5.19), is constant (with z) and is the same at the upper and lower walls (and this was discussed in Chapter 1):

$$\tau_{xz} = \mu \frac{U}{h}. \tag{5.25}$$

We calculate the friction coefficient on either the upper or lower surface by substituting the values of the shear stress and the average velocity:

$$C_f = \frac{\tau_{xz}}{\frac{1}{2}\rho U_{\text{av}}^2} = \frac{\mu \frac{U}{h}}{\frac{1}{2}\rho \frac{U^2}{4}} = \frac{8\mu}{\rho U h}. \tag{5.26}$$

This is a very important result because it introduces a nondimensional number, called the Reynolds number, Re, after the British fluid dynamicist Osborne Reynolds (1842–1912). It represents the ratio between the "inertia" and the viscosity effects:

$$\text{Re} = \frac{\rho U h}{\mu}. \tag{5.27}$$

Here Re is based on the clearance h between the plates (but in other cases, different quantities for the length scale may be used). The importance of this number will be discussed later, but now let us rewrite the friction coefficient for this case as

$$C_f = \frac{8}{\text{Re}}. \tag{5.28}$$

This equation is not only simple but also provides the basics for experimental evaluation of universal friction coefficients as a function of Re. Note that in this case (of laminar flow) the friction coefficient is reduced with increasing Re.

EXAMPLE 5.2. NUMERICAL EXAMPLE FOR THE FLOW BETWEEN TWO PARALLEL PLATES WITH NO PRESSURE GRADIENT. As a numerical example, following Example 5.1, let us consider the flow of water between two parallel plates. The upper plate moves at a velocity of $U = 5$ m/s and the clearance is $h = 1$ cm.

Solution: The average velocity is at the centerline,

$$U_{\text{av}} = \frac{U}{2} = \frac{5}{2} = 2.5 \text{ m/s},$$

and the volumetric flow rate per unit width is

$$Q = \frac{Uh}{2} = 0.01 \times 2.5 \times 1.0 = 0.025 \text{ m}^3/\text{s}.$$

The maximum velocity is at the contact with the upper plate ($z = h$),

$$U_{\max} = U = 5 \text{ m/s},$$

and the shear stress per unit width at the wall (upper or lower) is

$$\tau_{xz} = \mu \frac{U}{h} = 0.001 \frac{5}{0.01} \times 1.0 = 0.5 \frac{\text{N}}{\text{m}^2}.$$

Here the viscosity $\mu = 0.001$ for water was taken from Table 1.1. Suppose we want to calculate the force F required for pulling a unit area, $S = 1$ m^2, of the plate:

$$F = \tau_{xz} S = 0.5 \frac{\text{N}}{\text{m}^2} 1 \text{ m}^2 = 0.5 \text{ N}.$$

5.3.3 Flow between Two Infinite Parallel Plates — The Poiseuille Flow

This is another important case of the flow between plates because now both the upper and lower plates are stationary. Consequently the fluid motion is due to the pressure gradient. This flow is named the *Poiseuille flow* after Jean Louis Marie Poiseuille (1799–1869), a French physician who studied the pressure loss in small tubes.

Essentially this case was solved in the previous section, and information such as the average, maximum velocity and other properties can be obtained by simply setting the upper plate velocity to zero ($U = 0$). The velocity distribution, based on Eq. 5.12 is then

$$u(z) = -\frac{h^2}{2\mu} \frac{dp}{dx} \left(\frac{z}{h}\right) \left(1 - \frac{z}{h}\right). \tag{5.29}$$

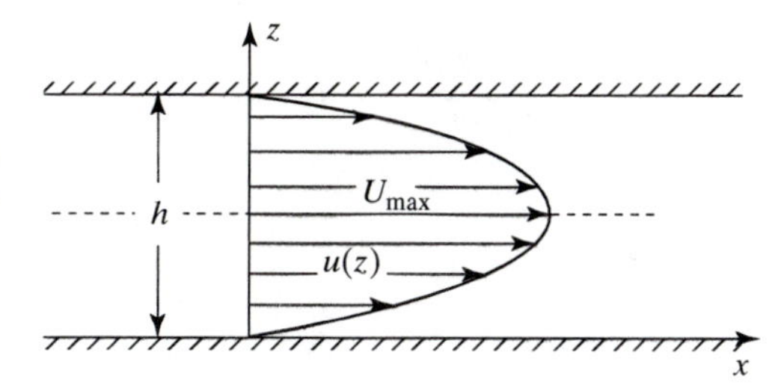

Figure 5.4. The flow that is due to a constant-pressure gradient between two parallel stationary plates.

This is a parabolic shape, as shown in Fig. 5.4, and $u(z) = 0$ at $z = 0$ and at $z = h$. Also this establishes the relation between pressure loss and viscous friction on the walls. The average velocity is obtained from Eq. (5.16),

$$U_{av} = -\frac{h^2}{12\mu}\frac{dp}{dx}, \tag{5.30}$$

and the volumetric flow rate (per unit width) from Eq. (5.17) is

$$Q = -\frac{h^3}{12\mu}\frac{dp}{dx}. \tag{5.31}$$

This indicates that the flow rate is directly (linearly) related to the pressure gradient. The maximum velocity location, based on Eq. (5.18), is found at the centerline:

$$\frac{z}{h} = \frac{1}{2}. \tag{5.32}$$

When this is substituted into the velocity distribution, Eq. (5.29), the maximum velocity is found:

$$U_{max} = -\frac{h^2}{8\mu}\frac{dp}{dx}. \tag{5.33}$$

Comparing this with the average velocity yields

$$U_{av} = \frac{2}{3}U_{max}. \tag{5.34}$$

Because of the symmetry (upper/lower), the shear stress on both walls is the same. To calculate the shear we use Eq. (5.19) (with $U = 0$):

$$\tau_{xz} = -\frac{h^2}{2}\frac{dp}{dx}\left(\frac{1}{h} - \frac{2z}{h^2}\right). \tag{5.35}$$

Note that the shear at the center ($z = h/2$) is zero and on the walls ($z = 0$, and $z = h$) is

$$\tau_{xz}|_{wall} = \pm\frac{h}{2}\frac{dp}{dx}, \tag{5.36}$$

and it seems to pull opposite to the pressure force. The friction coefficient C_f on one wall is then

$$C_f = \frac{-\frac{h}{2}\frac{dp}{dx}}{\frac{1}{2}\rho U_{av}^2}. \tag{5.37}$$

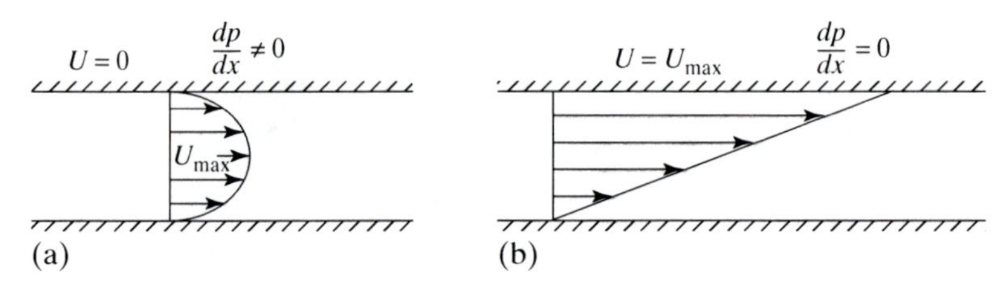

Figure 5.5. Two velocity distributions having the same shear stress at the wall.

This is not a very useful expression, but from Eq. (5.30) we can solve for the pressure gradient,

$$\frac{dp}{dx} = -\frac{12\mu}{h^2}U_{av},$$

and substitute this into Eq. (5.37):

$$C_f = \frac{\frac{h}{2}\frac{12\mu}{h^2}U_{av}}{\frac{1}{2}\rho U_{av}^2} = \frac{12\mu}{\rho U_{av} h}. \tag{5.38}$$

We see again that the Reynolds number appears in this equation. Using the Re definition from Eq. (5.27) we can write

$$C_f = \frac{12}{\text{Re}}. \tag{5.39}$$

It is interesting to compare the shear stress between this case and the basic shear flow (Example 5.1). First we need to rearrange the shear on the wall based on the maximum velocity. We can do this by solving Eq. (5.33) for dp/dx and substituting the result into Eq. (5.36):

$$\tau_{xz}\bigg|_{wall} = \frac{h}{2}\frac{dp}{dx} = \frac{h}{2}\frac{8\mu U_{max}}{h^2} = 4\mu\frac{U_{max}}{h}. \tag{5.40}$$

Now recall the shear term for the shear flow of Example 5.1:

$$\tau_{xz} = \mu\frac{U}{h} \tag{5.19}$$

Based on these results, to have the same shear at the wall, the flow created by the moving upper plate must have a maximum velocity four times larger than in the case in which the flow is moved by a pressure gradient (see Fig. 5.5).

We can also observe the force balance on the fluid in this flow, as depicted in Fig. 5.6. Because there is no acceleration in the x direction, the sum of the forces

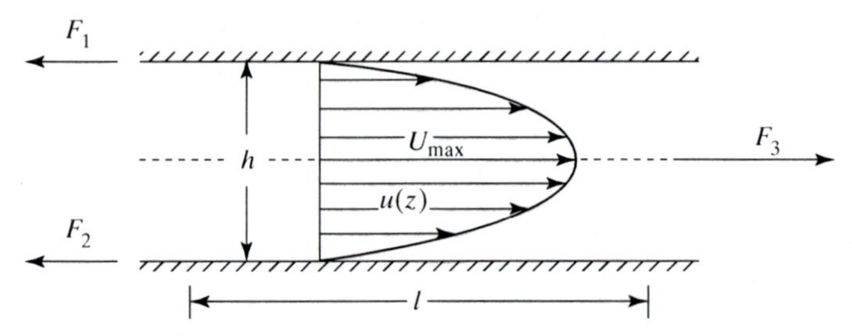

Figure 5.6. The shear flow on the walls must be equal to the force that is due to the pressure drop.

acting on the fluid between the two plates must be zero. To investigate this hypothesis let us select a segment of length l, as shown in the figure. The shear forces on the upper and lower walls for the selected segment (per unit width) are then

$$F_1 = F_2 = \tau_{xz} l \times 1 = -\frac{h}{2}\frac{dp}{dx} l \times 1.$$

Here we used the shear term from Eq. (5.36). The force F_3 that is due to the pressure gradient (per unit width) is

$$F_3 = \Delta p h \times 1 = -\frac{dp}{dx} l \times h \times 1,$$

and clearly

$$F_1 + F_2 + F_3 = 0.$$

EXAMPLE 5.3. WATER FLOW DUE TO PRESSURE GRADIENT BETWEEN PARALLEL PLATES. To demonstrate the applicability of the preceding formulation let us investigate the flow of water with an average velocity of 1 m/s, and a clearance between the plates of 1 cm. Using the viscosity value from Table 1.1, calculate the maximum velocity, the pressure gradient, the shear, and the friction coefficients.

First, from on Eq. (5.34), we can calculate the maximum velocity:

$$U_{\text{max}} = 1.5 U_{\text{av}} = 1.5 \text{ m/s}.$$

Next, we may use Eq. (5.30) to calculate the pressure drop:

$$\frac{dp}{dx} = \frac{-12\mu}{h^2} U_{\text{av}} = \frac{-12 \times 0.001 \times 1}{0.01^2} = -120 \frac{\text{N/m}^2}{\text{m}}.$$

The shear at the wall per unit width is calculated with Eq. (5.35):

$$\tau_{xz}|_{h=0} = -\frac{h}{2}\frac{dp}{dx} = -\frac{0.01}{2}(-120) = 0.6 \frac{\text{N}}{\text{m}^2}.$$

The Reynolds number is

$$\text{Re} = \frac{\rho U_{\text{av}} h}{\mu} = \frac{1000 \times 1 \times 0.01}{0.001} = 10^4,$$

and the friction coefficient [based on Eq. (5.39)] is

$$C_f = \frac{12}{\text{Re}} = \frac{12}{10^4} = 1.2 \times 10^{-3}.$$

5.3.4 The Hydrodynamic Bearing (Reynolds Lubrication Theory)

The potential of thin viscous fluid layers to reduce friction has been recognized and successfully used in many engineering applications. For example, rotating components such as the camshafts and crankshafts of the internal combustion engine are rolling on hydrodynamic bearings. The magnetic-reader mechanism of rotating computer disks is floating above the surface, based on the same principle. In all these cases, the moving surface is supported by the pressure, created by the viscous flow within the gap (also resulting in very low friction). The models for the rotating

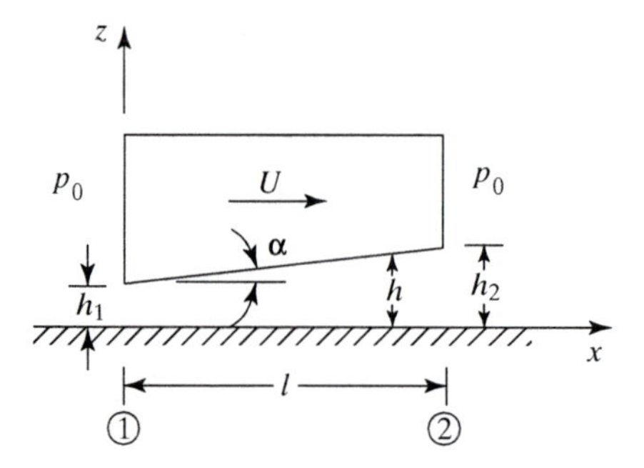

Figure 5.7. Model for the viscous laminar flow under a linear bearing.

shaft and the sliding block (slipper bearing) are quite similar, and here only the simpler linear slipper bearing is discussed. The theory was first developed by Osborne Reynolds in 1886, and the approach for laminar flow is shown schematically in Fig. 5.7.

Here a block is sliding at a velocity of U on a thin film of viscous fluid. The attempt is to describe a 2D steady-state case in which l is relatively very large and the gap h between the block and the solid surface is very small ($l \gg h$). An x–z coordinate system is attached to the upper sliding block, which moves at a velocity U to the right. Also the slope α is very small, and the velocity components in the z direction are small too:

$$u \gg w, \qquad (l \gg h). \tag{5.41}$$

With these assumptions only one term remains from the 2D incompressible continuity equation (5.1):

$$\frac{\partial u}{\partial x} \sim 0.$$

Because we expect a small change in the x direction we cannot conclude that the velocity profile in the x direction is constant. However, the flow rate Q is constant:

$$Q = \int_0^h u(z)dz = \text{const.} \tag{5.42}$$

Next we apply the previous assumptions to the momentum equations [Eqs. (5.2)]. For the x direction we get

$$u\frac{\partial u}{\partial x} = \frac{-1}{\rho}\frac{\partial p}{\partial x} + \frac{\mu}{\rho}\left(\frac{\partial^2 u}{\partial x^2} + \frac{\partial^2 u}{\partial z^2}\right), \tag{5.43}$$

and the momentum equation in the z direction becomes

$$0 = \frac{-1}{\rho}\frac{\partial p}{\partial z}. \tag{5.44}$$

Equation (5.44) indicates that there is no lateral or vertical change in the pressure. Hence the pressure can change only with x and is constant vertically:

$$p = p(x).$$

Considering the order of magnitude of the terms in the momentum equation in the x direction, we find that the left-hand inertia term is of the order of $O(l)$

$$u\frac{\partial u}{\partial x} \sim O\left(l\frac{l}{l}\right) \sim O(l)\,.$$

The order of the two second-order derivatives, based on inequalities (5.41), is

$$\frac{\partial^2 u}{\partial x^2} \sim O\left(\frac{l^2}{l^2}\right) = O(1),$$

$$\frac{\partial^2 u}{\partial z^2} \sim O\left(\frac{l^2}{h^2}\right),$$

and therefore this last second-order derivative term is clearly the largest:

$$\frac{\partial^2 u}{\partial z^2} \gg \frac{\partial^2 u}{\partial x^2}.$$

Consequently, the inertia term can be neglected too, and the momentum equation in the x direction becomes [exactly as in the Couette flow case – see Eq. (5.9)]

$$0 = \frac{-1}{\rho}\frac{\partial p}{\partial x} + \frac{\mu}{\rho}\left(\frac{\partial^2 u}{\partial z^2}\right). \tag{5.45}$$

The boundary conditions, based on Fig. 5.7, at $x = 0$ are

$$\begin{aligned} p &= p_0, \\ h &= h_1, \\ u(z=0) &= -U, \\ u(z=h_1) &= 0. \end{aligned} \tag{5.46}$$

Note that the coordinate system is attached to the sliding block and therefore the lower surface appears to be moving at a velocity of $-U$. The boundary conditions at the other end of the block at $x = l$ are

$$\begin{aligned} p &= p_0, \\ h &= h_2, \\ u(z=0) &= -U, \\ u(z=h_2) &= 0. \end{aligned} \tag{5.47}$$

Before integration, the momentum equation is rearranged,

$$\frac{\partial^2 u}{\partial z^2} = \frac{1}{\mu}\frac{\partial p}{\partial x},$$

and the velocity profile in the gap is found after two integrations:

$$u(z) = \frac{1}{\mu}\frac{\partial p}{\partial x}\frac{z^2}{2} + A(x)z + B(x) \tag{5.48}$$

We calculate the integration constant $B(x)$ by applying the boundary condition at $z = 0$ (for any x along the gap),

$$-U = B(x), \tag{5.49}$$

and we find $A(x)$ by applying the boundary condition at $z = h$,

$$0 = \frac{1}{\mu}\frac{\partial p}{\partial x}\frac{h^2}{2} + A(x)h - U. \tag{5.50}$$

Substituting $A(x)$ and $B(x)$ into Eq. (5.48) yields

$$u(z) = \frac{1}{\mu}\frac{\partial p}{\partial x}\frac{z^2}{2} + \left(\frac{-1}{\mu}\frac{\partial p}{\partial x}\frac{h}{2} + \frac{U}{h}\right)z - U,$$

and after rearranging, the velocity profile inside the gap becomes

$$u(z) = \frac{h^2}{2\mu}\frac{\partial p}{\partial x}\left(\frac{z^2}{h^2} - \frac{z}{h}\right) + U\left(\frac{z}{h} - 1\right). \tag{5.51}$$

The only unknown at this point is the shape of the pressure distribution, which we can find by applying continuity equation (5.42), which stated that the flow in the gap is constant:

$$\begin{aligned} Q = \int_0^h u(z)dz &= \int_0^h \left[\frac{h^2}{2\mu}\frac{\partial p}{\partial x}\left(\frac{z^2}{h^2} - \frac{z}{h}\right) + U\left(\frac{z}{h} - 1\right)\right]dz \\ &= -\frac{h^3}{12\mu}\frac{\partial p}{\partial x} - \frac{Uh}{2} = \text{const.} \end{aligned} \tag{5.52}$$

Now the pressure gradient as a function of the local h can be calculated:

$$\frac{\partial p}{\partial x} = -\frac{12\mu}{h^3}\left(\frac{Uh}{2} + Q\right). \tag{5.53}$$

At this point we can look at the boundary conditions and see that the pressure at both ends of the sliding block is p_0. This suggests that the pressure builds up toward the center and must have a maximum where $(\partial p/\partial x = 0)$. Let us identify the corresponding gap at this point (of maximum pressure) as h_0. With this definition, we can easily calculate the flow rate at h_0 [Eq. (5.52)] because Q is constant with x,

$$Q = -\frac{Uh_0}{2}, \tag{5.54}$$

and we can simplify the pressure gradient by substituting Eq. (5.54) into Eq. (5.53):

$$\frac{\partial p}{\partial x} = -\frac{6\mu U}{h^3}(h - h_0). \tag{5.55}$$

The pressure distribution can be found by integration with x; however, first, the gap geometry (based on Fig. 5.7) is obtained:

$$h = h_1 + (h_2 - h_1)\frac{x}{l} \approx h_1 + \alpha x. \tag{5.56}$$

We can simplify the integration more by exchanging the integration variable:

$$\frac{\partial p}{\partial x} = \frac{\partial p}{\partial h}\frac{\partial h}{\partial x} = \frac{\partial p}{\partial h}\alpha.$$

With this modification Eq. (5.55) becomes

$$\frac{\partial p}{\partial h} = -\frac{6\mu U}{\alpha}\left(\frac{1}{h^2} - \frac{h_0}{h^3}\right).$$

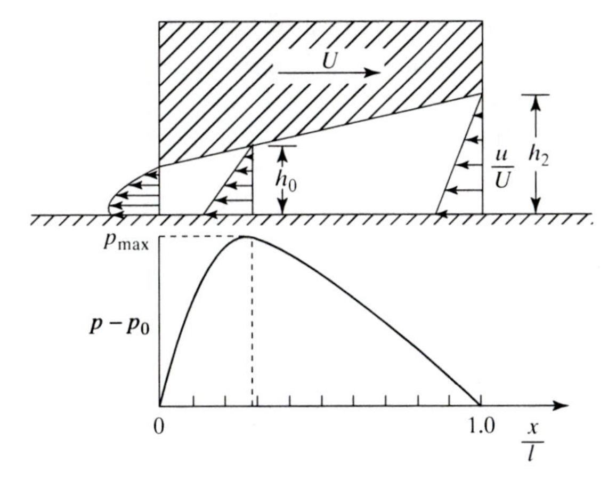

Figure 5.8. Schematic description of the pressure distribution under the sliding block [for the case of $(h_2/h_1) = 2.2$] and the shape of the velocity distribution in the gap.

This can be integrated with respect to h,

$$p = \frac{6\mu U}{\alpha}\left(\frac{1}{h} - \frac{h_0}{2h^2}\right) + C,$$

where C is the integration constant. We can find the two unknowns (C and h_0) by applying the boundary conditions at the two ends of the sliding block,

$$\text{at } x = 0,\; p = p_0,\; h = h_1,$$
$$\text{at } x = l,\; p = p_0,\; h = h_2,$$

and after some algebra we get

$$h_0 = \frac{2h_1 h_2}{h_1 + h_2}. \tag{5.57}$$

After we solve for the constant C, the longitudinal pressure distribution as a function of the gap h becomes

$$p - p_0 = \frac{6\mu U}{\alpha}\frac{(h - h_1)(h - h_2)}{h^2(h_1 + h_2)}. \tag{5.58}$$

The shape of this pressure distribution is shown schematically in Fig. 5.8. The maximum pressure occurs at $h = h_0$ and based on Eq. (5.58) [and recalling that $(h_2 - h_1)/l \approx \alpha$] it is

$$p_{\max} - p_0 = \frac{6\mu U}{\alpha}\frac{(h_0 - h_1)(h_0 - h_2)}{h_0^2(h_1 + h_2)} = \frac{3\mu U l}{2}\frac{(h_2 - h_1)}{h_1 h_2(h_1 + h_2)}. \tag{5.59}$$

We obtain the position of the maximum pressure in terms of the x coordinate by substituting h_0 from Eq. (5.57) into Eq. (5.56):

$$\left.\frac{x}{l}\right|_{p_{\max}} = \frac{h_1}{h_1 + h_2}. \tag{5.60}$$

The local velocity distribution is also shown in Fig. 5.8 and it is determined by Eq. (5.51). Initially, at the right-hand side of the block, it resembles the simple shear flow, as in Fig. 5.3, but in this case the pressure gradient is not constant. Near the left-hand side of the sliding block, the pressure gradient is positive and the parabolic term in the velocity profile is more dominant.

From the engineering point of view, the total force L lifting the sliding block and the friction drag D must be calculated. We can obtain the supporting force L (per unit width) by integrating the pressure along the block and again replacing the integration of x with the integration with h:

$$L = \int_0^l (p - p_0)dx = \frac{6\mu U}{\alpha^2}\int_{h_1}^{h_2} \frac{(h-h_1)(h-h_2)}{h^2(h_1+h_2)}dh$$
$$= \frac{6\mu U l^2}{h_1^2(h_2/h_1-1)^2}\left[\ln\frac{h_2}{h_1} - \frac{2(h_2/h_1-1)}{h_2/h_1+1}\right]. \tag{5.61}$$

We can find the drag force by integrating the shear force along the lower surface of the block, at $z = h$. Taking the velocity distribution from Eq. (5.51), we find that the shear stress is

$$\tau = \mu \left.\frac{du(z)}{dz}\right|_h = \frac{h^2}{2}\frac{\partial p}{\partial x}\left(\frac{2h}{h^2} - \frac{1}{h}\right) + \mu\frac{U}{h} = \frac{h}{2}\frac{\partial p}{\partial x} + \mu\frac{U}{h}. \tag{5.62}$$

Next, substituting $\partial p/\partial x$ from Eq. (5.53), we get

$$\tau = \mu\left(\frac{4U}{h} + \frac{6Q}{h^2}\right). \tag{5.63}$$

We now find the drag force (per unit width) by integrating the shear stress and exchanging the integration in the x direction with the h direction:

$$D = \int_0^l \tau dx = \frac{\mu}{\alpha}\int_{h_1}^{h_2}\left(\frac{4U}{h} + \frac{6Q}{h^2}\right)dh = \frac{\mu}{\alpha}\left(4U\ln h - \frac{6Q}{h}\right)_{h_1}^{h_2}. \tag{5.64}$$

Substituting Q from Eq. (5.54), α from Eq. (5.56), and h_0 from Eq. (5.57), we find that the drag force becomes

$$D = \frac{\mu U l}{h_1(h_2/h_1-1)}\left[4\ln\frac{h_2}{h_1} - \frac{6(h_2/h_1-1)}{h_2/h_1+1}\right]. \tag{5.65}$$

We can obtain the maximum lift by calculating $dL/d(h_2/h_1) = 0$, and after some algebra, the condition for the maximum lift occurs when $(h_2/h_1) = 2.2$. Calculating the lift [from Eq. (5.61)] and drag [from Eq. (5.65)] for the maximum lift condition, we get

$$L_{\max} = 0.16\mu U\left(\frac{l}{h_1}\right)^2,$$
$$D_{L\max} = 0.75\mu U\left(\frac{l}{h_1}\right), \tag{5.66}$$

and the lift-to-drag ratio (which is the mechanical friction coefficient) for this condition is

$$\left.\frac{D}{F}\right|_{L\max} = 4.7\frac{h_1}{l}. \tag{5.67}$$

Considering that the order of magnitude for a typical bearing is $\frac{l}{h_1} \sim 10^{-3}$ the friction coefficient of the hydrodynamic bearing is significantly lower than the dry friction!

EXAMPLE 5.4. LIFT OF A LINEAR BEARING. A 1-cm-long magnetic pickup is floating on a rotating disk where the linear velocity is 50 m/s. Assuming the allowed gap is $h_1 = 0.2$ mm and $(h_2/h_1) = 2.2$, calculate the weight of the magnetic head. Also calculate the drag-to-lift ratio.

Solution: For this 2D case we can use Eqs. (5.66) and assume that the disk operates in air. The lift is

$$L = 0.16\mu U\left(\frac{l}{h_1}\right)^2 = 0.16 \times 1.8 \times 10^{-5} \times 50\left(\frac{0.01}{0.0002}\right)^2 = 0.36\frac{\text{N}}{\text{m}}.$$

Suppose the pickup head is only 1 cm wide; then we can approximate the lift by taking 1/100 of the preceding value (e.g., the ratio between 1 cm and 1 m). We can calculate the drag-to-lift ratio by using Eq. (5.67),

$$\left.\frac{D}{F}\right| = 4.7\frac{0.0002}{0.01} = 0.094,$$

and this is at least one order of magnitude smaller than that of dry friction.

5.4 Laminar Flow in Circular Pipes (The Hagen–Poiseuille Flow)

Calculations, such as the pressure loss in long circular pipes, is of paramount importance in many engineering applications. It was extensively studied during the early 19th century by the German hydraulician G. H. L. Hagen (1797–1884) and the French physiologist J. L. M. Poiseuille (1799–1869). The mathematical formulation of this problem is similar to the one used for the flow between parallel plates; however, now a cylindrical coordinate system is used. The basic model is described in Fig. 5.9 and it is assumed that the flow is incompressible, viscous, laminar, and fully developed (far from entrance effects and no changes with x). The only body force we consider is gravitation (ρg), and the flow inside the pipeline moves only in the x direction and there are no velocity components in the other directions:

$$q_r = q_\theta = 0. \tag{5.68}$$

With these assumptions the continuity equation in cylindrical coordinates for an incompressible fluid [Eq. (2.45)] becomes

$$\frac{\partial q_x}{\partial x} = 0. \tag{5.69}$$

We assume the flow is axisymmetric (and no changes with θ) and we reach the conclusion that

$$q_x = q_x(r). \tag{5.70}$$

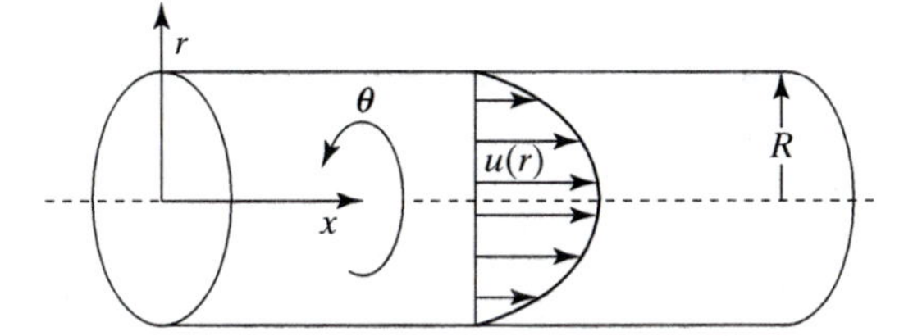

Figure 5.9. Fully developed laminar flow in a circular pipe.

The momentum equation in the r direction [Eq. (2.46)] is now reduced to

$$0 = -\frac{\partial p}{\partial r}. \tag{5.71}$$

The momentum equation [Eq. (2.47)] in the θ direction is

$$0 = -\frac{\partial p}{\partial \theta}, \tag{5.72}$$

and in the x direction [Eq. (2.48)] is

$$0 = \rho g - \frac{\partial p}{\partial x} + \mu \left(\frac{\partial^2 q_x}{\partial r^2} + \frac{1}{r}\frac{\partial q_x}{\partial r} \right). \tag{5.73}$$

From Eqs. (5.71), and (5.72) we conclude that the pressure varies only in the x direction:

$$p = p(x). \tag{5.74}$$

Because the term inside the parentheses in Eq. (5.73) is independent of x (or constant along the x axis), the term $\partial p/\partial x$ must be constant, too.

Now we can rearrange Eq. (5.73) as follows:

$$-\rho g + \frac{dp}{dx} = \mu \frac{1}{r}\frac{d}{dr}\left(r\frac{dq_x}{dr} \right).$$

A more convenient form is

$$\frac{d}{dr}\left(r\frac{dq_x}{dr} \right) = \frac{1}{\mu}\left(\frac{dp}{dx} - \rho g \right) r.$$

Now we can integrate with respect to r,

$$r\frac{dq_x}{dr} = \frac{1}{\mu}\left(\frac{dp}{dx} - \rho g \right)\frac{r^2}{2} + A,$$

where A is the constant of integration. We divide first by r,

$$\frac{dq_x}{dr} = \frac{1}{\mu}\left(\frac{dp}{dx} - \rho g \right)\frac{r}{2} + \frac{A}{r},$$

and perform the second integration

$$q_x = \frac{1}{2\mu}\left(\frac{dp}{dx} - \rho g \right)\frac{r^2}{2} + A \ln r + B. \tag{5.75}$$

To calculate the integration constant we must use the boundary conditions. At the center of the pipe, clearly there is symmetry and we can require that

$$\text{at} \qquad r = 0, \frac{dq_x}{dr} = 0,$$

$$\text{and at } r = R, q_x = 0. \tag{5.76}$$

And this second condition, at the pipe inner wall, requires that the velocity must be zero (based on the zero-slip boundary condition). Substituting the first boundary condition into Eq. (5.75) yields

$$\frac{dq_x}{dr} = \frac{1}{\mu}\left(\frac{dp}{dx} - \rho g \right)\frac{0}{2} + \frac{A}{0} = 0.$$

This is possible only if A is zero. Substituting the second boundary condition to solve for B, we get

$$0 = \frac{1}{2\mu}\left(\frac{dp}{dx} - \rho g\right)\frac{R^2}{2} + B,$$

and after rearranging, we get

$$B = -\frac{1}{2\mu}\left(\frac{dp}{dx} - \rho g\right)\frac{R^2}{2}. \tag{5.77}$$

By substituting this into Eq. (5.75), we obtain the velocity distribution at any x station:

$$q_x = \frac{R^2}{4\mu}\left(\frac{dp}{dx} - \rho g\right)\left[\left(\frac{r}{R}\right)^2 - 1\right]. \tag{5.78}$$

This velocity distribution has a parabolic shape, and maximum velocity is reached at the center. Next let us calculate quantities such as the maximum velocity, flow rate and shear on the wall. First, however, let us assume that the ρg term is similar to the pressure drop and use only one. In this case the velocity distribution is

$$q_x = \frac{R^2}{4\mu}\frac{dp}{dx}\left[\left(\frac{r}{R}\right)^2 - 1\right]. \tag{5.79}$$

The maximum velocity is at the center ($r = 0$)

$$U_{\max} = -\frac{R^2}{4\mu}\frac{dp}{dx}, \tag{5.80}$$

and the negative sign is a result of the flow in the positive x direction, for which the pressure gradient must be negative. The average velocity is related to the volumetric flow rate Q as

$$Q = U_{av}S = \int q_x dS;$$

therefore the average velocity is obtained:

$$U_{av} = \frac{1}{S}\int q_x dS, \tag{5.81}$$

where S is the pipe cross-section area and $dS = 2\pi r dr$. Substituting Eq. (5.79) yields

$$U_{av} = \frac{1}{\pi R^2}\int_0^R \frac{R^2}{4\mu}\frac{dp}{dx}\left[\left(\frac{r}{R}\right)^2 - 1\right]2\pi r dr = \frac{1}{2\mu}\frac{dp}{dx}\int_0^R\left[\frac{r^3}{R^2} - r\right]dr$$
$$= \frac{1}{2\mu}\frac{dp}{dx}\left[\frac{r^4}{4R^2} - \frac{r^2}{2}\right]_0^R,$$

and after substituting the constants at the two limits, we get

$$U_{av} = -\frac{R^2}{8\mu}\frac{dp}{dx}. \tag{5.82}$$

Comparing this with Eq. (5.80) reveals that

$$U_{\max} = 2U_{av}. \tag{5.83}$$

The volumetric flow is then

$$Q = U_{\text{av}} S = -\frac{R^2}{8\mu}\frac{dp}{dx}\pi R^2 = -\frac{\pi R^4}{8\mu}\frac{dp}{dx}. \tag{5.84}$$

This is a very important conclusion because it shows the relation between the pressure drop and the flow rate in a circular pipe. The relation is linear, with the pressure drop indicating that the higher pressure gradient will increase the flow rate. The effect of viscosity is also linear but opposite; for example, more pressure is needed to drive the same flow rate if the viscosity is increased. Next, we calculate the shear stress in the flow by deriving Eq. (5.78):

$$\tau_{xr} = -\mu\frac{dq_x}{dr} = -\mu\frac{R^2}{4\mu}\frac{dp}{dx}\frac{2r}{R^2} = -\frac{r}{2}\frac{dp}{dx}, \tag{5.85}$$

and the minus sign is a result of the coordinate system (e.g., the origin is at the center). Note that the shear stress is zero at the centers and increases linerly towards the wall. The shear at the wall is then

$$\tau_{xr}|_{\text{wall}} = -\frac{R}{2}\frac{dp}{dx}. \tag{5.86}$$

It is more useful to replace the pressure drop with an average velocity formulation. We do this by rearranging Eq. (5.82),

$$\frac{dp}{dx} = -\frac{8\mu U_{\text{av}}}{R^2},$$

and by substituting this into Eq. (5.86) we find that the wall shear term becomes

$$\tau_{rx}|_{\text{wall}} = \frac{2\mu}{R}U_{\max} = \frac{4\mu}{R}U_{\text{av}}. \tag{5.87}$$

It is interesting to compare this with the shear stress relation in the flow between parallel plates (with one plate moving at a velocity of $U_{\max}$, where $\tau_{\text{wall}} = \frac{\mu}{h}U_{\max}$). If the clearance h is compared with the pipe diameter ($2R$) then the shear in the pipe is four times larger! Once the shear stress is evaluated, the friction coefficient can be calculated:

$$C_f = \frac{\tau_{xr}}{\frac{1}{2}\rho U_{\text{av}}^2} = \frac{\frac{4\mu}{R}U_{\text{av}}}{\frac{1}{2}\rho U_{\text{av}}^2} = \frac{16\mu}{\rho U_{\text{av}}(2R)} = \frac{16}{\text{Re}}. \tag{5.88}$$

This is an amazingly good result that is validated by experiments and usually applicable up to $\text{Re} = 2000$. It also lays the foundation for more complex pipe flow calculations, and we discuss this in the next section. To conclude this section, let us evaluate the force balance between the shear and the pressure components. Suppose we consider a pipe with a length of l. The force pushing the fluid in the x direction is

$$F_1 = \Delta p(\pi R^2) = -\frac{dp}{dx}l(\pi R^2).$$

The shear on the wall acting into the opposite direction is

$$F_2 = -\tau_{xr}|_{\text{wall}}\, S = \frac{R}{2}\frac{dp}{dx}2\pi Rl = \frac{dp}{dx}\pi R^2 l,$$

and the two forces are equal (e.g., steady flow).

EXAMPLE 5.5. PRESSURE DROP IN A CIRCULAR PIPE. Estimate the pressure drop for a fully developed flow in a 3-m-long, 1.2-cm-diameter pipe, delivering 0.5 L/min of kerosene (at 20 °C).

Solution: To solve the problem we look up the viscosity and density values from Table 1.1. Because the flow rate is given, the pressure gradient and the average velocity are easily obtained [Eq. (5.84)]:

$$\frac{dp}{dx} = -\frac{Q8\mu}{\pi R^4} = -\frac{\frac{0.5}{60} \times 10^{-3}\ \mathrm{m^3/s} \times 8 \times 1.9 \times 10^{-2}\ (\mathrm{N\ s})/\mathrm{m}^2}{\pi 0.006^4\ \mathrm{m}^4} = -31.2\frac{\mathrm{N/m^2}}{\mathrm{m}}.$$

The average velocity is

$$U_{\mathrm{av}} = \frac{Q}{S} = \frac{\frac{0.5}{60} \times 10^{-3}\ \mathrm{m^3/s}}{\pi\ 0.006^2\ \mathrm{m^3}} = 0.0736\ \mathrm{m/s}.$$

Because it was mentioned that the preceding formula is valid up to Reynolds numbers of about 2000, let us calculate Re:

$$\mathrm{Re} = \frac{\rho U_{\mathrm{av}} D}{\mu} = \frac{814 \times 0.0736 \times 0.012}{1.9 \times 10^{-3}} = 378.8.$$

Therefore this calculation is within the range of this model.

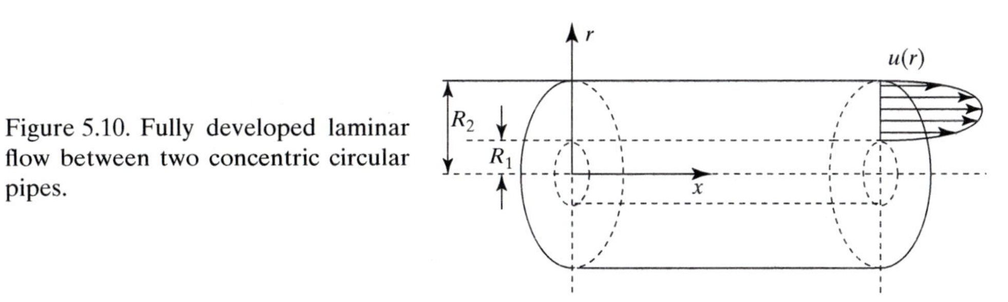

Figure 5.10. Fully developed laminar flow between two concentric circular pipes.

5.5 Fully Developed Laminar Flow between Two Concentric Circular Pipes

Let us consider the fully developed, viscous laminar flow between two concentric cylinders (or pipes), as depicted in Fig. 5.10. The assumptions here are the same as those in Section 5.3, and we can start with the solution obtained in Eq. (5.75):

$$q_x = \frac{1}{2\mu}\frac{dp}{dx}\frac{r^2}{2} + A \ln r + B. \tag{5.89}$$

The difference now is in the boundary conditions because the flow will stop at the wall of the inner cylinder. Consequently the boundary conditions are

$$\begin{aligned} &\text{at} \quad r = R_1, \quad q_x = 0, \\ &\text{and at} \quad r = R_2, \quad q_x = 0. \end{aligned} \tag{5.90}$$

Substituting these conditions into Eq. (5.75) provides two equations:

$$0 = \frac{1}{2\mu}\frac{dp}{dx}\frac{R_1^2}{2} + A\ln R_1 + B,$$

$$0 = \frac{1}{2\mu}\frac{dp}{dx}\frac{R_2^2}{2} + A\ln R_2 + B$$

After solving for the two constants we get

$$A = -\frac{\left[(R_2/R_1)^2 - 1\right]}{\ln(R_2/R_1)}R_1^2,$$

$$B = \frac{\left[(R_2/R_1)^2 - 1\right]}{\ln(R_2/R_1)}R_1^2\ln R_1 - R_1^2.$$

We now obtain the velocity distribution between the two concentric cylinders by substituting A and B into Eq. (5.89):

$$q_x = -\frac{1}{4\mu}\frac{dp}{dx}\left\{R_1^2 - r^2 + \frac{\left[(R_2/R_1)^2 - 1\right]}{\ln(R_2/R_1)}R_1^2\ln\frac{r}{R_1}\right\}. \tag{5.91}$$

We can obtain the flow rate by integrating the velocity between the two boundaries:

$$Q = \int_{R_1}^{R_2} q_x dS = -\int_{R_1}^{R_2}\frac{1}{4\mu}\frac{dp}{dx}\left\{R_1^2 - r^2 + \frac{\left[(R_2/R_1)^2 - 1\right]}{\ln(R_2/R_1)}R_1^2\ln\frac{r}{R_1}\right\}2\pi r\,dr$$

$$= \frac{\pi}{4\mu}\frac{dp}{dx}\left\{\frac{r^4}{2} - R_1^2r^2 - \frac{\left[(R_2/R_1)^2 - 1\right]}{\ln(R_2/R_1)}R_1^2\left(\ln\frac{r}{R_1} - \frac{1}{2}\right)r^2\right\}_{R_1}^{R_2}$$

$$= -\frac{\pi R_1^4}{8\mu}\frac{dp}{dx}\left\{(R_2/R_1)^4 - 1 - \frac{\left[(R_2/R_1)^2 - 1\right]^2}{\ln(R_2/R_1)}\right\}. \tag{5.92}$$

We now easily calculate the average velocity by dividing the flow rate by the area:

$$U_{\text{av}} = \frac{Q}{\pi\left(R_2^2 - R_1^2\right)} = -\frac{R_1^2}{8\mu}\frac{dp}{dx}\left[(R_2/R_1)^2 + 1 - \frac{(R_2/R_1)^2 - 1}{\ln(R_2/R_1)}\right]. \tag{5.93}$$

The shear stress on the outer wall is calculated as before:

$$\tau_{xr}\Big|_{R_2} = -\mu\frac{dq_x}{dr} = -\frac{R_1}{4}\frac{dp}{dx}\left[2(R_2/R_1) - \frac{(R_2/R_1)^2 - 1}{(R_2/R_1)}\frac{1}{\ln(R_2/R_1)}\right], \tag{5.94}$$

and the shear stress at the inner cylinder wall is

$$\tau_{xr}\Big|_{R_1} = \mu\frac{dq_x}{dr} = -\frac{R_1}{4}\frac{dp}{dx}\left[\frac{(R_2/R_1)^2 - 1}{\ln(R_2/R_1)} - 2\right]; \tag{5.95}$$

we dropped the minus sign for the inner cylinder because the velocity gradient there is positive.

5.6 Flow in Pipes: Darcy's Formula

Henry Philibert Gaspard Darcy (1803–1858) was a French scientist who made several important contributions to hydraulics. One of his most well-known contributions is the experimental development of the pressure-loss formula in pipes. Because we developed the exact laminar flow solution we should be able to arrive at the same formulation Darcy did. Let us start with the friction coefficient formula and substitute the shear stress from Eq. (5.86) into the definition of C_f [Eq. (5.26)]:

$$C_f = \frac{\tau_{xr}}{\frac{1}{2}\rho U_{\mathrm{av}}^2} = \frac{-\frac{R}{2}\frac{dp}{dx}}{\frac{1}{2}\rho U_{\mathrm{av}}^2} = \frac{-R\frac{dp}{dx}}{\rho U_{\mathrm{av}}^2}. \tag{5.96}$$

This relation basically implies that if the friction coefficient is known then we can calculate the pressure drop. Assuming long circular pipes, let us use $\frac{\Delta p}{L}$ instead of $-\frac{dp}{dx}$ where L is the length of the pipe:

$$\frac{\Delta p}{L} = C_f \rho \frac{U_{\mathrm{av}}^2}{R}.$$

This could be rearranged in a form used two decades ago:

$$\frac{\Delta p}{\rho} = C_f \frac{L}{R} U_{\mathrm{av}}^2.$$

Basically this is very close to Darcy's formula; however, he used the pipe diameter D (instead of $2R$) and a friction factor f instead of the friction coefficient, which is also four times larger

$$f = 4C_f. \tag{5.97}$$

If we use the results of Eq. (5.88) for laminar flow we get

$$f = 4C_f = \frac{64}{\mathrm{Re}}. \tag{5.98}$$

Now we can write Darcy's pressure-drop formula when the average velocity inside the pipeline is known:

$$\frac{\Delta p}{\rho g} = f \frac{L}{D} \frac{U_{\mathrm{av}}^2}{2g} \tag{5.99}$$

To follow the original spirit of the formula, g was added at both sides of the equation. Note that this is basically a *one-dimensional* model because only the average velocity is taken into account. Also, the dimensions on both sides of this equation are *length*. This is called the "head loss," h_f, and it can be measured in terms of the liquid-column height (the same liquid flowing in the pipeline):

$$h_f = \frac{\Delta p}{\rho g} = f \frac{L}{D} \frac{U_{\mathrm{av}}^2}{2g}, \tag{5.100}$$

and the friction factor f for laminar flow is calculated by Eq. (5.98). In many engineering applications the pressure drop is expressed versus the volumetric flow rate Q, and for a circular cross section we can write

$$Q = U_{\mathrm{av}} S = U_{\mathrm{av}} \frac{\pi D^2}{4}.$$

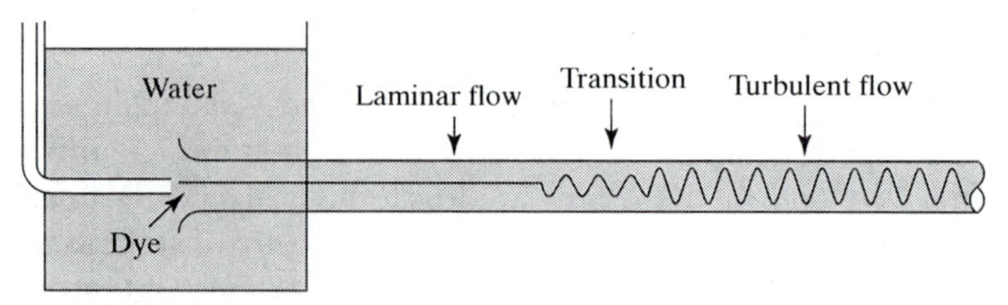

Figure 5.11. Schematic description of Reynolds' 1883 pipe flow experiment.

Replacing the average velocity in Eq. (5.99) with the volumetric flow rate yields

$$h_f = \frac{\Delta p}{\rho g} = f \frac{L}{D^5} \frac{8Q^2}{\pi^2 g}. \tag{5.101}$$

Another very important feature of Darcy's formulation [Eq. (5.99)] is that it can be used for turbulent flows as well. All that is needed now is to establish a database of the friction factors for various cases and we can calculate the pipeline pressure loss in many different situations. Prior to drawing practical conclusions, let us briefly discuss the effect of turbulence on the flow in pipes.

5.7 The Reynolds Dye Experiment, Laminar–Turbulent Flow in Pipes

One of Osborne Reynolds' (1842–1912) pipe flow experiments is described schematically in Fig. 5.11. Basically the fluid is flowing from a large container into a long pipe, and flow rates and other parameters can be changed. Near the center of the pipe inlet, colored dye is injected through a very thin tube (as shown). When the dye is observed as it flows inside the tube, initially a thin concentrated line is seen. This is expected in laminar flow, where the fluid particles move parallel (in the x direction in this case) so there is no reason (forget diffusion) for the dye to spread laterally. At a certain point this smooth flow is interrupted and the dye shows chaotic motions (laterally), and instead of a thin concentrated line, a rapid mixing of the dye is observed. This condition, in which the fluid particles chaotically move in all directions in addition to the main flow direction, is called turbulent flow. The region where the laminar flow turns into turbulent flow is called the transition region. Reynolds was able to show with his dye experiment that this condition occurs when the nondimensional number $\rho U_{av} D/\mu$ is about 2000 (and this is of course the Reynolds number).

Because of the additional motion of the fluid particles, turbulent flow losses are larger. In terms of our definitions, the shear stress and the friction coefficients are larger than in a laminar flow. Even if the average velocity has only one component, the one in the x direction, perturbations in the other directions exchange momentum and create larger losses. For example, the 2D shear stress at the wall will have an additional component, clearly increasing the shear stress:

$$\tau_{xr} = \mu \frac{\partial u}{\partial r} - \rho \overline{u'w'} \tag{5.102}$$

and here $u'w'$ are the perturbation velocities in the x and r directions. This also has an effect on the velocity distribution inside the pipe, as shown in Fig. 5.12, and both flows will have the same average velocity (and flow rate). Clearly the turbulent flow has larger shear ($\frac{\partial u}{\partial r}$) near the wall compared with that of the laminar solution.

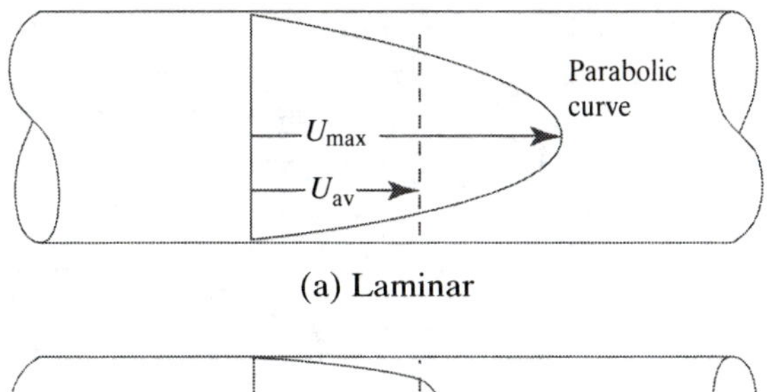

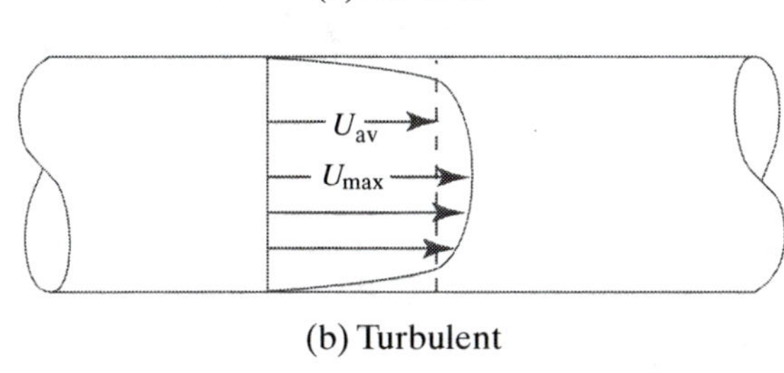

Figure 5.12. The difference between fully developed laminar and turbulent velocity distributions in a circular pipe flow.

Note that this is the first opportunity to observe the effect of turbulence in high-Reynolds-number flows. Because of the complex motion of the fluid, the engineering approach is to approximate the average quantities, based on the laminar flow model, as demonstrated next.

Now we can return to Darcy's formula and plot the friction parameter for a wide range of Reynolds numbers. For laminar flow the results are the same as in Eq. (5.88) and f is reduced with increasing Re:

$$f = 4C_f = \frac{64}{\mathrm{Re}}. \tag{5.103}$$

The chart in Fig. 5.13 is known as the "The Moody diagram," which was published by L. F. Moody in 1944. The turbulent curves for smooth pipes are significantly higher than for the laminar case, and for the higher Re only the turbulent flow case is present. Moody also included relative roughness, so that the effect of surface smoothness can be incorporated into the pressure-loss formula (the relative roughness is the average height of the roughness inside the pipe, k, divided by the pipe diameter).

EXAMPLE 5.6. PRESSURE LOSS IN TERMS OF HEAD LOSS. Water is flowing in a 10-m-long smooth pipe with a 0.02-m diameter and an average velocity of 0.15 m/s. Calculate the pressure drop.

Solution: First let us calculate the Reynolds number and use the values for the density and viscosity from Table 1.1:

$$\mathrm{Re} = \frac{1000 \times 0.15 \times 0.02}{10^{-3}} = 3000.$$

It is still possible to maintain laminar flow, and the friction factor, from Eq. (5.98) is

$$f = \frac{64}{3000} = 0.0213.$$

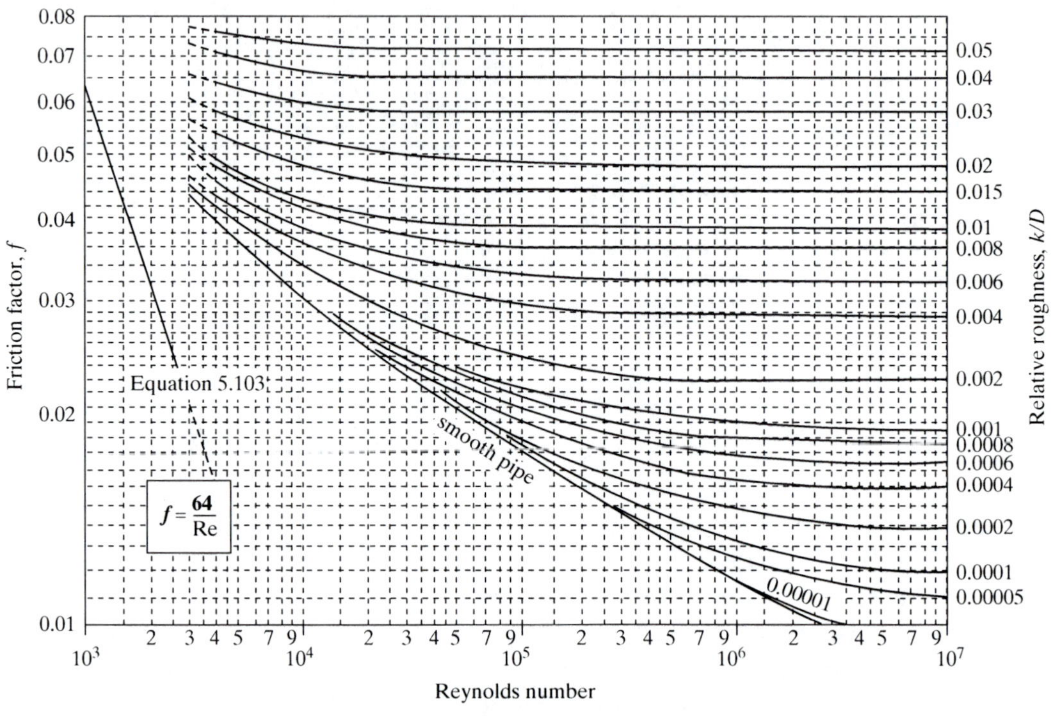

Figure 5.13. The Moody friction factor for fully developed flow in circular pipes. Note the effect of relative roughness (k = average height of inner surface roughness).

The head loss, based on Eq. (5.62), is then

$$h_f = \frac{\Delta p}{\rho g} = 0.0213 \frac{10}{0.02} \frac{0.15^2}{2 \times 9.8} = 0.0122 \text{ m} \,(\text{H}_2\text{O}),$$

so this is a little over 1 cm of water. Now if the flow is turbulent the friction factor is about 0.042 (based on Fig. 5.13) and the pressure drop will increase by more than two times.

5.8 Additional Losses in Pipe Flow

Actual pipelines include elbows, flanges, valves, and other devices; all may have an effect on the flow inside the pipe. For example, Fig. 5.14(a) shows the entrance into a pipe where the locally narrowing streamlines may create local flow recirculations and additional pressure losses. Similarly, Fig. 5.14(b) shows the flow in an elbow where the turning streamlines separate and create additional blockage and pressure loss. To accommodate such losses in the pressure-drop calculations, a loss coefficient must be defined.

Let us start with Darcy's formula; it was already pointed out that we can call the pressure loss a "head loss," h_f. Rewriting Eq. (5.99) we can see that the head loss measures the loss in terms of the pipe flow's liquid-column height:

$$h_f = \frac{\Delta p}{\rho g} = f \frac{L}{D} \frac{U_{\text{av}}^2}{2g}. \tag{5.104}$$

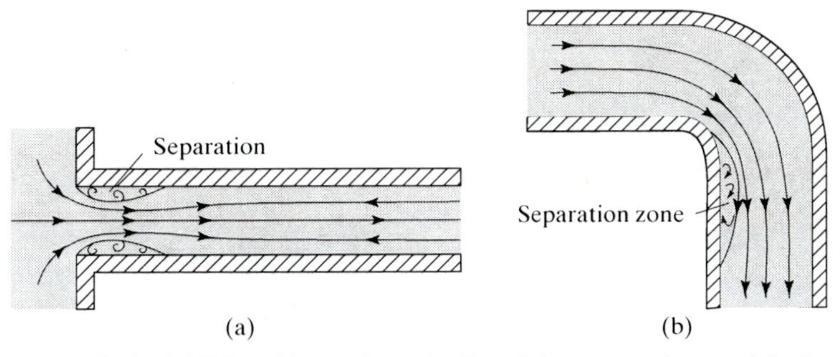

Figure 5.14. Additional losses in a pipeline: (a) entrance losses, (b) elbow losses.

Consequently the loss coefficients K must have the same units, and the head loss that is due to an elbow or entrance will be estimated by

$$h_f = \frac{\Delta p}{\rho g} = K \frac{U_{av}^2}{2g}. \tag{5.105}$$

A table showing the loss coefficient K values for various conditions is depicted in Fig. 5.15. Note that these coefficients depend on the geometry; for example, elbows with different diameters or turn radii will have different loss coefficients.

In summary, for a long pipeline (with the same diameter) but with elbows and other fittings, the total pressure loss (or head loss) can be summarized as

$$h_f = \frac{\Delta p}{\rho g} = \left(f \frac{L}{D} + \sum K \right) \frac{U_{av}^2}{2g}, \tag{5.106}$$

where the first term inside the parentheses is the friction inside the pipe and $\sum K$ is the sum of all the additional losses.

5.9 Summary of One-Dimensional Pipe Flow

Although we started this chapter with "exact solutions" in mind, Darcy's formula and the Bernoulli equation provide a simple 1D model for calculating the incompressible flow in pipes. The method of using these equations is depicted schematically in Fig. 5.16.

In principle we can write the Bernoulli equation [Eq. (4.7)] between two points, as shown. However if there are losses ($\sum h_f$) in the flow we can add those into the equation: as follows

$$z_1 + \frac{p_1}{\rho g} + \frac{u_1^2}{2g} = z_2 + \frac{p_2}{\rho g} + \frac{u_2^2}{2g} + \sum h_f; \tag{5.107}$$

the losses are summarized by Eq. (5.106):

$$h_f = \frac{\Delta p}{\rho g} = \left(f \frac{L}{D} + \sum K \right) \frac{U_{av}^2}{2g}. \tag{5.106}$$

Note that in terms of velocity we see here three variables (e.g., u_1, u_2, and U_{av} – if pipe-segment diameters are not equal), therefore, additional equations may be

Pipe entrance

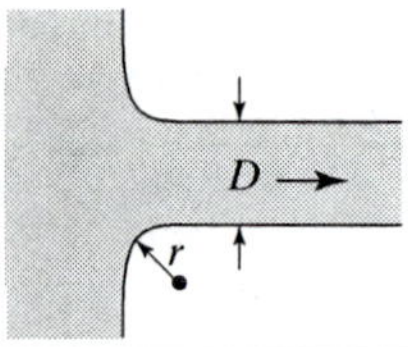

r/D	K
0.0	0.50
0.1	0.12
>0.2	0.03

Contraction

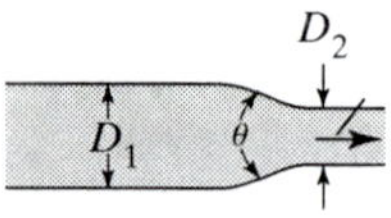

D_2/D_1	K $\theta = 60°$	K $\theta = 180°$
0.20	0.08	0.49
0.40	0.07	0.42
0.60	0.06	0.27
0.80	0.06	0.20
0.90	0.06	0.10

Expansion

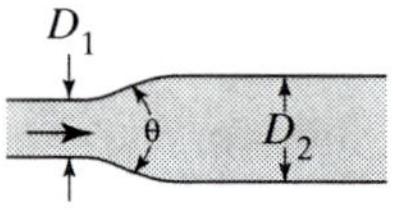

D_1/D_2	K $\theta = 20°$	K $\theta = 180°$
0.20	0.30	0.87
0.40	0.25	0.70
0.60	0.15	0.41
0.80	0.10	0.15

90° sharp bend (Turning vanes)

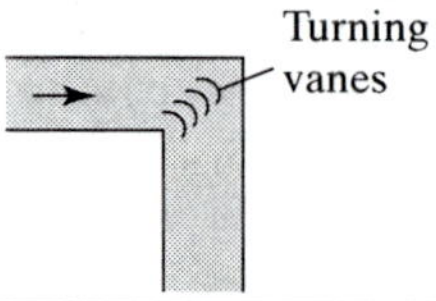

	K
No vanes	$K = 1.1$
With vanes	$K = 0.2$

90° smooth bend

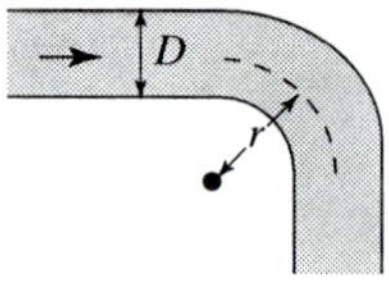

r/D	K
1	0.35
2	0.19
4	0.16
6	0.21
8	0.28
10	0.32

Gate valve

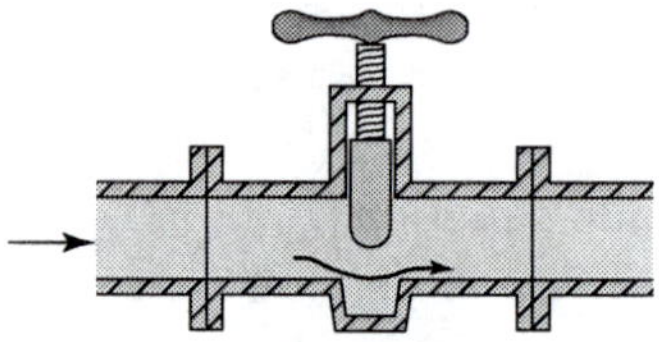

	K
full-open	0.20
half-open	2.20

Globe valve

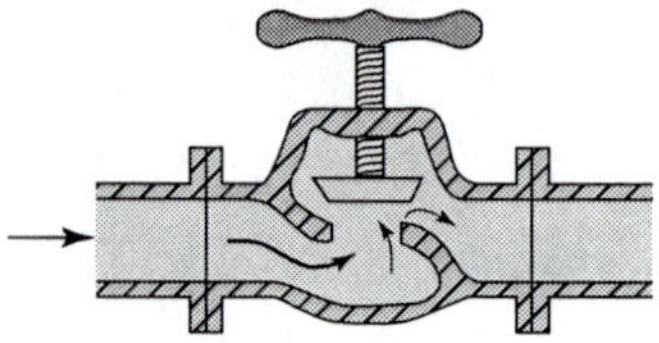

	K
full-open	10.00
half-open	5.60

Figure 5.15. Loss coefficients for various pipe-related fittings.

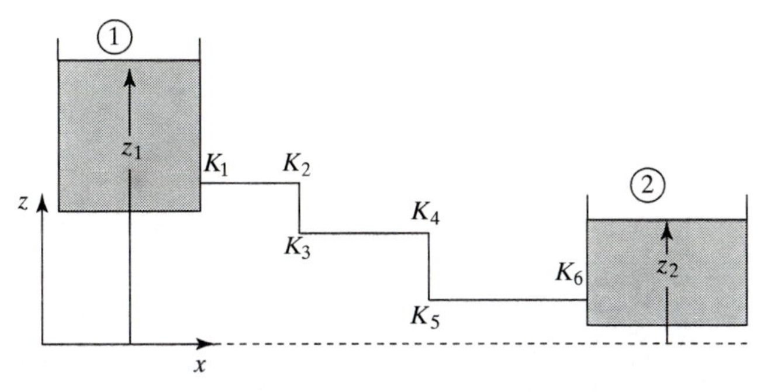

Figure 5.16. 1D model for pipe flow calculations.

needed (e.g., the continuity equation). Several examples to demonstrate the application of these formulas are presented later in this section.

Another important observation highlighting the simplicity of Darcy's equation can be demonstrated by a simple water-tower example (see Fig. 5.17). Such water reservoirs are placed at the highest point in a neighborhood, ensuring a pressurized water supply to the users below. Assuming that the water level inside the reservoir (point 1) is not changing then the exit velocity at station 2, without losses, is given by the simple 1D model of Eq. (4.12):

$$u_2 = \sqrt{2g(z_1 - z_2)} = \sqrt{2gh},$$

and here we use h for the height difference. The losses along the pipeline that are due to the fluid flow can be summarized by use of the head-loss term h_f as given in Eq. (5.106). Therefore a simple estimate for the average velocity in the pipe at point 2 can be obtained by reducing the available water column height for accelerating the fluid:

$$u_2 = \sqrt{2g(h - h_f)}. \tag{5.108}$$

This example not only depicts the meaning of the term *head loss* but it also highlights the ingenious formulation of Darcy's formula.

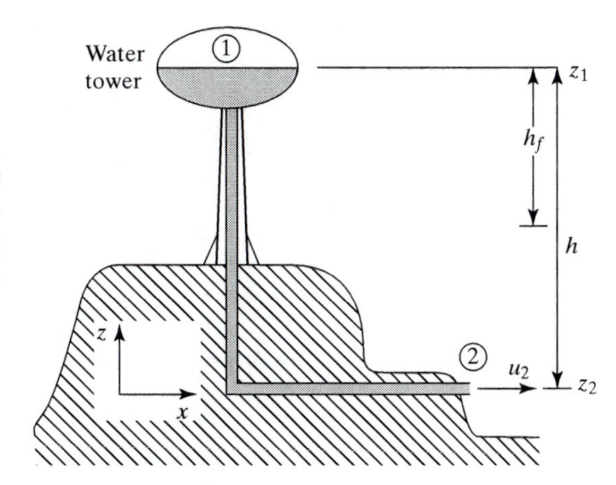

Figure 5.17. Schematic description of a water tower. By placing the water reservoir at the highest point in a neighborhood the problem of a pressurized water supply is solved.

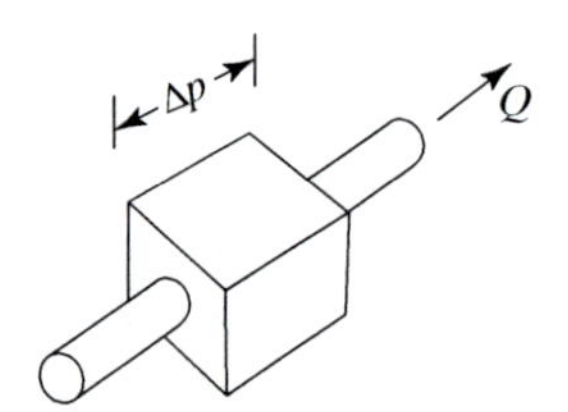

Figure 5.18. Simple pump model.

5.9.1 Simple Pump Model

Because we are dealing with pipelines, one important engineering question relates to the power required to pump the fluid. Let us examine a simple incompressible case in which the pump creates a pressure jump Δp and the flow rate is Q (see Fig. 5.18).

In general, the definition of work W is force F times distance l. But the force is equal to the pressure difference times the cross-section area of the pipe ($F = \Delta p S$) and the work then is the pressure times the fluid volume V.

$$W = F \times l = \Delta p S \times l = \Delta p \times V,$$

The power is the work per unit time and is measured in watts (W):

$$\text{Power} = \frac{d}{dt} W = \Delta p \frac{d}{dt} V = \Delta p Q,$$

and here Q is the volumetric flow rate. In conclusion, the power required to move the fluid is

$$\text{Power} = \Delta p \times Q. \tag{5.109}$$

If the pump efficiency is denoted as η, the power supplied to the pump is

$$\text{Power}_{\text{to pump}} = \frac{\Delta p \times Q}{\eta}. \tag{5.110}$$

Sometimes, the pump performance is measured as a *head gain* h_{pump}, which, based on the preceding relation, becomes

$$h_{\text{pump}} = \frac{\Delta p}{\rho g} = \eta \frac{\text{Power}_{\text{pump}}}{\rho g Q}.$$

With this definition, Eq. (5.107) can be modified to account for the pump head gain:

$$z_1 + \frac{p_1}{\rho g} + \frac{u_1^2}{2g} \pm h_{\text{pump}} = z_2 + \frac{p_2}{\rho g} + \frac{u_2^2}{2g} + \sum h_f, \tag{5.111}$$

and the $\pm$ sign depends on the direction the pump is operating (the same applies to the $\sum h_f$ term). A more detailed discussion on pumps and their internal flows is presented in Chapter 11.

5.9.2 Flow in Pipes with Noncircular Cross Sections

Based on Eq. (5.98), the formula for the friction factor f can be modified to include the flow in pipes with different cross sections. For the laminar case we can do this

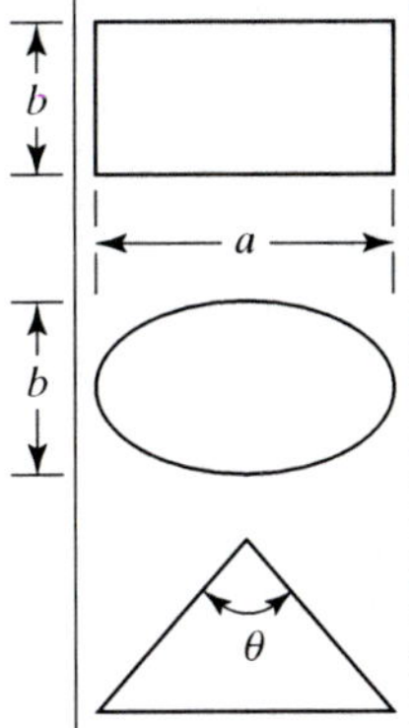

Cross-section shape		C_K
Rectangle (sides a, b)	$\frac{a}{b} =$ 1	56.9
	2	62.2
	4	72.9
	8	82.3
	16	93.0
Ellipse (axes a, b)	$\frac{a}{b} =$ 1	64.0
	2	67.3
	4	72.9
	8	76.6
	16	78.2
Triangle (apex angle θ)	$\theta =$ 10°	50.8
	30°	52.3
	60°	53.3
	90°	52.6
	120°	51.0

Figure 5.19. Friction factors for several generic pipe cross sections for fully developed laminar flows.

by simply replacing the number 64 in the numerator with a constant C_K (and for circular pipes $C_K = 64$). The modified formula then becomes

$$f = \frac{C_K}{\text{Re}}. \tag{5.112}$$

However, the Reynolds number is now based on the hydraulic diameter D_h and the hydraulic diameter is based on the wetted perimeter P_h and the cross-section area S:

$$D_h = \frac{4S}{P_h}. \tag{5.113}$$

With these definitions the Reynolds number becomes

$$\text{Re} = \frac{\rho U_{\text{av}} D_h}{\mu}. \tag{5.114}$$

Of course, for a circle the hydraulic diameter remains the circle's diameter and for a pipe with a square cross section the hydraulic diameter remains its height.

With these modifications and based on the geometry of the pipe cross section, the friction factor f for the cases shown in Fig. 5.19 can be estimated (for fully developed laminar flows).

For example, the hydraulic diameter for a rectangular cross-section pipe where $a = 2b$ is

$$D_h = \frac{4S}{P_h} = \frac{4 \times a \times 2a}{2(a + 2a)} = \frac{4}{3}a.$$

To estimate the pressure drop for high-Reynolds-number turbulent flows in noncircular pipes, Darcy's formula [Eq. (5.100)] can still be used. However, in this case the hydraulic diameter based on Eq. (5.113) must be used to estimate the friction factor from the Moody diagram.

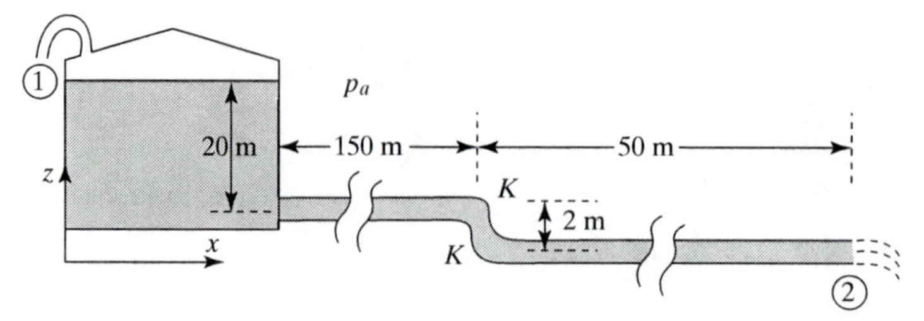

Figure 5.20. One dimensional flow in a long pipe line.

5.9.3 Examples for One-Dimensional Pipe Flow

EXAMPLE 5.7. HEAD LOSS IN AN ELBOW. The purpose of this example is to demonstrate the use of the head loss formulation to determine the corresponding pressure loss and associated units. Let us consider a 6-L/min flow of water in a 3-cm-diameter pipe that has a 90° elbow. Calculate the head loss that is due to the elbow that has a loss coefficient: $K = 0.32$.

Solution: The head loss is calculated by Eq. (5.105), but the average velocity in the pipe must be calculated first:

$$U_{\mathrm{av}} = \frac{Q}{S} = \frac{6/60 \times 10^{-3}\ \mathrm{m}^3/\mathrm{s}}{\pi 0.015^2\ \mathrm{m}^2} = 0.141\frac{\mathrm{m}}{\mathrm{s}}.$$

Now we can use Eq. (5.105) to calculate the head loss:

$$h_f = K\frac{U_{\mathrm{av}}^2}{2g} = 0.32\frac{0.141^2}{2 \times 9.8} = 3.24 \times 10^{-4}\ \mathrm{m}.$$

The unit of the head loss is in the length (meters in this case) or height of the fluid inside the pipeline. To return to pressure units we must look at the left-hand side of Eq. (5.105):

$$h_f = \frac{\Delta p}{\rho g}.$$

Consequently the pressure drop is

$$\Delta p = h_f \rho g = 3.24 \times 10^{-4} \times 1000 \times 9.8 = 3.18\frac{\mathrm{N}}{\mathrm{m}^2}.$$

EXAMPLE 5.8. PRESSURE LOSSES IN A LONG PIPE. Kerosene is flowing through the pipeline shown in Fig. 5.20. The fluid level in the main tank is 20 m above the exit and along the line there is an additional elevation drop of 2 m, as shown. The pipe inner diameter is 0.3 m and the inside resistance coefficient is $f = 0.015$. Assume a loss coefficient for the two elbows at $K = 0.3$, and the properties of kerosene are $\rho = 804\ \mathrm{kg/m^3}$ and $\mu = 1.9 \times 10^{-3}\ \mathrm{N\ s/m^2}$. Based on this, calculate

(a) The average discharge velocity at the end of the pipe.
(b) The Re of the flow.

Solution: Let us use Eq. (5.107) for this case. The coordinate system is set at the lowest elevation of the pipeline; the upper level of the fluid in the tank is considered as point 1 and the pipe exit at the bottom is point 2. The velocity at point 1 is zero and the pressure at both points is ambient (e.g., $p_1 = p_2 = p_a$ because at point 2 the pressure is evaluated outside the pipe); therefore the pressure terms cancel and Eq. (5.107) becomes

$$(20+2) + \frac{p_a}{\rho g} + \frac{0}{2g} = 0 + \frac{p_a}{\rho g} + \frac{u_2^2}{2g} + \sum h_f. \qquad (*)$$

So if the flow has no friction, the potential energy ($\rho g z_1$) will be converted into velocity at point 2; however, the friction in this case will reduce the exit velocity. Next we must calculate the friction losses in the pipe and in the elbows:

$$\sum h_f = f\frac{L}{D}\frac{u_2^2}{2g} + 2K\frac{u_2^2}{2g}.$$

Substituting this into (*) we get

$$22 = \frac{u_2^2}{2 \times 9.8}\left(1 + 0.015\frac{202}{0.3} + 2 \times 0.3\right).$$

Note that the units are meters – because the calculation is for head loss! Basically we can solve now for the velocity, and we get

$$u_2 = 6.07\frac{\text{m}}{\text{s}}.$$

The corresponding flow rate is

$$Q = u_2 S = 6.07 \times \pi \times 0.15^2 = 0.429\frac{\text{m}^3}{\text{s}}.$$

In this case the friction coefficient was given (if not, we have an iterative process with the Moody diagram). In both cases it is desirable to calculate Re (and to see if the friction factor f selection is reasonable):

$$\text{Re} = \frac{804 \times 6.07 \times 0.3}{1.9 \times 10^{-3}} = 0.773 \times 10^6,$$

and from Fig. 5.13 it appears that the friction coefficient selection was reasonable. Also, now that the average velocity was calculated it is interesting to evaluate $\sum h_f$ and demonstrate the meaning of *head loss*:

$$\sum h_f = f\frac{L}{D}\frac{u_2^2}{2g} + 2K\frac{u_2^2}{2g} = \left(0.015\frac{202}{0.3} + 2 \times 0.3\right)\frac{6.07^2}{2 \cdot 9.8} = 20.12 \text{ m}.$$

This means that, instead of the potential height of 22 m, only 1.88 m is available at the end of the pipeline (in the form of jet kinetic energy)! Thus we can use the ideal flow of Eq. (4.12) to calculate the exit velocity,

$$u_2 = \sqrt{2gh} = \sqrt{2 \times 9 \times 8 \times 1.88} = 6.07 \text{ m/s},$$

and this is the same result!

EXAMPLE 5.9. LAMINAR FLOW IN A VERTICAL PIPE. In this example we combine the friction formula with Eq. (5.107), resulting in a quadratic equation: Consider the vertical flow of glycerin at 20 °C down an $h_2 = 0.2$-m-long circular tube

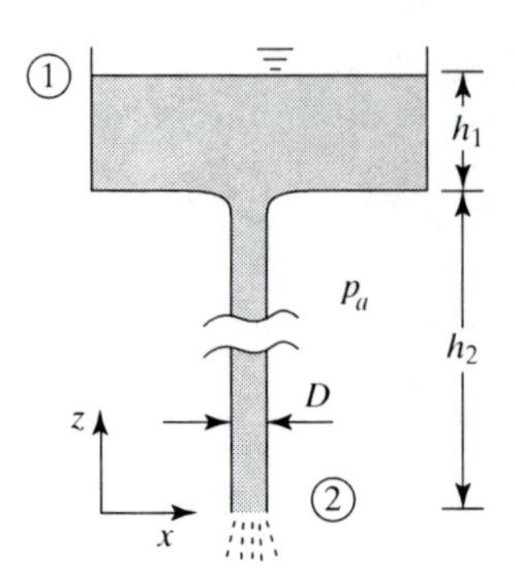

Figure 5.21. Laminar flow in a vertical pipe.

of diameter $D = 0.01$ m, as shown in Fig. 5.21. The liquid height in the large container is $h_1 = 0.1$ m. Estimate the average discharge velocity at the bottom. (From Table 1.1 for glycerin we get $\rho = 1254$ kg/m^3 and $\mu = 0.62$ N s/m^2.)

Solution: If we place our coordinate system at the bottom exit then the liquid column height is 0.3 m. Again we consider point 1 at the top of the liquid (at $z_1 = 0.3$ m) and point 2 at the exit (a small distance after the fluid left the pipe) and the ambient pressures are the same. Writing Eq. (5.107) for this case results in

$$0.3 + \frac{p_a}{\rho g} + \frac{0}{2g} = 0 + \frac{p_a}{\rho g} + \frac{u_2^2}{2g} + \sum h_f.$$

Assuming laminar flow (which must be verified at the end), we can use Eq. (5.98) for the pressure loss:

$$f = \frac{64}{\text{Re}}.$$

Now we can calculate the head loss as

$$\sum h_f = f \frac{L}{D} \frac{u_2^2}{2g} = \frac{64}{\text{Re}} \frac{L}{D} \frac{u_2^2}{2g} = \frac{64\mu}{\rho} \frac{L}{D^2} \frac{u_2}{2g}.$$

Substituting the head loss into Eq. (5.107) results in a quadratic equation:

$$0.3 = \frac{u_2^2}{2 \times 9.8} + \frac{32 \times 0.62}{1254} \frac{0.2}{0.01^2} \frac{u_2}{9.8}.$$

Solving for the exit velocity we get

$$u_2 = 0.93 \text{ m/s}.$$

Now, to make sure that we have a laminar flow, let us check the Re:

$$\text{Re} = \frac{1254 \times 0.93 \times 0.01}{0.62} = 18.8.$$

The Re is much less than 2000 so clearly the flow must be laminar.

EXAMPLE 5.10. FLOW IN A PIPE WITH TWO DIFFERENT DIAMETERS. Water is flowing from a container through the pipeline, as shown in Fig. 5.22. The fluid level in the main tank is 20 m above the exit and the pipe diameter is 0.3 m for the first 50 m. At this point, the diameter is reduced to 0.2 m and the loss coefficient for the contraction is $K = 0.1$. The smaller-diameter pipeline is 100 m long and at

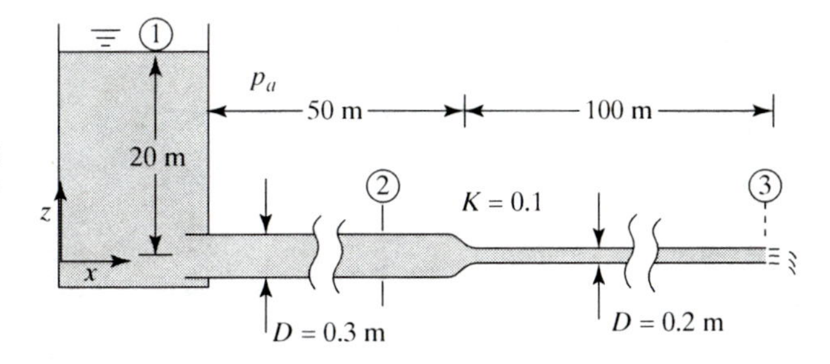

Figure 5.22. 1D flow in a long pipeline with a change in diameter.

the end the water is released. Assume the same friction factor of $f = 0.015$ for the two pipe segments and the properties of kerosene are taken from Table 1.1. Calculate the average discharge velocity at the end of the pipe.

Solution: Let us again use Eq. (5.107) for this case. The coordinate system is set at the lowest elevation of the pipeline; the upper level of the fluid in the tank is considered to be point 1 and the pipe exit at the bottom is point 3. The velocity at point 1 is zero, and the pressure at both points is ambient (e.g., $p_1 = p_3 = p_a$). Because of the change in pipe diameter, station 2 is added in the larger-diameter pipe section. Writing Eq. (5.107) between point 1 and point 3 yields

$$20 + \frac{p_a}{\rho g} + \frac{0}{2g} = 0 + \frac{p_a}{\rho g} + \frac{u_3^2}{2g} + \sum h_f.$$

We calculate the friction losses $\sum h_f$ by using Eq. (5.106):

$$\sum h_f = f \frac{L_2}{D_2}\frac{u_2^2}{2g} + K\frac{u_2^2}{2g} + f\frac{L_3}{D_3}\frac{u_3^2}{2g}.$$

At this point we use the continuity equation between point 2 and point 3

$$\rho u_2 S_2 = \rho u_3 S_3.$$

Knowing the pipe diameters, we can write

$$u_2 = u_3 \frac{S_3}{S_2} = u_3 \left(\frac{0.2}{0.3}\right)^2.$$

Substituting the friction losses and the velocity ratio into the Bernoulli equation results in

$$20 = \frac{u_3^2}{2 \times 9.8}\left[1 + 0.015\frac{50}{0.3}\left(\frac{0.2}{0.3}\right)^4 + 0.1\left(\frac{0.2}{0.3}\right)^4 + 0.015\frac{100}{0.2}\right].$$

As noted earlier, the units of this equations are in meters. Solving for the velocity, we get

$$u_3 = 6.59 \text{ m/s}$$

The corresponding flow rate is

$$Q = u_3 S_3 = 6.59 \times \pi \times 0.1^2 = 0.20 \text{ m/s}.$$

Let us calculate Re to verify that the flow is within the turbulent range and that the estimate of f is reasonable. The density and viscosity are taken from

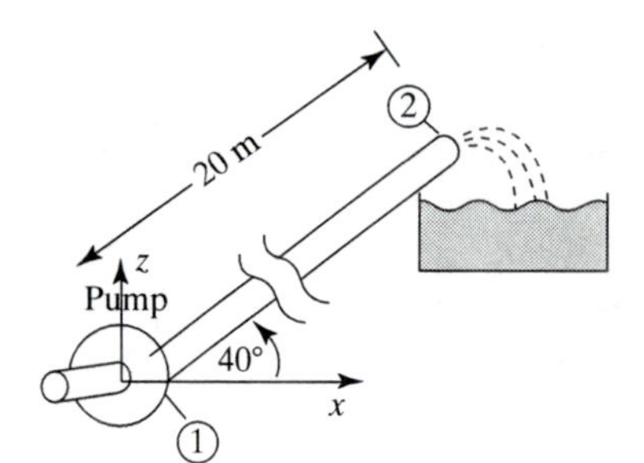

Figure 5.23. Power and pressure requirement for uphill pumping.

Table 1.1 as $\rho = 1000$ kg/m^3 and $\mu = 1.0 \times 10^{-3}$ N s/m^2:

$$\text{Re} = \frac{1000 \times 6.73 \times 0.2}{1.0 \times 10^{-3}} = 1.346 \times 10^6,$$

and clearly this is within the turbulent flow range and the friction factor selection is reasonable.

EXAMPLE 5.11. PUMP POWER REQUIREMENTS. Consider a 20-m-long, 6-cm-diameter pipe discharging into an open container, as shown in Fig. 5.23. The pipe inclination is 40°. The desirable flow rate is 7.63 L/s, and the liquid (oil) properties are $\rho = 900$ kg/m^3 and $\mu = 0.18$ N s/m^2. Calculate the power required to pump the liquid.

Solution: From Eq. (5.109), we need to calculate the pressure drop in the pipe. Let us select point 1 at the pump exit and set $z_1 = 0$ there. Next, place point 2 at the discharge exit of the pipe (as shown in the figure). Because the required flow rate and the pipe diameter are known, we can calculate the average velocity:

$$u_1 = u_2 = \frac{Q}{S} = \frac{7.63 \times 10^{-3}}{\pi 0.03^2} = 2.7 \text{ m/s}.$$

Next we can calculate the Reynolds number so that we can select the proper friction coefficient (and evaluate if it is laminar or turbulent):

$$\text{Re} = \frac{900 \times 2.7 \times 0.06}{0.18} = 810.$$

This is within the laminar region, and we can use the laminar flow formula:

$$f = \frac{64}{\text{Re}} = 0.079.$$

The head loss in the pipe is therefore

$$h_f = f \frac{L}{D} \frac{u_2^2}{2g} = 0.079 \frac{20}{0.06} \frac{2.7^2}{2 \times 9.8} = 9.79 \text{ m}.$$

Now we can return to Eq. (5.107):

$$0 + \frac{p_1}{\rho g} + \frac{u_1^2}{2g} = 20 \sin 40 + \frac{p_2}{\rho g} + \frac{u_2^2}{2g} + 9.79.$$

But $u_1 = u_2$, and we get

$$\frac{p_1 - p_2}{\rho g} = 20 \sin 40 + 9.79 = 22.65 \text{ m};$$

after multiplying by ρg, we get

$$\Delta p = p_1 - p_2 = 22.65 \times 900 \times 9.8 = 199{,}773\,\mathrm{N/m^2},$$

which is about 2 atm. Now we can calculate the power requirement:

$$\text{Power} = \Delta p \times Q = 199{,}773 \times 7.63 \times 10^{-3} = 1524.3\,\mathrm{W},$$

which is close to 2 hp.

5.9.4 Network of Pipes

Complex networks of pipes are used in places such as chemical plants or water distribution or treatment plants. A typical engineering requirement is to estimate the pressure loss or the flow rates in the pipe network. To demonstrate the generic approach for solving such problems let us use the simple 1D pipe flow model developed in Section 5.6. We can use Darcy's formula [Eq. (5.99)] and apply it to the simple network shown in Fig. 5.24.

Let us assume incompressible fluid and that the input pressure p_1 and the exit pressure p_0 are known. The diameters D_i and lengths L_i of the four segments are also known. The problem now is to find the flow rates in each of the pipe branches. We can apply Darcy's formula for the four branches as follows (let us simplify and assume the same friction factor f for all pipes):

$$\Delta p_i = \rho f \frac{L_i}{D_i} \frac{u_i^2}{2}, \quad i = 1, 2, 3, 4. \tag{5.115}$$

At this point the four velocities and the resulting pressure drops are unknown. Three equations can be constructed, stating the total pressure drop in the system:

$$\begin{aligned} p_1 - p_0 &= \Delta p_1 + \Delta p_2, \\ p_1 - p_0 &= \Delta p_1 + \Delta p_3, \\ p_1 - p_0 &= \Delta p_1 + \Delta p_4. \end{aligned} \tag{5.116}$$

An additional equation is based on the continuity equation, stating that the inflow in pipe 1 is equal to the flow leaving through the three branches:

$$Q_1 = Q_2 + Q_3 + Q_4. \tag{5.117}$$

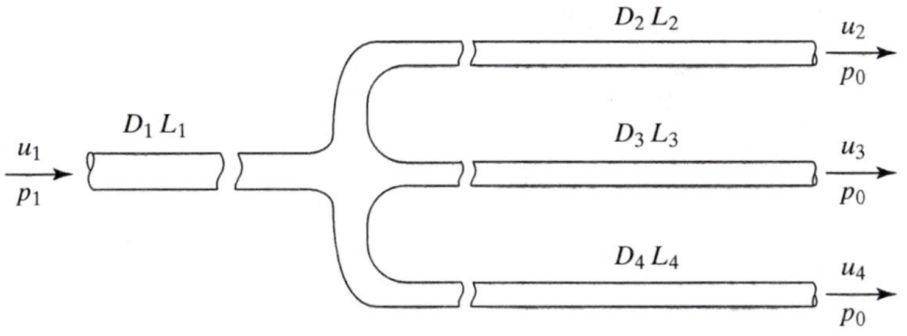

Figure 5.24. A simple network of pipes.

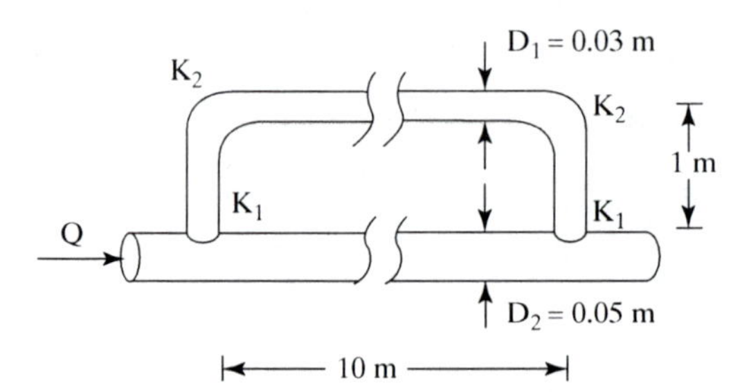

Figure 5.25. Flow in two parallel pipes.

The flow rate is related to the average velocity as

$$Q_i = \frac{\pi D_i^2}{4} u_i, \quad i = 1, 2, 3, 4, \tag{5.118}$$

and the four equations for the four unknown velocities are

$$\begin{aligned} \frac{2(p_1 - p_0)}{\rho f} &= \frac{L_1}{D_1} u_1^2 + \frac{L_2}{D_2} u_2^2, \\ \frac{2(p_1 - p_0)}{\rho f} &= \frac{L_1}{D_1} u_1^2 + \frac{L_3}{D_3} u_3^2, \\ \frac{2(p_1 - p_0)}{\rho f} &= \frac{L_1}{D_1} u_1^2 + \frac{L_4}{D_4} u_4^2, \\ D_1^2 u_1 &= D_2^2 u_2 + D_3^2 u_3 + D_4^2 u_4. \end{aligned} \tag{5.119}$$

In principle these four equations can be solved for the four velocities u_i. Usually the number of pipe segments is very large and (iterative) numerical techniques are used to solve the system of equations.

EXAMPLE 5.12. PARALLEL FLOW IN TWO PIPES. The simplest example for a network of pipes is when the flow splits between two pipes, as shown in the Fig. 5.25. Assume water at a flow rate of 5 L/min entering a 10-m-long, and 5-cm diameter pipe at the left. The pipe then splits into two, as shown, but the upper pipe diameter is only 3 cm. The total length of the upper pipe $L_1 = 12$ m, the entrance/exit losses are $K_1 = 0.9$, and the loss coefficients in the two bends are $K_2 = 0.2$. Assuming $f = 0.025$ for both pipes, calculate the flow rates for the upper and lower pipes.

Solution: The pressure drop along the two pipes (between the two junctions) is the same. Therefore the pressure loss (in terms of the head loss) is calculated for the two segments. The head loss in the upper pipe is

$$h_u = \left(f \frac{L_u}{D_u} + 2K_1 + 2K_2 \right) \frac{u_u^2}{2g},$$

and the head loss in the lower pipe is

$$h_L = f \frac{L_L}{D_L} \frac{u_L^2}{2g}.$$

Since the pressure drop is the same, we can compare these two equations:

$$\left(f \frac{L_u}{D_u} + 2K_1 + 2K_2 \right) \frac{u_u^2}{2g} = f \frac{L_L}{D_L} \frac{u_L^2}{2g}.$$

Solving for the velocity ratio, we get

$$\frac{u_u^2}{u_L^2} = \frac{f\dfrac{L_L}{D_L}}{f\dfrac{L_u}{D_u} + 2K_1 + 2K_2} = \frac{0.025\dfrac{10}{0.05}}{0.025\dfrac{12}{0.03} + 2 \times 0.9 + 2 \times 0.2} = 0.4098,$$

and the velocity ratio is

$$\frac{u_u}{u_L} = 0.64.$$

The two flow rates are equal to the incoming flow of 5 L/s:

$$\pi 0.025^2 u_L + \pi 0.015^2 u_u = 5 \times 10^{-3} \text{ L/s}.$$

Solving for the velocity and the flow rate in the lower pipe, we get:

$$u_L = 2.07 \text{ m/s},$$

$$Q_L = \pi 0.025^2 u_L = 4.06 \text{ L/s}.$$

5.10 Open Channel Flows

Liquids flowing in open channels can be found in rivers, irrigation canals, and in numerous industrial processes. The main difference between such flows and the flow inside enclosed pipes is that flow rates cannot be increased by simply applying an arbitrary pressure gradient. Moreover, in a typical case the flow is driven by gravity (elevation change) and there is no pressure gradient.

Exact solutions of the flow details are significantly complicated by the free-surface effects. Over the years, instead of complex multidimensional exact solutions, a much simpler but practical approach evolved, based on the 1D pipe flows discussed earlier in this chapter.

The objective of this section is to familiarize the reader with the basic concepts of open channel flow (at an introductory level). Therefore the surface-wave concept is introduced first, resulting in a classification of such flows. Next the uniform flow case (which is somewhat similar to the pipe flow) is demonstrated. Finally, some complex flows such as those found in hydraulic jumps and weirs are discussed, again using overly simplified first-order approximations.

5.10.1 Simple Models for Open Channel Flows

The first task at this point is to simplify the open channel flow model without losing the dominant physics, so that practical engineering solutions can be obtained. One approach is to apply the methodology used earlier in this chapter, leading to simple "one-dimensional" models. The circular pipe flow problem of Section 5.7 provided a smooth transition between the 2D axisymmetric flow and the 1D model relying on experimental friction coefficients. Before the same approach is adapted, some complications in the case of open channel flow are discussed. Of course, it is clear that parameters such as the channel slope, the cross section, the shape (top view),

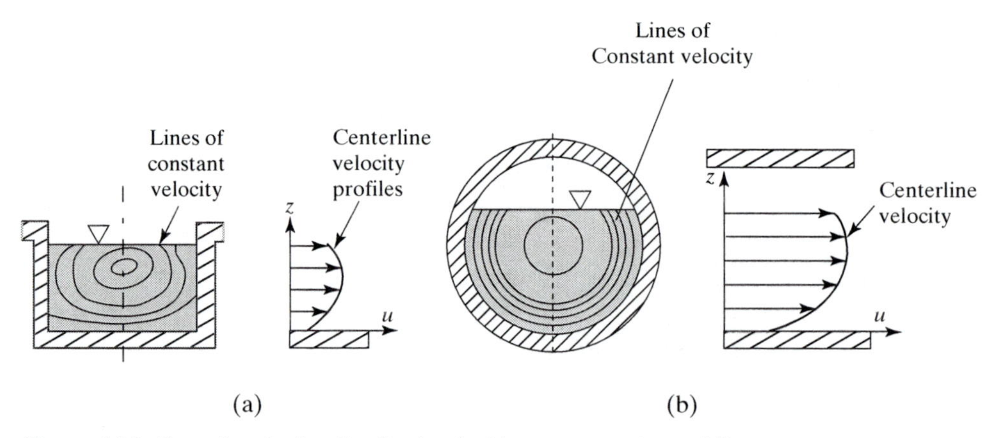

Figure 5.26. Generic velocity distribution inside two open channel flows.

or the wall's surface roughness will have a strong effect on engineering quantities such as flow speed and flow rate. However, even a simple 2D velocity profile cannot be established by closed-form solutions. Consequently, from the 2D solution of the laminar flow inside pipes (Section 5.7) we can speculate about the expected velocity profiles, as shown in Fig. 5.26. Here rectangular and circular cross-section channels are shown, and when the liquid level is quite high (as shown) the velocity profiles are similar. Those velocity profiles (based on experimental results) show that the maximum velocity is not necessarily at the free surface, and the velocity distribution also depends on the level of the liquid in the channel.

These generic velocity distributions (shown in Fig. 5.26) can be further complicated by 3D effects resulting from complex cross-section shapes or rapid turns in channel direction. Even when significant simplifications are introduced, such as 1D flow, the problem remains complicated and empirical data are used for practical engineering solutions (see [1–3]).

When open channel flows are compared with pipe flows, one additional complication results from the fact that the open surface shape may not be known. Let us start with a simple example, investigating the speed at which a small surface perturbation moves. This could be viewed as an approximation for surface-wave velocity.

Assume that a liquid column of height h is flowing at a velocity u, as shown in Fig. 5.27. Because of a downstream disturbance, such as a wall or any other obstruction (far to the right), a small discontinuity Δh in the liquid-column height is observed. A control volume is attached to this traveling surface wave so that the sketch in Fig. 5.27 appears to be stationary. For the simplest model we can use the 1D continuity and momentum (or Bernoulli) equations. It is also clear that, for an

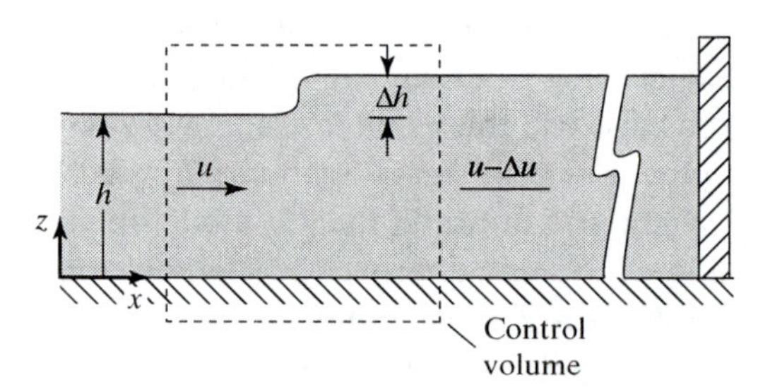

Figure 5.27. Simple model for a surface-wave propagation over shallow water.

incompressible liquid, the exit velocity will be lower by Δu. For example, the continuity equation [Eq. (2.26)] applied to the incoming and the exiting flows is

$$\rho u h = \rho(u - \Delta u)(h + \Delta h), \tag{5.120}$$

and the Bernoulli equation [Eq. (4.3)] is

$$h + \frac{u^2}{2g} = (h + \Delta h) + \frac{(u - \Delta u)^2}{2g}. \tag{5.121}$$

Our objective is to solve for u, which is the perturbation propagation velocity. From Eq. (5.120) we get

$$\Delta u = \frac{u \Delta h}{h + \Delta h} \approx \frac{u \Delta h}{h}, \tag{5.122}$$

and because $\Delta h \ll h$, Δh is neglected. Similarly, by expanding the square term on the right-hand side of Eq. (5.121) we get

$$\frac{u^2}{2g} = \Delta h + \frac{u^2 - 2u\Delta u + \Delta u^2}{2g}.$$

Again $\Delta u \ll u$ and we get

$$\frac{u^2}{2g} \approx \Delta h + \frac{u^2}{2g} + \frac{2u\Delta u}{2g}.$$

Solving for Δu we get

$$\Delta u = \frac{\Delta h g}{u}. \tag{5.123}$$

Comparing Δu in Eqs. (5.122) and (5.123) results in

$$u^2 = gh. \tag{5.124}$$

The velocity of this surface disturbance is similar to the speed of sound, and we use the notation c instead (e.g., $c \equiv u$). So the surface-wave speed is

$$c = \sqrt{gh} \tag{5.125}$$

This is a simple approximation; a more accurate solution taking the wavelength into account is presented in [4, Chapter 6]. Now that the surface disturbance was developed, we can consider other open channel flows with an average velocity of u. The ratio between this flow speed u and the speed of the surface-wave velocity c is called the Froude number, Fr (after William Froude, 1810–1879, an English hydrodynamicist):

$$\mathrm{Fr} = \frac{u}{c} = \frac{u}{\sqrt{gh}}. \tag{5.126}$$

This number plays an important role in open channel flows. For example, if the flow is slower than c then perturbations such as surface waves can travel upstream. This condition is called subcritical. If the average speed u is equal to c (called critical condition) then a standing wave is observed, and no information can pass from the elevated side of the flow (in Fig. 5.27). In the case in which the flow is much faster

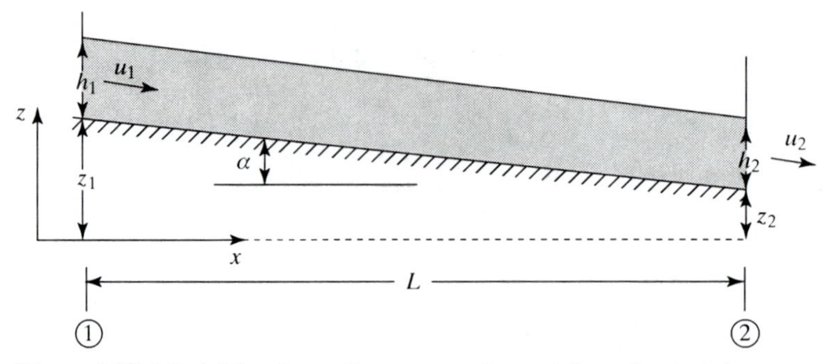

Figure 5.28. Model for the uniform open channel flow, flowing down a moderate slope.

than c, then clearly no wave can travel upstream and therefore the flow is called supercritical. Using Fr, we can categorize the flow regimes as

$$\begin{aligned} &\mathrm{Fr} < 1, \quad \text{subcritical flow,} \\ &\mathrm{Fr} = 1, \quad \text{critical flow,} \\ &\mathrm{Fr} > 1, \quad \text{supercritical flow.} \end{aligned} \tag{5.127}$$

The first example deals with the most common form of channel flow in which liquid such as water is flowing slowly because of gravity. In most cases the conditions are subcritical and $\mathrm{Fr} < 1$. On the other hand, if the flow is very fast, and $\mathrm{Fr} > 2$ (supercritical), a hydraulic jump is observed; this case will be discussed later.

5.10.2 Uniform Open Channel Flows

One of the simplest open channel models is the 1D flow, down a moderate grade, driven by gravitational forces. Most irrigation canals or rivers can be included within this category. It is assumed that there are no sudden slope changes resulting in free-surface irregularities or even waterfalls. The flow is fully developed, the channel cross section is fixed, and therefore the average velocity is also unchanged. As noted, instead of the complex velocity profiles (see Fig. 5.26), a uniform (1D) velocity is assumed. This average velocity assumption was also used for the calculation of the pressure drop in pipe flows and allowed the inclusion of experimental data for a wide range of Reynolds numbers (Section 5.6). With the preceding assumptions the following 1D model is proposed and the corresponding nomenclature is depicted in Fig. 5.28.

Here the side view of the channel of length L is shown in which the liquid is flowing from the left (station 1) to the right (station 2) because of the slope α, as shown. The liquid depth in the channel may vary (at this point) and is represented by h_1 and h_2, respectively. From the continuity equation we can write that the flow rate Q is the same at both sections:

$$Q = u_1 S_1 = u_2 S_2, \tag{5.128}$$

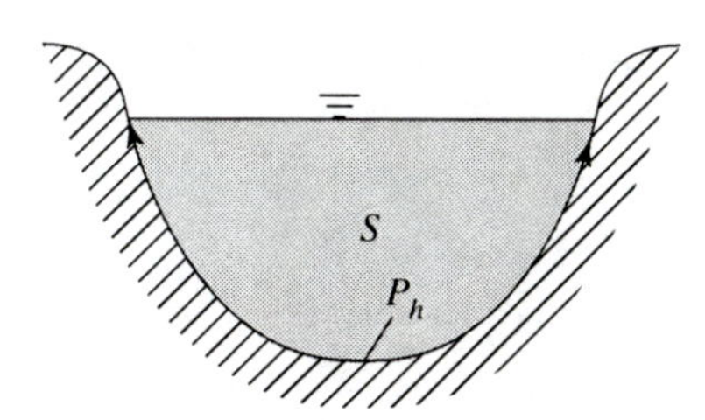

Figure 5.29. Nomenclature used to define the hydraulic radius.

where S_1 and S_2 are channel cross-section areas. Next, the Bernoulli equation, Eq. (5.107), can be modified for this model and because we assume open-air conditions the outside pressure is unchanged between section 1 and section 2:

$$h_1 + z_1 + \frac{u_1^2}{2g} = h_2 + z_2 + \frac{u_2^2}{2g} + \sum h_f, \tag{5.129}$$

and here $\sum h_f$ represents the losses (e.g., friction on the walls) that are due to the flow. In principle, we can use Darcy's formula [Eq. (5.99)] to estimate the losses; however, a slightly different form, based on the hydraulic radius R_h, is used for this problem. This term is clarified in Fig. 5.29, where S is the liquid cross-section area and P_h is the wetted perimeter (representing the friction between the fluid and the channel wall).

The hydraulic radius R_h is then defined as

$$R_h = \frac{S}{P_h}. \tag{5.130}$$

To use Darcy's formula, the fully wetted diameter D must be replaced with a quantity based on the hydraulic perimeter. A simple approach is to consider a rectangular pipe with a width of D. If the liquid is flowing inside the pipe, then the wetted perimeter $P_h = 4D$ and the cross-section area is $S = D^2$. The hydraulic radius for this case is then

$$R_h = \frac{D}{4}.$$

Now, recall Darcy's formula [Eq. (5.99)] and replace the pipe diameter with the hydraulic radius (e.g., $D = 4R_h$):

$$\sum h_f = \frac{\Delta p}{\rho g} = f \frac{L}{4R_h} \frac{U_{av}^2}{2g}. \tag{5.131}$$

At this point we can execute calculations similar to the pipe flow with friction. However, hydraulic engineers studying long irrigation channels and river flows usually follow a somewhat simpler model. In this case we assume that the flow cross-section area is the same and therefore, based on the continuity equation [Eq. (5.128)], the velocity is constant. Consequently Eq. (5.129) reduces to

$$z_1 = z_2 + \sum h_f. \tag{5.132}$$

Combining Eq. (5.132) with head-loss formula (5.131) and assuming that the friction coefficient represents the average quantity for the whole length of the channel, we get

$$z_1 - z_2 = f \frac{L}{4R_h} \frac{U_{av}^2}{2g}.$$

Solving for the average velocity, we get

$$U_{av} = \sqrt{\frac{z_1 - z_2}{L} \frac{8R_h g}{f}}, \tag{5.133}$$

and the flow rate Q is the average velocity multiplied by the cross-section area S:

$$Q = U_{av} S. \tag{5.134}$$

Note that the slope of the channel can be represented by the slope angle α:

$$\frac{z_1 - z_2}{L} = \tan \alpha. \tag{5.135}$$

This formulation provides an approximate method for calculating the average velocity and the flow rate in the open channel, based on the slope, cross-section geometry, and the friction coefficient. Now let us go back in history: Antoine Chezy (1718–1798), a French engineer studying the water supply of Paris, arrived empirically at a similar formula, stating that

$$U_{av} = C\sqrt{R_h \frac{z_1 - z_2}{L}}. \tag{5.136}$$

Naturally the coefficient C is called the Chezy coefficient:

$$C = \sqrt{\frac{8g}{f}}. \tag{5.137}$$

This coefficient includes the friction effects, and Robert Manning (1816–1897) a French-born Irish engineer experimentally modified this relation for actual river and water conduits. He proposed the following formula for the coefficient C:

$$C = \frac{R_h^{1/6}}{n}, \tag{5.138}$$

where n is a nondimensional friction parameter, sometimes called the Manning roughness coefficient. A short list for the values of this coefficient applicable to various surface conditions is given in Table 5.1. Substituting this coefficient into Eq. 5.131 yields

$$U_{av} = \frac{1}{n} R_h^{2/3} \sqrt{\frac{z_1 - z_2}{L}}, \tag{5.139}$$

which is called the Manning equation. Because this formula is based on an empirical correlation, the units require special attention. In the preceding formula, the units for the hydraulic diameter and the resulting velocity are in meters (the parameter n has no units). When the hydraulic diameter is given in feet and the resulting velocity is in feet per second, then the formula changes slightly to

$$U_{av} = \frac{1.486}{n} R_h^{2/3} \sqrt{\frac{z_1 - z_2}{L}}. \tag{5.140}$$

Once the average velocity is calculated, the flow rate can be calculated with Eq. (5.134).

The following examples demonstrate the applicability of this simple model to various open channel flows.

Table 5.1. *The Manning roughness coefficient (or friction parameter) for several channel surfaces (after Ven Te Chow,* Open Channel Hydraulics, *McGraw-Hill, New York, 1959)*

Surface	n	Average roughness height ε mm
Artificially lined channels		
Glass	0.010±0.002	0.3
Brass	0.011±0.002	0.6
Steel, smooth	0.012±0.002	1.0
Painted	0.014±0.003	2.4
Riveted	0.015±0.002	3.7
Cast iron	0.013±0.003	1.6
Cement, finshed	0.012±0.002	1.0
Unfinished	0.014±0.002	2.4
Planed wood	0.012±0.002	1.0
Clay tile	0.014±0.003	2.4
Brickwork	0.015±0.002	3.7
Asphalt	0.016±0.003	3.4
Corrugated metal	0.022±0.005	37
Rubble masonry	0.025±0.005	80
Excavated earth channels		
Clean	0.022±0.004	37
Gravelly	0.025±0.005	80
Weedy	0.030±0.005	240
Stony, cobbles	0.035±0.010	500
Natural channels		
Clean and straight	0.030±0.005	240
Sluggish, deep pools	0.040±0.010	900
Major rivers	0.035±0.010	500
Floodplains		
Pasture, farmland	0.035±0.010	500
Light brush	0.05±0.02	2000
Heavy brush	0.075±0.025	5000
Trees	0.15±0.05	10000

EXAMPLE 5.13. FLOW IN A TRAPEZOIDAL CHANNEL. Calculate the fully developed average velocity and the water flow rate in a concrete wall, trapezoidal channel, with the dimensions shown in Fig. 5.30. The slope of the channel is 0.3°.

Solution: First let us calculate the cross-section area and the hydraulic radius. The area S is

$$S = \left(1 + \frac{0.8}{\tan 60}\right) 0.8 = 1.17\,\text{m}^2,$$

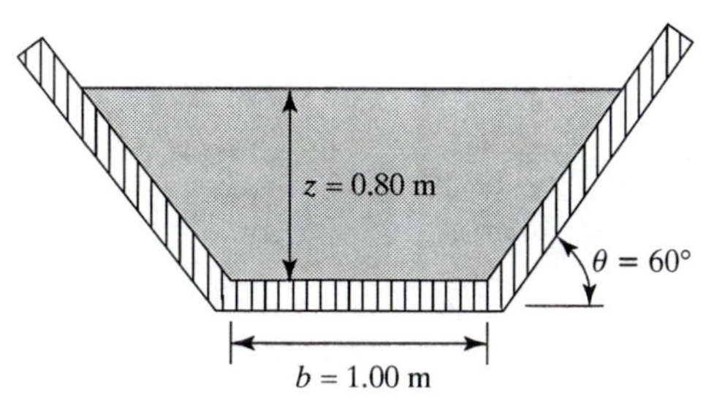

Figure 5.30. Dimensions of the trapezoidal channel for Example 5.13.

and the hydraulic perimeter is

$$P_h = 1 + 2\frac{0.8}{\sin 60} = 2.84\,\text{m}.$$

The hydraulic radius can now be calculated:

$$R_h = \frac{S}{P_h} = 0.41\text{ m}.$$

The Manning coefficient for unfinished concrete is taken from Table 5.1 as $n = 0.012$. With this information, we calculate the average velocity by using Eq. (5.139),

$$U_{\text{av}} = \frac{1}{n} R_h^{2/3} \sqrt{\frac{z_1 - z_2}{L}} = \frac{1}{0.012} 0.41^{2/3} \sqrt{\tan 0.3} = 3.33\text{ m/s},$$

and the flow rate is

$$Q = U_{\text{av}} S = 3.33 \times 1.17 = 3.89\text{ m}^3\text{/s}.$$

EXAMPLE 5.14. THE BEST RECTANGULAR CROSS SECTION OF A RECTANGULAR CHANNEL. Assuming a fully developed steady flow in a rectangular channel, calculate the flow rate versus liquid heights z, for a given slope and Manning factor n (see Fig. 5.31).

Solution: For a fixed slope and Manning coefficient n, the velocity in Eq. (5.139) depends on the hydraulic radius. So the best design is one in which the flow cross section is the largest for a given wetted area P_h. Consequently for the highest average velocity the largest hydraulic diameter is desirable. For the rectangular cross section the area and the wetted perimeter are

$$S = bz,$$

$$P_h = b + 2z.$$

Replacing b in the second equation with S/z results in

$$P_h = \frac{S}{z} + 2z.$$

So for a constant cross-section area the best condition is

$$\frac{dP_h}{dz} = -\frac{S}{z^2} + 2 = 0;$$

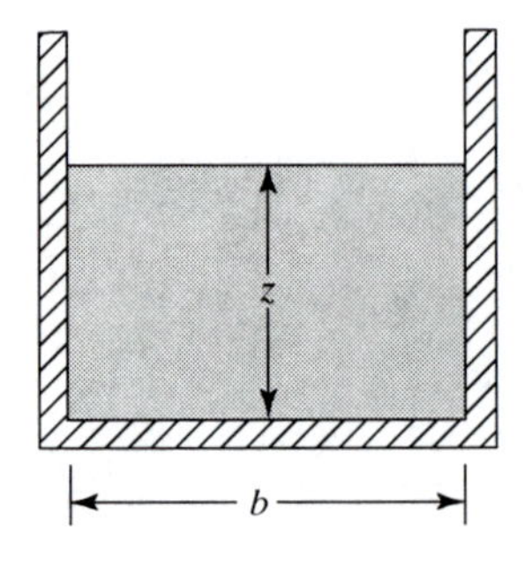

Figure 5.31. Nomenclature of the rectangular channel for Example 5.14.

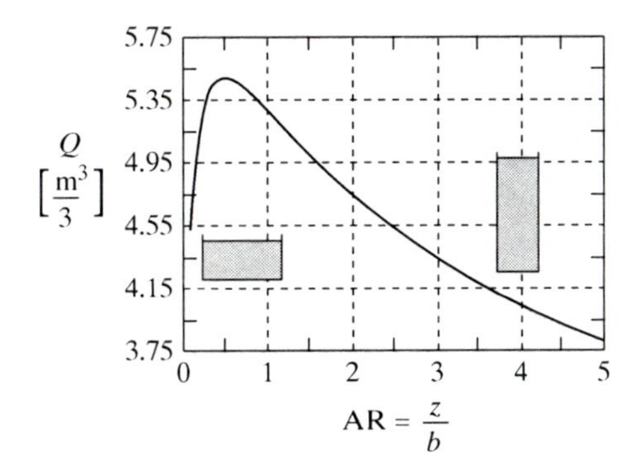

Figure 5.32. Effect of cross-section aspect ratio (AR) on the flow rate in a rectangular channel.

or, by solving this equation, we get

$$z^2 = \frac{S}{2} = \frac{bz}{2} \rightarrow z = \frac{b}{2}.$$

If we define a cross-section aspect ratio AR as

$$\mathrm{AR} = z/b,$$

then the conclusion is that AR $= 1/2$ is desirable. To provide a more complete variation of flow rates with AR, a set of calculations is performed with a fixed cross-section area of 1 m^2, a slope of 1°, and $n = 0.012$. The calculations result in the graph in Fig. 5.32.

So the wider and shallower cross section is usually better, but below AR $=$ 0.5 the flow rate drops fast (because of the rapid increase in the wetted–friction area).

EXAMPLE 5.15. FLOW RATES IN A CIRCULAR CHANNEL (OR PARTIALLY FILLED PIPE). Consider a steady uniform flow in a partially filled circular pipe, as shown in Fig. 5.33. Calculate the variation of flow rate and the average velocity versus liquid height z using the Manning formula (the pipe diameter is D and the Manning friction parameter is n).

Solution: First let us establish parameters such as the wetted perimeter and the cross-section area. It is easier to use a view-angle parameter θ, as shown in the figure. and calculate the liquid height z as a function of θ:

$$z = R(1 - \cos\theta).$$

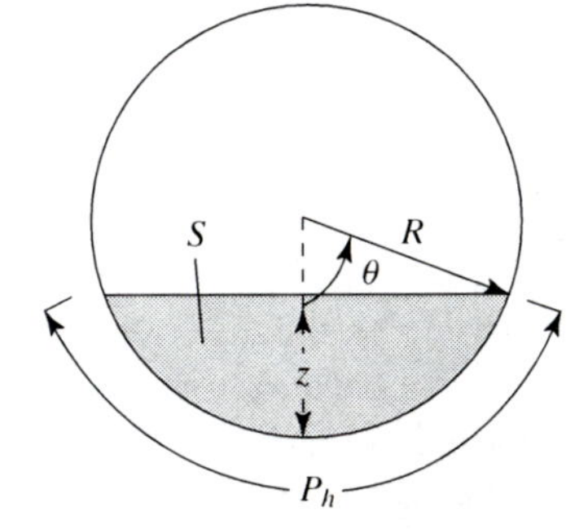

Figure 5.33. Flow in a partially filled pipe.

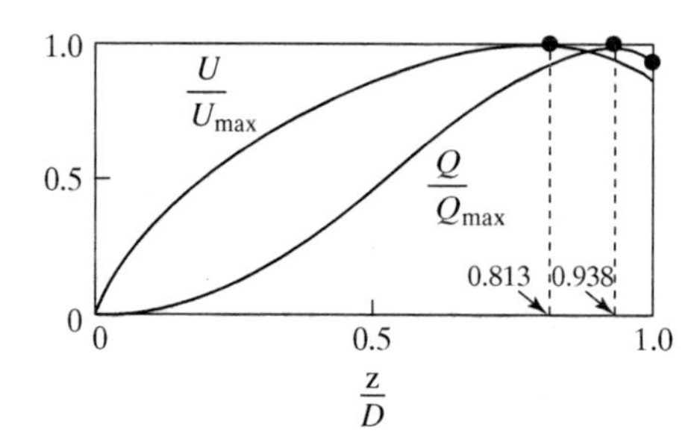

Figure 5.34. Variation of the average velocity and flow rate versus fluid height inside a circular pipe (steady uniform flow).

The wetted perimeter, in terms of the angle θ, is

$$P_h = 2\pi R \frac{\theta}{180}.$$

Next, the cross-section area S is calculated:

$$S = \pi R^2 \frac{\theta}{180} - R\cos\theta R\sin\theta = \pi R^2 \frac{\theta}{180} - \frac{R^2}{2}\sin 2\theta, \tag{5.141}$$

and the hydraulic radius is

$$R_h = \frac{S}{P_h} = \frac{\pi R^2 \frac{\theta}{180} - \frac{R^2}{2}\sin 2\theta}{2\pi R \frac{\theta}{180}} = \frac{R}{2} - \frac{45 R \sin 2\theta}{\pi \quad \theta}.$$

The average velocity is now calculated with Eq. (5.139)

$$U_{av} = \frac{1}{n} R_h^{2/3}\sqrt{\tan\alpha}, \tag{5.142}$$

and the flow rate Q is

$$Q = U_{av} S. \tag{5.134}$$

Suppose we assume $D = 1$ m, a slope of $\alpha = 0.5^\circ$, and a Manning friction parameter of $n = 0.012$. With these values, the average velocity [Eq. (5.142)] and flow rate can be calculated, as shown in Fig. 5.34. The results are normalized by the maximum value (which was at $z/D = 0.813$ for the velocity and $z/D = 0.938$ for the flow rate). These data show that, as the circular pipe fills up, both average speed and flow rate increase. At a certain point, however (before filling up), the velocity reaches a maximum and then slightly drops before the pipe is full. The flow rate follows a similar trend but reaches its maximum a bit later, because U_{av} is multiplied by the cross-section area S, which always increases with z.

5.10.3 Hydraulic Jump

When a rapid flow merges with a larger body of still liquid or some obstacle suddenly slows down the flow, a hydraulic jump can form. We call that condition supercritical ($\mathrm{Fr} > 1$) because disturbances behind the jump cannot move upstream. The actual fluid dynamics of such a flow is very complex, but with a simple 1D model, some low-order estimates are possible.

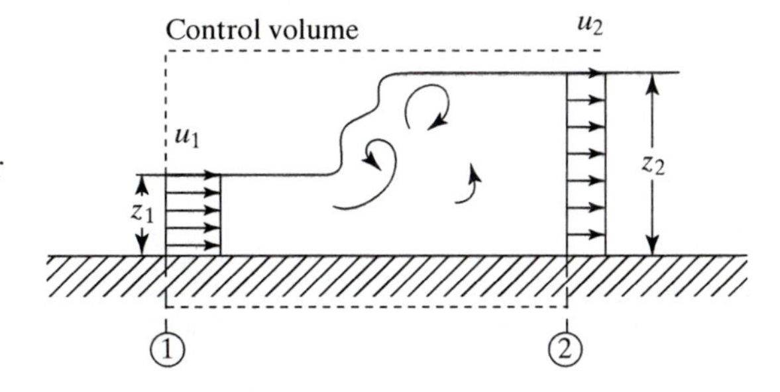

Figure 5.35. Nomenclature used for the hydraulic jump.

A typical hydraulic jump is described schematically in Fig. 5.35. A high-speed steady flow from the left (Fr $\gg$ 1) is slowed down (by an obstacle an the right), and as a result the velocity slows down and the liquid height increases. To estimate this head increase, let us use a control volume, as shown in the figure (width b – normal to the page), and apply the basic conservation laws. The mass flow rate of the liquid entering the control volume at station 1 is then

$$\dot{m} = \rho_1 u_1 z_1 b, \tag{5.143}$$

and it must be equal to the flow of the liquid leaving station 2:

$$\rho_1 u_1 z_1 b = \rho_2 u_2 z_2 b.$$

Because the flow is incompressible the continuity equation reduces to

$$u_1 z_1 = u_2 z_2. \tag{5.144}$$

We can use the 1D momentum equation [as in Eq. (2.29)], and by neglecting the external force F_x (representing the shear component) in such a short distance, we get

$$\rho u_2^2 A_2 - \rho u_1^2 A_1 = -(p_2 - p_a)A_2 + (p_1 - p_a)A_1.$$

We can simplify the left-hand side by using the continuity equation to yield

$$\rho u_2^2 A_2 - \rho u_1^2 A_1 = \dot{m}(u_2 - u_1),$$

and we can approximate the pressure on both sides by the average liquid-column height:

$$(p_1 - p_a) \approx \rho g \frac{z_1}{2},$$

$$(p_2 - p_a) \approx \rho g \frac{z_2}{2}.$$

With these modifications and recalling that $A_1 = z_1 b$ and $A_2 = z_2 b$, we find that the momentum equation becomes

$$\dot{m}(u_2 - u_1) = \rho g \left(-\frac{z_2}{2} z_2 b + \frac{z_1}{2} z_1 b\right). \tag{5.145}$$

Now, substituting $\dot{m} = \rho_1 u_1 z_1 b$ from Eq. (5.107), we get

$$\frac{u_1 z_1}{g}(u_2 - u_1) = \frac{z_1^2}{2} - \frac{z_2^2}{2}. \tag{5.146}$$

Next we use the Bernoulli (or energy) equation (5.107) between the two stations, assuming that the outside pressure is unchanged between station 1 and station 2:

$$z_1 + \frac{u_1^2}{2g} = z_2 + \frac{u_2^2}{2g} + h_f, \tag{5.147}$$

where h_f represents the head loss that is due to the rapid mixing. At this point there are three equations that can be solved for the head loss h_f. Let us start with continuity equation (5.144) and solve for u_2:

$$u_2 = \frac{u_1 z_1}{z_2}.$$

Next we substitute this into momentum equation (5.140):

$$\frac{u_1 z_1}{g}\left(\frac{u_1 z_1}{z_2} - u_1\right) = \frac{z_1^2}{2} - \frac{z_2^2}{2}.$$

This could be rearranged as:

$$\frac{u_1^2}{g}\left(\frac{z_1}{z_2}\right)(z_1 - z_2) = \frac{1}{2}(z_1 + z_2)(z_1 - z_2).$$

After canceling $(z_1 - z_2)$ from both sides of the preceding equation, we can rearrange it as

$$\left(\frac{z_2}{z_1}\right)^2 + \frac{z_2}{z_1} - 2\frac{u_1^2}{g z_1} = 0, \tag{5.148}$$

which is a simple quadratic equation for the liquid-column height (ratio). The third term in this equation contains the Froude number at station 1,

$$\mathrm{Fr}_1^2 = \frac{u_1^2}{g z_1}, \tag{5.149}$$

and we can rewrite Eq (5.148) as

$$\left(\frac{z_2}{z_1}\right)^2 + \frac{z_2}{z_1} - 2\mathrm{Fr}_1^2 = 0, \tag{5.150}$$

the solution of which is

$$\frac{z_2}{z_1} = \frac{-1 \pm \sqrt{1 + 8\mathrm{Fr}_1^2}}{2}. \tag{5.151}$$

For the present problem only the positive value of the square root is considered. Also for Froude numbers larger than zero, $(z_2/z_1) > 1$ is expected. To calculate the head loss we return to the energy equation [Eq. (5.147)],

$$\frac{h_f}{z_1} = 1 - \frac{z_2}{z_1} + \frac{u_1^2}{2g z_1} - \frac{u_2^2}{2g z_1} = 1 - \frac{z_2}{z_1} + \frac{u_1^2}{2g z_1}\left[1 - \left(\frac{z_1}{z_2}\right)^2\right],$$

and here we use the continuity equation, namely, $u_1 z_1 = u_2 z_2$. Finally, using the Froude number, we can write:

$$\frac{h_f}{z_1} = 1 - \frac{z_2}{z_1} + \frac{\mathrm{Fr}_1^2}{2}\left[1 - \left(\frac{z_1}{z_2}\right)^2\right]. \tag{5.152}$$

Fr	$\frac{z_2}{z_1}$	Description	Sketch
<1	1	Smooth flow	
1 to 1.7	1 to 2.0	Standing-wave or undulant jump	z_1 z_2
1.7 to 2.5	2.0 to 3.1	Weak jump	z_1 z_2
2.5 to 4.5	3.1 to 5.9	Oscillating jump	z_1 z_2
4.5 to 9.0	5.9 to 12	Stable, well-balanced steady jump	z_1 z_2
>9.0	>12	Strong, unsteady jump	z_1 z_2

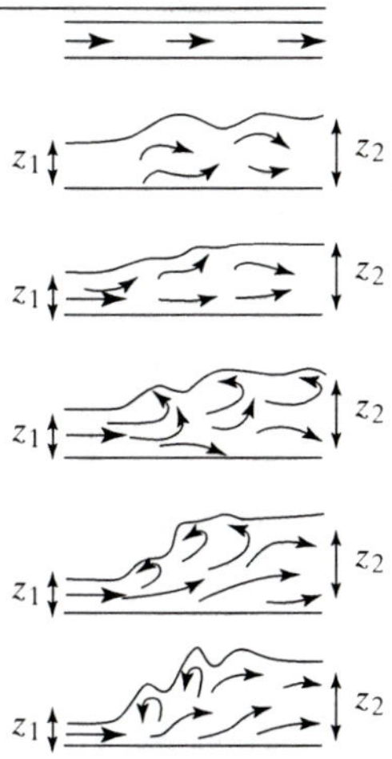

Figure 5.36. Schematic classification of the types of hydraulic jumps versus Fr.

This equation estimates the head loss as a function of the Froude number. To calculate the percentage of head loss we can compare the total heads at both sides of the hydraulic jump. The incoming total head is then h_1 and can be calculated by

$$h_1 = z_1 + \frac{u_1^2}{2g}. \tag{5.153}$$

The percentage of loss is then defined as

$$\%\text{Loss} = \frac{h_f}{h_1} = \frac{h_f}{z_1 + \frac{u_1^2}{2g}}. \tag{5.154}$$

Figure 5.36 provides a pictorial description for the relation between the hydraulic jump and Fr. Of course, for Fr < 1, a uniform flow is expected and a hydraulic jump is not possible. Near critical conditions (Fr $\sim$ 1.0–1.7) a standing wave is expected with up to 7% head loss. For higher Fr (1.7–2.5) a clear hydraulic jump is forming, with head losses of up to 15%. As the flow speed increases (Fr $\sim$ 2.5–4.5) a strong hydraulic jump is observed, leading to head losses of up to 45%. For much faster flows (like jets) the losses are even higher and can reach 80% head loss.

To demonstrate the applicability of the formulation developed for the hydraulic jump, let us solve the following example.

EXAMPLE 5.16. HEAD LOSS IN A HYDRAULIC JUMP. Water is flowing in a wide channel at a velocity of 8 m/s. If the water-column height before the hydraulic jump is 1 m, then calculate Fr and the head loss.

Solution: Let us first calculate the Fr of the incoming flow:

$$\text{Fr}_1 = \frac{u_1}{\sqrt{g z_1}} = \frac{8}{\sqrt{9.8 \times 1}} = 2.555.$$

Based on Fig 5.36, this can be classified as a weak hydraulic jump. To calculate the height of the jump, we use Eq. (5.151):

$$\frac{z_2}{z_1} = \frac{-1 \pm \sqrt{1 + 8\mathrm{Fr}_1^2}}{2} = 3.148.$$

Consequently the height of the column because of the hydraulic jump is

$$z_2 = 3.148\,\mathrm{m}.$$

We estimate the average velocity after the jump by using the continuity equation,

$$u_2 = \frac{u_1 z_1}{z_2} = \frac{8 \times 1.00}{3.148} = 2.541\ \mathrm{m/s},$$

and the Froude number behind the jump is

$$\mathrm{Fr}_2 = \frac{2.541}{\sqrt{9.8 \times 3.148}} = 0.457,$$

which is a subcritical number. We then calculate the head loss by using Eq. (5.152):

$$\frac{h_f}{z_1} = 1 - \frac{3.148}{1} + \frac{2.555^2}{2}\left[1 - \left(\frac{1}{3.148}\right)^2\right] = 0.787,$$

and $h_f = 0.787$ m. The percentage of loss is then calculated with Eq. (5.154),

$$\%\mathrm{Loss} = \frac{h_f}{z_1 + \frac{u_1^2}{2g}} = \frac{0.787}{1 + \frac{8^2}{2 \times 9.8}} = 0.184,$$

so the loss is 18.4%.

5.10.4 Flow Discharge through Sharp-Crested Weirs

A sharp-crested weir is usually a flat plate placed across the open end channel, as shown schematically in Fig. 5.37. Such devices can be used in open channel flows to control and measure flow rates (e.g., in a field irrigation system). A similar flow-control device, the sluice gate, was discussed in Subsection 4.5.4.

The shape of the weir (from the front view) can vary, and two generic cases are shown in Fig. 5.38. In general, the flow on the left is moving slowly and its height

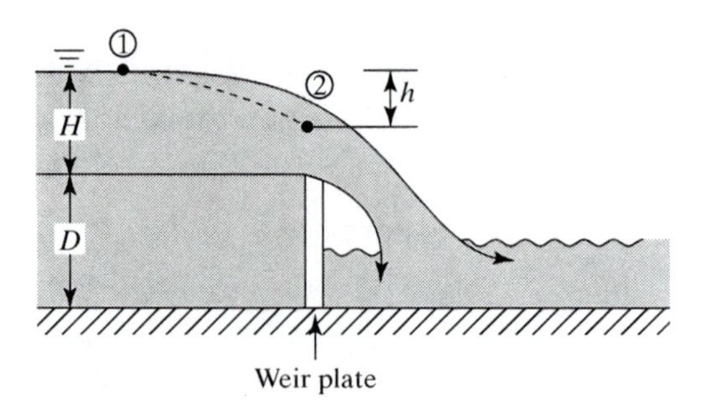

Figure 5.37. Nomenclature used for the flow over a sharp-crested weir.

is $H + D$, as shown in Fig. 5.37. As the flow approaches the weir it accelerates and then drops sharply at the right-hand side, mixing with the fluid below.

The flow field across the weir and particularly behind it and through the lower mixing zone is extremely complex. However, the flow rate is mostly determined by the upstream conditions, which can be approximated by ideal flow. This is an approach similar to the method used to estimate the flow rates through an orifice (Subsection 4.5.3), and the losses are accounted for by an empirical loss coefficient. Along the same lines let us observe a fluid particle far upstream (point 1) where its velocity is negligible and its head, relative to the weir edge, is H. As the particle reaches the weir its velocity increases, depending on its height h above the weir. Assuming no losses between these two points we can write the Bernoulli equation [Eq. (5.107)] as

$$H + D + \frac{u_1^2}{2g} = H + D - h + \frac{u_2^2}{2g}.$$

Assuming that the velocity at point 1 is negligible, then the velocity at point 2 is

$$u_2 = \sqrt{2gh},$$

which is the free-fall equation. Also note that the velocity above the weir varies with h according to this relation (so the velocity is larger when point 2 is deeper). To calculate the flow rate Q we must integrate the velocity. Assuming a rectangular weir as in Fig. 5.38(a), we have

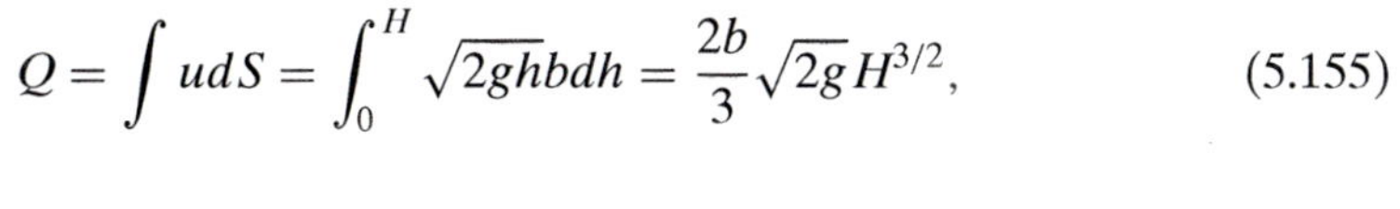

$$Q = \int u dS = \int_0^H \sqrt{2gh} b dh = \frac{2b}{3}\sqrt{2g} H^{3/2}, \tag{5.155}$$

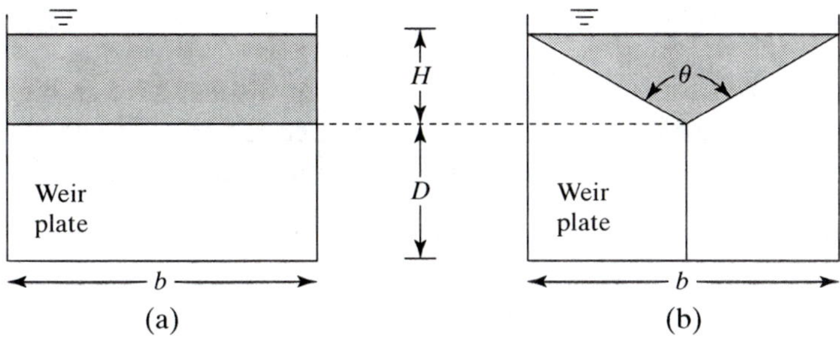

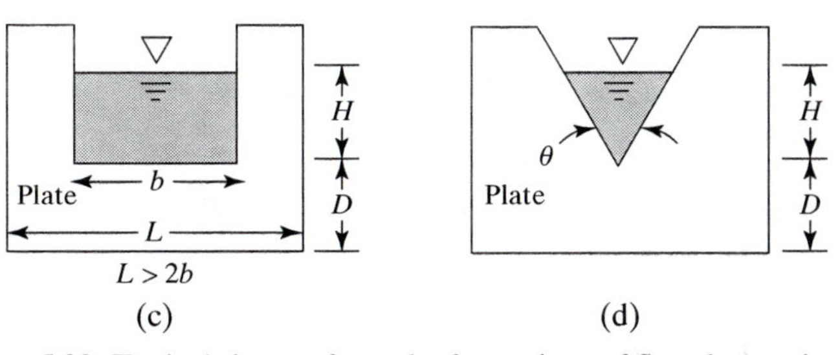

Figure 5.38. Typical shapes, from the front view, of flat-plate weirs: (left) rectangular, (right) triangular.

where b is the weir width. In the case of the triangular weir, the area element is

$$dS = h \tan \frac{\theta}{2} dh$$

and the flow rate is

$$Q = \int_0^H \sqrt{2gh} \tan \frac{\theta}{2} h dh = \frac{2}{5} \tan \frac{\theta}{2} \sqrt{2g} H^{5/2}. \quad (5.156)$$

Defining a discharge coefficient that will account for the losses, we can rewrite Eq. (5.155) for the rectangular weir as

$$Q = C_D b \sqrt{2g} H^{3/2} \quad (5.157)$$

and we approximate the experimental discharge coefficient C_D for the weir in Fig 5.38(a) as

$$C_D = 0.399 + 0.0598 \frac{H}{D}. \quad (5.158)$$

For a partial-width weir s, shown in Fig. 5.38(c), the flow is reduced by the contraction on both sides and the discharge coefficient can be approximated as

$$C_D = 0.410 \left(1 - 0.100 \frac{H}{b}\right), \quad (5.159)$$

and for the triangular weir case

$$Q = C_D \tan \frac{\theta}{2} \sqrt{2g} H^{5/2}. \quad (5.160)$$

The discharge coefficient for both weir shapes [shown in Figs. 5.38(b) and 5.38(d)] is estimated as

$$C_D = 0.31. \quad (5.161)$$

EXAMPLE 5.17. FLOW MEASUREMENT BY A WEIR. A 0.5-m-wide and 0.5-m-high weir [as in Fig. 5.38(a)] is measuring the flow from a larger channel, where the depth ahead of the weir is 1 m. Calculate the flow rate across the weir.

Solution: The flow rate is given by Eq. (5.157), and the discharge coefficient [from Eq. (5.158)] is

$$C_D = 0.399 + 0.0598 \frac{0.5}{0.5} = 0.4588.$$

The flow rate is then

$$Q = 0.4588 \times 0.5 \sqrt{2 \times 9.8} \times 0.5^{3/2} = 0.358 \, \text{m}^3/\text{s}.$$

5.11 Advanced Topics: Exact Solutions; Two-Dimensional Inviscid Incompressible Vortex Flow

This chapter presented several "exact solutions," and the case of a simple vortex flow can clearly fit in this category. This simple flow establishes important relations between solid-body-type rotation and terms such as *vorticity* and *circulation*, which

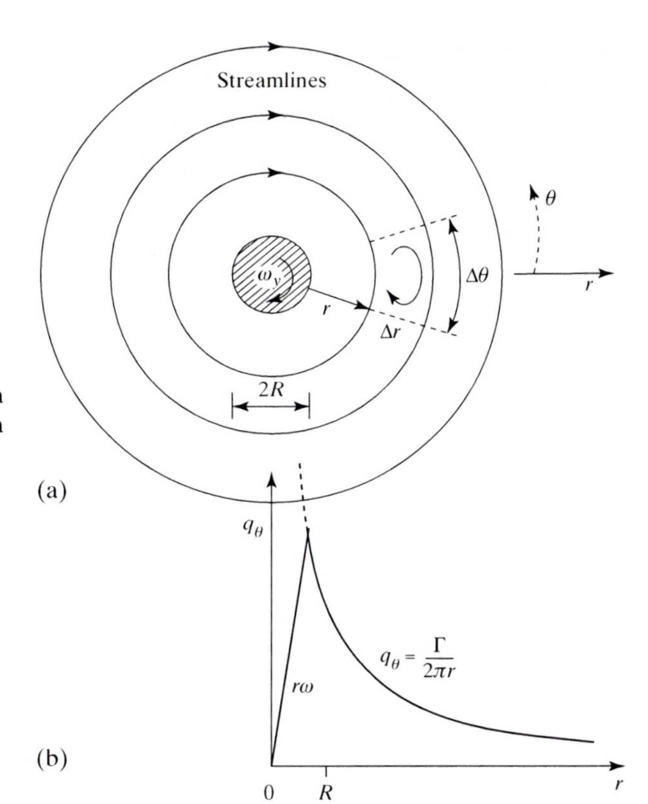

Figure 5.39. 2D flow near a cylindrical core rotating as a rigid body.

will be used in the next three chapters dealing with high-Reynolds-number flows. Vortex flows are often seen in nature, and the basic properties can be calculated by a 2D model. To illustrate the flow field of a 2D vortex, consider a 2D rigid cylinder of radius R rotating in a viscous fluid at a constant angular velocity of ω, as shown in Fig. 5.39(a) (think about mixing paint with a rotating rod). The no-slip boundary condition dictates that the particles near the wall (or surface) of the rotating cylinder will move at the same velocity. Because of the axisymmetric nature of this flow, we can assume circular streamlines with a zero-radial-velocity component.

We select the cylindrical coordinate system with the rotation ω (see direction in Fig. 5.39) about the x axis. Consequently the continuity equation [Eq. (2.45)] in the $r - \theta$ plane becomes

$$\frac{\partial q_\theta}{\partial \theta} = 0. \tag{5.162}$$

Integrating this equation suggests that q_θ is a function of the other coordinate:

$$q_\theta = q_\theta(r). \tag{5.163}$$

With the previous assumptions, the Navier–Stokes equation in the r direction [Eq. (2.46)], neglecting the body-force terms, becomes

$$-\rho\frac{q_\theta^2}{r} = -\frac{\partial p}{\partial r}. \tag{5.164}$$

Because q_θ is a function of r only and because of the axial symmetry of the problem, the pressure must be either a function of r or a constant. Therefore its derivative will not appear in the momentum equation in the θ direction [Eq. (2.47)],

$$0 = \mu\left(\frac{\partial^2 q_\theta}{\partial r^2} + \frac{1}{r}\frac{\partial q_\theta}{\partial r} - \frac{q_\theta}{r^2}\right), \tag{5.165}$$

and because q_θ is a function of r only, ordinary differentials are used. After rearranging the terms in Eq. (5.165) we get

$$0 = \frac{d^2 q_\theta}{dr^2} + \frac{d}{dr}\left(\frac{q_\theta}{r}\right). \tag{5.166}$$

Integrating with respect to r yields

$$\frac{dq_\theta}{dr} + \frac{q_\theta}{r} = A,$$

where A is the constant of integration. Rearranging this again yields

$$\frac{1}{r}\frac{d}{dr}(rq_\theta) = A,$$

and after an additional integration

$$q_\theta = \frac{A}{2}r + \frac{B}{r}. \tag{5.167}$$

The boundary conditions are

$$q_\theta = -R\omega, \quad \text{at } r = R, \tag{5.168a}$$

$$q_\theta = 0, \quad \text{at } r = \infty. \tag{5.168b}$$

and the minus sign is a result of the rotation being opposite to the direction of θ in Fig. 5.39. The second boundary condition is satisfied only if $A = 0$, and by use of the first boundary condition, the velocity becomes

$$q_\theta = -\frac{R^2\omega}{r}. \tag{5.169}$$

This velocity distribution is plotted in Fig. 5.39(b). The velocity within the solid core increases with the radius ($r\omega$ – as in solid-body rotation) but in the fluid it decreases at a rate inverse to the distance r. This is exactly the conservation of angular momentum, because we can write for two points in the fluid that

$$q_\theta r\Big|_1 = q_\theta r\Big|_2. \tag{5.170}$$

When vortex flows are discussed, the term "circulation" is often mentioned. The basic definition of the circulation Γ is

$$\Gamma \equiv \oint_c \vec{q}\cdot dl, \tag{5.171}$$

where c is a closed curve. To investigate the meaning of this term, let us perform the closed-loop integral in the same direction of ω. The path of integration is a circle enclosing the rotating core at a distance $r > R$,

$$\Gamma = \int_{2\pi}^{o} q_\theta r\,d\theta = \int_{2\pi}^{o} \frac{-R^2\omega}{r} r\,d\theta = 2\omega\pi R^2, \tag{5.172}$$

and this quantity is independent of r and is constant. With the help of the circulation Γ, the tangential velocity can be rewritten as

$$q_\theta = -\frac{\Gamma}{2\pi r}. \tag{5.173}$$

This velocity distribution is shown in Fig. 5.39(b) and is called vortex flow. If $r \to 0$ then the velocity becomes very large near the core, as shown by the dashed lines in Fig. 5.39. More important, it has been demonstrated that Γ is the circulation generated by the rotating cylinder. Based on Eq. (5.172), the circulation is

$$\Gamma = 2\omega S, \tag{5.174}$$

and its magnitude is twice the solid-body rotation of the cylinder multiplied by the area S of the rotating core.

5.11.1 Angular Velocity, Vorticity, and Circulation

Because the term "angular velocity" has appeared, let us elaborate on this topic a little bit more.

In general, the arbitrary motion of a fluid element consists of translation, rotation, and deformation. To illustrate the rotation of a moving fluid element, consider at $t = t_0$ the control volume shown in Fig. 5.40(a). Here, for simplicity, an infinitesimal rectangular element is selected that is being translated in the $z = 0$ plane by a velocity (u, v) of its corner 1. The lengths of the sides, parallel to the x and y directions, are Δx and Δy, respectively. Because of the velocity variations within the fluid the element may deform and rotate, and, for example, the x component of the velocity at the upper-left corner (4) of the element will be $(u + \frac{\partial u}{\partial y}\Delta y)$, where higher-order terms in the small quantities Δx and Δy are neglected. At a later time (e.g., $t = t_0 + \Delta t$), this will cause the deformation shown in Fig. 5.40(b). We can obtain the angular velocity component ω_z (note that the positive direction in the figure follows the right-hand rule) of the fluid element can be obtained by averaging the instantaneous angular velocities of segments 1–2 and 1–4 of the element. The instantaneous angular velocity of segment 1–2 is the difference in the linear velocities of the two

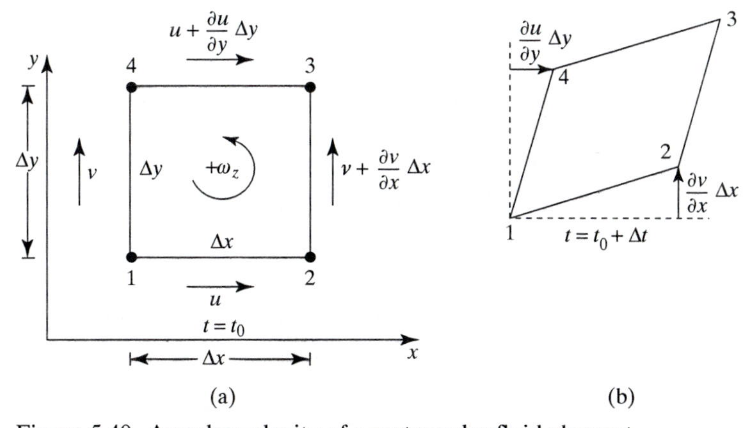

Figure 5.40. Angular velocity of a rectangular fluid element.

edges of this segment, divided by the distance (Δx):

$$\text{angular velocity of segment } 1-2 \approx \frac{\text{relative velocity}}{\text{radius}} = \frac{v + \frac{\partial v}{\partial x}\Delta x - v}{\Delta x} = \frac{\partial v}{\partial x}.$$

The angular velocity of the 1–4 segment is

$$\frac{-(u + \frac{\partial u}{aly}\Delta y) + u}{\Delta y} = -\frac{\partial u}{\partial y}.$$

The z component of the angular velocity of the fluid element is then the average of these two components:

$$\omega_z = \frac{1}{2}\left(\frac{\partial v}{\partial x} - \frac{\partial u}{\partial y}\right). \tag{5.175}$$

The two additional components of the angular velocity can be obtained similarly, and in vector form the angular velocity becomes

$$\vec{\omega} = \frac{1}{2}\nabla \times \vec{q}. \tag{5.176}$$

It is convenient to define the vorticity $\vec{\zeta}$ as twice the angular velocity:

$$\vec{\zeta} \equiv 2\vec{\omega} = \nabla \times \vec{q}. \tag{5.177}$$

In Cartesian coordinates the vorticity components are

$$\begin{aligned}
\zeta_x &= 2\omega_x = \left(\frac{\partial w}{\partial y} - \frac{\partial v}{\partial z}\right), \\
\zeta_y &= 2\omega_y = \left(\frac{\partial u}{\partial z} - \frac{\partial w}{\partial x}\right), \\
\zeta_z &= 2\omega_z = \left(\frac{\partial v}{\partial x} - \frac{\partial u}{\partial y}\right).
\end{aligned} \tag{5.178}$$

Because we already introduced the term *circulation* [in Eq. (5.171)] let us investigate its relation to vorticity. This relation can be illustrated again with the simple fluid element of Fig. 5.40. The circulation Γ is obtained by the evaluation of the closed-line integral of the tangential velocity component around the fluid element (Fig. 5.40a). Note that the *positive* direction corresponds to the *positive* direction of $\vec{\omega}$:

$$\begin{aligned}
\Gamma &= \oint_c \vec{q} \cdot dl = u\Delta x + \left(v + \frac{\partial v}{aly}\Delta x\right)\Delta y - \left(u + \frac{\partial u}{aly}\Delta y\right)\Delta x - v\Delta y \\
&= \left(\frac{\partial v}{\partial x} - \frac{\partial u}{\partial y}\right)\Delta x \Delta y = \int_s \zeta_z dS.
\end{aligned} \tag{5.179}$$

The circulation is therefore somehow tied to the rotation in the fluid (e.g., to the angular velocity of a solid-body-type rotation). To estimate the vorticity in the vortex flow of Fig. 5.39 we observe the results of Eq. (5.174),

$$\Gamma = 2\omega S, \tag{5.180}$$

which is exactly the same result if we use $\zeta_x = 2\omega$. Now let us calculate the same integral along a line (not including the origin) shown by the dashed lines in Fig. 5.39(a). Integrating the velocity in a clockwise direction and recalling that $q_r = 0$ results in

$$\oint \vec{q} \cdot d\vec{l} = 0 \cdot \Delta r + \frac{\Gamma}{2\pi(r + \Delta r)}(r + \Delta r)\Delta\theta - 0 \cdot \Delta r - \frac{\Gamma}{2\pi r} r \Delta\theta = 0. \tag{5.181}$$

This indicates that this vortex flow is irrotational (vorticity free) everywhere, but at the core where *all* the vorticity is generated. When the core size approaches zero ($R \to 0$) then this flow is called an *irrotational vortex* (excluding the core point, where the velocity approaches infinity). Some important conclusions can be drawn from this example:

1. Vorticity and rotation are generated near solid boundaries, and in this case we call the flow in these regions *rotational* because

$$\nabla \times \vec{q} \neq 0. \tag{5.182}$$

2. However, even in viscous flows, but not close to the solid-surface boundary, the fluid will not be rotated by the shear force of the neighboring fluid elements. In this case the flow is considered as *irrotational*:

$$\nabla \times \vec{q} = 0. \tag{5.183}$$

3. It appears that, when the flow is irrotational, there are no viscous flow losses (or friction) and we can define a conservative flow field. A possible benefit of that condition is that without viscous losses the Bernoulli equation can be applied between any two points in the flow. This will be clarified in the discussion about potential-flow in Chapter 8.

5.12 Summary and Concluding Remarks

This chapter demonstrated several simple solutions showing the effects of viscosity in a moving fluid. At the same time the engineering approach for treating practical fluid-flow-related problems was established. For example, the analytical solutions for laminar flow clearly showed that the friction coefficient depends on the Reynolds number. Consequently, similar relations were established for the high-Reynolds-number turbulent flow regime, and the principal coefficients were based on empirical data. This rationale is followed in this chapter for the open channel flows and also is used in the following chapters when the force coefficients (e.g., on vehicles) are calculated for a wide range of Reynolds numbers.

REFERENCES

[1] Henderson, F. M., *Open Channel Flow,* Macmillan, New York, 1966.
[2] French, R. H., *Open Channel Hydraulics,* McGraw-Hill, New York, 1985.

[3] Chaudhry, M. H., *Open Channel Flow,* Prentice-Hall, Upper Saddle River, NJ, 1993.
[4] Newman, J. N., *Marine Hydrodynamics,* MIT Press, Cambridge, MA, 1977.

PROBLEMS

5.1. A flat plate is moving at a velocity of $U = 10$ m/s on top of a 1-cm-thick oil film. The density of the oil is $\rho = 920$ kg/m^3 and viscosity $\mu = 0.4$ kg/(m s). Calculate the average forward velocity of the oil film, the flow rate, the force required to pull the plate (per 1 m^2), and the friction coefficient on the plate.

5.2. Consider the viscous laminar flow of oil between two stationary parallel plates as shown in the figure (2D, $\rho = 920$ kg/m^3, $\mu = 0.4$ N s/m^2). Assuming that the pressure difference between station (1) and (2) is $\Delta p = 1$ atm, calculate the following values.

(a) the velocity distribution,
(b) the average and maximum velocities,
(c) the flow rate (per unit width),
(d) the shear force on the lower plate and the upper plate,
(e) the Reynolds number and the friction coefficient on the lower wall.

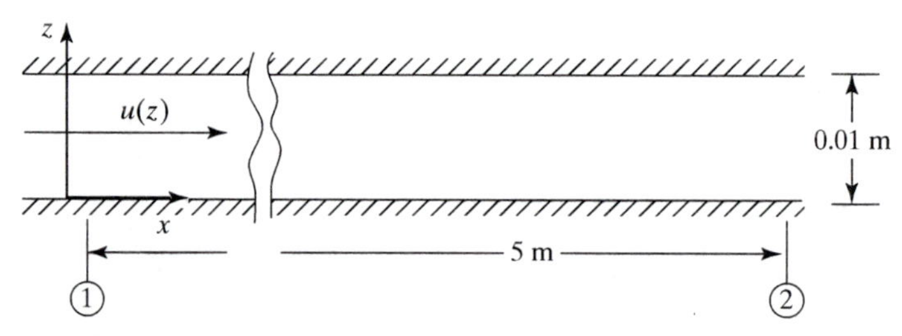

Problem 5.2.

5.3. Consider the laminar viscous flow between two infinite parallel plates (the lower is stationary). Assuming that fluid viscosity μ, the distance between the plates h, and the pressure gradient dp/dx are known, provide an expression for a zero-shear-stress condition on the upper moving plate.

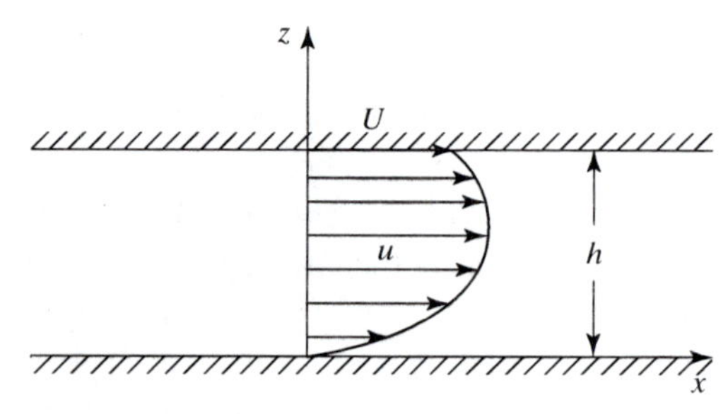

Problem 5.3.

5.4. A flat plate is moving at a velocity of $U = 5$ m/s on top of a 1-cm-thick oil film. The density of the oil is $\rho = 920$ kg/m^3 and viscosity $\mu = 0.4$ kg/(m s). Also, there is a favorable pressure gradient of $dp/dx = -2\mu U/h^2$. Calculate the average forward velocity of the oil film, the flow rate, the force required to pull the upper plate (per 1 m^2), and the friction coefficient on the lower surface (at $z = 0$).

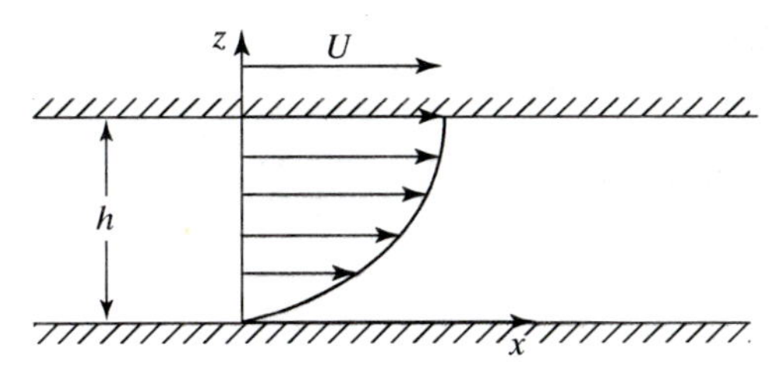

Problem 5.4.

5.5. A flat plate is moving at a velocity of $U = 10$ m/s on top of a 1-cm-thick oil film. The density of the oil is $\rho = 920$ kg/m^3 and viscosity $\mu = 0.4$ kg/(m s). Next, a favorable pressure gradient is applied, resulting in no shear on the upper surface (see figure). Calculate the average forward velocity of the oil film, the pressure gradient, the flow rate, the force required to pull the upper plate (per 1 m^2), and the friction coefficient on the lower surface ($z = 0$).

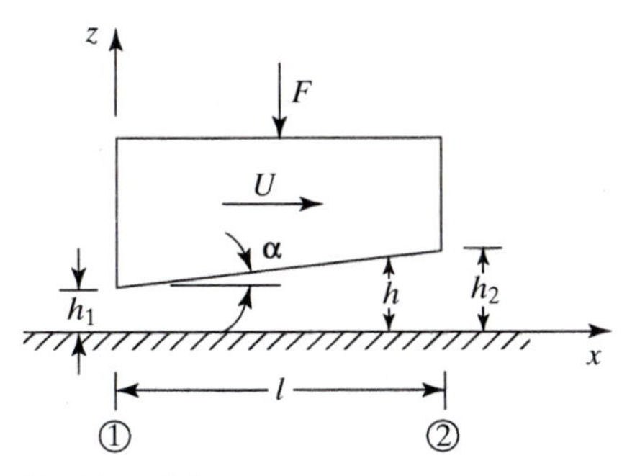

Problem 5.5.

5.6. A 1-cm-long, 2-cm-wide slipper bearing is floating on an oil film with a viscosity of 0.29 N s/m^2. Assuming the front gap is $h_1 = 0.1$ mm and $(h_2/h_1) = 2.2$, calculate the allowed vertical F load on the slider at a forward speed of 10 m/s. Also estimate the friction drag.

5.7. The laminar flow velocity profile inside a 0.1-m-diameter pipe is given by

$$u(x) = 5\left[\left(\frac{r}{R}\right)^2 - 1\right] \text{ m/s.}$$

(a) Calculate the shear force on the wall ($\mu = 0.3$).
(b) Calculate the flow rate and the average velocity.
(c) Calculate the friction coefficient ($\rho = 800$).

5.8. Oil with a density of $\rho = 920$ kg/m^3 and viscosity $\mu = 0.4$ kg/(m s) is flowing at an average velocity of 0.5 m/s in a 5-m-long, 0.02-m-inner-diameter smooth pipe. Calculate the Reynolds number, the pressure gradient, and the power required to pump the flow.

5.9. Water is flowing down from a container, between two parallel vertical plates, as shown in the figure. Assuming the density of the water is $\rho = 1000$ kg/m^3, viscosity $\mu = 0.001$ kg/(m s), and the spacing between the plates is 1 cm, calculate the following volues:

(a) maximum velocity of the water film,
(b) average velocity of the water film,
(c) the flow rate,
(d) shear force on the plate (per 1 m^2),
(e) the friction coefficient C_f.

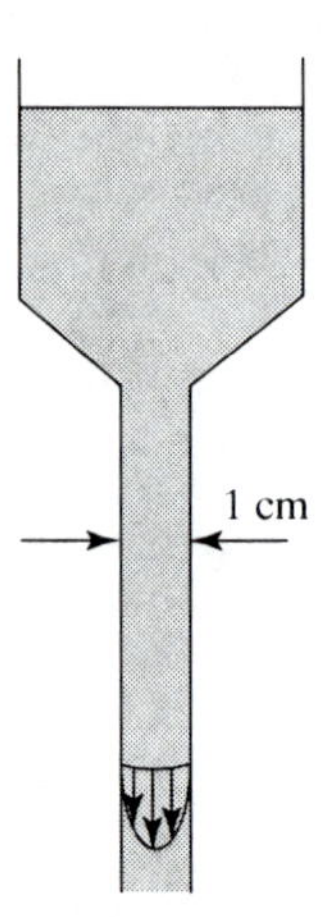

Problem 5.9.

5.10. Oil from a large container is flowing vertically down in a 3-cm inner-diameter circular pipe, as shown in the figure. Assuming the flow is fully developed, calculate the shear stress at the pipe inner wall. Also calculate the shear at the center of the pipe. Estimate the shear force acting on the inner surface of a 1-m-long pipe segment. What is the flow rate and the Reynolds number? Use motor-oil properties listed in Table 1.1.

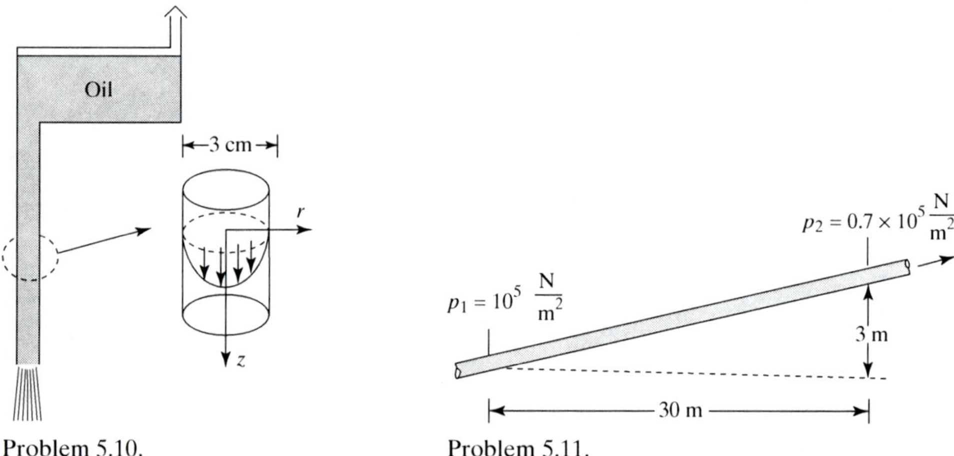

Problem 5.10.

Problem 5.11.

5.11. A fluid with density $\rho = 800$ kg/m^3 and viscosity $\mu = 0.8 \times 10^{-3}$ is flowing upward through an 8-cm-diameter galvanized iron pipe. If the relative roughness inside the pipe is $k/D = 0.001875$, calculate the flow rate between station 1 and station 2.

5.12. Estimate the diameter of a cast iron pipe required for carrying water at a discharge rate of 85 L/s and with a head loss of 1.2 m per 300 m of pipe (assume $f = 0.015$). Calculate Re and check on the Moody diagram if the value used for f is reasonable.

5.13. A fluid with a density of 800 kg/m^3 is flowing in a long 0.3-m-diameter pipe. If the maximum velocity along the centerline is 1.5 m/s and the pressure drop along a 100-m pipe segment is 1900 N/m^2, then calculate the value of the viscosity μ and the kinematic viscosity υ (assume laminar flow).

5.14. Oil flows at an average velocity of 0.3 m/s in a 3-cm-diameter pipe. Calculate the flow rate and the pressure drop for a 4-m-long pipe. Also calculate the Reynolds number and check if the laminar flow formula is valid. How much will the flow increase if the pressure gradient is increased by a factor of two.

5.15. A fluid is flowing in a 2-cm-diameter pipe at an average velocity of 0.4 m/s. The pressure drop along a 5-m-long segment is 10,000 N/m^2. Assuming laminar flow, calculate the fluid viscosity.

5.16. Water is flowing in a 3-cm-diameter pipe at an average velocity of $U_{\mathrm{av}} = 2$ m/s. Assuming water density of $\rho = 1000$ kg/m^3 and viscosity $\mu = 10^{-3}$ N s/m^2, calculate the velocity at the center of the pipe, the shear τ at the wall, and the Reynolds number. Assuming laminar flow, calculate friction coefficient C_f and pressure drop dp/dx.

5.17. Water is flowing between the two reservoirs, as shown in the figure. The difference in water surface elevation between the two reservoirs is 5 m, and the horizontal distance between them is 300 m. Determine the size (inner diameter) of steel pipe needed for a discharge of 2 m^3/s. Use the following loss coefficients for the pipe: entrance $K_e = 0.5$, valve loss $K_v = 0.2$, and exit loss, $K_{ex} = 1.0$, and $f = 0.0116$. Calculate the Re number and check on the Moody diagram if the value used for f is reasonable? ($\upsilon = 10^{-6}$ m^2/sec and $\rho = 1000$ kg/m^3).

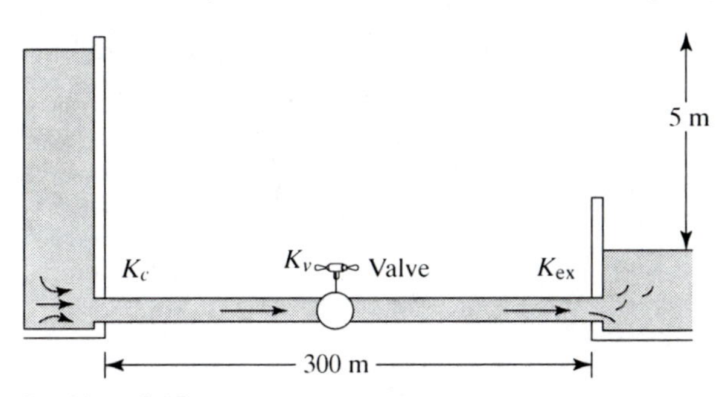

Problem 5.17.

5.18. A 100-m-long and 0.1-m-diameter pipe is used to water the garden near a large dam, as shown. The water level in the reservoir is 30 m above the horizontal pipe ($\rho = 1000$ kg/m^3, $\mu = 10^{-3}$ N s/m^2, and assume a friction coefficient $f = 0.012$).

(a) Calculate the average discharge velocity.
(b) Calculate the Reynolds number in the pipe.

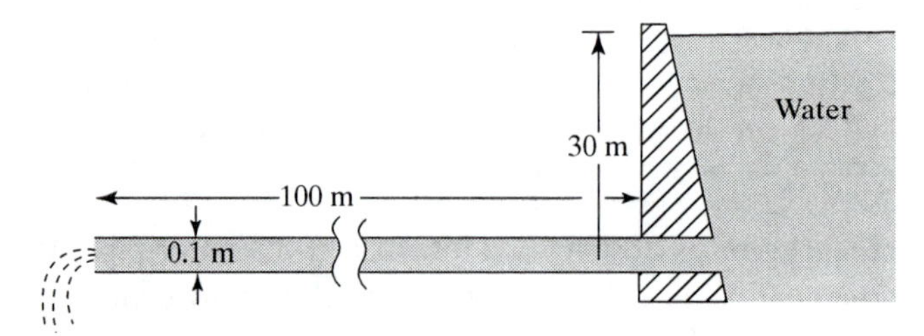

Problem 5.18.

5.19. Water is flowing through a 0.05-m-diameter pipe from a large reservoir, as shown. The water level in the reservoir is 25 m above the horizontal pipe segment at the exit from the tank. The pipe initially is bent down and later up, emerging above ground level. The loss coefficient in the three elbows is $K = 0.25$, and we can assume a Darcy friction coefficient of $f = 0.02$. Also, for water, $\rho = 1000$ kg/m^3, $\mu = 10^{-3}$ N s/m^2.

(a) Calculate the average discharge velocity and the Reynolds number in the pipe.
(b) Calculate the height h of the water fountain above the end of the pipe.

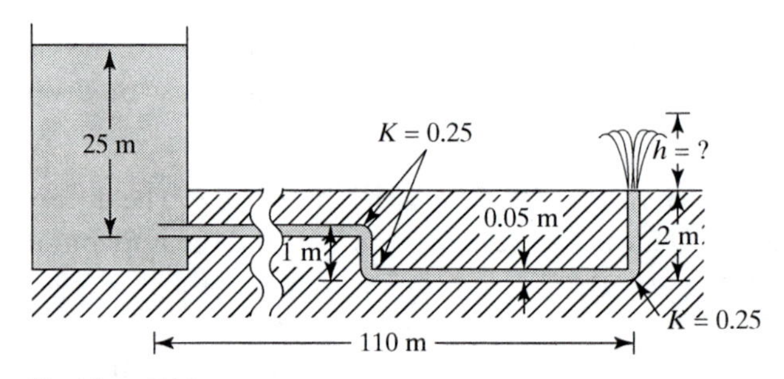

Problem 5.19.

5.20. Kerosene is flowing through the pipeline shown in the figure. The fluid level in the main tank is 20 m above the exit, and along the line there is an additional elevation drop of 2 m, as shown. The pipe diameter is 0.3 m and the inside resistance coefficient $f = 0.015$. Assume the loss coefficients for the two elbows at $K = 0.3$ and the properties of kerosene are $\rho = 804$ kg/m^3 and $\mu = 1.9 \times 10^{-3}$. Calculate the Re of the flow and the average discharge velocity at the end of the pipe.

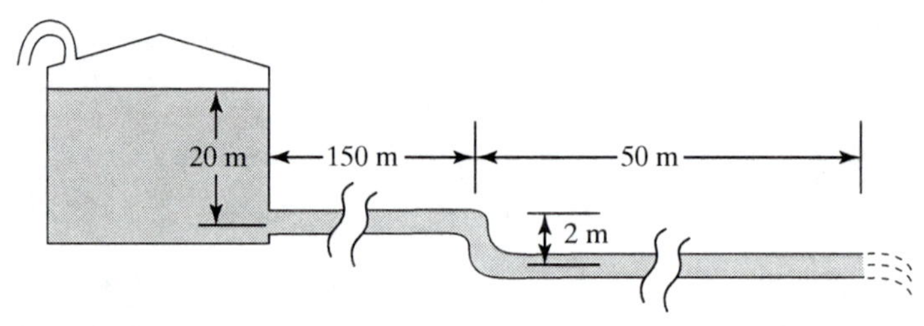

Problem 5.20.

5.21. The flow in the previous problem is reversed by placing a pump at the end of the pipeline. Consequently the kerosene is now flowing into the large reservoir at an average velocity of 6 m/s. Calculate the power required to pump the flow (pump efficiency is about 0.8).

5.22. Water is flowing from a large container along an inclined pipe (20°), as shown in the figure. The fluid level in the main tank is $h = 10$ m, the pipe inner diameter is 0.1 m and its horizontal length is $x = 200$ m. Assuming an inside resistance coefficient

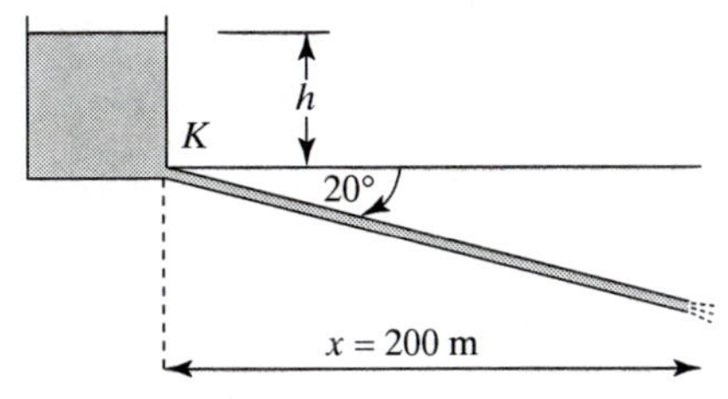

Problem 5.22.

of $f = 0.025$ and an entrance loss coefficient of $K = 0.5$, calculate the following values:

(a) The average discharge velocity at the end of the pipe.
(b) The Re of the flow ($\rho = 1000$ kg/m^3 and $\mu = 10^{-3}$ N s/m^2).

5.23. Water is flowing from a large container along an inclined pipe as shown in the figure of the previous problem. If the water is discharging at the pipe end at a rate of 80 L/s, then calculate the distance x in the figure ($f = 0.025$).

5.24. A conical funnel of diameter $D = 30$ cm is filled with oil and the fluid exits at the bottom of a long vertical pipe, as shown in the figure. The oil level in the upper conical part is $h_1 = 20$ cm, the pipe inner diameter is $d = 1$ cm, and its length is $h_2 = 60$ cm. Neglecting the losses in the conical section, and assuming a resistance coefficient of $f = 0.9$ in the pipe and an entrance loss coefficient of $K = 0.5$, calculate the average discharge velocity at the end of the pipe (assuming h_1 is not changing fast).

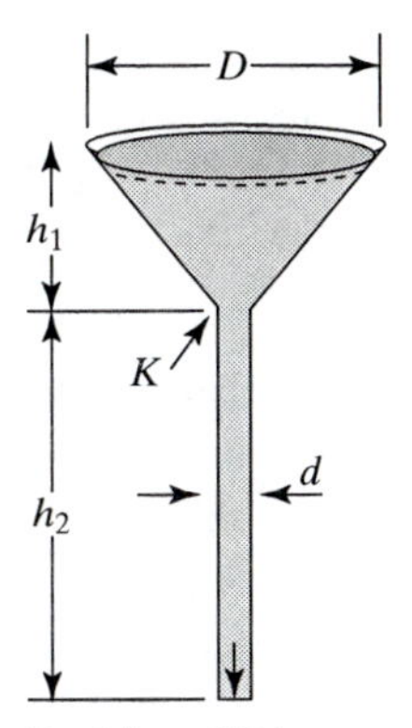

Problem 5.24.

5.25. A water tower supplies water to a tap through an 8-cm inner-diameter pipe. Dimensions of the pipeline are given in the figure and the friction coefficient is $f = 0.018$. Assuming the three loss coefficients are $K_1 = K_2 = K_3 = 0.25$, calculate the exit velocity and the flow rate.

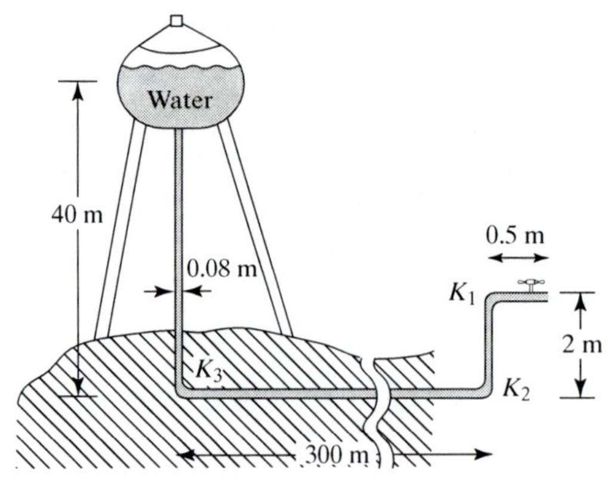

Problem 5.25.

5.26. Motor oil is siphoned from a large container, as shown in the figure. The tube inner diameter is 0.01m, the oil density is 919 kg/m^3, and its viscosity is 0.29 N s/m^2. Calculate the exit velocity at the bottom of the tube and the flow rate (volume per second). Also calculate the Reynolds number and the friction factor f.

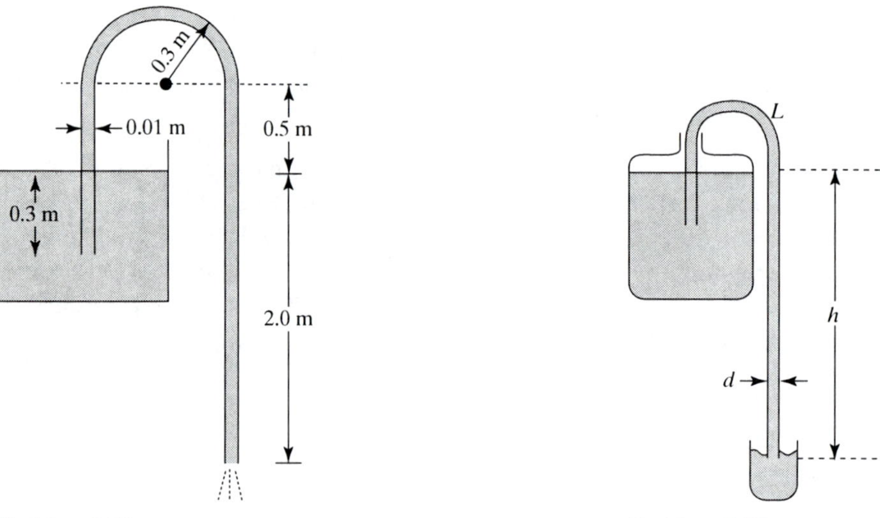

Problem 5.26.

Problem 5.27.

5.27. Wine is siphoned from a large container, as shown in the figure. The tube inner diameter is $d = 0.01$ m and its total length is $L = 1.5$ m. The elevation difference is $h = 1$ m, the wine density is 1000 kg/m^3, and its viscosity is 1.9×10^{-3} N s/m^2. Calculate the exit velocity at the bottom of the tube, the flow rate, the Reynolds number, and the friction factor f. (*Hint*: Use the Moody diagram for smooth tubes and iterate for the friction coefficient.)

5.28. A motorist is helping another motorist in a stranded car by siphoning gasoline from his automobile tank, as shown. The 1.5-m-long tube inner diameter is 0.7 cm and the friction coefficient is estimated at $f = 0.018$.

(a) How long does it takes to siphon 2 liters of gasoline (assuming no effect on the fluid level in the tank and in the container)?

(b) Repeat the calculation but assuming no friction in the pipe. How realistic is this solution?

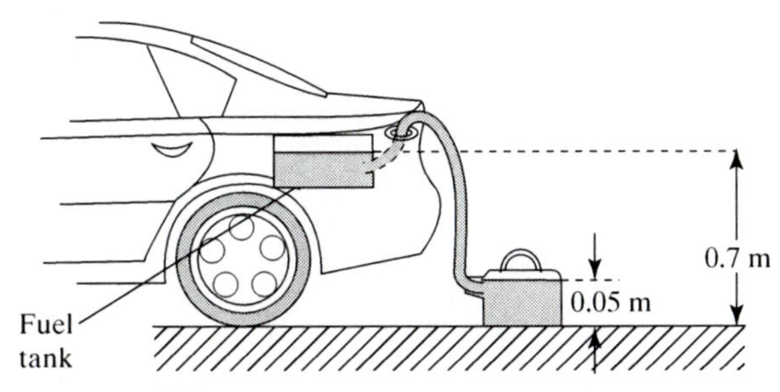

Problem 5.28.

5.29. An orifice with a friction loss coefficient of $C_D = 0.7$ is placed in a pipeline. How does this loss factor compare with the head-loss coefficient K, as defined in this chapter?

$$h_f = K\frac{u^2}{2g}.$$

Note that C_D was defined for an orifice–Venturi tube in this equation:

$$\dot{m} = C_D\sqrt{\frac{2\rho(p_1 - p_2)}{\frac{1}{A_2^2} - \frac{1}{A_1^2}}}.$$

5.30. Water is flowing out of the taller container through 0.05-m-diameter pipe, as shown in the figure. Assuming a friction factor of $f = 0.03$ in the pipe system, $K_1 = K_2 = 0.3$, $K_3 = 0.2$, and that the flow rate is not affecting the water level in the containers, calculate the flow rate in the pipes ($\Delta x = 70$ m, $\Delta z = 10$ m, $z_1 = 12$ m, $z_2 = 5$ m).

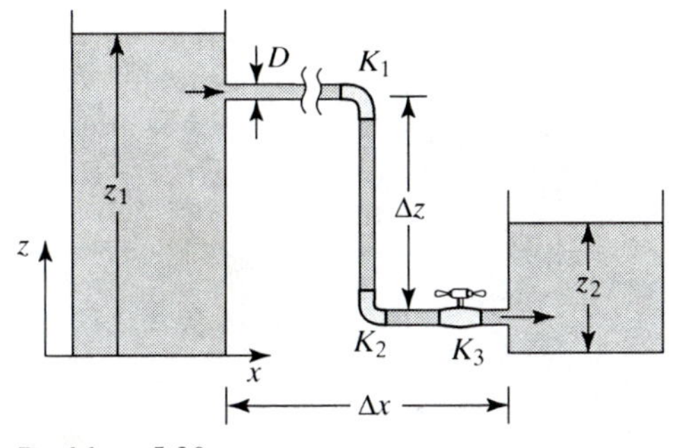

Problem 5.30.

5.31. Water is flowing out of the taller container through a 0.05-m-diameter pipe, as shown in the figure. Assuming a friction factor of $f = 0.03$ in the pipe system, $K_1 = K_2 = 0.3$, $K_3 = 0.2$, and that the flow rate is not affecting the water level in the containers, calculate the required z_1 so that the flow rate in the pipes will be 10 L/s. ($\Delta x = 70$ m, $\Delta z = 10$ m, $z_2 = 2$ m).

5.32. Water is flowing from the taller container through a long pipe that has two segments, as shown in the figure. The inner diameter of the thicker pipe is 6 cm and its length is 30 m, whereas the length of the thinner pipe is 20 m and its inner diameter is 4 cm. The loss that is due to the transition between the two pipe diameters is $K_1 = 0.2$ and the friction factor for both pipes is $f = 0.03$. Calculate the flow rate between the two containers when $z_1 = 3$ m and $z_2 = 5$ m.

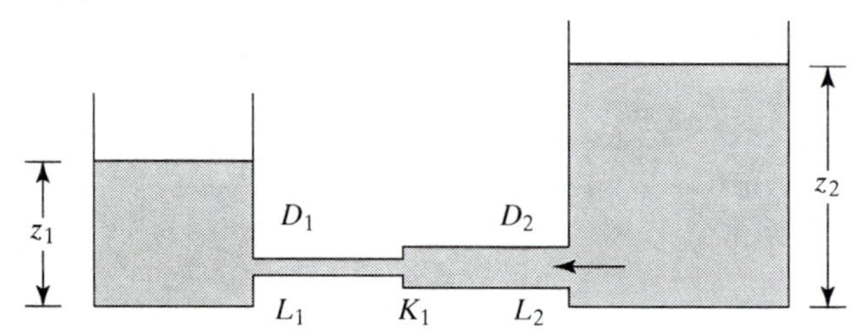

Problem 5.32.

5.33. Water is pumped through a 12-m long pipe (inner diameter $D = 0.08$ m) from a lower reservoir into a tank. The elevation difference is $z_1 - z_2 = 10$ m and $K_1 = 0.2$. Estimate the power required to drive the pump for a flow rate of 10 L/s. Assume the pump efficiency is 0.8; the fluid properties for water are listed in Table 1.1 (assume a smooth pipe when using the Moody diagram).

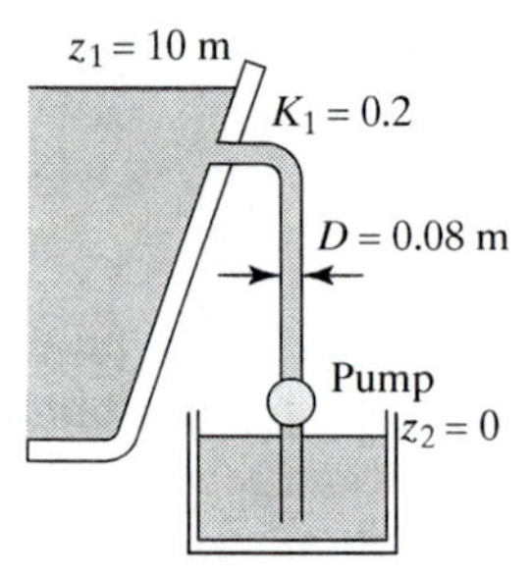

Problem 5.33.

5.34. Water is pumped from the lower container through 0.05-m-diameter pipe, as shown in the figure. Assuming a friction factor of $f = 0.03$ in the pipe system, $K_1 = K_2 = 0.3$, $K_3 = 0.2$, and that the flow rate is not affecting the water level in the two containers, calculate the power required to pump a flow rate of 10 L/s ($\Delta x = 70$ m, $\Delta z = 10$ m, $z_1 = 12$ m, $z_2 = 5$ m).

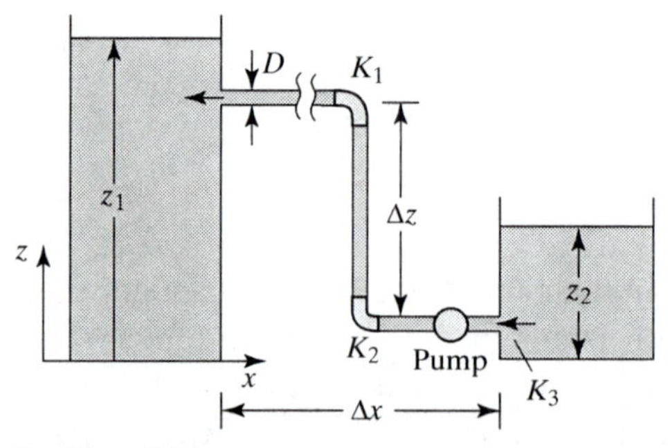

Problem 5.34.

5.35. Under normal conditions the left-hand side of the human heart pumps blood at a rate of about 5 L/min into the systemic circulation, which creates a pressure drop of about 10,600 N/m^2. The right-hand side pumps at the same flow rate but at a lower pressure of 3400 N/m^2 into the pulmonary circulation.

(a) Using the preceding average values for the pressure and flow rate, calculate the power required to pump the fluid in both sides of the human heart.
(b) Some estimate that, at the same time, the heart consumes energy at a rate of 12 W. Estimate the pumping efficiency.
(c) If the flow exits the left-hand side into the aorta with a diameter of 22 mm, then calculate the local Reynolds number (ρ = 1060 kg/m^3, μ = 0.003 N s/m^2).

5.36. Oil flows at an average velocity of 0.3 m/s in a 4-cm-wide and 2-cm-high elliptical pipe. Calculate the flow rate and the pressure drop for a 4-m-long pipe. Also calculate the Reynolds number and check if the laminar flow formula is valid (assume density = 919 kg/m^3, viscosity = 0.29 N s/m^2, and an ellipse circumference of $P_h = 2\pi\sqrt{0.5(a^2 + b^2)}$.

5.37. Oil flows at an average velocity of 0.3 m/s in a 4-cm-wide and 2-cm-high rectangular-cross-section pipe. Calculate the flow rate and the pressure drop for a 4-m-long pipe. Also calculate the Reynolds number and check if the laminar flow formula is valid (assume density = 919 kg/m^3 and viscosity = 0.29 N s/m^2).

5.38. Oil flows at an average velocity of 0.3 m/s in a 4-cm-wide, 2-cm-high triangular-cross-section pipe. Calculate the flow rate and the pressure drop for a 4-m-long pipe. Also calculate the Reynolds number and check if the laminar flow formula is valid (assume density = 919 kg/m^3 and viscosity = 0.29 N s/m^2).

5.39. Compare the pressure drop for two rectangular 5-m-long pipes, with oil flowing at an average velocity of 0.3 m/s (assume density = 919 kg/m^3 and viscosity = 0.29 N s/m^2). The first pipe is 4 cm wide and 4 cm high and the other is 8 cm wide and 2 cm high (so both have the same cross-section area).

5.40. Air at 300 K and 1 atm is flowing at an average speed of 10 m/s through a 10-m-long air conditioning duct having a square cross section of 0.3 × 0.3 m. Calculate the Reynolds number, based on the hydraulic diameter (assume viscosity is 1.8×10^{-5} N s/m^2). Estimate the power needed to pump the air, using the Moody diagram to calculate the friction coefficient (assume a smooth pipe).

5.41. Water is flowing at 5 L/min into the pipe on the left. The pipe then splits into two, as shown, but the upper valve is partially closed. All pipes have the same inner diameter, 2 cm. Assuming a loss coefficient of K_1 = 0.2 at the two junctions, K_2 = 0.3 at the elbows, and K_3 = 10 at the upper valve, calculate the flow rate at the upper and lower pipes.

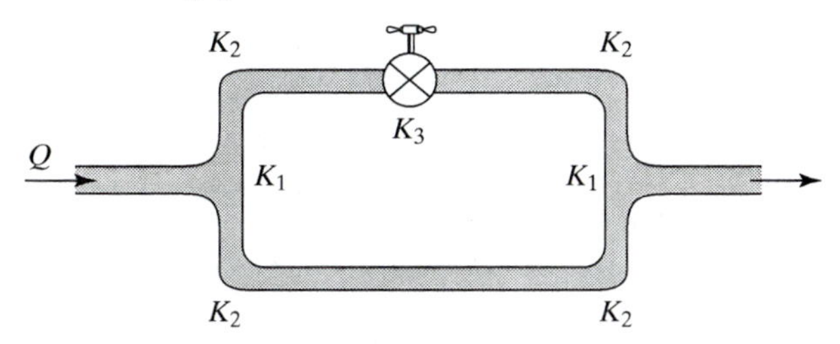

Problem 5.41.

5.42. Water is flowing at 5 L/min into a 5-cm-diameter pipe at the left. The pipe then splits into two, as shown, but the upper pipe diameter is only 3 cm. The total length of the upper pipe $L_1 = 15$ m, and it has two bends with a loss coefficient of $K_1 = 0.2$. The length of the lower pipe between points A and B is 10 m. Calculate the flow rates for the upper and lower pipes (assume $f = 0.025$ for both pipes).

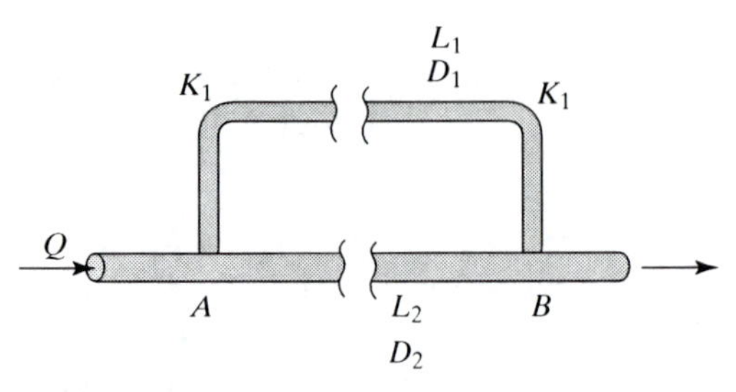

Problem 5.42.

5.43. Water is flowing at 4 L/min into a 5-cm inner-diameter pipe at the left. The pipe then splits into two, as shown, and the upper pipe inner diameter is also 5 cm. The total length of the upper pipe $L_1 = 10$ m and it has two bends with a loss coefficient of $K_1 = 0.2$. The length of the lower pipe between points A and B is 8 m. Calculate the pressure loss between point A and point B (assume $f = 0.025$ for both pipes).

5.44. Water is flowing in a V-shaped channel, as shown in the figure. The channel slope is 1.5-m drop per 1000-m length and the Manning coefficient can be estimated as $n = 0.012$. Calculate the flow rate Q.

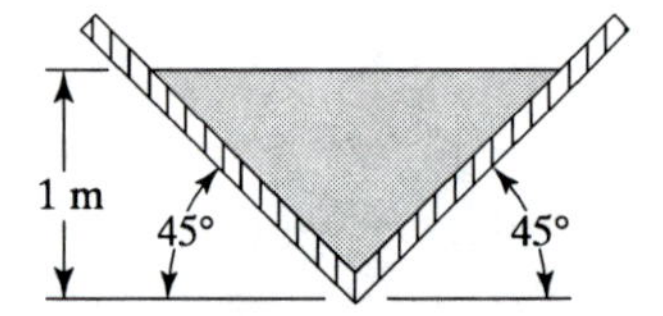

Problem 5.44.

5.45. A 60° triangular weir (as in Fig. 5.38b) is measuring the flow in a water channel, where the head ahead of the weir is $H = 0.5$ m. Calculate the flow rate across the weir.

5.46. A 2-m-wide rectangular weir (shown in the figure) is controlling the water flow in a 5-m-wide water channel. Calculate the flow rate across the weir.

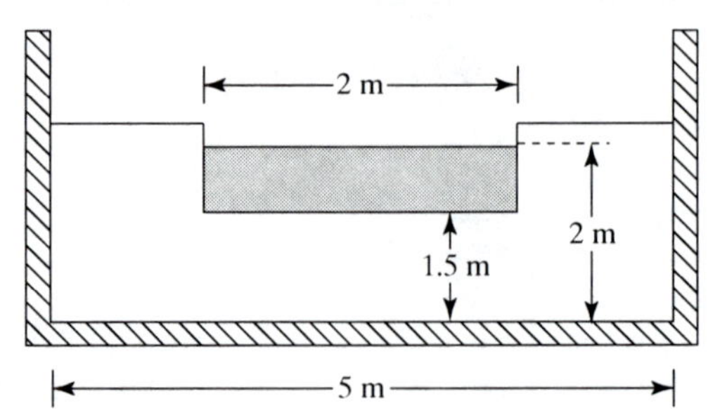

Problem 5.46.

5.47. A rectangular weir (as shown in the figure) is measuring the flow in a water channel. Calculate the flow rate across the weir

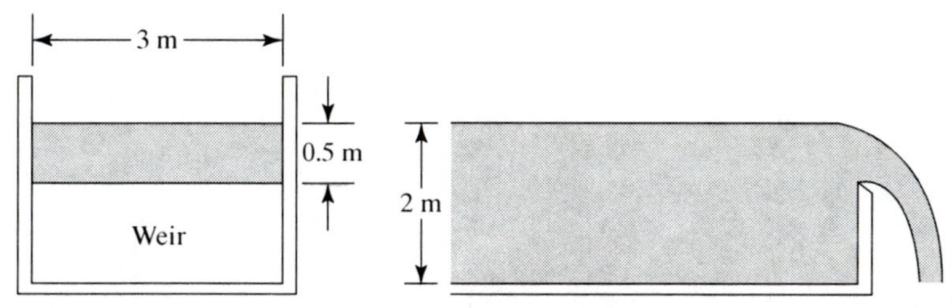

Problem 5.47.

5.48. Water is flowing in a 1-m inner-diameter concrete pipe in a slope of 2 m per 1 km. If water level inside the pipe is $h = 0.3$ m, estimate the flow rate (assume a Manning roughness coefficient of 0.012).

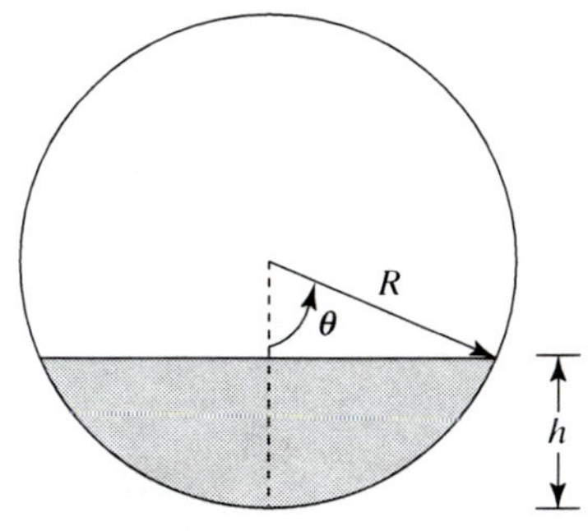

Problem 5.48.

5.49. Water is flowing in a 1-m inner-diameter concrete pipe in a slope of 2 m per 1 km (as in the previous problem). However, this time the pipe is half full, or $h = 0.5$ m. Estimate the flow rate, assuming a Manning roughness coefficient of 0.012.

5.50. Calculate the fully developed average velocity and the water flow rate in a concrete wall, trapezoidal channel, with the dimensions shown in the figure. The slope of the channel is 0.2°. Assume a Manning coefficient for unfinished concrete of $n = 0.012$.

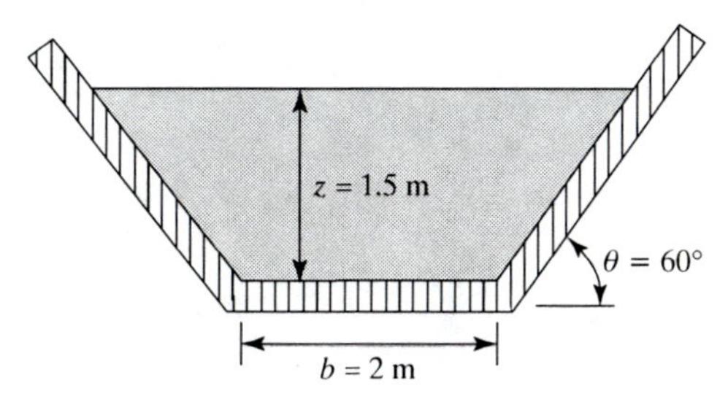

Problem 5.50.

5.51. Water flow from a reservoir is controlled by a $\theta = 60°$ weir, and the water level is $H + D = 5$ m. If the opening height is $H = 1$ m, calculate the flow rate across the weir.

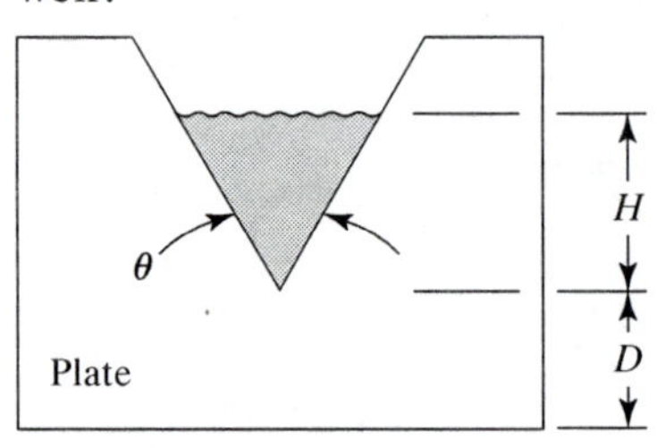

Problem 5.51.

5.52. A small rotating rod creates a 2D vortex flow at the origin (x–z coordinates) with an intensity of $\Gamma = 5$ m^2/s. Calculate the velocity components at a point (5,5).

5.53. A small rotating rod creates a 2D vortex flow at the origin (x–z coordinates) with an intensity of $\Gamma = 5$ m^2/s. Calculate the value of the integral $\int q dl$ for a circle of radius 5 m around the origin.

6 Dimensional Analysis and High-Reynolds-Number Flows

6.1 Introduction

We have seen in the previous chapter that for specific cases certain terms of the fluid dynamic equations can be neglected, yet these simplified solutions still contain the dominant physical elements. The flow in pipes was an excellent example for developing exact solutions for the low-Reynolds-number case and then the formulation was extended to the higher-Reynolds-number cases, based on experimental observations. Also, the examples presented in the previous chapter can be considered as *internal flows*. The discussion in this chapter extends the modeling capability to include *external flows* as well. One of the objectives of this chapter is to demonstrate that neglecting certain terms in the governing equations can be done systematically. The use of dimensional analysis allows a logical approach for simplifying the governing equations and provides relative scaling for the various terms. A secondary objective of this chapter is to introduce a flow regime called the high-Reynolds-number flow (discussed in Chapters 7 and 8) and to explain the success of incompressible flow models. In spite of the complex equations, a reasonable solution for the flow over bodies (and the resulting forces such as lift and drag) can be obtained and their physical origins explained. This approach of treating high-Reynolds-number flows provides the basis for modern aerodynamics and hydrodynamics.

The process of simplifying the fluid dynamic equations (and neglecting certain terms) is not arbitrary. In fact, it is based on rigorous assumptions. To demonstrate this process, let us apply dimensional analysis in the following section.

6.2 Dimensional Analysis of the Fluid Dynamic Equations

The governing equations that were developed in Chapter 2 are very complex and, as noted, their solution, even by numerical methods, is difficult for many practical applications. If some of the more complex terms can be neglected in certain regions of the flow field and at the same time the dominant physical features retained, then a set of simplified equations will result (and probably solved with less effort). This rationale, for simplifying the governing equations, is based on comparing the magnitude of the various terms in the equations. To determine the relative magnitude of the various components, the following dimensional analysis is performed. For simplicity, consider the fluid dynamic equations with constant properties ($\mu = \text{const.}$, and $\rho = \text{const.}$). The continuity equation [Eq. (2.33)] in vector form is

$$\nabla \cdot \vec{q} = 0, \tag{6.1}$$

and the momentum equation [Eq. (2.40)] is

$$\rho\left(\frac{\partial \vec{q}}{\partial t} + \vec{q} \cdot \nabla \vec{q}\right) = \rho \vec{f} - \nabla p + \mu \nabla^2 \vec{q}. \tag{6.2}$$

The first step is to define some characteristic or reference quantities, relevant to the physical problem to be studied:

L, reference length (e.g., length of a car).
V, reference speed (e.g., the free-stream speed).
T, characteristic time (e.g., one cycle of a periodic process, or simply L/V).
p_0, reference pressure (e.g., the free-stream pressure, p_∞)
f_0, body force (e.g., magnitude of earth's gravitation, g)

With the aid of these characteristic quantities we can define the following nondimensional variables:

$$\begin{aligned} x^* &= \frac{x}{L}, \quad y^* = \frac{y}{L}, \quad z^* = \frac{z}{L}, \\ u^* &= \frac{u}{V}, \quad v^* = \frac{v}{V}, \quad w^* = \frac{w}{V}, \\ t^* &= \frac{t}{T}, \quad p^* = \frac{p}{p_0}, \quad f^* = \frac{\vec{f}}{f_0}, \end{aligned} \tag{6.3}$$

If the magnitudes of the characteristic quantities are properly selected, then all the nondimensional variables in Eqs. (6.3) will be of the order of one, $O(1)$. Next, the governing equations need to be rewritten using the quantities of Eqs. (6.3) . As an example, the first term of the continuity equation becomes

$$\frac{\partial u}{\partial x} = \frac{\partial u}{\partial u^*}\frac{\partial u^*}{\partial x^*}\frac{\partial x^*}{\partial x} = \frac{V}{L}\left(\frac{\partial u^*}{\partial x^*}\right),$$

and applying this method to all three terms in the incompressible continuity equation yields

$$\frac{V}{L}\left(\frac{\partial u^*}{\partial x^*} + \frac{\partial v^*}{\partial y^*} + \frac{\partial w^*}{\partial z^*}\right) = 0. \tag{6.4}$$

It appears that all terms have the same magnitude (none of the terms can be neglected). Consequently, the continuity equation remain unchanged. As an example for the momentum equation, only the equation in the x direction is shown. By applying the chain derivatives to the various terms, it appears that the multiplier can readily be written, based on the method used for the continuity equation. For example, the first (unsteady) term becomes

$$\frac{\partial u}{\partial t} = \frac{\partial u}{\partial u^*}\frac{\partial u^*}{\partial t^*}\frac{\partial t^*}{\partial t} = \frac{V}{T}\left(\frac{\partial u^*}{\partial t^*}\right),$$

and the momentum equation in the x direction is

$$\rho\left(\frac{V}{T}\frac{\partial u^*}{\partial t^*}+V\frac{V}{L}u^*\frac{\partial u^*}{\partial x^*}+V\frac{V}{L}v^*\frac{\partial u^*}{\partial y^*}+V\frac{V}{L}w^*\frac{\partial u^*}{\partial z^*}\right) \tag{6.5}$$

$$=\rho f_0 f_x^* - \frac{p_0}{L}\frac{\partial p^*}{\partial x^*}+\mu\frac{V}{L^2}\left(\frac{\partial^2 u^*}{\partial x^{*2}}+\frac{\partial^2 u^*}{\partial y^{*2}}+\frac{\partial^2 u^*}{\partial z^{*2}}\right).$$

The corresponding equations in the y and z directions can be obtained by the same procedure, and they will have the same multipliers. Now, by multiplying Eq. (6.4) by L/V and Eq. (6.5) by $L/\rho V^2$, we end up with

$$\frac{\partial u^*}{\partial x^*}+\frac{\partial v^*}{\partial y^*}+\frac{\partial w^*}{\partial z^*}=0, \tag{6.6}$$

$$\left(\frac{L}{TV}\right)\frac{\partial u^*}{\partial t^*}+u^*\frac{\partial u^*}{\partial x^*}+v^*\frac{\partial u^*}{\partial y^*}+w^*\frac{\partial u^*}{\partial z^*} \tag{6.7}$$

$$=\left(\frac{Lf_0}{V^2}\right)f_x^*-\left(\frac{p_0}{\rho V^2}\right)\frac{\partial p^*}{\partial x^*}+\left(\frac{\mu}{\rho VL}\right)\left(\frac{\partial^2 u^*}{\partial x^{*2}}+\frac{\partial^2 u^*}{\partial y^{*2}}+\frac{\partial^2 u^*}{\partial z^{*2}}\right).$$

If all the nondimensional variables in Eqs. (6.3) are of the order of one, then all terms appearing with an asterisk (*) will also be of the order of one, and the relative magnitude of each group in the equations is fixed by the nondimensional numbers appearing inside the parentheses. In the continuity equation [Eq. (6.6)], all terms have the same order of magnitude and for an arbitrary 3D flow all terms are equally important (so we cannot neglect any term). In the momentum equation, however, several nondimensional numbers multiply the various terms. The first nondimensional number is

$$\Omega=\frac{L}{TV}. \tag{6.8}$$

Ω is often called a time constant and it signifies the importance of time-dependent phenomena. A more frequently used form of this nondimensional number is the *Strouhal* number, in which the characteristic time is the inverse of the frequency ω of a periodic occurrence (e.g., wake shedding frequency behind a separated plate):

$$\mathrm{St}=\frac{L}{(1/\omega)V}=\frac{\omega L}{V}. \tag{6.9}$$

If the Strouhal number is very small, perhaps because of very low frequencies, then the time-dependent first term in Eq. (6.7) can be neglected compared with the other terms of the order of one. The second group of nondimensional numbers, in which gravity is the body force and f_0 is the gravitational acceleration (g), is called the *Froude* number (already introduced in Section 5.10), and stands for the ratio of inertial force to gravitational force (actually the square root is used):

$$\mathrm{Fr}=\frac{V}{\sqrt{Lg}}.\left(=\sqrt{\frac{V^2}{Lf_0}}\right) \tag{6.10}$$

Small values of Fr [note that Fr^{-2} appears in Eq. (1.50)] will mean that body forces such as gravity should be included in the equations, as in the case of free-surface river flows, waterfalls, ship hydrodynamics, etc. The third nondimensional number is the *Euler* number, which represents the ratio between the pressure and the inertia forces:

$$\mathrm{Eu} = \frac{p_0}{\rho V^2}. \tag{6.11}$$

A frequently used quantity that is related to the Euler number is the pressure coefficient C_p, which measures the nondimensional pressure difference, relative to a reference pressure p_0:

$$C_p \equiv \frac{p - p_0}{\frac{1}{2}\rho V^2} \tag{6.12}$$

The last nondimensional group in Eq. (6.7) represents the ratio between the inertial and the viscous forces and is called the *Reynolds* number, which was introduced in the previous chapter:

$$\mathrm{Re} = \frac{\rho V L}{\mu} = \frac{V L}{\nu}. \tag{6.13}$$

Here ν is the kinematic viscosity, which is often used for sake of brevity:

$$\nu = \mu/\rho. \tag{6.14}$$

For the flow of gases, from the kinetic theory point of view (see [4, p. 257] in Chapter 2), the viscosity can be connected to the average velocity of the molecules c and to the mean distance λ that they travel between collisions (mean free path) by

$$\mu \approx \rho \frac{c\lambda}{2}.$$

Substituting this into Eq. (6.13) yields

$$\mathrm{Re} \approx 2(V/c)(L/\lambda).$$

This formulation shows that the Reynolds number represents the scaling of the velocity times length compared with the molecular scale. Note that c is larger than the speed of sound (see Fig. 1.16). The conditions for neglecting the viscous terms when $\mathrm{Re} \gg 1$ is discussed in more detail in the next section. For simplicity, at the beginning of this analysis an incompressible fluid was assumed. However, if compressibility is to be considered, an additional nondimensional number appears that is called the *Mach* number, and it is the ratio of the velocity to the speed of sound a [see Eq. (1.33) for evaluating the speed of sound for an ideal gas]:

$$M = V/a. \tag{6.15}$$

Usually when the characteristic velocity V is much less than the speed of sound, then the flow can be considered as incompressible (e.g., a car traveling at 150 km/h).

6.3 The Process of Simplifying the Governing Equations

The most important outcome of this dimensional analysis of the governing equations is that now the relative magnitude of the terms appearing in the equations can be determined and compared. If desired, small terms can be neglected, resulting in simplified equations that are easier to solve but still contain the dominant physical effects. As noted, in the case of the continuity equation, all terms have the same magnitude and none can be neglected. For the momentum equation we can obtain the relative magnitude of the terms by substituting the nondimensional numbers into Eq. (6.7) , and for the x direction we get

$$\Omega\frac{\partial u^*}{\partial t^*}+u^*\frac{\partial u^*}{\partial x^*}+v^*\frac{\partial u^*}{\partial y^*}+w^*\frac{\partial u^*}{\partial z^*}=\frac{1}{F^2}f_x^*-\mathrm{Eu}\frac{\partial p^*}{\partial x^*}+\frac{1}{\mathrm{Re}}\left(\frac{\partial^2 u^*}{\partial x^{*2}}+\frac{\partial^2 u^*}{\partial y^{*2}}+\frac{\partial^2 u^*}{\partial z^{*2}}\right). \tag{6.16}$$

The first term on the left-hand side is multiplied by the time constant, and the rest of the terms are of the order of $O(1)$. In the case in which time-dependent changes are small (or the time scale is very large), Ω is also small. Consequently for $\Omega \ll 1$ the first term can be neglected and we can solve the steady-state problem only. Similarly, if the Froude number Fr is large, then body forces have negligible effects, and the body-force term can be neglected. The Euler number Eu is usually not negligible because the pressure is responsible for the changes in the inertia of the moving fluid. This leaves us with the last nondimensional group, the Reynolds number. Note that it multiplies a second-order differential term, and, if neglected, only a first-order differential equation remains.

One important flow regime is the so-called *creeping flow*, in which fluid viscosity is very high, fluid motion is slow and the inertia terms are negligible. It is usually assumed that $\mathrm{Re} < 1$ and therefore the viscous terms cannot be neglected. With these considerations in mind, and without body forces, we find that the momentum equation, Eq. (6.2), is reduced to

$$0=-\nabla p+\mu\nabla^2\vec{q}. \tag{6.17}$$

The general solution of this equation is beyond the scope of the present text, but the viscous flow in pipes (Chapter 5), for example, represents a similar case.

6.4 Similarity of Flows

Another interesting aspect of the process of nondimensionalizing the equations in the previous section is that two different flows are considered to be similar if the nondimensional numbers of Eq. (6.16) are the same. For most practical cases, in which gravity and unsteady effects are negligible, only the Reynolds and the Mach numbers need to be matched. For example, many airplanes are tested in small scale (e.g. 1/5 scale). To keep the Reynolds number the same, either the airspeed or the air density must be increased (e.g., by a factor of 5). This is a typical conflict that test engineers face, because increasing the airspeed five times will bring the Mach number to an unreasonably high range. The second alternative of reducing the kinematic viscosity ν by compressing the air is possible in only a very few wind tunnels, and in most cases matching both of these nondimensional numbers is difficult. Another possibility of applying the similarity principle is to exchange fluids between the actual and the test conditions (e.g., water with air in which the ratio of kinematic viscosity is about 1:15). Thus a 1/15-scale model of a submarine can be tested in a wind tunnel at true speed conditions. Usually it is better to increase the speed in the wind tunnel, and then even a smaller-scale model can be tested (of course the Mach number is not always matched but for such low Mach number applications this is less critical).

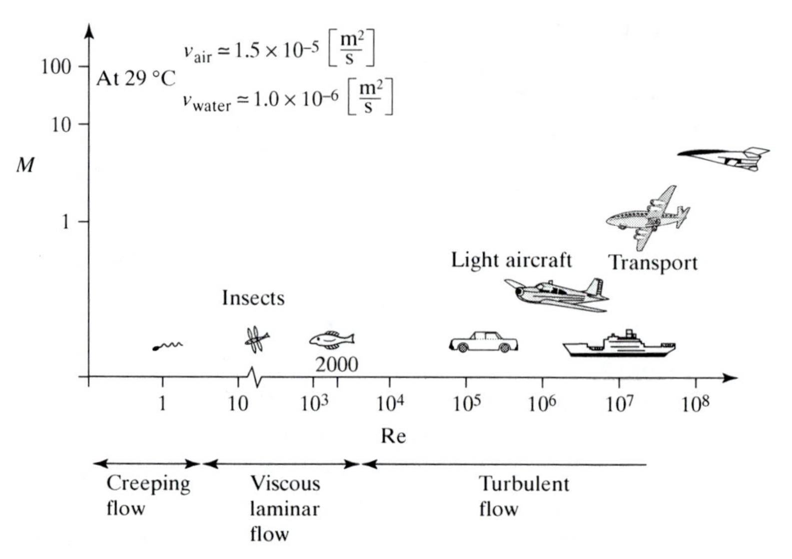

Figure 6.1. Some typical fluid flows and their Reynolds numbers. High speed sometimes means high Mach number, and therefore the relevant Mach number range is also presented.

6.5 Flow with High Reynolds Number

One of the objectives of this discussion is to demonstrate why incompressible inviscid models are successfully used to predict the lift of airplanes or the pressure field around ship hulls. This is quite obvious if the Reynolds number is high and the viscosity terms in the momentum equations can be neglected. However, when high-Reynolds-number flows are discussed, the effect of turbulence must be considered, and this is done briefly in the next section.

Before proceeding further, let us examine the range of Reynolds number and Mach number for some typical engineering problems. Because the viscosity of typical fluids such as air and water is very small, a large variety of practical engineering problems (aircraft low-speed aerodynamics, hydrodynamics of naval vessels, etc.) fall within the large-Reynolds-number range, as shown in Fig. 6.1.

From the large variety of cases shown in Fig. 6.1, we can conclude that, for high-Reynolds-number flows, the viscous terms become small compared with the other terms of the order of $O(1)$ in Eq. (6.16) . But before we neglect these terms, a closer look at the high-Reynolds-number-flow condition is needed. As an example, consider the flow near a streamlined body (e.g., a fish in this case), as shown in Fig. 6.2. In general, based on the assumption of a high Reynolds number, the viscous terms in the momentum equations can be neglected in the outer flow regions (outside the immediate vicinity of a solid surface where $\nabla^2\vec{q}$ may be large). Therefore, in this outer flow region, we can approximate the solution by solving the incompressible continuity and the Euler equations:

$$\nabla \cdot \vec{q} = 0, \tag{6.18}$$

$$\frac{\partial \vec{q}}{\partial t} + \vec{q} \cdot \nabla \vec{q} = \vec{f} - \frac{\nabla p}{\rho}. \tag{6.19}$$

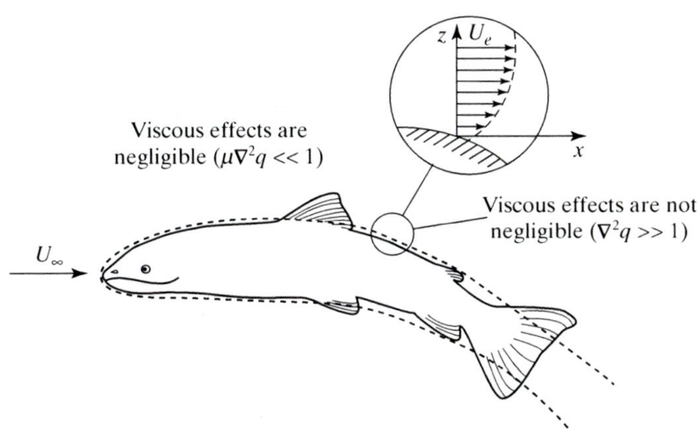

Figure 6.2. Two major flow regimes in a high-Reynolds-number attached flow: (a) The outer mostly inviscid and (b) the thin inner region, shown by the dashed curves, dominated by viscous effects.

Equation (6.18) is a first-order partial differential equation that requires a boundary condition on the normal velocity component on a solid surface compared with a boundary condition on all velocity components (including tangential) for the viscous equations. Because the flow is assumed to be inviscid, there is no physical reason for the tangential velocity component to be zero on a stationary solid surface, and therefore what remains from the no-slip boundary condition is that the normal component of velocity must be zero:

$$q_n = 0 \text{ (on a solid surface).} \tag{6.20}$$

However, a closer investigation of such flow fields reveals that, near the solid boundaries in the fluid, shear flow derivatives such as $\nabla^2\vec{q}$ become large and the viscous terms cannot be neglected even for high values of the Reynolds number (see Fig. 6.2). This thin region is usually called the *boundary layer* and is discussed in the next chapter. So, in conclusion, for high-Reynolds-number attached flows we can identify two dominant regions in the flow field:

1. The outer flow (away from the solid boundaries) where the viscous effects are negligible. A solution for the inviscid flow in this region provides information about the pressure distribution and the related forces. Only the zero-normal-velocity boundary condition is used [e.g., Eq. (6.19)].
2. The thin boundary layer (near the solid boundaries) where the viscous effects cannot be neglected. Solutions of the boundary-layer equations will provide information about the shear stress distribution, the related (friction) forces, and the thickness of this layer. For the solution of the boundary-layer equations, the no-slip boundary condition is applied on the solid boundary. The tangential velocity profile inside the boundary layer is shown schematically in the inset to Fig. 6.2, and it is seen that, as the outer region is approached, the tangential velocity component reaches the speed of the outer flow, U_e. The interface between the boundary-layer region and the outer flow region is not precisely defined and occurs at a distance δ, the boundary-layer thickness, from the wall.

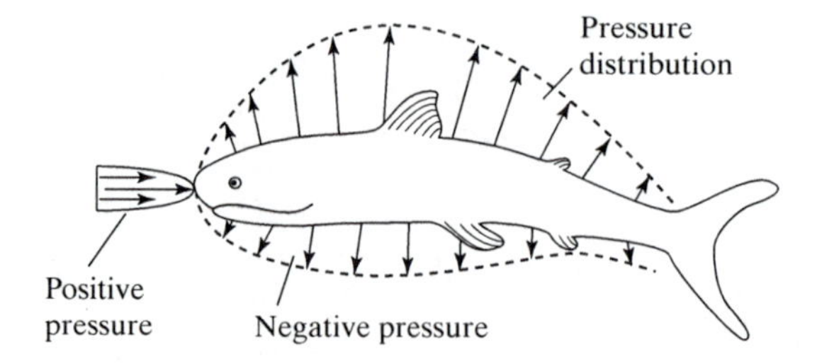

Figure 6.3. The pressure distribution can be obtained by the solution of the outer (irrotational) flow. The arrows show the direction of the pressure force.

Consequently, the classical approach for modeling the complete solution for a high-Reynolds-number, attached flow-field proceeds as follows:

1. First, A solution is found for the inviscid flow past the body (as in Fig. 6.2). For this solution the boundary condition of zero velocity normal to the solid surface is applied at the surface of the body (which is indistinguishable from the edge of the boundary layer on the scale of the body's length). Note that the required velocity field can be calculated with the continuity equation only [Eq. (6.17)]. The tangential velocity component on the body surface U_e is then obtained as part of the inviscid solution and the pressure distribution along the solid surface is then determined. For irrotational flows the Bernoulli equation can be used to calculate the pressure field (instead of solving the Euler equation). Figure 6.3 depicts a schematic pressure distribution obtained from the outer solution. Integrating these pressure forces on the body will provide the so-called pressure lift and pressure drag.
2. Next, the surface pressure distribution is taken from the inviscid flow solution and inserted into the boundary-layer equations (see next chapter). Also, U_e is taken from the inviscid solution as the tangential component of the velocity at the outer edge of the boundary layer and is used as a boundary condition in the solution of the boundary-layer equations. The solution for a high-Reynolds-number-flow field with the assumption of an inviscid fluid is therefore the first step toward the solution of the complete physical problem. In terms of results, this solution will provide information on the boundary-layer thickness and the skin friction, integration of which will give the viscous drag. Note that the pressure field is dictated entirely by the outer solution because there are no pressure changes across the boundary layer.

This method is very effective in modeling attached flows, and led to major developments in the fields of aero and hydrodynamics. For separated flows, however, this modeling technique is less effective.

6.6 High-Reynolds-Number Flows and Turbulence

The concept of turbulent flow was mentioned briefly in Subsection 1.4.2 and its effect on the flow inside pipes was discussed in Section 5.7. The internal pipe flow example clearly demonstrates that with increased Reynolds number the flow turns turbulent and the friction near solid surfaces increases. Because the next two chapters focus on this high-Reynolds-number-flow region, the subject of turbulent flow is revisited here briefly. Consider a velocity-measuring probe that is inserted into the turbulent free stream shown in Fig. 1.1. Although the average speed is in one direction only, small fluctuations in the other directions can be detected, as well. As

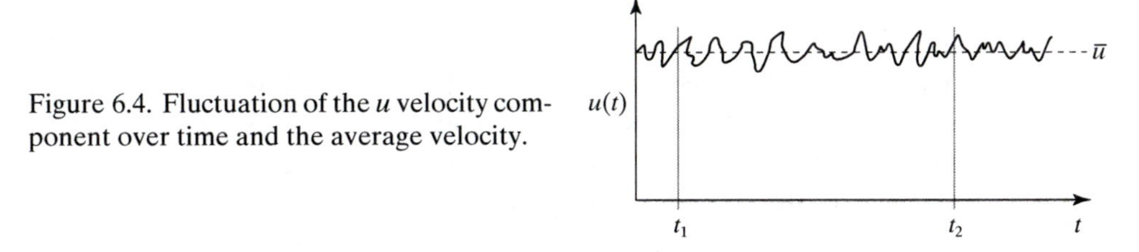

Figure 6.4. Fluctuation of the u velocity component over time and the average velocity.

an example, Fig. 6.4 shows the time-dependent recording of the momentary velocity in the x direction.

Based on the data in Fig. 6.4, the average velocity $\bar{u}$ for the time period between t_1 and t_2 is defined as

$$\bar{u} = \frac{1}{t_2 - t_1} \int_{t_1}^{t_2} u dt. \tag{6.21}$$

Similarly the momentary velocity vector will have perturbation components in all directions (u', v', w'), and will have a form

$$\vec{q} = (\bar{u} + u', v', w'), \tag{6.22}$$

and the average velocity $\vec{q}_{av}$ is in one direction only

$$\vec{q}_{av} = (\bar{u}, 0, 0).$$

This short introduction suggests that turbulent flows are *time dependent and three-dimensional*! This feature significantly complicates the fluid mechanic model and usually average-flow-based models are used (as in the case of the pipe flow in Section 5.4). The next question is related to how and under which conditions turbulence evolves. Usually strong shear conditions near solid surfaces or in open jets can lead to turbulence, but most laminar flows may turn turbulent if the Reynolds number is increased (up to a point called the transition point). The Reynolds experiment (see Section 5.7) suggested that transition from laminar to turbulent flow inside a smooth pipe occurs at about Re = 2000. For external flows over streamlined shapes, such as airfoils, transition occurs at much higher Reynolds numbers (e.g., Re–10^5–10^7) and depends on the body's shape (streamlining) and its surface roughness.

The illustration in Fig. 6.5 provides an oversimplified model for the transition from laminar to turbulent flow near the surface of a flat plate. Initially at the front of a body, the Reynolds number is low and the flow is laminar. Near the surface the

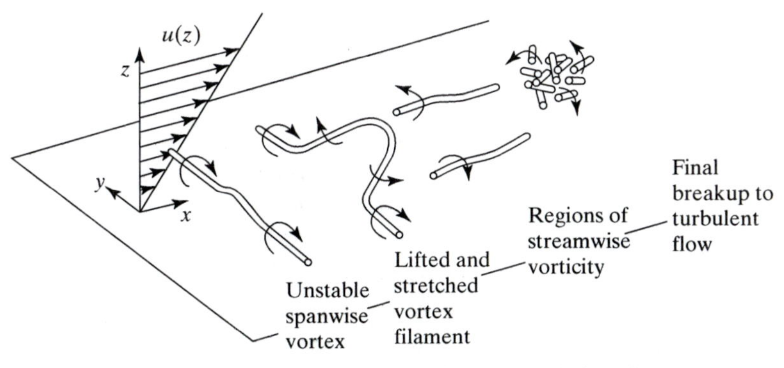

Figure 6.5. Schematic description of transition to turbulent flow.

fluid particles come to a halt (the no-slip boundary condition) and the shear flow near the wall is described by the velocity diagram on the left. The shear naturally creates angular momentum, or vorticity, as illustrated by the single spanwise vortex. As the angular momentum accumulates, an instability develops, partially lifting the vortices above the wall. Of course the velocity above the wall is faster, resulting in the stretching and eventually the breaking up of the vortices into small segments. These vortices are now responsible for the perturbations of the velocity vector in the other directions.

This brief introduction suggests that in turbulent flow more momentum is lost (larger friction), but diffusion is enhanced. So if a red drop of dye is placed on the surface it will smear fast and will not clearly follow a streamline, as expected in laminar flow.

One possible approach for modeling turbulent flows was proposed by Osborne Reynolds and is based on averaging the velocity components. Based on Eq. (6.22) the velocity vector is

$$\vec{q} = (\bar{u} + u', \bar{v} + v', \bar{w} + w'). \tag{6.23}$$

For this example let us assume steady-state (average) flow and no body forces; the incompressible momentum equation [Eq. (6.2)] in the x direction becomes

$$u\frac{\partial u}{\partial x} + v\frac{\partial u}{\partial y} + w\frac{\partial u}{\partial z} = \frac{-1}{\rho}\frac{\partial p}{\partial x} + \frac{\mu}{\rho}\left(\frac{\partial^2 u}{\partial x^2} + \frac{\partial^2 u}{\partial y^2} + \frac{\partial^2 u}{\partial z^2}\right). \tag{6.24}$$

By introducing the turbulent velocity vector of Eq. (6.23), this equation may end up in the following form; the ($^-$) sign indicates average quantities:

$$\bar{u}\frac{\partial \bar{u}}{\partial x} + \bar{v}\frac{\partial \bar{u}}{\partial y} + \bar{w}\frac{\partial \bar{u}}{\partial z} = \frac{-1}{\rho}\frac{\partial \bar{p}}{\partial x} + \frac{\mu}{\rho}\left(\frac{\partial^2 \bar{u}}{\partial x^2} + \frac{\partial^2 \bar{u}}{\partial y^2} + \frac{\partial^2 \bar{u}}{\partial z^2}\right) - \left(\frac{\partial \bar{u}'^2}{\partial x} + \frac{\partial \overline{u'v'}}{\partial y} + \frac{\partial \overline{u'w'}}{\partial z}\right). \tag{6.25}$$

This demonstrates the level of complication that is due to the addition of more unknowns. The additional perturbation-based Reynolds stresses (the terms in the right-hand parentheses) are responsible for the additional momentum transfer. In recent years large efforts were invested in modeling turbulent flows. Because of the highly complex nature of such models, the following two chapters are based on an empirical approach in which turbulent flows are discussed (similar to the treatment of turbulent flows in pipes). Therefore the discussion usually starts with a laminar flow model, which is then extended into the turbulent flow range, based on experimental data.

6.7 Summary and Conclusions

The dimensional analysis presented have provides a rational approach for simplifying the governing equations. As a prelude to the following two chapters, simplified models for the high Reynolds number flows are presented. In fact, very large percentage of fluid mechanic problems can be classified as high-Reynolds-number flows (e.g., subsonic aerodynamics or hydrodynamics). From the historical perspective, inviscid solutions evolved rapidly and were used to calculate the flow and the resulting pressure field over objects (without starting with the Navier–Stokes equations). One important example is the explanation of an airplane wing's lift through the use of simple vortex models. The analysis of this chapter attempts to explain why these methods produced reasonably accurate results.

PROBLEMS

6.1. Write the Euler equations in the 2D r–θ coordinate system. [*Hint:* Start with Eqs. (2.46) and (2.47)].

6.2. The incompressible, steady-state, 2D momentum equation in the x direction (without body forces) is

$$u\frac{\partial u}{\partial x} + w\frac{\partial u}{\partial z} = \frac{-1}{\rho}\frac{\partial p}{\partial x} + \frac{\mu}{\rho}\left(\frac{\partial^2 u}{\partial x^2} + \frac{\partial^2 u}{\partial z^2}\right).$$

Derive the nondimensional form of this equation by using the following characteristic quantities for the dimensional analysis:

$$x^* = \frac{x}{L}, \quad z^* = \frac{z}{\delta} \quad u^* = \frac{u}{U}, \quad w^* = \frac{w}{W}, \quad p^* = \frac{p}{p_0}.$$

Suppose that $\delta \ll L$ and $U \gg W$; then which terms can be neglected?

6.3. Use the characteristic quantities from the previous problem to simplify the 2D momentum equation in the z direction

$$u\frac{\partial w}{\partial x} + w\frac{\partial w}{\partial z} = \frac{-1}{\rho}\frac{\partial p}{\partial z} + \frac{\mu}{\rho}\left(\frac{\partial^2 w}{\partial x^2} + \frac{\partial^2 w}{\partial z^2}\right).$$

Again, assume that $\delta \ll L$ and $W \ll U$.

6.4. Perform a dimensional analysis on the 2D incompressible Navier–Stokes equations:

$$u\frac{\partial u}{\partial x} + w\frac{\partial u}{\partial z} = f_x - \frac{1}{\rho}\frac{\partial p}{\partial x} + \frac{\mu}{\rho}\left(\frac{\partial^2 u}{\partial x^2} + \frac{\partial^2 u}{\partial z^2}\right),$$

$$u\frac{\partial w}{\partial x} + w\frac{\partial w}{\partial z} = f_z - \frac{1}{\rho}\frac{\partial p}{\partial z} + \frac{\mu}{\rho}\left(\frac{\partial^2 w}{\partial x^2} + \frac{\partial^2 w}{\partial z^2}\right).$$

Use the following nondimensional parameters,

$$x^* = \frac{x}{L}, \quad z^* = \frac{z}{L} \quad u^* = \frac{u}{U}, \quad w^* = \frac{w}{U}, \quad f^* = \frac{f}{g} \quad p^* = \frac{p}{p_0},$$

but now consider the creeping flow assumptions where $U^2 \gg Lg$ and $\mu \gg \rho UL$.

6.5. In the discussion about the viscous terms in the fluid dynamic equations it was mentioned that the Reynolds number represents the ratio between actual and molecular scalings of length times velocity. To check this model, consider a bird with a characteristic length of $L = 0.2$ m, flying at a speed of 14 m/s. Assuming standard conditions, $c = 468$ m/s and $\lambda = 6.8 \times 10^{-8}$ m, calculate the Reynolds number by the formula

$$\text{Re} = 2\frac{VL}{c\lambda}$$

and by using the definition in Eq. (6.13) (take the density and viscosity values for air from Table 1.1).

6.6. Prove that the Reynolds and the Strouhal numbers are nondimensional by substituting engineering units into their definitions.

6.7. The vortex-shedding frequency ω behind a large truck traveling at a speed of U can be estimated by the Strouhal number as

$$\mathrm{St} = \frac{\omega}{2\pi}\frac{L}{U} = 0.2,$$

where L is the truck width.

This can be felt by smaller vehicles (such as a motorcycle) traveling behind the truck. Calculate the shedding frequency for a 2-m-wide truck traveling at 100 km/h. Also estimate the distance D between two vortex cycles.

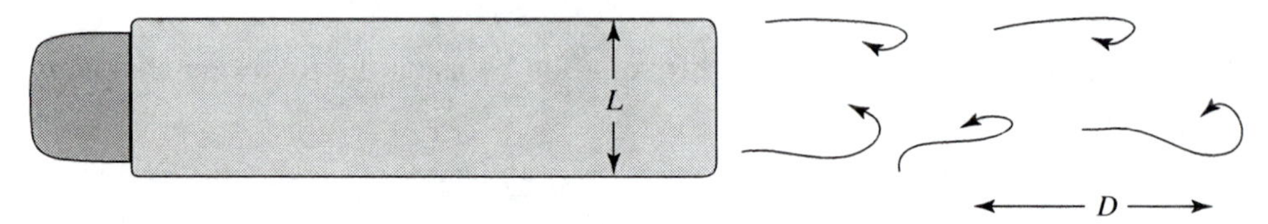

Problem 6.7.

6.8. A quarter-scale model of the truck in the preceding problem was placed in a wind tunnel. If the model is tested at a true speed of 100 km/hr, calculate the vortex-shedding frequency and the distance between two shedding cycles. If the intention is to have the same frequency as in full scale, then what is the desirable air speed in the wind tunnel?

6.9. The quarter-scale model of a tall cylindrical chimney was tested in a wind tunnel at a speed of 25 km/h and the vortex-shedding frequency was measured at 3 Hz.

(a) At what wind speed will the vortex-shedding frequency be the same for the full-scale chimney? (*Hint:* Keep St constant).

(b) Can you keep both the Reynolds number and the Strouhal number constant in such a quarter-scale test?

6.10. The Froude number for an open channel flow of velocity U and a depth of h was defined in Chapter 5 as

$$\mathrm{Fr} = \frac{U}{\sqrt{gh}}.$$

(a) Calculate the value of the Froude number for a 1-m-deep channel in which the water is flowing at an average speed of 0.5 m/s.

(b) The channel shape was tested in a 1/10-scale model. What is the desirable average velocity in order to keep the same Froude number (and not change the surface-wave behavior)?

6.11. The length of a proposed submarine is 100 m and the planned speed is 20 km/h. To estimate the power requirements for the propulsion system, the vehicle resistance must be evaluated. It was proposed to build a wind-tunnel model of the submarine and test it in air. If the similitude assumption is used, what is the required length of the wind-tunnel model if the test speed is 80 m/s?

6.12. A 0.3-m-long fish is swimming in a river at a speed of 0.2 m/s. Calculate the Reynolds number based on its length.

6.13. A 20-m-long boat sails at a sped of 20 km/h. Calculate the Reynolds number (based on its length), using the seawater properties listed in Table 1.1.

6.14. To study the velocity profile of motor-oil flow in a circular pipe, it was proposed to conduct a test using water (instead of oil). The oil-conducting pipe diameter is 0.4 m and the expected average velocity is about 1.5 m/s. For the properties of these two fluids, use the values listed in Table. 1.1.

(a) Suppose the same diameter pipe is used for the test with the water. What is the average water velocity for the two flows to be similar?
(b) Would you change the pipe diameter, too, for the test with the water?

6.15. Near the radiator inlet of a car moving at a speed of $U = 100$ km/h the air is stopped and the pressure coefficient is

$$C_p \equiv \frac{p - p_0}{\frac{1}{2}\rho U^2} = 1.0.$$

Calculate the pressure rise compared with the undisturbed surrounding pressure p_0 ($\rho = 1.2$ kg/m^3).

Convert your results to pressure in terms of water head (water density is 1000 kg/m^3).

6.16. At the roof of the car in the previous problem, a pressure coefficient of -1.0 was measured. Calculate the airspeed at that point.

6.17. Suppose a quarter-scale model of the same automobile was tested but at a true speed of 100 km/h. Calculate the pressure in front of the radiator, where $C_p = 1$, and on the roof, where $C_p = -1$.

6.18. The average chord of an airplane's wing is 4 m and it flies at sea level ($\rho =$ 1.225 kg/m^3, $\mu = 1.78 \times 10^{-5}$ N s/m^2, $T = 288$ K, $\gamma = 1.4$) at a speed of 1000 km/h. Calculate the Mach number and the Reynolds number based on the wing average chord.

6.19. The airplane of the previous problem flies at the same speed of 1000 km/h, but at an altitude of 10,000 m, where $\rho = 0.413$ kg/m^3, $\mu = 1.46 \times 10^{-5}$ N s/m^2, $T =$ 223 K, $\gamma = 1.4$.

(a) Calculate the Mach number and the Reynolds number based on the wing average chord.
(b) What percentage of speed reduction is required if the airplane must fly at the same Mach number as calculated for sea level (in the previous problem)?

6.20. Water is flowing in a 2.5-cm inner-diameter pipe. Calculate the minimum flow rate at which the flow is expected to be laminar (assume that transition occurs at Re = 2000. Use fluid properties from Table 1.1).

6.21. Oil flows through a 3-mm inner-diameter tube to lubricate the bearings of a large machine.

(a) Taking the properties of SAE 30 oil from Table 1.1, calculate the flow rate for a laminar Reynolds number of Re = 1000.
(b) This flow is modeled by use of a larger tube of 2.5-cm inner diameter. Calculate the flow rate for the same Reynolds number.

6.22. An airplane flies at a speed of 900 km/h, at an altitude of 10,000 m, where $\rho = 0.413$ kg/m^3, $\mu = 1.46 \times 10^{-5}$ N s/m^2, $T = 223$ K, $\gamma = 1.4$. If the wing average chord is 4 m and transition from laminar to turbulent flow is expected at Re $= 10^6$, calculate the percentage of chord with laminar flow.

6.23. The drag of a football is investigated in a water tunnel with a quarter-scale model and matching the full-scale Reynolds number. If the actual expected speed is assumed at 95 km/h, estimate the ratio of the drag force between the model and the full-size football. Use the fluid properties (for air and water) as listed in Table 1.1 and assume that the force F on the ball varies as $F \sim \rho U^2 D^2$, where D is the ball diameter.

6.24. A 5-m-wide, 2-m-deep irrigation water channel is simulated by a 1/10–scale model. To maintain smooth flow, without strong surface waves, the Froude number $(u/\sqrt{gh})$ must be kept below 1. Estimate the average speed and the water flow rate in full scale and for the small-scale model for Fr $= 0.8$.

6.25. A 4-m-wide road sign oscillates heavily in a 15-km/h crosswind. To study the problem a 1/20-scale model was placed in the water-tunnel. Estimate the water-tunnel speed if both the Reynolds number and the Strouhal number must be kept the same (at 0.2) as in full scale, as well as the oscillation frequency. (Use fluid properties from Table 1.1.)

6.26. To study a small insect's aerodynamics, helium was used instead of air in a wind tunnel. If flow speed remains the same as in actual scale, estimate how much the insect model size can increase and still maintain the true Reynolds number (for helium use $\rho = 0.179$ kg/m^3, $\mu = 1.9 \times 10^{-5}$ N s/m^2).

6.27. It is proposed to build a wind tunnel using compressed air so that smaller models can be tested (but at the same Reynolds number). Compare the power requirements for two wind tunnels operating at the same speed; one is using atmospheric condition and the other (one half-width and height) has 2-atm compressed air inside. (Assume the required power is proportional to the mass flow rate and air viscosity is independent of pressure.)

6.28. The free-stream velocity in the test section during an early morning wind-tunnel test was 100 km/h and the temperature was 280 K (use $\rho = 1.2$ kg/m^3, $\mu = 1.8 \times 10^{-5}$ N s/m^2). The same test is repeated in the afternoon but now the temperature was 310 K (to calculate ρ, use the ideal-gas relation, $\mu = 2.0 \times 10^{-5}$ N s/m^2).

(a) Calculate the desirable free-stream speed to maintain the Reynolds number of the morning test.

(b) At what speed will the forces (e.g., lift and drag) be the same as in the morning?

7 The (Laminar) Boundary Layer

7.1 Introduction

The previous chapter discussed the physical aspects of high-Reynolds-number flow and concluded that far from a solid surface the viscous effects are negligible (and the flow is irrotational). The models presented in this chapter and in the next are aimed at streamlined shapes without flow separation. Based on experimental data, these results can be later generalized to include effects of flow separation. It was also concluded in Chapter 6 that near a solid surface the fluid particles must adhere to the zero-slip boundary condition and therefore viscous effects cannot be neglected. This region is called the *boundary layer* and is assumed to be thin compared with the dimensions (e.g., length) of the body around which the fluid flows. To clarify the high-Reynolds-number-flow assumption, consider a 4-m-long streamlined automobile traveling at $U_\infty = 100$ km/h (as in Fig. 7.1). Taking the density and viscosity values for air from Table 1.1, we can calculate the Reynolds number:

$$\mathrm{Re} = \frac{\rho U_\infty L}{\mu} = \frac{1.22 \times 100/3.6 \times 4}{1.8 \times 10^{-5}} = 7.5 \cdot 10^6.$$

This is a very large number, considering that all terms in Eq. (6.7) were of the order of $O(1)$, and clearly indicates that the viscous terms are negligible. The conclusion at this point is that indeed many flows of interest fall within this high-Reynolds-number-flow category. Figure 7.1 demonstrates the nature of the two flow regimes. As the air moves around the body, its velocity changes and at any point we can call this (outer) velocity U_e (how to calculate this velocity is discussed in the next chapter). Within the thin boundary layer, denoted as δ, the velocity must change from zero to U_e, as shown in the inset to Fig. 7.1.

Consequently the discussion in this chapter is aimed at deriving a basic boundary-layer model and the elements necessary to explain the concept of combining the inner viscous and the outer inviscid flows. With these two (inner and outer flow) high-Reynolds-number-flow models in mind, the information sought from the viscous boundary-layer solution in this chapter is as follows:

1. The scale, or thickness, of the boundary layer and its streamwise growth.
2. Displacement effects (to the outer flow model) that are due to the slower velocity inside the viscous layer.

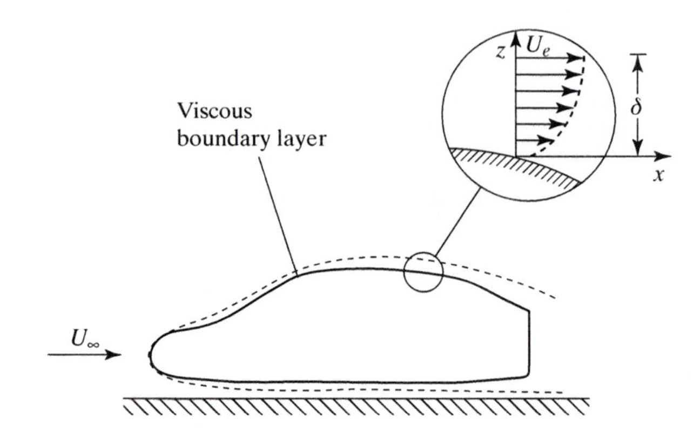

Figure 7.1. Schematic description of a high-Reynolds-number-flow boundary layer on the upper surface of a car. The inset depicts the velocity distribution near the solid surface.

3. The skin-friction and resulting drag estimates that cannot be calculated by the outer inviscid flow.
4. Based on the laminar boundary-layer model, parameters such as the boundary-layer thickness or the skin friction will be extended into the turbulent flow range, mostly by use of empirical correlations (similar to the method used for the friction coefficient in pipes – with the Moody diagram).

It is interesting to compare the *internal* flow between two parallel plates in Section 5.33 (the Poiseuille flow – between the wall and the centerline) with the current *external* boundary-layer flow. In both cases the no-slip boundary condition is applied to the solid surface resulting in a velocity profile that increases with the distance from the solid surface. Consequently we may expect similar conclusions regarding the skin friction and its dependence on the Reynolds number. We shall see later that indeed the laminar friction coefficient decreases with increased Reynolds number (see also the Moody diagram for pipe flows), as well as the trends in the experiment-based turbulent friction coefficient.

From the historical perspective, the German scientist Ludwig Prandtl (1874–1953) was the first to develop the 2D boundary-layer equations, circa 1904. One of his first students, Paul Richard Heinrich Blasius (1883–1970), provided the first analytical solution (circa 1908) for these equations. This solution is quite complex and beyond the scope of this text, but its results are described in the advanced topics subsection, Subsection 7.6.1. Of course, nowadays we can solve the boundary-layer equations numerically, and the results should duplicate the Blasius solution. A third approach is called the integral approach that is quite simple to explain and is presented next.

7.2 Two-Dimensional Laminar Boundary-Layer Flow over a Flat Plate – (The Integral Approach)

Let us propose a simple flat plate model, as shown in Fig. 7.2, in which the plate is parallel to the free stream U_∞. In actual flows, as in the case described in Fig. 7.1, the surface is curved and the outer velocity changes locally [e.g., $U_e = U_e(x, t)$ where x is the coordinate along the surface]. In this case, however, we assume a thin plate and therefore the outer velocity remains ($U_e = U_\infty$). We can select several vertical points and inject tracers (or colored dye) at $t = 0$. After a while (1 s in Fig. 7.2) we observe the location of the tracers and we can see the expected velocity profile inside the boundary layer, so an additional objective would be to determine the shape of

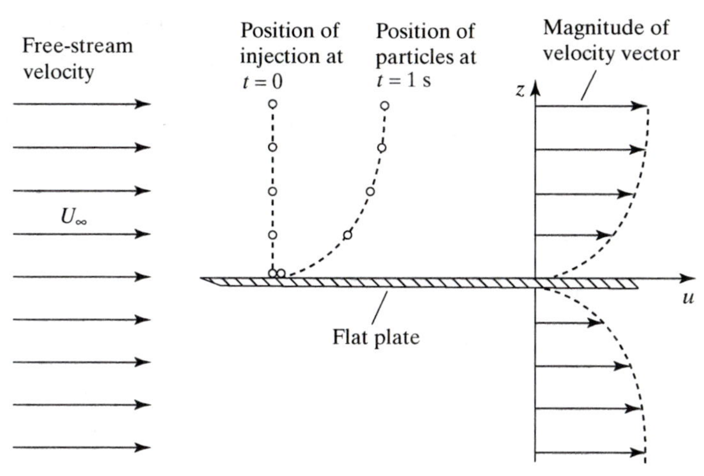

Figure 7.2. Boundary-layer model on a flat plate.

this velocity distribution. Of course, in the case shown in Fig. 7.2 such a boundary layer will form both above and under the plate, but our discussion will focus on one (the upper) layer only.

As noted in Chapter 2, an analytical solution of the complex fluid dynamic equations is next to impossible, but through careful modeling we may simplify them into a form that can be solved. This case is no exception, and we can start with the assumptions that properties are constant (μ = const. and ρ = const.) and that body forces such as gravity are negligible. Next we limit ourselves to a 2D laminar flow case (as in Fig. 7.2). Another very important assumption, based on the exact solution, is that velocity profiles $u(z)$ at any station along the plate are similar in shape. This means that a universal velocity profile can be assumed (when there is no pressure gradient). Also, as noted, a control-volume approach for developing the boundary-layer integral formulation is more intuitive and much simpler to present.

This basic model is described schematically in Fig. 7.3, where a laminar boundary layer develops along a flat plate, placed in a parallel free stream with a velocity of U_e. An x–z coordinate system is placed at the leading edge of the plate, as shown in the figure. From this point and on, a boundary layer of thickness $\delta(x)$ develops, having a velocity profile (not yet known) with zero velocity at $z = 0$. A rectangular control volume (actually a 2D control surface) of length dx is placed such that it is bounded by the four corners 1–2–3–4. The plane 2–4 is placed above the boundary layer, at $z = z_e$, where there are no transverse changes in the u component of the velocity; and a constant free-stream speed U_e prevails. Our objective is to evaluate

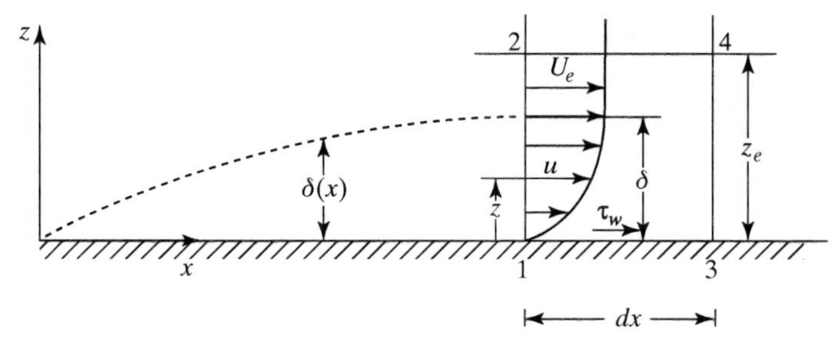

Figure 7.3. The control-volume model used to develop the integral boundary-layer equations.

the change in the momentum entering and leaving the control volume and to balance this change with the external forces (namely the shear stress on the plate and the pressure gradient in the x direction).

Let us start by evaluating the mass flow rate entering the control element through plane 1–2 and assume the flow is incompressible:

$$\rho \int_0^{z_e} u dz.$$

The flow rate leaving through plane 3–4 can be approximated by use of the first term of a Taylor series:

$$\rho \int_0^{z_e} u dz + \rho \frac{d}{dx}\left(\int_0^{z_e} u dz\right) dx.$$

Because there is no flow across the wall (plane 1–3), the net change in the mass flow rate must have occurred through plane 2–4 and is

$$\rho \frac{d}{dx}\left(\int_0^{z_e} u dz\right) dx \tag{7.1}$$

In a similar manner (simply multiplying by the velocity), the momentum in the x direction entering across plane 1–2 is

$$\rho \int_0^{z_e} u^2 dz \tag{7.2}$$

and leaving through plane 3–4 is

$$\rho \int_0^{z_e} u^2 dz + \rho \frac{d}{dx}\left(\int_0^{z_e} u^2 dz\right) dx. \tag{7.3}$$

The mass flow rate change through plane 2–4, expressed by Eq. (7.1), has a constant speed of U_e outside the control surface. Therefore the momentum crossing this plane is

$$\rho U_e \frac{d}{dx}\left(\int_0^{z_e} u dz\right) dx. \tag{7.4}$$

The time rate of the momentum change within the control surface is

$$\rho \frac{\partial}{\partial t}\left(\int_0^{z_e} u dz\right) dx. \tag{7.5}$$

Thus the net rate of change of the momentum in the x direction is due to the change with time [Eq. (7.5)] and due to the difference between the momentum leaving [Eq. (7.3)] and entering [Eqs. (7.2) and (7.4)] the control surface;

$$\begin{aligned}
&\rho \frac{\partial}{\partial t}\left(\int_0^{z_e} u dz\right) dx + \rho \int_0^{z_e} u^2 dz + \rho \frac{d}{dx}\left(\int_0^{z_e} u^2 dz\right) dx \\
&-\rho \int_0^{z_e} u^2 dz - \rho U_e \frac{d}{dx}\left(\int_0^{z_e} u dz\right) dx \\
&= \rho \frac{\partial}{\partial t}\left(\int_0^{z_e} u dz\right) dx + \rho \frac{d}{dx}\left(\int_0^{z_e} u^2 dz\right) dx - \rho U_e \frac{d}{dx}\left(\int_0^{z_e} u dz\right) dx.
\end{aligned} \tag{7.6}$$

At this point, Eq. (7.6) provides the information on the fluid acceleration. To complete the momentum relation (e.g., $\rho a_x = \sum F$), the forces acting on the control element must be added. Thus, according to the momentum principle, this change in the linear momentum [as in Eq. (7.6)] must be equal to the forces acting on the control volume in Fig. 7.3. Because the body forces were neglected, the only forces acting are the pressure and the laminar shear stress on the wall. Using Eq. (1.14), we find that the shear force on the wall, along segment 1–3, is

$$-\tau_w dx = -\mu \frac{\partial u}{\partial z}\bigg|_{z=0} dx. \tag{7.7}$$

Because this is a thin layer and the vertical velocity components are small, all terms in the momentum equation [Eq. (5.2c)] vanish and we conclude that the pressure is independent of z (inside the boundary layer); then $p(x)$ is a function of x only. Consequently the pressure force on segment 1–2 is simply pz_e and on segment 3–4 is

$$-\left(p + \frac{dp}{dx}dx\right) z_e;$$

the net force acting on the control surface is the sum of the shear and pressure forces:

$$\sum F = -\tau_w dx - \frac{dp}{dx} dx z_e. \tag{7.8}$$

Equating the forces in Eq. (7.8) with the change in the momentum in Eq. (7.6) results in

$$\rho \frac{\partial}{\partial t}\left(\int_0^{z_e} u dz\right) dx + \rho \frac{d}{dx}\left(\int_0^{z_e} u^2 dz\right) dx - \rho U_e \frac{d}{dx}\left(\int_0^{z_e} u dz\right) dx$$
$$= -\tau_w dx - \frac{dp}{dx} dx z_e.$$

Now if we let $z_e \to \delta$ and divide by ρdx, we obtain the von Kármán integral equation for momentum change in the boundary layer:

$$\frac{\partial}{\partial t}\int_0^{\delta} u dz + \frac{d}{dx}\int_0^{\delta} u^2 dz - U_e \frac{d}{dx}\int_0^{\delta} u dz = -\frac{\tau_w}{\rho} - \frac{1}{\rho}\frac{dp}{dx}\delta. \tag{7.9}$$

This equation is named after the Hungarian-born and later U.S. scientist Theodore von Kármán (1881–1963) who completed his doctoral dissertation under Prandtl.

7.3 Solutions Based on the von Kármán Integral Equation

The shear on the wall, for example, in integral momentum equation (7.9) depends on the velocity distribution; therefore this equation cannot be solved without additional information. The necessary information can be obtained from an assumed similar velocity profile family $f(\eta)$:

$$\frac{u(x, z)}{U_e} = f(\eta), \tag{7.10}$$

where the nondimensional vertical parameter η is defined as

$$\eta = \frac{z}{\delta(x)}. \tag{7.11}$$

This means that at any x station along the plate, the shape of the velocity distribution [Eq. (7.10)] is uniform. For example, at $z = \delta$, the velocity is equal to the outer flow, $u(z) = U_e$, as well as the first and second derivatives of $u(z) = 0$. This idea was actually presented by Blasius in his 1908 solution of the boundary-layer equations. He too assumed that the shape of the velocity profile is uniform and will grow along x and its nondimensional shape [as in (Eq. 7.10)] is unchanged (if there is no pressure gradient). Consequently, the first step in the process of solving this problem is to guess a velocity profile inside the boundary layer.

We can demonstrate the simplicity of the integral approach by suggesting an approximate velocity distribution within the boundary layer. Then, parameters such as the boundary-layer thicknesses and skin-friction coefficient can be readily calculated. For example, even simple polynomial velocity profiles can be used:

$$\frac{u}{U_e} = f(\eta) = a_0 + a_1\eta + a_2\eta^2 + \cdots + \{0 \le \eta \le 1\}.$$

Let us limit the discussion to the case for the boundary layer along a flat plate, in steady state and without a pressure gradient. For the flow between two plates (section 5.3.3) we have seen linear and parabolic velocity distributions, so clearly the a_1 and a_2 terms will be included. However, the a_0 term cannot be included because the velocity must be zero on the plate (at $z = 0$). In an effort to expand the number of terms used, let us propose the following velocity distribution:

$$\frac{u}{U_e} = a_1\eta + a_2\eta^2 + a_3\eta^3 + a_4\eta^4. \tag{7.12}$$

The boundary conditions for the original boundary layer problem are

$$\text{at } z = 0, \quad u = w = 0;$$

$$\text{at } z = \delta, \quad u = U_e, \quad \frac{\partial u}{\partial z} = 0, \tag{7.13}$$

and the requirement for smooth transition at the outer edge of the boundary-layer dictate: $\partial u/\partial z = 0$ at $z = \delta$. We can generate additional boundary conditions by observing the change of the streamwise momentum inside the boundary layer, but without the pressure gradient. Thus assuming that $(\partial U_e/\partial x) = 0$, combined with the previous boundary conditions, results in

$$\text{at } z = 0, \quad \frac{\partial^2 u}{\partial z^2} = 0;$$

$$\text{at } z = \delta, \quad \frac{\partial^2 u}{\partial z^2} = 0. \tag{7.14}$$

Applying at $z = \delta$, $u = U_e$ to Eq. (7.12), we get

$$1 = a_1 + a_2 + a_3 + a_4; \tag{*}$$

at $z = \delta$, $(\partial u/\partial z) = 0$,

$$0 = a_1 + 2a_2 + 3a_3 + 4a_4; \tag{**}$$

at $z = \delta$, $(\partial^2 u/\partial z^2) = 0$,

$$0 = 2a_2 + 6a_3 + 12a_4, \tag{***}$$

and the last boundary condition at $z = 0$, $(\partial^2 u/\partial z^2) = 0$,

$$0 = 2a_2, \tag{****}$$

Solving these four equations for the four coefficients a_1, a_2, a_3, a_4 results in the following equation for the velocity profile:

$$\frac{u}{U_e} = 2\eta - 2\eta^3 + \eta^4. \tag{7.15}$$

As noted, this is a universal velocity distribution, which is similar along the plate.

The next step is to solve for the evolution of the boundary-layer thickness along the plate and for the associated frictional losses. For simplicity, let us consider a steady-state case, without a pressure gradient (e.g., $\partial/\partial t = 0$ and $dp/dx = 0$). To solve the problem for δ we substitute this velocity profile into von Kármán's integral equation 7.9, but first we calculate the wall shear stress by using Eq. (7.15):

$$\tau_w = \mu\left(\frac{\partial u}{\partial z}\right)_{z=0} = \mu\frac{2U_e}{\delta}. \tag{7.16}$$

Substituting this and the velocity profile into Eq. (7.9), without the pressure term, yields

$$\frac{d}{dx}\int_0^\delta U_e^2\left[2\left(\frac{z}{\delta}\right) - 2\left(\frac{z}{\delta}\right)^3 + \left(\frac{z}{\delta}\right)^4\right]^2 dz$$
$$-U_e\frac{d}{dx}\int_0^\delta U_e\left[2\left(\frac{z}{\delta}\right) - 2\left(\frac{z}{\delta}\right)^3 + \left(\frac{z}{\delta}\right)^4\right] dz = -\frac{\mu}{\rho}\frac{2U_e}{\delta}.$$

Evaluating the two integrals, we get

$$U_e^2\frac{d}{dx}(0.5825\delta) - U_e^2\frac{d}{dx}(0.7000\delta) = -\frac{\mu}{\rho}\frac{2U_e}{\delta},$$

and, after rearranging,

$$\delta\frac{d\delta}{dx} = 17.021\frac{\mu}{\rho U_e}.$$

Integrating with x and recalling that $\delta = 0$ at $x = 0$, we get

$$\frac{\delta^2}{2} = 17.021\frac{\mu x}{\rho U_e},$$

and, after rearranging again, we obtain the growth of the boundary layer with x:

$$\delta = 5.836\sqrt{\frac{\mu x}{\rho U_e}}. \tag{7.17}$$

This is an important result because it demonstrates that the boundary layer grows at a rate of $\sqrt{x}$. This relation can further be rearranged as

$$\delta = 5.836\sqrt{\frac{\mu x}{\rho U_e}} = 5.836x\sqrt{\frac{\mu}{\rho U_e x}}.$$

Now, if we define a length-based Reynolds number

$$\mathrm{Re}_x = \frac{\rho U_e x}{\mu}, \tag{7.18}$$

then we can write

$$\delta = 5.836 \frac{x}{\sqrt{\mathrm{Re}_x}}. \tag{7.19}$$

With this solution for δ and with the velocity profile of Eq. (7.15), $u(x, z)$ is known everywhere and the values for the boundary-layer thickness can be calculated. It is customary to rewrite Eq. (7.19) in a nondimensional form as

$$\frac{\delta}{x} = \frac{5.836}{\sqrt{\mathrm{Re}_x}}. \tag{7.20}$$

At this point the question about the boundary-layer thickness was answered; next we need to develop the displacement-thickness relation. This quantity should provide "thickness-loss" information for the outer flow. Consider the velocity profile in the boundary layer, as depicted on the left-hand side of Fig. 7.4, where the outer velocity U_e is reduced to zero near the plate. If the plate (and resulting skin friction) weren't there, then the velocity would remain U_e everywhere. So the net effect of placing the plate in the free stream is a loss of fluid flow with a thickness of δ^* – a quantity called the displacement thickness (as shown in the figure).

To evaluate the displacement thickness, the flow rate loss that is due to the reduced velocity $(U_e - u)$ in the boundary layer [see Fig. 7.4(a)] must be calculated. This loss is

$$\int_0^\infty (U_e - u) dz,$$

and of course the integration is from 0 to δ, but for added confidence the infinity sign is used. Figure 7.4(b) shows the flow rate loss – which is due to not having any flow up to a height of δ^*. This loss is $U_e\delta^*$, per unit width, and therefore

$$U_e\delta^* = \int_0^\infty (U_e - u) dz.$$

Consequently the definition of the displacement thickness is

$$\delta^* = \int_0^\infty \left(1 - \frac{u}{U_e}\right) dz. \tag{7.21}$$

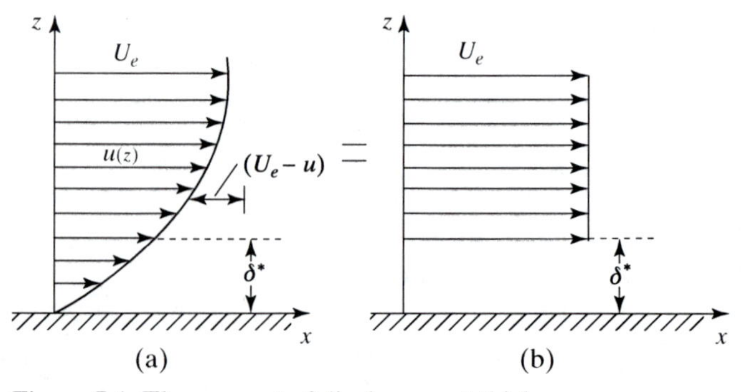

Figure 7.4. The concept of displacement thickness.

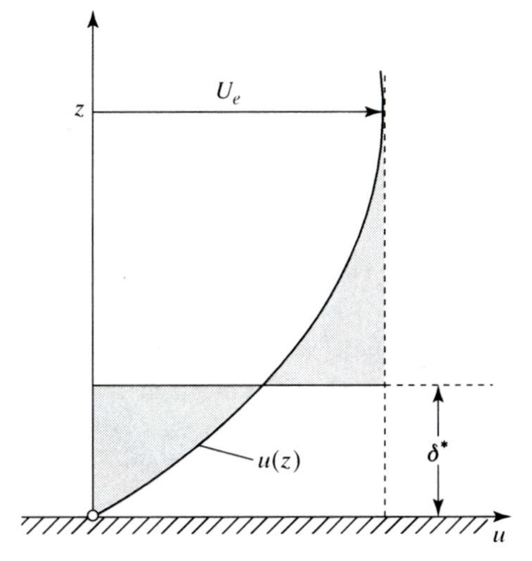

Figure 7.5. The presence of the boundary layer reduces the flow rate (of velocity U_e) by the displacement thickness δ^*.

Using the velocity profile of Eq. (7.15), performing the integration from 0 to δ, and substituting the boundary-layer thickness from Eq. (7.18), we get

$$\delta^* = \int_0^\delta \left(1 - \frac{u}{U_e}\right) dz = 1.751 \frac{x}{\sqrt{\mathrm{Re}_x}}.$$

Or, in nondimensional form, we have

$$\frac{\delta^*}{x} = \frac{1.751}{\sqrt{\mathrm{Re}_x}}. \tag{7.22}$$

Now, returning to Fig. 7.1, the displacement thickness means that the body shape effectively gained thickness and the outer flow must be solved over a thicker body shape. Sometimes the displacement thickness at a particular x location is described schematically, as in Fig. 7.5. The solid curve shows the actual velocity profile. However, eliminating the flow below $z = \delta^*$ and transferring it above this line will result in a constant-velocity profile of U_e (so the flow rate inside the two "triangular-shaped" shaded areas is the same).

It is possible to define another boundary-layer thickness θ, representing the deficiency in the momentum that is due to the presence of the boundary layer. Referring to Fig. 7.4, we find that the calculation is similar, but now we simply multiply by the local velocity u. Consequently the equation for calculating the *momentum thickness* θ is

$$\theta = \int_0^\infty \left(1 - \frac{u}{U_e}\right) \frac{u}{U_e} dz, \tag{7.23}$$

and by substituting the velocity profile of Eq. (7.15), performing the integration from 0 to δ, and substituting the boundary-layer thickness from Eq. (7.19), we get

$$\theta = \int_0^\delta \left(1 - \frac{u}{U_e}\right) \frac{u}{U_e} dz = 0.685 \frac{x}{\sqrt{\mathrm{Re}_x}},$$

and in nondimensional form

$$\frac{\theta}{x} = \frac{0.685}{\sqrt{\mathrm{Re}_x}}. \tag{7.24}$$

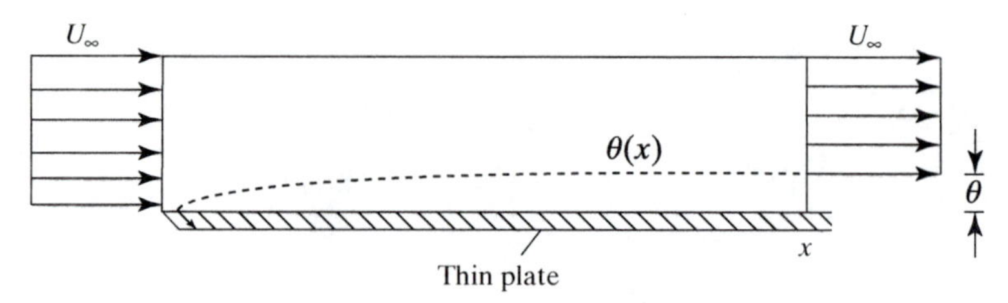

Figure 7.6. The momentum thickness represents the momentum loss up a particular point on the plate.

Sometimes the ratio between these two boundary-layer thicknesses is defined as H, the *shape factor*, and in this case it equals 2.556:

$$H = \frac{\delta^*}{\theta} = 2.556. \tag{7.25}$$

The importance of the shape factor is in identifying the transition to turbulent flows where this ratio is much smaller.

With the aid of the momentum thickness we can demonstrate the effect of friction in the boundary layer; see the schematic diagram in Fig. 7.6. At the leading edge of the plate, the velocity is uniform and equal to the free stream. After a while the viscous flow losses on the plate create a loss of thickness θ, as shown. Therefore the drag force (per unit width) on the plate must be equal to this loss:

$$D = \dot{m}U_e = \rho U_e \theta \cdot U_e = \rho U_e^2 \theta. \tag{7.26}$$

We can obtain the same result by applying the 1D momentum equation [Eq. (2.29)] between the plate's leading edge and station x on the figure.

More specifically, we can show this by integrating the loss of momentum across the boundary layer,

$$D = \rho \int_0^\delta (U_e - u)u dz = \rho U_e^2 \int_0^\infty \left(1 - \frac{u}{U_e}\right) \frac{u}{U_e} dz = \rho U_e^2 \theta, \tag{7.27}$$

and this is the same result. The nondimensional form of the drag force is called the *drag coefficient* and is defined as

$$C_D \equiv \frac{D}{1/2 \rho U^2 S}, \tag{7.28}$$

where S is the reference area. Substituting the drag from Eq. (7.26) and the momentum thickness from Eq. (7.24) yields

$$C_D(x) = \frac{\rho U_e^2 \theta}{1/2 \rho U_e^2 S} = \frac{2}{S} \frac{0.685}{\sqrt{\mathrm{Re}_x}} x,$$

where $C_D(x)$ represent the drag up to a point x on the plate. For a plate of length $x = L$ and unit width (so $S = l \times 1$) we get

$$C_D = \frac{1.370}{\sqrt{\mathrm{Re}_L}}. \tag{7.29}$$

Next, let us calculate the shear stress along the plate. Using the result of Eq. (7.16) and substituting δ from Eq. (7.20), we get

$$\tau_w = \mu \frac{2U_e}{\delta} = \frac{2U_e}{5.836} \sqrt{\frac{\rho \mu U_e}{x}} = 0.3427 \sqrt{\frac{\rho \mu U_e^3}{x}}. \tag{7.30}$$

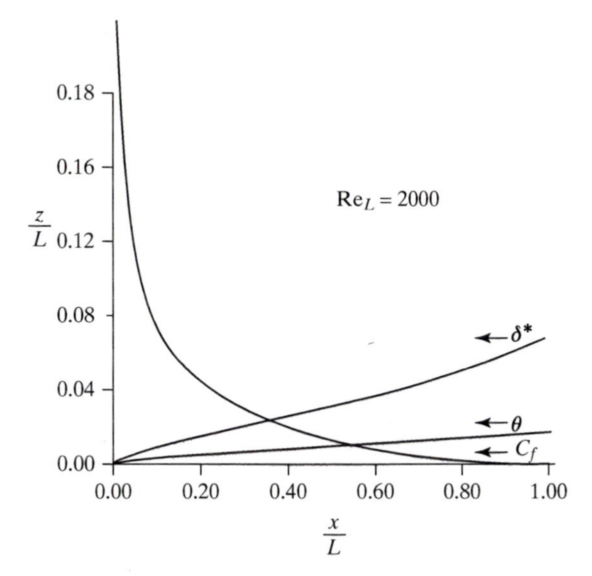

Figure 7.7. Variation of the displacement and momentum thicknesses and the skin-friction coefficient along the plate of length L (laminar flow).

This is again an important result because it shows that the shear force is the largest at the plate's leading edge and decays at the rate of $\sqrt{1/x}$. Next, we can use this result to calculate the skin-friction coefficient:

$$C_f = \frac{\tau_w}{\frac{1}{2}\rho U_e^2} = \frac{0.3427}{\frac{1}{2}}\sqrt{\frac{\mu}{\rho U_e x}} = \frac{0.685}{\sqrt{\mathrm{Re}_x}}. \tag{7.31}$$

We can calculate the drag D on the upper surface by integrating the shear stress along the plate [and it must verify Eq. (7.27)]

$$D = \int_0^L \tau_w dx = \int_0^L \mu \frac{2U_e}{\delta} dx = \int_0^L \mu \frac{2U_e}{5.836\frac{x}{\sqrt{\mathrm{Re}_x}}} dx$$

$$= \frac{2}{5.836}\sqrt{\rho\mu U_e^3}\int_0^L \sqrt{\frac{1}{x}} dx = \frac{2}{5.836}\sqrt{\rho\mu U_e^3} \times 2\sqrt{L} = 0.685\sqrt{\rho\mu U_e^3 L}. \tag{7.32}$$

To make sure that this is the same result for the drag as obtained in Eq. (7.29), we use the nondimensional definition of the drag coefficient:

$$C_D = \frac{D}{1/2\rho U_e^2 S} = \frac{0.685\sqrt{\rho\mu U_e^3 L}}{1/2\rho U_e^2 L \times 1} = \frac{1.370}{\sqrt{\mathrm{Re}_L}} = 2C_f. \tag{7.33}$$

This is the same result of Eq. (7.29). Also note that the drag of the whole plate (one side) is twice the local friction coefficient C_f. Finally Fig. 7.7 shows the variation of some of the boundary-layer parameters along a flat plate of length L. Clearly the boundary-layer thicknesses increase with x and the momentum thickness is much smaller than the displacement thickness. The skin-friction coefficient and shear stress are the largest at the leading edge because the boundary layer is thin near the leading edge.

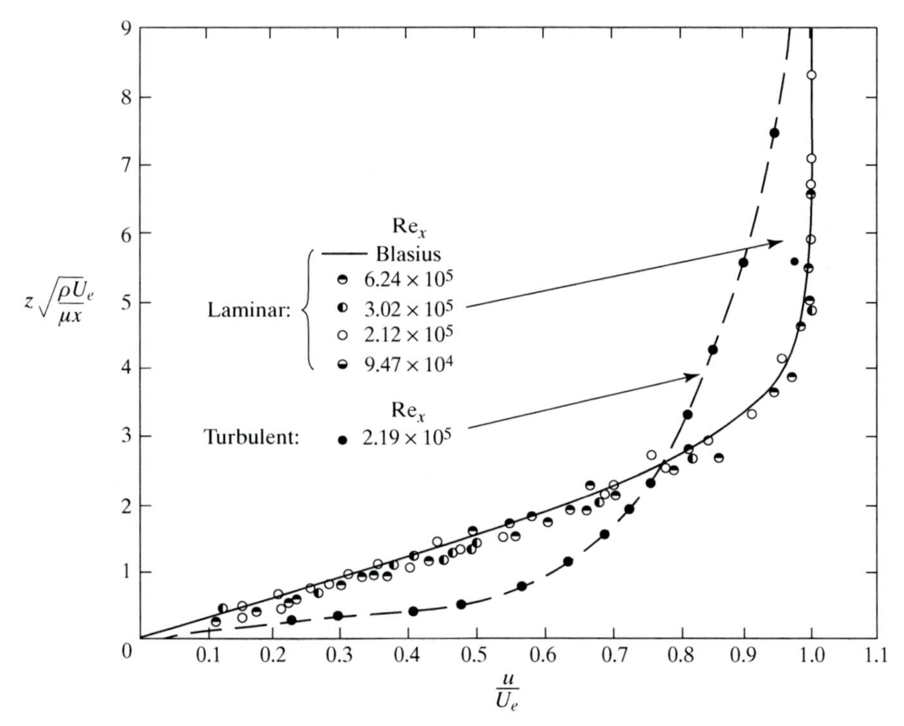

Figure 7.8. Velocity profiles in both laminar (Blasius solution) and turbulent boundary layers and comparison with experimental results (dots).

7.4 Summary and Practical Conclusions

Although we started with an approximate solution and an approximate and similar velocity profile, the results generated are quite satisfactory. The exact solution presented by Blasius (in 1908) was based on a more detailed velocity distribution, and experimental results (see the laminar flow curve in Fig. 7.8) show excellent agreement with his model.

As we discussed earlier for the case of laminar flow in pipes, at higher Reynolds number the flow can be turbulent. Even on a flat plate, as the Reynolds number increases a transition to turbulent flow takes place. The velocity profile inside the turbulent boundary layer is shown in Fig. 7.8 as well, and clearly the shear near the wall is larger, suggesting significantly larger friction coefficients (also a thicker boundary layer). An approximate formula describing the velocity profile inside the turbulent boundary layer is

$$\frac{\bar{u}}{U_e} = \eta^{\frac{1}{7}}, \tag{7.34}$$

and the turbulent shear stress based on the model discussed in Section 6.6 is

$$\tau_{xz} = \mu \frac{\partial u}{\partial z} - \rho \overline{u'w'}. \tag{7.35}$$

So not only is the velocity gradient larger, but there is an additional loss of momentum through the average turbulent components u' and w' (e.g., momentum loss in the other directions). A quite typical situation in high-Reynolds-number flows is

Table 7.1. *Comparison among laminar, Blasius, and turbulent (based on the $\frac{1}{7}$-power law) flat-plate, integral boundary-layer properties. Here δ_{99} is used instead of δ, representing 99% of the boundary-layer thickness*

Property	Approximate solution	Exact solution (Blasius)	Turbulent flow
$\frac{\delta_{99}}{x}$	$\frac{5.83}{\sqrt{\text{Re}_x}}$	$\frac{5.00}{\sqrt{\text{Re}_x}}$	$\frac{0.37}{\sqrt[5]{\text{Re}_x}}$
$\frac{\delta^*}{x}$	$\frac{1.751}{\sqrt{\text{Re}_x}}$	$\frac{1.721}{\sqrt{\text{Re}_x}}$	$\frac{0.046}{\sqrt[5]{\text{Re}_x}}$
$\frac{\theta}{x}$	$\frac{0.685}{\sqrt{\text{Re}_x}}$	$\frac{0.664}{\sqrt{\text{Re}_x}}$	$\frac{0.036}{\sqrt[5]{\text{Re}_x}}$
C_f	$\frac{0.685}{\sqrt{\text{Re}_x}}$	$\frac{0.664}{\sqrt{\text{Re}_x}}$	$\frac{0.0576}{\sqrt[5]{\text{Re}_x}}$
C_D	$\frac{1.370}{\sqrt{\text{Re}_x}}$	$\frac{1.328}{\sqrt{\text{Re}_x}}$	$\frac{0.074}{\sqrt[5]{\text{Re}_x}}$
H	2.56	2.59	1.28

shown schematically in Fig. 7.9. Here the simplest case of the parallel flow above a thin plate is shown, and, as expected, at the front of the plate, a laminar boundary layer forms. As the local Reynolds number increases (within the range of 10^5–10^7) a transition takes place, perhaps through the sequence shown in Fig. 6.5. From this point and on, the boundary layer is thicker (because of the lateral perturbations) and the momentum losses increase.

The boundary-layer parameters for our approximate laminar flow solution and for the exact Blasius solution, along with data for turbulent flow, are presented in Table 7.1. First note the quite good agreement of the approximate model with the exact solution for the laminar flow case. Also, the turbulent boundary layer is thicker (Fig. 7.9) and has a larger friction coefficient (so look closer at the formulas). The shape factor, on the other hand, is much smaller for turbulent flow, and because of this feature, it can be used to identify transition.

From the engineering point of view, estimating the friction drag is very important. Therefore it is desirable to provide information on the friction coefficient for a wide range of Reynolds numbers (similar to the flow in pipes). Such data are presented in Fig. 7.10, and the laminar flow curve for the drag coefficient is the Blasisus solution from Table 7.1. The curve for turbulent boundary layers is based on experimental data and can be used reliably for engineering calculations. Also, there is a large overlap between the laminar and turbulent flow regions (Re $\sim 10^5$–10^7), presenting an engineering opportunity to reduce skin friction by retaining longer

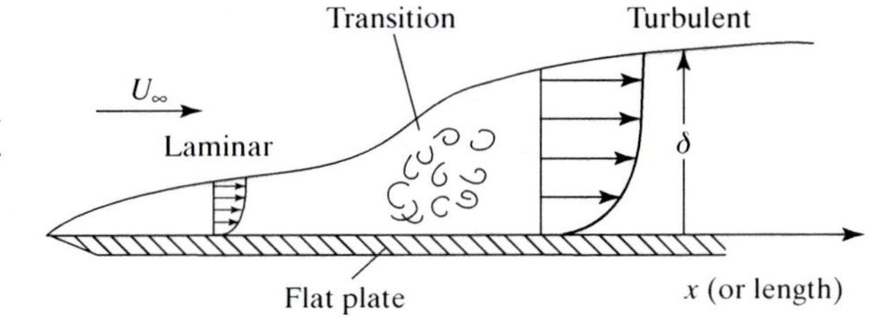

Figure 7.9. Schematic description of transition from laminar to turbulent flow on a flat plate.

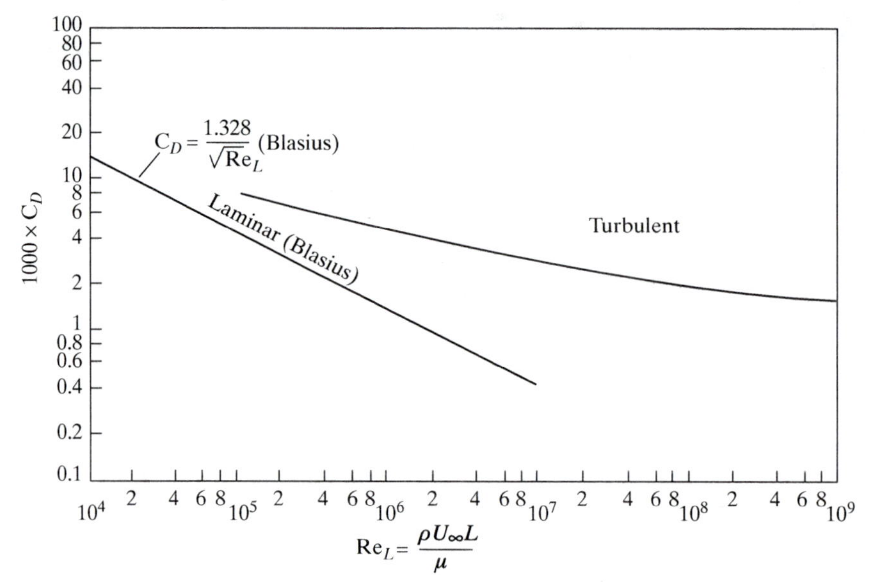

Figure 7.10. Drag coefficients on a smooth flat plate of length L for laminar and turbulent flows (at $M \to 0$).

laminar flow regions. Note that C_f is the local friction coefficient and the drag coefficient is for a plate of length L. According to Eq. (7.33),

$$C_D = 2C_f. \tag{7.33}$$

EXAMPLE 7.1. SKIN-FRICTION DRAG OF A FLAT PLATE. A 1-m-wide, 0.5-long-thin plate is towed in a water tank at a speed of 2 m/s. Assuming laminar flow, calculate the drag force.

Solution: We use the Blasius solution for this case. First we calculate the Reynolds number at the end of the plate:

$$Re_L = \frac{1000 \times 2 \times 0.5}{10^{-3}} = 10^6.$$

The drag coefficient, per side, is then

$$C_D = 2C_f = 2\frac{0.664}{\sqrt{Re_L}} = 1.328 \times 10^{-3}.$$

The total drag on the plate (two sides) is

$$D = 2C_D\frac{1}{2}\rho U^2 S = 2 \times 1.328 \times 10^{-3}\frac{1}{2}1000 \times 4(0.5 \times 1) = 2.656 \text{ N}.$$

Next let us calculate the boundary-layer thicknesses:

$$\delta^* = \frac{1.721}{\sqrt{Re_L}}L = 0.86 \times 10^{-3}\text{ m} = 0.86\text{ mm},$$

and the momentum thickness is

$$\theta = \frac{0.664}{\sqrt{Re_L}}L = 0.33 \times 10^{-3}\text{ m} = 0.33\text{ mm}.$$

Now let us repeat the same calculations but with the plate turned by 90°, so $L = 1$ m. The Reynolds number at the end of the plate is

$$\mathrm{Re}_L = \frac{1000 \times 2 \times 1}{10^{-3}} = 2 \times 10^6,$$

and the drag coefficient per side is

$$C_D = 2C_f = 2\frac{0.664}{\sqrt{\mathrm{Re}_L}} = 0.939 \times 10^{-3}.$$

The total drag on the plate (two sides) is

$$D = 2C_D\frac{1}{2}\rho U^2 S = 2 \times 0.939 \times 10^{-3}\frac{1}{2}1000 \times 4(0.5 \times 1) = 1.878\,\mathrm{N}.$$

Next let us calculate the boundary-layer thicknesses:

$$\delta^* = \frac{1.721}{\sqrt{\mathrm{Re}_L}}L = 1.22 \times 10^{-3}\,\mathrm{m} = 1.22\,\mathrm{mm},$$

and the momentum thickness is

$$\theta = \frac{0.664}{\sqrt{\mathrm{Re}_L}}L = 0.47 \times 10^{-3}\,\mathrm{m} = 0.47\,\mathrm{mm}.$$

Therefore, although the boundary layer at the trailing edge is thicker in the second case, the drag is lower. This is because the skin friction is larger near the leading edge and a wider leading edge results in more drag.

EXAMPLE 7.2. TURBULENT DRAG OF A FLAT PLATE. A 12-m/s wind is blowing parallel to the roof of an exposed carport. Its width (relative to the wind) is 10 m and its length is 4 m. Because of the surface roughness, assume the flow is fully turbulent. Calculate the drag force on the roof.

Solution: First we calculate the Reynolds number at the end of the plate (the properties of air are taken from Table 1.1):

$$\mathrm{Re}_L = \frac{1.22 \times 12 \times 4}{1.8 \times 10^{-5}} = 3.25 \times 10^6.$$

Based on Fig. 7.10, both laminar and turbulent boundary layers are possible. However, as stated, because of the not-so-smooth surface, we assume turbulent flow and the value of the drag coefficient per side (from Fig. 7.10) is $C_D = 0.0032$. The total drag on the plate (two sides) is

$$D = 2C_D\frac{1}{2}\rho U^2 S = 2 \times 0.0032\frac{1}{2}1.22 \times 12^2(10 \cdot 4) = 22.49\,\mathrm{N}.$$

7.5 Effect of Pressure Gradient

The boundary-layer solution in the previous section (for the sake of simplicity) assumed a flat plate with no pressure gradient. In most practical applications, however, the surface is not flat and there is a pressure gradient. Typically in most flow fields the pressure gradient is a function of the surface shape and both positive and negative pressure gradients are possible. If the pressure outside the boundary layer (along the surface) changes from high pressure to low ($dp/dx < 0$), then we call

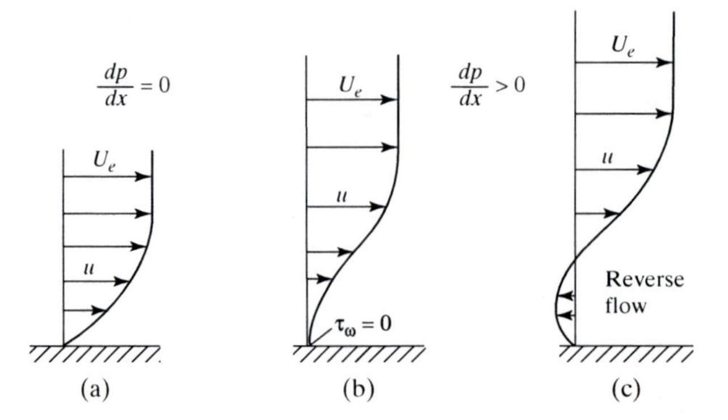

Figure 7.11. Effect of pressure gradient on velocity profile inside the boundary layer.

this a favorable pressure gradient. Such a pressure gradient energizes the boundary layer and usually transition to turbulent flow is delayed.

Figure 7.11 depicts the case in which the pressure gradient ($dp/dx > 0$) is unfavorable or adverse. In this case the pressure increases with x along the surface and in effects slows down the flow. For example, Fig. 7.11(a) shows the velocity profile inside the boundary layer with no pressure gradient; Fig. 7.11(b) depicts schematically the effect of an adverse pressure field. In the particular case shown here the adverse pressure slows down the flow near the wall, and the shear stress there is zero. This condition is usually the borderline ahead of flow separation, which is described schematically in Fig. 7.11(c). In this case the adverse pressure forces the flow backward and flow recirculation is observed. It is possible to have all the preceding cases in one flow. For example, the boundary layer starts as in Fig. 7.11(a) and gradually, because of the adverse pressure, transitions into zero shear stress and then to flow separation.

In addition to the velocity profile inside the boundary layer, we can observe the effect of pressure gradient on other parameters. As noted, if the pressure gradient is favorable (as on the left-hand side of Fig. 7.12) then an initially laminar boundary layer stays laminar up to Reynolds numbers of several millions (see Fig. 7.10). Even if the conditions without a pressure gradient would determine transition to turbulent flow, the presence of a favorable pressure gradient will delay the transition [and also flow separation – as depicted in Fig. 7.11(c)]. If the outer conditions dictate an adverse pressure gradient, as shown on the right-hand side of Fig. 7.12, then transition to turbulent flow will be soon triggered, as shown schematically in this figure. Using these principles can lead to designs with less drag. An example demonstrating the redesign of an airfoil such that the outer flow will have a long

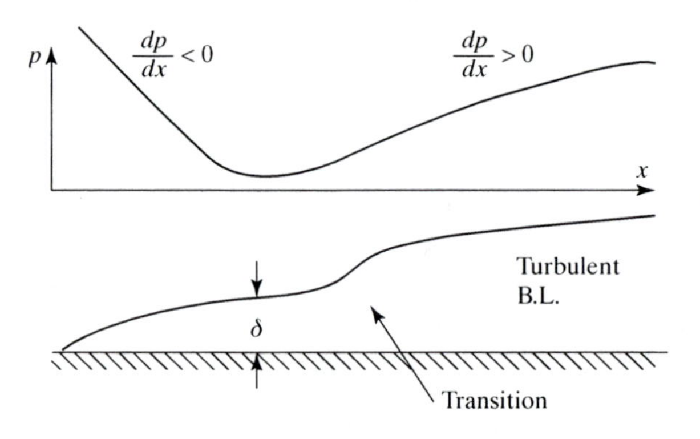

Figure 7.12. Effect of pressure gradient on the boundary layer.

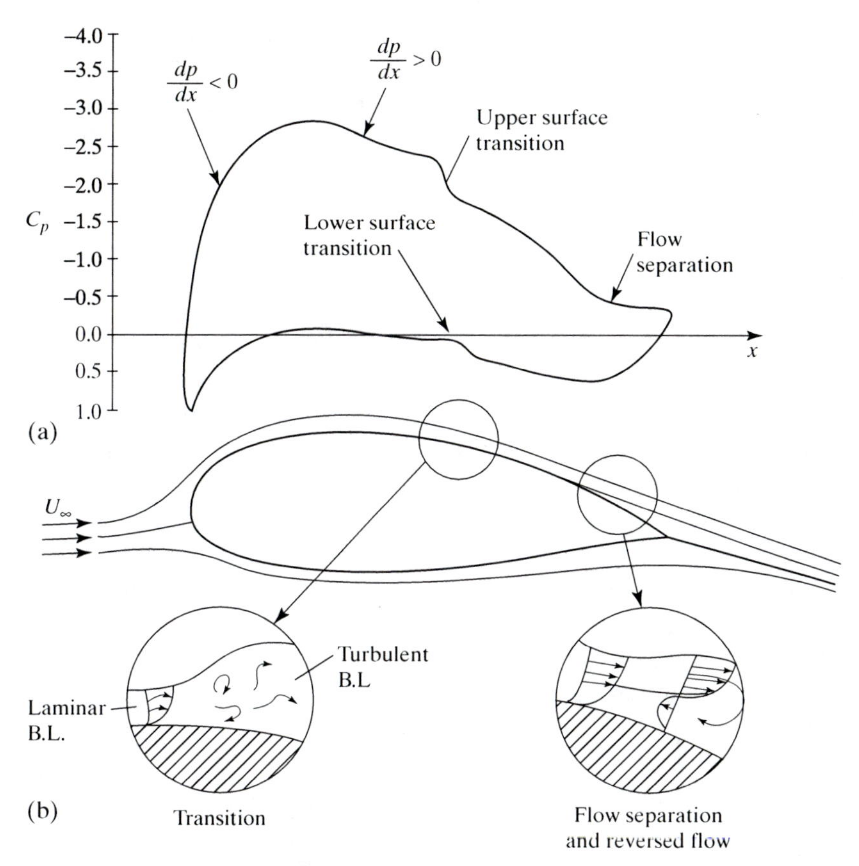

Figure 7.13. High-Reynolds-number flow over a thick airfoil and effect of boundary-layer transition and separation on pressure distribution ($\text{Re} = 10^6$).

favorable pressure distribution is shown later in Fig. 8.38). Consequently the longer laminar flow resulted in significant reduction in the drag.

Figure 7.13 summarizes the preceding principles for a thick airfoil. The shape of the airfoil and the nearby streamlines are shown at the center, and above it the expected pressure distribution (which is due to the external flow). The stagnation streamline is shown in front and it stops at the leading edge, resulting in a high pressure there. From that point on (on both the upper and lower surfaces) the flow accelerates, velocity increases, and the pressure is reduced (recall the Bernoulli equation). Throughout this initial region a favorable pressure gradient exist and the boundary layer stays laminar! However, from the geometrical point of view, the airfoil surface curvature must change toward the trailing edge – so that a closed body is formed. Consequently the pressure gradient becomes positive, indicating that the velocity slows down toward the trailing edge. In the case shown in Fig. 7.13, the initial downward slope of dp/dx is moderate and the boundary-layer transition to turbulent flow is delayed a bit. The transition can be detected by the small kinks in the pressure distribution curve (on both upper and lower surfaces), as indicated in the figure. This kink is a result of a localized laminar separation (called *laminar bubble*), which is responsible for the sudden increase in boundary-layer thickness. Behind these transition points, the skin friction increases, increasing the airfoil skin-friction drag. As the upper surface curvature decreases, closing in on the trailing edge, dp/dx is increased (see upper surface pressure distribution), leading to boundary-layer separation near the trailing edge. The reverse flow inside the boundary layer is shown by the inset. Also, the pressure inside the separated flow region is almost constant (negative), resulting in increased pressure drag on the airfoil. Of course, this figure was drawn to highlight the effects of pressure gradient on the boundary layer. However, in a good design, flow separation must be eliminated (or reduced) for less drag and better performance.

7.6 Advanced Topics: The Two-Dimensional Laminar Boundary-Layer Equations

From the historical point of view, the differential form of the boundary-layer equations was developed first (by Prandtl in 1904). We can derive these equations by neglecting smaller terms in the Navier–Stokes equations. Also, this topic is included in the advanced topics section for two reasons. First, the solution is beyond the scope of an introductory course; the integral approach used earlier requires much simpler mathematics. The second reason is that time is limited when one is presenting a comprehensive introduction to fluid flow, and a more detailed treatment usually belongs in a more advanced viscous flow course.

For simplicity, we limit ourselves to a 2D case for the flow over a flat plate (as in Fig. 7.2) without body forces; the incompressible fluid dynamic equations for this case [Eqs. (5.1), (5.2)] are

$$\frac{\partial u}{\partial x} + \frac{\partial w}{\partial z} = 0, \tag{7.36}$$

$$\frac{\partial u}{\partial t} + u\frac{\partial u}{\partial x} + w\frac{\partial u}{\partial z} = \frac{-1}{\rho}\frac{\partial p}{\partial x} + \frac{\mu}{\rho}\left(\frac{\partial^2 u}{\partial x^2} + \frac{\partial^2 u}{\partial z^2}\right), \tag{7.37}$$

$$\frac{\partial w}{\partial t} + u\frac{\partial w}{\partial x} + w\frac{\partial w}{\partial z} = \frac{-1}{\rho}\frac{\partial p}{\partial z} + \frac{\mu}{\rho}\left(\frac{\partial^2 w}{\partial x^2} + \frac{\partial^2 w}{\partial z^2}\right). \tag{7.38}$$

Let us reiterate that the boundary-layer thickness $\delta = \delta(x)$ is much smaller than the characteristic length, L (e.g., the plate length), along the solid surface:

$$\delta \ll L. \tag{7.39}$$

This is probably true everywhere except at the leading edge, where both the length and the boundary-layer thicknesses are zero. Equations (7.36)–(7.38) can be simplified by a dimensional analysis, similar to the one used in Chapter 6, and a consideration of the order of magnitude of the terms. For this analysis, a set of characteristic quantities may be defined again: However, now we make a distinction between the streamwise and the normal directions:

$$\begin{aligned} x^* &= \frac{x}{L}, \quad z^* = \frac{z}{\delta}, \\ u^* &= \frac{u}{U}, \quad w^* = \frac{w}{W}, \\ p^* &= \frac{p}{p_0}, \\ t^* &= \frac{t}{T} = \frac{t}{L/U}. \end{aligned} \tag{7.40}$$

Here U (instead of U_∞) is the characteristic velocity in the x direction and W is the characteristic velocity in the z direction. The flow equations are now rewritten in terms of the new variables and the continuity equation becomes

$$\frac{U}{L}\frac{\partial u^*}{\partial x^*} + \frac{W}{\delta}\frac{\partial w^*}{\partial z^*} = 0. \tag{7.41}$$

If we assume that all nondimensional variables of Eq. (7.41) are of the order of $O(1)$ inside the boundary layer, then for both terms in the continuity equation to be of the same order it is necessary that U/L be of the order of W/δ. Therefore, if $\delta \ll L$, then it follows that $W \ll U$, and the order of magnitude of W is determined as

$$\frac{W}{U} = O\left(\frac{\delta}{L}\right). \tag{7.42}$$

Consequently we conclude that the continuity equation is unchanged (and we cannot neglect either of the two terms). Introducing the nondimensional variables into the momentum equation in the x direction [similar to the treatment of Eq. (6.7)] results in

$$\Omega\frac{\partial u^*}{\partial t^*} + u^*\frac{\partial u^*}{\partial x^*} + \frac{W}{U}\frac{L}{\delta}w^*\frac{\partial u^*}{\partial z^*} = -\text{Eu}\frac{\partial p^*}{\partial x^*} + \frac{1}{\text{Re}}\left[\frac{\partial^2 u^*}{\partial x^{*2}} + \left(\frac{L^2}{\delta^2}\right)\frac{\partial^2 u^*}{\partial z^{*2}}\right]. \tag{7.43}$$

All three terms on the left-hand side of this equation appear to have the same order of magnitude; the first (viscous) term can be clearly neglected in comparison with the second viscous term. If we recall our basic assumption that, inside the boundary layer, the inertia terms [left-hand side of Eq. (7.43)] are of the same order of magnitude as the viscous terms, then the remaining viscous term is of the order of 1; therefore,

$$\frac{1}{\text{Re}}\left(\frac{L^2}{\delta^2}\right) \approx O(1),$$

and it follows that

$$(\delta/L) = O(\text{Re}^{-1/2}), \quad (W/U) = O(\text{Re}^{-1/2}). \tag{7.44}$$

Consequently, only one term, the first viscous term [from left in Eq. (7.43)] is neglected in this equation! Substituting the nondimensional quantities into the momentum equation in the z direction results in

$$\frac{W}{U}\Omega\frac{\partial w^*}{\partial t^*} + \frac{W}{U}u^*\frac{\partial w^*}{\partial x^*} + \frac{W^2}{U^2}\frac{L}{\delta}w^*\frac{\partial w^*}{\partial z^*}$$
$$= -\frac{L}{\delta}\text{Eu}\frac{\partial p^*}{\partial z^*} + \frac{1}{\text{Re}}\frac{W}{U}\left[\frac{\partial^2 w^*}{\partial x^{*2}} + \left(\frac{L^2}{\delta^2}\right)\frac{\partial^2 w^*}{\partial z^{*2}}\right] \tag{7.45}$$

Again, all inertia terms on the left-hand side are of the same order of magnitude $[O(\delta/L)]$ and are considerably smaller than the pressure term, which is multiplied by L/δ. The viscous terms are of the same order as the inertia terms $[O(\delta/L)]$ because, according to Eqs. (7.44), $1/\text{Re} = O(\delta^2/L^2)$. Therefore all inertia and viscous terms appearing in this equation are much smaller than the pressure term and can be neglected. Rearranging the remaining terms in the three equations yields for the continuity

$$\frac{\partial u}{\partial x} + \frac{\partial w}{\partial z} = 0. \tag{7.46}$$

For the momentum equation in the x direction, only one viscous term is neglected,

$$\frac{\partial u}{\partial t} + u\frac{\partial u}{\partial x} + w\frac{\partial u}{\partial z} = -\frac{1}{\rho}\frac{\partial p}{\partial x} + \frac{\mu}{\rho}\frac{\partial^2 u}{\partial z^2}, \tag{7.47}$$

whereas, in the z direction, all terms but the normal pressure gradient become negligible, implying that the normal pressure gradient itself is equal to zero as well:

$$0 = -\frac{\partial p}{\partial z}. \tag{7.48}$$

So the pressure is dictated by the outer flow and there are no changes in p across the boundary layer. Equations (7.46)–(7.48) define the classical 2D boundary-layer equations proposed by Prandtl (German scientist, 1874–1953) in 1904. At the wall, the no-slip boundary condition remains,

$$z = 0, \quad u = w = 0, \tag{7.49a}$$

and at the edge of the boundary layer ($z = \delta$) the tangential velocity component must approach the inviscid surface value of $U_e(x, t)$,

$$z = \delta, \quad u = U_e(x, t). \tag{7.49b}$$

We can simplify the momentum equation in the x direction by rewriting it at the outer edge of the boundary layer (at $z > \delta$), where there are no changes with z:

$$\frac{\partial U_e}{\partial t} + U_e\frac{\partial U_e}{\partial x} = \frac{-1}{\rho}\frac{\partial p}{\partial x}. \tag{7.50}$$

This value can be inserted into Eq. (7.47) so that the pressure p is no longer an unknown in the problem. With the preceding assumptions and for the case of steady-state flow, Eq. (7.47) reduces to

$$u\frac{\partial u}{\partial x} + w\frac{\partial u}{\partial z} = U_e\frac{\partial U_e}{\partial x} + \frac{\mu}{\rho}\frac{\partial^2 u}{\partial z^2}. \tag{7.51}$$

An exact solution for the boundary-layer equations is quite complex (and beyond the scope of this introductory text), but it was solved by Heinrich Blasius, one of Prandtl's first students, circa 1908.

7.6.1 Summary of the Blasius Exact Solution for the Laminar Boundary Layer

The key to the analytic solution of the boundary-layer equation was a uniform velocity distribution proposed by Heinrich Blasius. The assumption is that along the flat plate (as in Fig. 7.3) the boundary layer will grow but the shape of the velocity profile will stay similar:

$$\frac{u(x, z)}{U_e} = f\left[\frac{z}{\delta(x)}\right] = f(\eta). \tag{7.52}$$

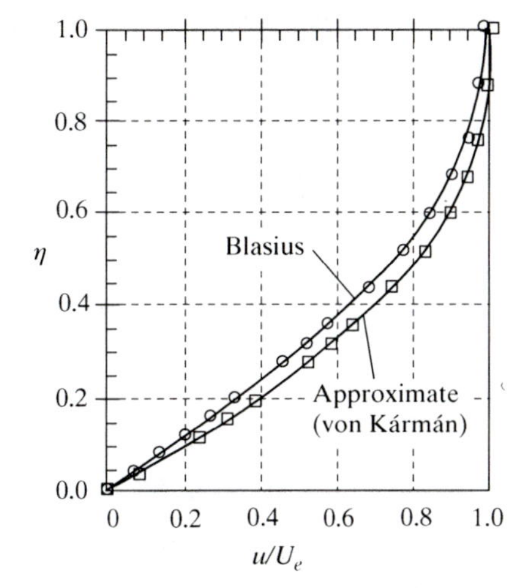

Figure 7.14. The Blasius velocity distribution inside the boundary layer and the simple approximation of Eq. (7.15).

With this assumption he was able to solve the boundary-layer equations, and the resulting velocity distribution is depicted in Fig. 7.14 (also in Fig. 7.8). Note that at the edge of the boundary layer ($z = \delta$) the velocity is equal to the outer velocity (U_e), and the slope near the wall is determined by the solution for the velocity profile. This velocity profile is plotted in Fig. 7.14, which also presents the approximate polynomial profile of Eq. (7.15).

Once the velocity profile is determined, the boundary-layer growth is calculated. Because the changes in the Blasius velocity profile near the edge of the boundary layer are very small δ_{99} was calculated, a point at which 99% of the outer velocity is reached:

$$\frac{\delta_{99}}{x} = \frac{5.00}{\sqrt{\mathrm{Re}_x}}. \tag{7.53}$$

This equation shows that the boundary-layer thickness increases with $\sqrt{x}$, as shown schematically in Fig. 7.15. This figure also shows the growth of the displacement thickness δ^*, which we calculate by using the definition of Eq. (7.22):

$$\frac{\delta^*}{x} = \frac{1.721}{\sqrt{\mathrm{Re}_x}}. \tag{7.54}$$

The momentum thickness is calculated by Eq. (7.24); for the Blasius solution it is

$$\frac{\theta}{x} = \frac{0.664}{\sqrt{\mathrm{Re}_x}}. \tag{7.55}$$

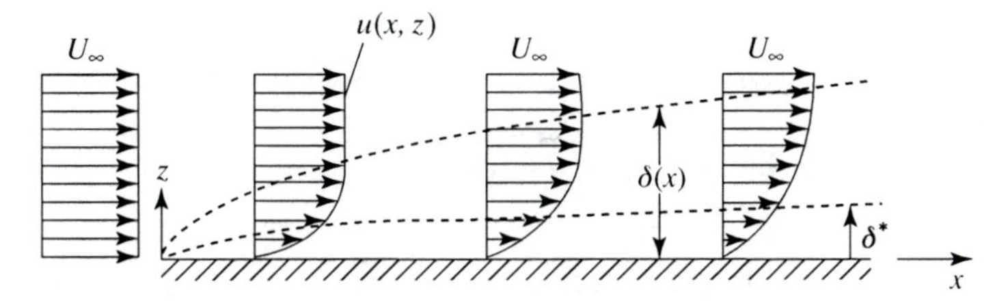

Figure 7.15. Schematic description of the Blasius solution for a laminar boundary layer.

The friction coefficient is calculated next [as in Eq. (7.31)]:

$$C_f = \frac{0.664}{\sqrt{\text{Re}_x}}. \tag{7.56}$$

We obtain the shape factor H by dividing the displacement thickness by the momentum thickness:

$$H = 2.59. \tag{7.57}$$

and these results are quite close to the results generated by the integral approach of Section 7.3.

7.7 Concluding Remarks

This chapter follows the rationale developed in Chapter 5. Resembling the approach for solving the laminar pipe flow, the solution of the shear stress and friction coefficients were calculated first for the laminar case. Then, based on experimental results, the formulation was extended into the turbulent flow regime, so that the shear stress and resulting drag can be estimated for a wide range of Reynolds numbers.

The present analysis indicates that the boundary layers in an attached flow are thin compared with the length scale. Also, experimental results indicate that laminar flow friction drag is significantly less than the drag in turbulent flow. Therefore the possibility of delaying boundary-layer transition provides an opportunity for drag reduction.

PROBLEMS

7.1. A 15-km/h wind is blowing parallel to a 5-m-long flat plate. Calculate the shear stress at a distance of 5 cm from the leading edge, at 1 m from the leading edge, and at the end of the plate (e.g., 5 m from the leading edge). Calculate Re_L first to ensure the use of the laminar flow formulas. Assume that $\mu = 1.8 \times 10^{-5}$ N s/m^2 and $\rho = 1.2$ kg/m^3.

7.2. A 15-km/h wind is blowing parallel to a 5-m-long flat plate. Calculate the boundary-layer thickness at a distance of 5 cm from the leading edge, at 1 m from the leading edge, and at the end of the plate (e.g., 5 m from the leading edge). Calculate Re_L first to ensure the use of the laminar flow formulas. Assume that $\mu = 1.8 \times 10^{-5}$ N s/m^2 and $\rho = 1.2$ kg/m^3.

7.3. The wing of a sail plane is mostly laminar and its wingspan is 8 m. If the average chord is 0.4 m, then calculate the laminar, skin-friction drag force on the wing at a speed of 60 km/h (use air properties from Table 1.1). Do you think the actual drag of the wing is higher?

7.4. The chord of an airplane's thin wing is 2 m and it is flying at 300 km/h. Calculate the friction drag for unit width (a) assuming laminar flow on both the upper and lower surfaces and (b) assuming turbulent flow on both surfaces. Assume that $\mu = 1.8 \times 10^{-5}$ N s/m^2 and $\rho = 1.2$ kg/m^3.

7.5. The chord of an airplane's thin wing is 2 m and it is flying at 200 km/h. Using the flat-plate model, calculate the displacement thickness and friction coefficient at the wing's trailing edge (a) assuming laminar flow and (b) assuming turbulent flow. Assume that $\mu = 1.8 \times 10^{-5}$ N s/m^2 and $\rho = 1.2$ kg/m^3.

7.6. A typical setup for automobile model testing is shown in the figure. The model length is $L = 1.5$ m, the ground clearance is $h = 3$ cm, and the test speed is 100 km/h ($\mu = 1.8 \times 10^{-5}$ N s/m^2, $\rho = 1.2$ kg/m^3). Ahead of the model, on the floor there is a suction slot and it is assumed that the boundary layer is eliminated there. Calculate the Reynolds number and the boundary-layer thickness on the floor at a distance $l = 2$ m behind the car nose for non moving floor. Also calculate the boundary-layer thickness on the lower surface of the model (of length L). Assume laminar flow and use the Blasius formulas. Can these two boundary layers affect the test results?

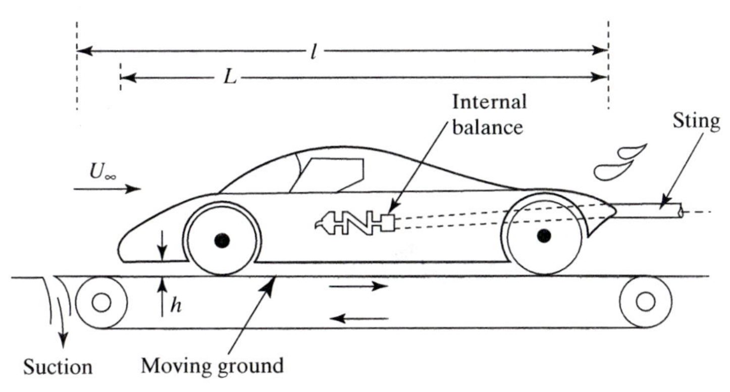

Problem 7.6.

7.7. The boundary layer on the bottom of the vehicle is tripped by a stripe of coarse sandpaper and it is turbulent along the whole surface. Estimate the boundary-layer thickness on the lower surface of the vehicle and the drag on this surface if model width is 0.5 m. Would you recommend using the moving-ground simulation?

7.8. A 1-m-long, 3-m-wide plate is towed at a velocity of 2 m/s in a fluid. Assuming laminar flow, find the boundary-layer thickness δ, displacement thickness δ^*, and momentum thickness θ at the end of the plate. Also calculate the drag force on one side of the plate ($\mu = 10^{-3}$ N s/m^2, $\rho = 1000$ kg/m^3).

7.9. A 1.5-m-long, 1-m-wide flat plate is towed in water at a speed of 0.2 m/s. Calculate the drag force (resistance) of the plate and the boundary-layer thickness at its end. Is this a laminar boundary layer?

7.10. A flat plate is towed in a resting fluid. The density of the fluid is 1.5 kg/m^3 and its viscosity is 10^{-5} N s/m^2. Find the skin-friction drag on the plate per unit width if the plate is 2 m long and the free-stream velocity is 20 m/s. Also calculate the wall velocity gradient at the center of the plate (1 m from the leading edge). Assume turbulent flow where $\delta = 0.16/\mathrm{Re}^{1/7}$, $C_D = 0.523/\ln^2(0.06\mathrm{Re})$.

7.11. Assuming a laminar boundary layer exists over a submerged flat plate (in water, $\rho = 1000$ kg/m^3, $\mu = 10^{-3}$ N s/m^2). If the free-stream velocity is $U_\infty = 10$ m/s, then calculate the displacement thickness δ^* at a distance of $L = 2$ m from the plate's leading edge. Also calculate the total friction force (on one side of the plate) per unit width for the 2-m-long section. Use the Blasius laminar flow formulas.

7.12. A 1-m-long, thin symmetric airfoil is placed at a free stream as shown in the figure. Calculate Re and the drag force for the speed of $U_\infty = 50$ m/s and $U_\infty = 150$ m/s. Assume that the average friction coefficient per side is $C_D = 0.002$ ($\mu = 1.8 \times 10^{-5}$ N s/m^2, $\rho = 1.2$ kg/m^3).

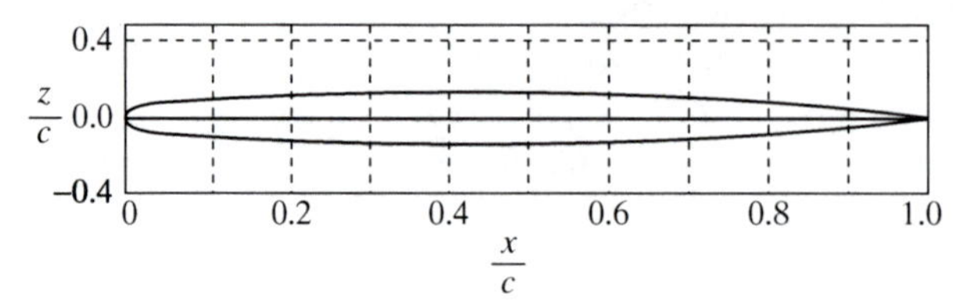

Problem 7.12.

7.13. If the flow at 40 m/s would be entirely laminar on the airfoil in the previous problem, then calculate the boundary-layer and displacement thicknesses at the trailing edge. Calculate the shear force at the leading edge and at the trailing edge.

7.14. The airfoil shown in Problem 7.12 has a chord of 2 m, its maximum thickness of 12% is at the center ($c/2$), and it is towed in water at a free-stream speed of 5 m/s. Assuming laminar flow, estimate how thick the airfoil appears at the point of its maximum thickness (*Hint: Think* of the displacement thickness as an effective increase in the body's thickness).

7.15. Repeat the previous problem, but now the boundary-layer flow is tripped and is turbulent along the whole length.

7.16. The drag coefficient of a NACA 0012 airfoil is 0.006 at a Reynolds number of 0.9×10^6. Estimate the drag based on both the laminar and turbulent models and determine which is applicable (or a combination thereof)?

7.17. Suppose the wind is blowing over a mirror-smooth lake at 15 km/h. Using the air density and viscosity values from Table 1.1, calculate Re and boundary-layer thickness at a distance of 10 km. (Use the laminar flat-plate model with $\delta_{99} = 5.0/\mathrm{Re}^{1/2}$.)

7.18. Several 2-m-wide, 4-m-long thin metal plates are towed for an underwater construction. Assuming the tow vehicle moves at a speed of 2 m/s and the plates are parallel to the free stream, calculate the drag force for a single plate positioned as in cases a and b in the figure. Calculate the boundary-layer thickness at the end of the plate (for the two cases). Assume laminar flow (Blasius solution) and use seawater properties from Table 1.1.

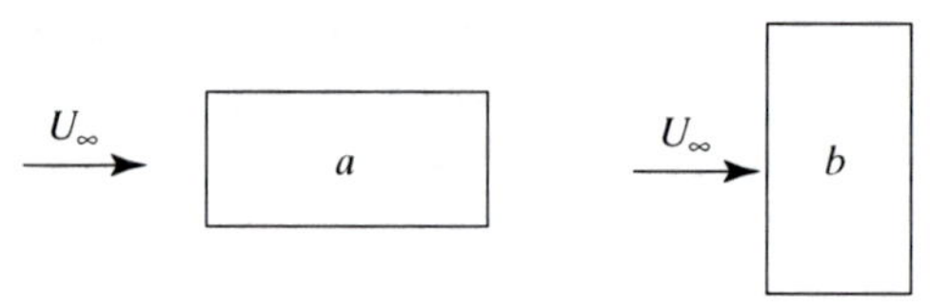

Problem 7.18.

7.19. An airplane tows a 2-m-tall advertising banner at a speed of 160 km/h. If the banner is flat and 5 m long, then estimate its drag assuming fully turbulent flow ($\mu = 1.8 \times 10^{-5}$ N s/m^2, $\rho = 1.2$ kg/m^3).

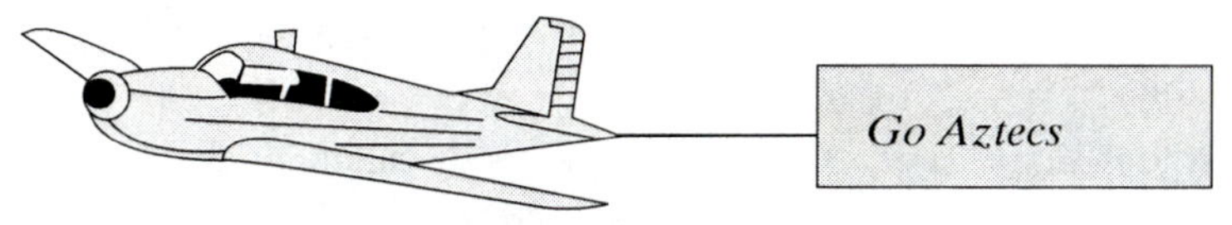

Problem 7.19.

7.20. How thick is the boundary layer at the trailing edge of the banner in the figure of Problem 7.19? Also calculate the shear force at the trailing edge. Use turbulent boundary-layer properties from Table 7.1.

7.21. A carport is covered by a 20-m long, 4-m-wide flat roof (its length is parallel to the prevailing wind). The boundary layer thickness was measured at the end of the (20-m) roof and found to be 25 cm thick. Assuming turbulent flow along the whole surface, estimate the wind speed and the drag force on the roof (from both upper and lower boundary layers). Use air properties from Table 1.1.

7.22. The velocity profile above a flat plate placed parallel to a stream is

$$\frac{u}{U_e} = 2\left(\frac{z}{\delta}\right) - 2\left(\frac{z}{\delta}\right)^3 + \left(\frac{z}{\delta}\right)^4.$$

Obviously when $z > \delta$ the velocity is constant at U_e. The incoming flow rate (at $x > 0$) between $z = 0$ and $z = \delta$, per unit width (without the plate), is then $\rho U_e \delta$. However, when the plate is inserted, then, because of the previously described boundary layer, the mass flow will be reduced. This new mass flow rate is simply the integral of the velocity deficit:

$$\rho \int_0^\delta (U_e - u) dz.$$

Calculate the percentage loss of the mass flow rate that is due to the plate (by inserting the above velocity profile into this integral).

7.23. If the boundary layer thickness is given [as in Eq. (7.20)] by

$$\frac{\delta}{x} = \frac{5.836}{\sqrt{\mathrm{Re}_x}},$$

then, by substituting the polynomial from Problem 7.22 into the definition of δ^* [Eq. (7.21)], prove Eq. (7.22) for δ^*/x.

7.24. A laminar boundary layer develops along a flat plate and its thickness at a particular point is $\delta = 0.1$ m. If we approximate the boundary-layer velocity distribution by

$$u/U_e = 2(z/\delta) - 2(z/\delta)^3 + (z/\delta)^4,$$

then calculate the local shear τ and friction coefficient C_f. Assume that $U_\infty = 30$ m/s, $\rho = 1.2$ kg/m^3, and $\mu = 1.8 \times 10^{-5}$ N s/m^2.

7.25. The sinusoidal shape is one of the simplest velocity profiles proposed for the boundary layer. Suppose we approximate the velocity distribution as

$$\frac{u}{U_e} = \sin\left(\frac{\pi z}{2\delta}\right), \qquad z \le \delta;$$

Calculate the displacement thickness and the momentum thickness in terms of the local boundary-layer thickness δ.

7.26. Using the sinusoidal shape $(u/U_e) = \sin(\pi z/2\delta)$, $z \leq \delta$, calculate the expression for the boundary-layer thickness δ by evaluating the von Kármán integral equation.

7.27. Use the results from the previous problem to calculate the local shear on the wall and the total drag for a plate with a length L. How does this approximation compare with the exact result of Blasius?

7.28. Calculate the local skin friction and the local drag coefficient for a plate of length L, using the previous sinusoidal velocity profile.

7.29. Assuming a velocity profile for the laminar flat plate in the form

$$\frac{u}{U_e} = 2\left(\frac{z}{\delta}\right) - \left(\frac{z}{\delta}\right)^2,$$

develop a relation for the boundary layer (δ/x) as a function of the local Reynolds number by evaluating the von Kármán integral equation. Compare with the exact results of Blasius.

7.30. Calculate the displacement and momentum thickness for the preceding velocity profile and compare with the exact results.

7.31. Calculate the local shear stress and friction coefficient for the preceding velocity profile and compare with the exact results.

7.32. Suppose that the velocity profile proposed for a turbulent boundary layer has the form

$$\frac{u}{U_e} = \left(\frac{z}{\delta}\right)^{\frac{1}{9}}.$$

Calculate the displacement thickness and the momentum thickness in terms of the local boundary-layer thickness δ.

7.33. The schematic growth of the boundary layer along a flat plate is shown in the figure. Can the Bernoulli equation provide the pressure difference between point 1 and point 2 (the latter is on the surface where we apply the no-slip boundary condition)? What is the pressure difference between point 2 and point 3 and between point 2 and point 1?

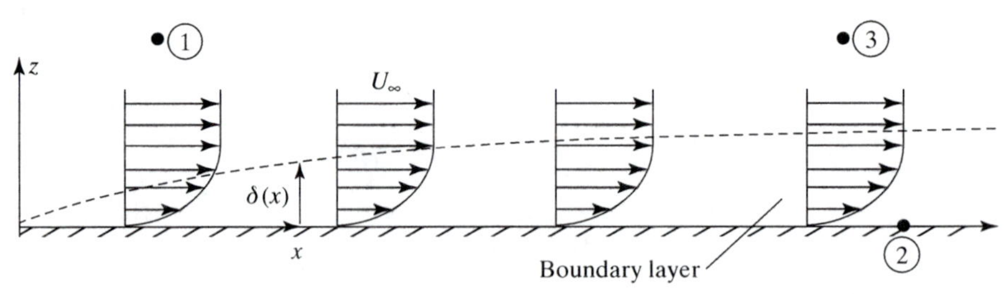

Problem 7.33.

7.34. Air is entering a small 2D laminar flow wind tunnel at a speed of $u_1 = 10$ m/s ($L = 1$ m, $h = 0.2$ m). Calculate the boundary-layer thickness at the exit of the wind tunnel at section 2. Also calculate the exit velocity u_2.

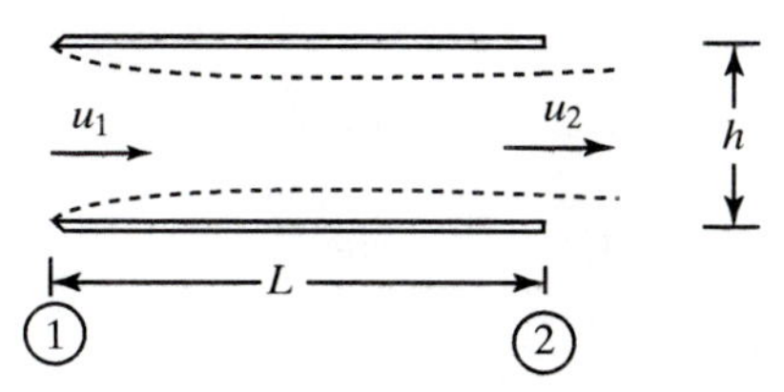

Problem 7.34.

7.35. Repeat the previous problem, but now the flow is forced to be fully turbulent by placing sandpaper strips at the entrance (use Table 7.1 for the boundary-layer properties).

8 High-Reynolds-Number Flow over Bodies (Incompressible)

8.1 Introduction

The concept of high-Reynolds-number flows was discussed in Chapter 6 and it was concluded that near a solid-body surface for an attached flow a thin boundary layer exists. In Chapter 7 this boundary layer was investigated, and the small-thickness assumption was verified. It was also concluded that the pressure distribution around a vehicle could be obtained by the solution of the inviscid flow outside the thin boundary layer. These modeling conclusions, along with some general features of such flow fields, is summarized in Fig. 8.1.

In term of forces, the boundary layer solution provides the skin-friction estimate and the resulting skin-friction-related drag-force component. However, viscous effects, in addition to the boundary layer, can be present in the wake and in areas of flow separation. For example, we can see the effects of viscous flow-momentum loss by comparing the velocity distribution ahead and behind the vehicle (as shown in the figure). Clearly, in the wake the flow is slower and there is a loss of linear momentum (which is the drag, as was shown in the previous chapter).

Solution of the flow outside this viscous layer should provide information on the velocity and pressure distributions, as depicted by the centerline pressure distribution shown in the upper part of Fig. 8.1 (recall that there is no change in the pressure across the boundary layer). In the case in which the flow is attached, we can define an irrotationl flow model and solve for the velocity distribution. We can then calculate the corresponding pressure field by using the Bernoulli equations (instead of the full Euler equation). This process is demonstrated in this chapter. It is expected that the integration of the pressure will result in a force (in addition to the skin friction) that may act in other directions (e.g., lift, side force, or drag). However, when the flow is separated (and unsteady) as in the wake shown in Fig. 8.1, then these models may not be accurate. The engineering approach is then to define force coefficients for a wide range of Reynolds numbers (similar to the approach taken for pipe flows in Chapter 5) and with such a "database" approach we can estimate the forces even in separated, turbulent, or both types of flows.

The first task therefore is to demonstrate a solution that will provide the pressure distribution for a particular geometry (even a very simple one). This model should demonstrate the process and allow the extrapolation for treating more

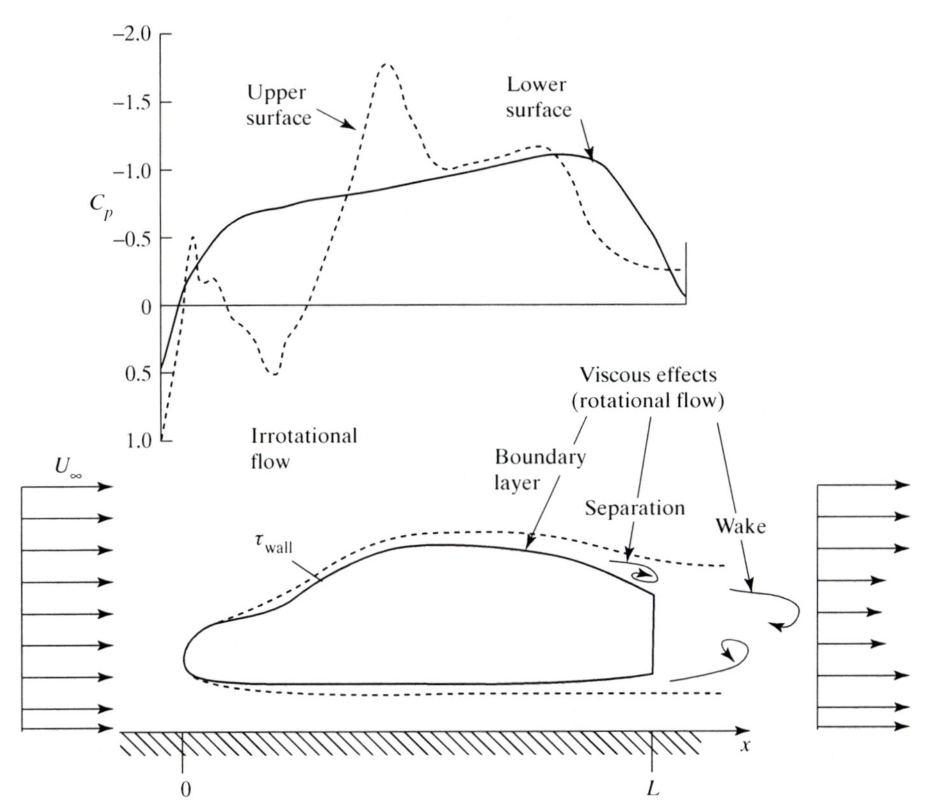

Figure 8.1. Summary of the expected results from a high-Reynolds-number-flow solution. The pressure distribution (from the outer solution) is shown in the upper diagram and the viscous flow effects (inner solution) are depicted in the lower diagram.

complex engineering problems. However, prior to solving the flow over a simple shape, we must regroup the mathematical tools required for treating this problem.

8.2 The Inviscid Irrotational Flow (and Some Math)

The conclusions from the dimensional analysis in Chapter 6 is that the flow outside the boundary layer is mostly inviscid and the simplified equations are the continuity and the Euler equations:

$$\nabla \cdot \vec{q} = 0, \tag{6.17}$$

$$\frac{\partial \vec{q}}{\partial t} + \vec{q} \cdot \nabla \vec{q} = \vec{f} - \frac{\nabla p}{\rho}. \tag{6.18}$$

Because in this model the viscous terms are neglected, only the boundary condition requiring zero velocity, normal to a solid surface, remains.

This set of equations is still quite complex (for exact analytic solutions) and additional simplifications are required. For example, if the flow is irrotational, there are no frictional losses, and we can use potential (or conservative) models. This is similar to mechanics, when we claim that the work done by the gravitational force is independent of the path (as long as there is no friction). So the first task is to convince ourselves that the outer flow is vorticity $\vec{\zeta}$ free (e.g., irrotational). The vorticity

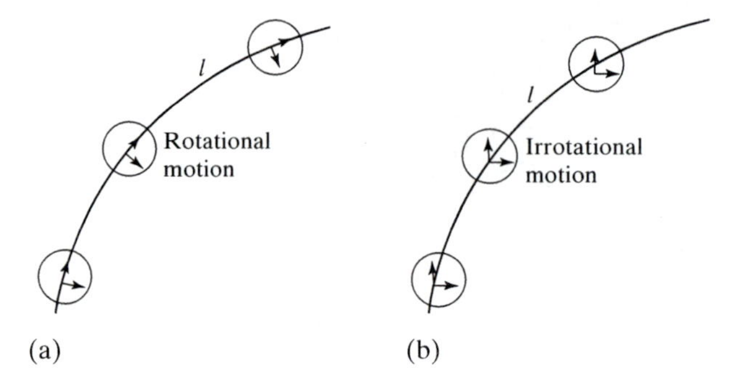

Figure 8.2. Roatational and irrotational motions of a fluid element.

was defined in Chapter 5 [Eq. (5.177)], and it is twice the solid-body rotation $\vec{\omega}$

$$\vec{\zeta} = 2\vec{\omega} = \nabla \times \vec{q}. \tag{5.171}$$

Solid-body rotation is possible only when strong viscous forces are present. To illustrate the motion of a fluid with rotation, consider the control volume shown in Fig. 8.2(a), moving along the path l. Let us assume that the viscous forces are very large (e.g., think of it as a cup filled with jelly and the cup also rotates as it moves along the circular path), and the fluid will rotate as a rigid body while following the path l. In this case $\nabla \times \vec{q} \neq 0$, and the flow is called rotational. For the fluid motion described in Fig. 8.2(b), the shear forces in the fluid are negligible (think of it as a cup filled with water), and the fluid will not be rotated by the shear force of the neighboring fluid elements. In this case $\nabla \times \vec{q} = 0$, and the flow is considered irrotational.

To further clarify our argument, let us select a fluid element of length dx, in a boundary layer, as shown in Fig. 8.3. Recall the definition of vorticity in Cartesian coordinates [Eq. (5.177)]:

$$\zeta_y = 2\omega_y = \left(\frac{\partial u}{\partial z} - \frac{\partial w}{\partial x}\right).$$

Note that inside the boundary layer $w \sim 0$ and we can approximate the change in the u velocity component between point 1 and 2 as

$$\zeta_y = \left(\frac{\partial u}{\partial z} - \frac{\partial w}{\partial x}\right) \approx \frac{U_e}{\delta},$$

where δ is the local boundary-layer thickness. We arrived at a similar conclusion in the example about the flow near a rotating cylinder (the vortex in Section 5.11), namely, that near the surface where the zero-slip boundary condition is fulfilled, the vorticity is nonzero! However, away from the surface, the flow is irrotational. So, to conclude this short discussion: Vorticity is generated near a solid surface (or in

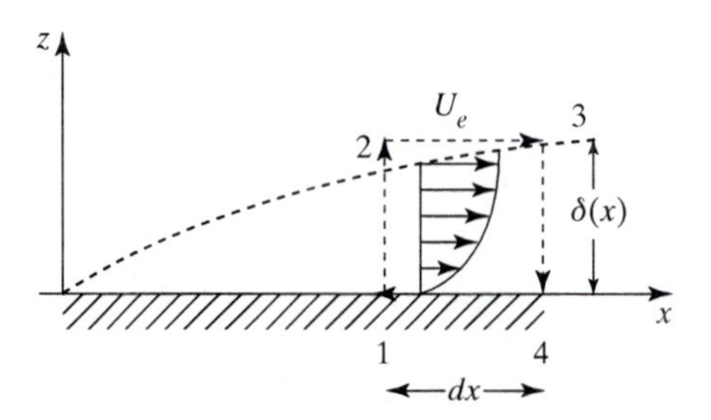

Figure 8.3. Vorticity is created in the boundary layer.

the wakes) and the rest of the flow can be considered irrotational (a more rigorous proof can be found in [1, Chapter 2]).

Once it is accepted that the vorticity in the high-Reynolds-number flow fields is confined to the boundary layer and wake regions where the influence of viscosity is not negligible, we can assume an irrotational as well as inviscid flow outside these confined regions. Now we can return to our analogy to work done by a (conservative) gravitational force (e.g., without friction) in which the force times distance integral is independent of path. In a similar way we can consider the line integral (of the velocity instead of the force vector) along the line C:

$$\int_C \vec{q} \cdot d\vec{l} = \int_C u dx + v dy + w dz. \tag{8.1}$$

If the flow is irrotational in this region then $udx + vdy + wdz$ is an exact differential (see [2, p. 475]) of a potential Φ that is independent of the integration path C. The potential at a point $P(x, y, z)$ is therefore

$$\Phi(x, y, z) = \int_{P_0}^{P} u dx + v dy + w dz \tag{8.2}$$

where P_0 is an arbitrary reference point and P is the point where the potential is evaluated. The result of the integration, Φ, is called the velocity potential and the velocity at each point can be obtained as its gradient,

$$\vec{q} = \nabla \Phi, \tag{8.3}$$

and in Cartesian coordinates

$$u = \frac{\partial \Phi}{\partial x}, \qquad v = \frac{\partial \Phi}{\partial y}, \qquad w = \frac{\partial \Phi}{\partial z}. \tag{8.4}$$

The substitution of Eq. (8.3) into the continuity equation [Eq. (6.17)] leads to the differential equation for the velocity potential (which is really the continuity equation):

$$\nabla \cdot \vec{q} = \nabla \cdot \nabla \Phi = \nabla^2 \Phi = 0 \tag{8.5}$$

Of course, this is the *Laplaces equation* (named after the French mathematician Pierre S. De Laplace, 1749–1827). It is a statement of the incompressible continuity equation for an irrotational fluid. Note that Laplace's equation is a linear differential equation. Because the fluid's viscosity has been neglected, the no-slip boundary condition on a solid–fluid boundary cannot be enforced and only the normal velocity on a solid surface is set to zero:

$$q_n = 0. \tag{8.6a}$$

In a more general form (as shown in Fig. 8.4), we obtain this boundary condition by simply multiplying the unit vector normal to the surface ($\vec{n}$) by the local velocity vector $\vec{q}$

$$\vec{q} \cdot \vec{n} = 0. \tag{8.6b}$$

It now appears that the velocity field can be obtained from a solution of Laplace's equation for the velocity potential. This is a major simplification for the solution procedure! Instead of solving for a velocity vector field (with three

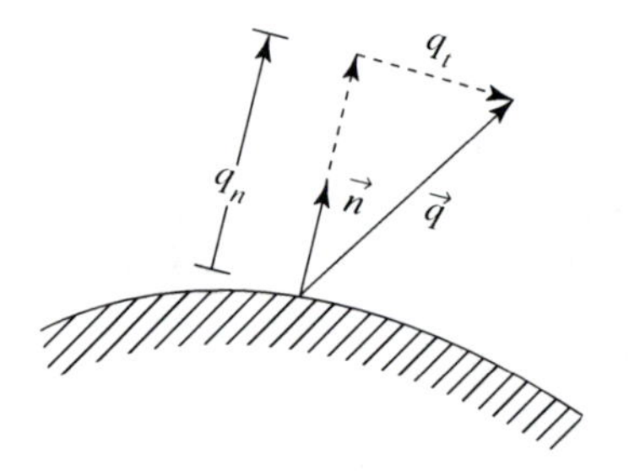

Figure 8.4. In an irrotational fluid only the "zero-normal-velocity" boundary condition remains.

unknowns: u, v, w, at any point) we must find a scalar function Φ with only one unknown per point.

Note that we have not yet used the Euler equation that connects the velocity to the pressure. Once the velocity field is obtained it is necessary to also obtain the pressure distribution on the body surface to allow for a calculation of the fluid dynamic forces and moments.

For completeness (and we shall need this later) the 2D continuity and Laplace equations are rewritten in cylindrical coordinates:

1. The continuity equation, based on Eq. (2.45), is:

$$\frac{\partial q_r}{\partial r} + \frac{1}{r}\frac{\partial q_\theta}{\partial \theta} + \frac{\partial q_x}{\partial x} + \frac{q_r}{r} = 0. \tag{8.8}$$

2. The Laplace equation in cylindrical coordinates [using Eq. (2.43)] is

$$\nabla^2\Phi = \frac{\partial^2\Phi}{\partial r^2} + \frac{1}{r}\frac{\partial \Phi}{\partial r} + \frac{1}{r^2}\frac{\partial^2\Phi}{\partial \theta^2} + \frac{\partial^2\Phi}{\partial x^2} = 0. \tag{8.9}$$

8.3 Advanced Topics: A More Detailed Evaluation of the Bernoulli Equation

The objective of this section is to arrive at the Bernoulli equation directly from the Euler equation [Eq. (6.19)]. Now that we understand the connection between the effects of viscosity and vorticity, it will be easier to understand the limitation that applies to the usage of this relation. Simple vector algebra can show that the steady-state inertia term in the Navier–Stokes equation can be rewritten by use of the following vector identity:

$$\vec{q}\cdot\nabla\vec{q} = -\vec{q}\times\nabla\times\vec{q} + \nabla\frac{q^2}{2} = -\vec{q}\times\vec{\zeta} + \nabla\frac{q^2}{2}. \tag{8.10}$$

The incompressible Euler equation can now be rewritten with the use of Eq. (8.10) as

$$\frac{\partial\vec{q}}{\partial t} - \vec{q}\times\vec{\zeta} + \nabla\frac{q^2}{2} = \vec{f} - \nabla\frac{p}{\rho}. \tag{8.11}$$

For irrotational flow $\vec{\zeta} = 0$! This is a major simplification of the momentum equation. Instead of a differential equation an algebraic relation results! Next, the time derivative of the velocity can be written as

$$\frac{\partial\vec{q}}{\partial t} = \frac{\partial}{\partial t}\nabla\Phi = \nabla\left(\frac{\partial\Phi}{\partial t}\right). \tag{8.12}$$

Let us also assume that the body force is conservative with a potential gh:

$$\vec{f} = -\nabla(gh). \tag{8.13}$$

The Euler equation for incompressible irrotational flow with this conservative body force [by substituting Eqs. (8.12) and (8.13) into Eq. (8.11)] then becomes

$$\nabla\left(gh + \frac{p}{\rho} + \frac{q^2}{2} + \frac{\partial\Phi}{\partial t}\right) = 0.$$

This is true only if the quantity in parentheses is a function of time only:

$$gh + \frac{p}{\rho} + \frac{q^2}{2} + \frac{\partial\Phi}{\partial t} = C(t).$$

This is the more general *Bernoulli equation* for inviscid, incompressible, irrotational flow. We obtain a more useful form of the Bernoulli equation by comparing the quantities on the left-hand side of this equation at two points in the fluid; the first is an arbitrary point and the second is a reference point at infinity. The equation then becomes

$$\left(gh + \frac{p}{\rho} + \frac{q^2}{2} + \frac{\partial\Phi}{\partial t}\right) = \left(gh + \frac{p}{\rho} + \frac{q^2}{2} + \frac{\partial\Phi}{\partial t}\right)_\infty. \tag{8.14}$$

At this point we can conclude that

1. the Bernoulli equation is valid between two arbitrary point in an incompressible irrotational fluid, and
2. If the flow is steady and incompressible but rotational, the Bernoulli equation [Eq. (8.14)] is still valid with the time-derivative term set equal to zero if the constant $C(t)$ on the right-hand side is now allowed to vary from streamline to streamline [see Eq. (2.8)]. This is because the product $\vec{q} \times \vec{\zeta}$ is normal to the streamline $d\vec{l}$ and their dot product vanishes along the streamline. Consequently, Eq. (8.14) can be used in a rotational fluid between two points lying on the same streamline.

For the cases discussed here we use only the steady-state form of the Bernoulli equation (as in Section 4.2) and we can write

$$gh + \frac{p}{\rho} + \frac{q^2}{2} = \text{const.} \tag{8.15}$$

8.4 The Potential Flow Model

In the last three chapters we have established the notion that outside viscous regions (such as the boundary layer) the flow can be considered irrotational and, for subsonic flows, also incompressible. With the definition of the velocity potential, this model is called *potential flow* and the governing equations can be summarized (in Cartesian coordinates) as follows:

$$\frac{\partial^2\Phi}{\partial x^2} + \frac{\partial^2\Phi}{\partial y^2} + \frac{\partial^2\Phi}{\partial z^2} = 0. \tag{8.16}$$

This is of course the continuity equation and its solution will yield the velocity field. The boundary condition, as stated in Eq. (8.6b) is

$$\nabla\Phi \cdot \vec{n} = 0. \tag{8.17}$$

Once the velocity field is known the pressure can be calculated at any point by use of the Bernoulli equation (which is now replacing the momentum equation):

$$gh + \frac{p}{\rho} + \frac{q^2}{2} = \text{const.} \tag{8.15}$$

Note that the 2D potential flow problem can be stated in terms of the stream function, which immediately depicts the streamlines. This method is not easily extended to three dimensions and therefore is not discussed here (recall that this is an introductory text). More information on this attractive approach can be found in [1, Section 2.13] and in the 2D discussion in Chapter 3.

8.4.1 Methods for Solving the Potential Flow Equations

There are various approaches to solve the Laplace equation. Of course, a simple trial-and-error approach may work, and by substituting various functions into the Laplace equation, feasible solutions can be found. Plotting the corresponding streamlines can show the nature of the solutions, and polynomial functions are probably the first candidates for such an exercise. A more systematic method is based on the Green's identity [after the British physicist George Green (1793–1841)], by which generic flow fields can be constructed by the superposition of several basic solutions. From the fluid dynamic point of view, the need to solve the flow for an arbitrary geometry (e.g., the flow over a car) is probably the top priority. Such an approach, in the spirit of the Green's identity, is described in [1] and in Subsection 8.5.5 of this chapter. Therefore, prior to attempting the solution of any practical problem, some of the elementary solutions to Eq. (8.16) are sought, We already have seen that the flow near a vortex (see Section 5.11) is irrotational, and therefore the vortex could qualify as an elementary solution. We shall develop two additional elements (the source and the doublet) and, combined with a free-stream model, the flow over many practical shapes can be solved (we can view those basic solutions as individual tools in a toolbox). The basic approach then is to combine these elementary solutions in a manner such that the zero normal flow on a solid surface boundary condition is satisfied. In fact, this process can be automated and several computer codes work using these principles. The first step, however, is to establish that several basic (or elementary) solutions of the Laplace equation can be added or combined.

8.4.2 The Principle of Superposition

If $\Phi_1, \Phi_2, \ldots, \Phi_n$ are solutions of the Laplace equation [Eq. (8.16)], which is linear, then a linear combination of those solutions,

$$\Phi = \sum_{k=1}^{n} c_k \Phi_k, \tag{8.18}$$

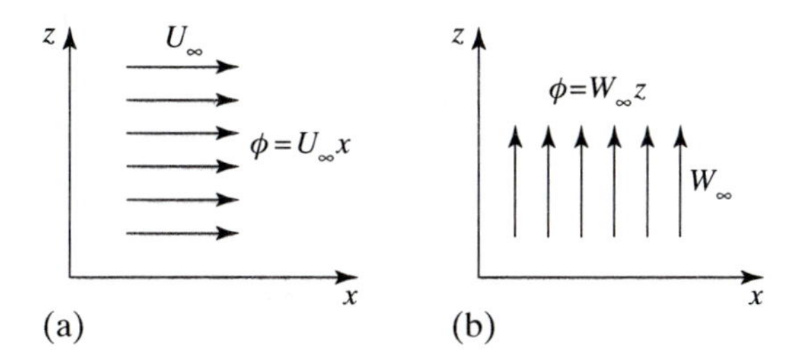

Figure 8.5. Velocity potential of a free stream (a) parallel to the x axis and (b) parallel to the z axis.

is also a solution for that equation in that region. Here $c_1, c_2, \ldots, c_n$ are arbitrary constants and therefore

$$\nabla^2 \Phi = \sum_{k=1}^{n} c_k \nabla^2 \Phi_k = 0.$$

This principle is a very important property of the Laplace equation, paving the way for solutions of the flow field near complex boundaries. In principle, by using a set of elementary solutions, the solution process (of satisfying a set of given boundary conditions) can be reduced to an algebraic search for the right linear combination of these elementary solutions.

8.5 Two-Dimensional Elementary Solutions

The principle of superposition allows us to combine elementary solutions and calculate the flow over various body shapes. The next step therefore is to develop those solutions, and for simplicity we limit ourselves to 2D flows only. A polynomial series can be a good start and clearly low-order terms should work – so let us suggest such a solution next.

8.5.1 Polynomial Solutions

Because Laplace's equation is a second-order differential equation, a linear function of position will be a solution, too. So let us propose a first-order polynomial:

$$\Phi = Ax + Bz. \tag{8.19}$$

The velocity components that are due to such a potential are

$$u = \frac{\partial \Phi}{\partial x} = A \equiv U_\infty, \qquad w = \frac{\partial \Phi}{\partial z} = B \equiv W_\infty, \tag{8.20}$$

where U_∞ and W_∞ are constant-velocity components in the x and z directions. Hence the velocity potential that is due to a constant free-stream flow in the x direction is

$$\Phi = U_\infty x. \tag{8.21}$$

The velocity field that is due to the potential of Eq. (8.21) is described in Fig. 8.5(a). Similarly, a free-stream in the z direction can be defined by the velocity potential,

$$\Phi = W_\infty z, \tag{8.22}$$

which is shown in Fig. 8.5(b). From the principle of superposition we can combine the two, and in general

$$\Phi = U_\infty x + W_\infty z. \tag{8.23}$$

This potential describes a free stream in the combined directions. For example, if $U_\infty = W_\infty$, then the free stream will be at 45° between the x and z axes.

Along the same lines, additional polynomial solutions can be sought. As an example, let's consider the second-order polynomial, with A and B being constants:

$$\Phi = Ax^2 + Bz^2. \tag{8.24}$$

To satisfy the continuity equation,

$$\nabla^2 \Phi = A + B = 0.$$

The solution to this equation is

$$A = -B, \tag{8.25}$$

and by substituting this result into Eq. (8.24) the velocity potential becomes

$$\Phi = A(x^2 - z^2). \tag{8.26}$$

The velocity components for this 2D flow in the x–z plane are the derivatives of Eq. (8.26):

$$\begin{aligned} u &= 2Ax, \\ w &= -2Az. \end{aligned} \tag{8.27}$$

To visualize this flow, the streamlines can be plotted. Recall Eq. (2.9), indicating that the flow is parallel to the streamline,

$$\frac{dx}{u} = \frac{dz}{w}, \tag{2.9}$$

and substituting the velocity components yields

$$\frac{dx}{2Ax} = \frac{dz}{-2Az}.$$

Integration by separation of variables results in

$$xz = \text{const.} = D. \tag{8.28}$$

The streamlines for different constant values of $D = 1, 2, 3\ldots$are plotted in Fig. 8.6 and, for example, if only the first quadrant of the x–z plane is considered, then this potential describes the flow around a corner. If the upper half of the x–z plane is considered, then this flow describes a stagnation flow against a wall. Note that when $x = z = 0$, the velocity components $u = w = 0$ vanish too – which means that a stagnation point is present at the origin, and the coordinate axes x and z are also the stagnation streamlines.

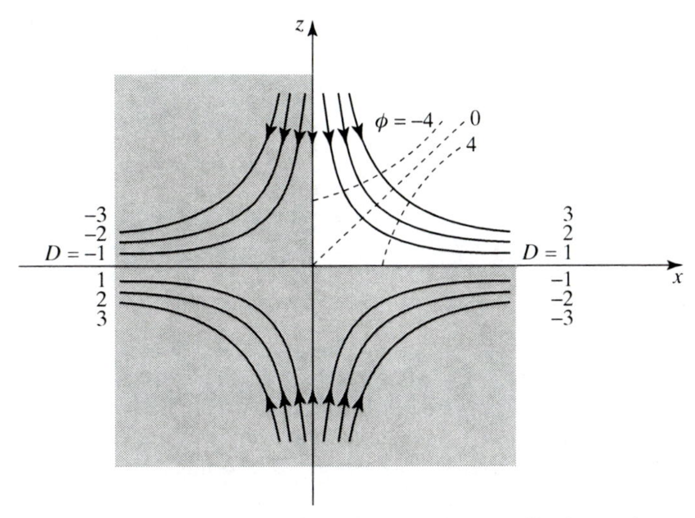

Figure 8.6. Streamlines defined by xz = const. Each quadrant describes the flow in a 90° corner.

8.5.2 Two-Dimensional Source (or Sink)

A source is a point from which fluid flows in the radial direction in straight lines. Thus, in the 2D r–θ coordinate system the tangential velocity component vanishes (e.g., $q_\theta = 0$). Because we are searching for a potential flow solution, this flow field must be irrotational! Therefore we can start with the definition of vorticity (representing rotation), and require it to be zero!

$$\zeta_y = 2\omega_y = -\frac{1}{r}\left[\frac{\partial}{\partial r}(rq_\theta) - \frac{\partial}{\partial \theta}(q_r)\right] = \frac{1}{r}\frac{\partial}{\partial \theta}(q_r) = 0. \tag{8.29}$$

Thus the velocity component in the r direction is a function of r only $[q_r = q_r(r)]$. Also, the remaining radial velocity component must satisfy the continuity equation [Eq. (8.8)], which in the r–θ coordinate system (without the tangential velocity component) is

$$\nabla \cdot \vec{q} = \frac{dq_r}{dr} + \frac{q_r}{r} = \frac{1}{r}\frac{d}{dr}(rq_r) = 0. \tag{8.30}$$

This indicates that

$$rq_r = \text{const.} = A.$$

Using this constant, the velocity components are

$$q_\theta = 0, \qquad q_r = \frac{A}{r}. \tag{8.31}$$

Let us call the flow rate passing through an arbitrary circle at a distance r_1 as σ, and according to the continuity equation it is the same as the flow rate at a different distance r_2:

$$\sigma = q_r 2\pi r = \frac{A}{r_1}2\pi r_1 = \frac{A}{r_2}2\pi r_2 = 2\pi A.$$

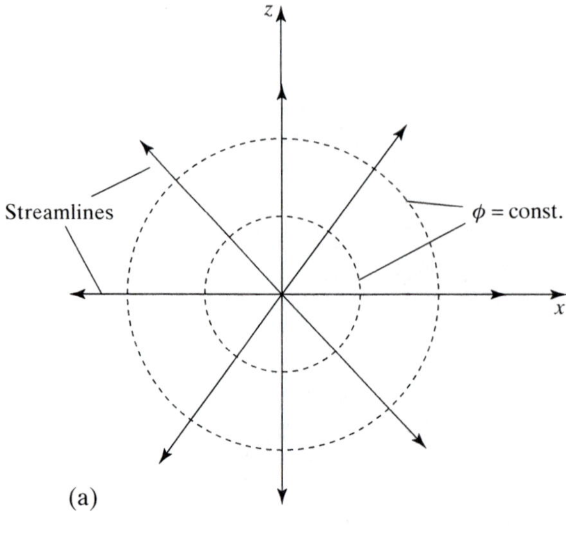

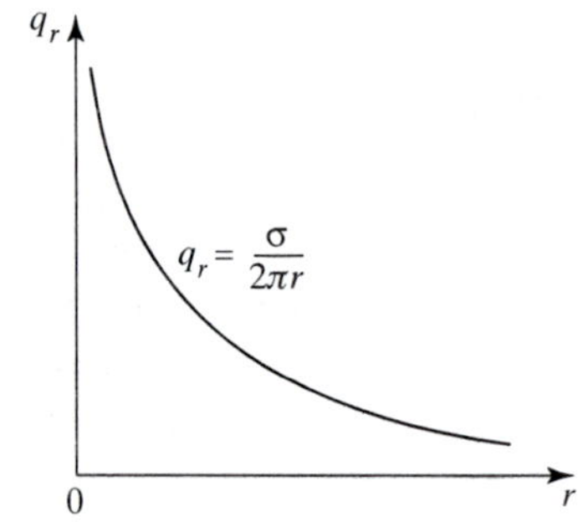

Figure 8.7. Streamlines and equipotential lines for a 2D source placed at the origin.

From this we conclude that the constant in Eq. (8.31) is

$$A = \frac{\sigma}{2\pi}, \tag{8.32}$$

where σ represents the volume flow rate introduced by the source. The resulting velocity components for a source element at the origin are

$$q_\theta = 0, \qquad q_r = \frac{\sigma}{2\pi r}. \tag{8.33}$$

This element is described in Fig. 8.7; the streamlines are clearly straight radial lines and the velocity decays as a function of $1/r$ [see Fig. 8.7(b)]. At the center of the source (origin in this case) the velocity goes to infinity, and these types of elements are called *singular solutions*. In terms of the velocity potential in the r–θ coordinate system,

$$q_r = \frac{\partial \Phi}{\partial r} = \frac{\sigma}{2\pi r}, \tag{8.34}$$

$$q_\theta = \frac{1}{r}\frac{\partial \Phi}{\partial \theta} = 0. \tag{8.35}$$

By integrating these equations, we find the velocity potential,

$$\Phi = \frac{\sigma}{2\pi} \ln r + C, \tag{8.36}$$

and the constant C can be set to zero. Also a source is emitting fluid while a sink is removing fluid at a rate of σ. The only difference is the sign, and in general we can write

$$\Phi = \pm \frac{\sigma}{2\pi} \ln r, \tag{8.37}$$

where the minus sign is for a sink.

Equations (8.34)–(8.37) describe a source–sink placed at the origin, and for any other location (as in r_0) we can replace the distance r with $r - r_0$. Similarly, in Cartesian coordinates the distance r between two points (x, z) and (x_0, z_0) is

$$r = \sqrt{(x - x_0)^2 + (z - z_0)^2}.$$

Based on this, the corresponding equations for a source located at (x_0, z_0) are

$$\Phi(x, z) = \frac{\sigma}{2\pi} \ln \sqrt{(x - x_0)^2 + (z - z_0)^2}, \tag{8.38}$$

and the velocity components obtained by deriving the velocity potential are

$$u = \frac{\partial \Phi}{\partial x} = \frac{\sigma}{2\pi} \frac{x - x_0}{(x - x_0)^2 + (z - z_0)^2}, \tag{8.39}$$

$$w = \frac{\partial \Phi}{\partial z} = \frac{\sigma}{2\pi} \frac{z - z_0}{(x - x_0)^2 + (z - z_0)^2}, \tag{8.40}$$

Note that the source here is placed at (x_0, z_0) and the velocity is evaluated at (x, z). If placed at the origin then clearly $x_0 = z_0 = 0$.

A sink is the same as a source but the flow direction is reversed. The only difference is that instead of positive flux a negative sign is added $(-\sigma)$ in the source equations! In this case σ represents the flow disappearing at the point sink.

EXAMPLE 8.1. VELOCITY INDUCED BY A SOURCE. A source of strength $\sigma = 5\ \mathrm{m^2/s}$ is located at a point (1,1). Calculate the velocity at $(0, -1)$

Solution: Using the velocity equations for the source and substituting the values for the two points we get

$$u = \frac{5}{2\pi} \frac{0 - 1}{(0 - 1)^2 + (-1 - 1)^2} = -0.159 \frac{\mathrm{m}}{\mathrm{s}},$$

$$w = \frac{5}{2\pi} \frac{-1 - 1}{(0 - 1)^2 + (-1 - 1)^2} = -0.318 \frac{\mathrm{m}}{\mathrm{s}}.$$

8.5.3 Two-Dimensional Doublet

A doublet is like a small jet engine emitting fluid in one direction and sucking the same amount of flow from behind (so no fluid is introduced as in the case of a source or sink). Consequently we can obtain the 2D doublet by letting a point source and a point sink (of equal strength) approach each other, as depicted in Fig. 8.8.

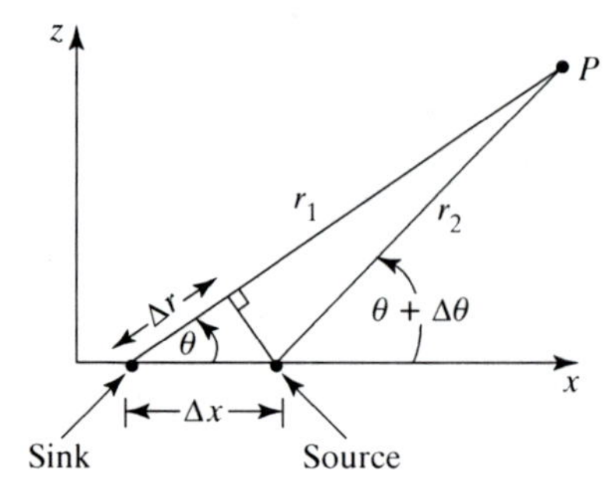

Figure 8.8. A doublet is a product of a source and a sink placed next to each other.

The velocity potential at an arbitrary point P because of the two point elements is [based on Eq. (8.37)]

$$\Phi = \frac{\sigma}{2\pi}(\ln r_2 - \ln r_1) = -\frac{\sigma}{2\pi}\ln\frac{r_1}{r_2}, \tag{8.41}$$

and the source is located forward. So the fluid is ejected into the $+x$ direction. Because r_1 is only a bit longer in the figure (e.g., Δx is small) we can write

$$\frac{r_1}{r_2} = 1 + \frac{\Delta r}{r},$$

where r is the average distance to P and $\frac{\Delta r}{r} \ll 1$. Consequently we can rewrite Eq. (8.41) as

$$\Phi = -\frac{\sigma}{2\pi}\ln\left(1 + \frac{\Delta r}{r}\right),$$

and expanding the log term, assuming Δr is small, we get

$$\Phi = -\frac{\sigma}{2\pi}\ln\left(1 + \frac{\Delta r}{r}\right) = -\frac{\sigma}{2\pi}\left[\frac{\Delta r}{r} - \frac{1}{2}\left(\frac{\Delta r}{r}\right)^2 + \frac{1}{3}\left(\frac{\Delta r}{r}\right)^3 - \cdots\right].$$

Next, observing the geometry in Fig. 8.8, it appears that

$$\Delta r = \Delta x \cos\theta.$$

Taking the limit process as $\Delta x \to 0$, neglecting smaller terms, and assuming that the source–sink strength multiplied by their separation distance becomes the constant μ (e.g., $\sigma\,\Delta x \to \mu$), we get

$$\Phi = \lim_{\Delta x \to 0}\frac{-\sigma}{2\pi}\left[\frac{\Delta x\cos\theta}{r} - \frac{1}{2}\left(\frac{\Delta x\cos\theta}{r}\right)^2 + \frac{1}{3}\left(\frac{\Delta x\cos\theta}{r}\right)^3 - \cdots\right]$$
$$= \frac{-\mu\cos\theta}{2\pi r}.$$

Consequently the velocity potential for a doublet at the origin becomes

$$\Phi(r,\theta) = \frac{-\mu}{2\pi}\frac{\cos\theta}{r}. \tag{8.42}$$

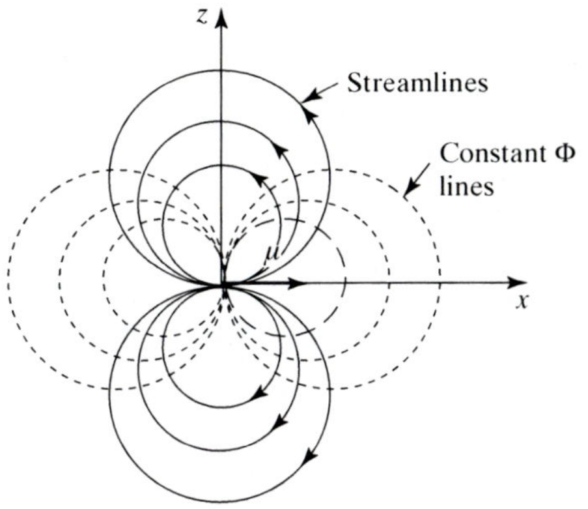

Figure 8.9. Streamlines and equipotential lines for a 2D doublet placed at the origin. Note that this doublet points in the positive x direction.

We can obtain the velocity field that is due to this element by differentiating the velocity potential:

$$q_r = \frac{\partial \Phi}{\partial r} = \frac{\mu \cos\theta}{2\pi r^2}, \tag{8.43}$$

$$q_\theta = \frac{1}{r}\frac{\partial \Phi}{\partial \theta} = \frac{\mu \sin\theta}{2\pi r^2}. \tag{8.44}$$

Again, the preceding doublet is placed at the origin. In Cartesian coordinates the sin and cos functions are (also see Fig. 8.8)

$$\cos\theta = \frac{x - x_0}{\sqrt{(x - x_0)^2 + (z - z_0)^2}}$$

$$\sin\theta = \frac{z - z_0}{\sqrt{(x - x_0)^2 + (z - z_0)^2}}$$

and recall the expression for the distance r:

$$r = \sqrt{(x - x_0)^2 + (z - z_0)^2}.$$

Using the preceding expressions, we find that the velocity potential in Cartesian coordinates for such a doublet at a point (x_0, z_0) is

$$\Phi(x, z) = \frac{-\mu}{2\pi}\frac{x - x_0}{(x - x_0)^2 + (z - z_0)^2}, \tag{8.45}$$

and the velocity components are

$$u = \frac{\partial \phi}{\partial x} = \frac{\mu}{2\pi}\frac{(x - x_0)^2 - (z - z_0)^2}{[(x - x_0)^2 + (z - z_0)^2]^2}, \tag{8.46}$$

$$w = \frac{\partial \phi}{\partial z} = \frac{\mu}{2\pi}\frac{2(x - x_0)(z - z_0)}{[(x - x_0)^2 + (z - z_0)^2]^2}. \tag{8.47}$$

The velocity field and constant-potential lines for this doublet are depicted schematically in Fig. 8.9. The streamlines are circles originating at the doublet front and returning at its back. The constant-potential lines are normal to the streamlines and consist of similar circles, but rotated by 90°, as shown in the figure.

The arrow in Fig. 8.9 indicates that a doublet is directional, and the one formulated here points in the positive x direction. Doublets pointing into other directions can be derived by simple (rotational) transformation of the preceding equations. The simplest case is the one in which the doublet formulas are multiplied by –1 to get a doublet pointing in the $-x$ direction.

EXAMPLE 8.2. VELOCITY INDUCED BY A DOUBLET. Calculate the velocity induced by a doublet of strength $\mu = 1$ m³/s located at the origin at a point $x = 1$, $z = 0$.

For the solution let us use the Cartesian form of the velocity formulas:

$$u = \frac{\mu}{2\pi}\frac{(x-x_0)^2-(z-z_0)^2}{[(x-x_0)^2+(z-z_0)^2]^2} = \frac{1}{2\pi}\frac{1^2-0^2}{[1^2+0^2]^2} = \frac{1}{2\pi}\frac{\text{m}}{\text{s}},$$

$$w = \frac{\mu}{2\pi}\frac{2(x-x_0)(z-z_0)}{[(x-x_0)^2+(z-z_0)^2]^2} = \frac{1}{2\pi}\frac{2\times 1\times 0}{[1^2+0^2]^2} = 0.$$

Suppose we want to find the velocity at a point located at $x = 1$, $z = 1$. Then, from Fig. 8.9, we suspect that the velocity vector will point straight up (because we are at a 90° position on the streamline):

$$u = \frac{1}{2\pi}\frac{(1^2-1^2)}{[1^2+1^2]^2} = 0,$$

$$w = \frac{1}{2\pi}\frac{2\times 1\times 1}{[1^2+1^2]^2} = \frac{1}{4\pi}\frac{\text{m}}{\text{s}},$$

and this result verifies the expected direction of the velocity vector.

8.5.4 Two-Dimensional Vortex

The velocity field that is due to a 2D vortex was developed in Section 5.11. It was also demonstrated that the flow that is due to a rotating vortex core is vorticity free when the vortex core is excluded. We would like to develop the same 2D vortex element, but using the approach used for the source element (in this section). We can start by searching for a singularity element with only a tangential velocity component, as shown in Fig. 8.10(a), whose velocity will decay in a manner similar to the decay of the radial velocity component of a 2D source (e.g., it will vary with $1/r$).

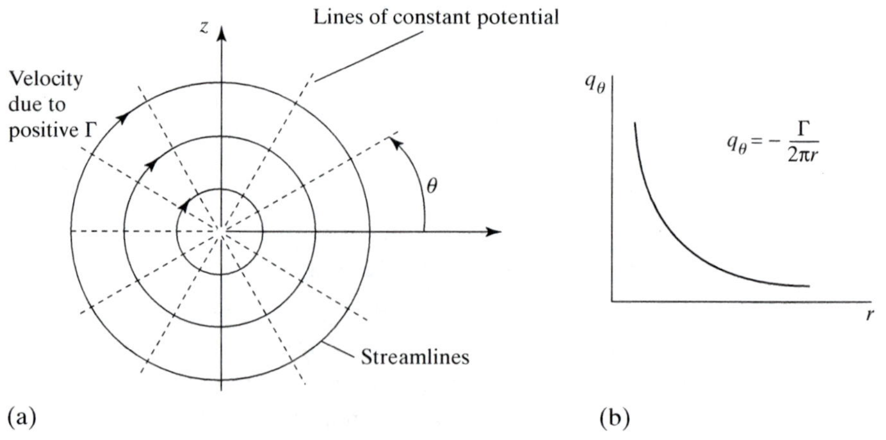

Figure 8.10. Streamlines and equipotential lines for a 2D vortex placed at the origin. Note that the tangential velocity decays as $1/r$.

The expected velocity components are then

$$q_r = 0,$$
$$q_\theta = q_\theta(r, \theta).$$

Substituting these velocity components (e.g., $q_r = 0$) into the continuity equation [Eq. (2.45)] results in q_θ being a function of r only:

$$q_\theta = q_\theta(r).$$

Similar to the approach used to develop the source, we can substitute these relations into the vorticity expression and require that the rotation be zero:

$$\zeta_y = 2\omega_y = -\frac{1}{r}\left[\frac{\partial}{\partial r}(rq_\theta) - \frac{\partial}{\partial \theta}(q_r)\right] = -\frac{1}{r}\frac{\partial}{\partial r}(rq_\theta) = 0.$$

By integrating with respect to r, we get

$$rq_\theta = \text{const.} = A. \tag{8.48}$$

This, of course, is the conservation of angular momentum, and the magnitude of the velocity varies with $1/r$, similar to the radial velocity component of a source. The value of the constant A can be calculated by use of the definition of the circulation Γ, as in Eq. (5.171):

$$\Gamma = \oint_c \vec{q} \cdot d\vec{l} = \int_{2\pi}^{0} q_\theta \times r d\theta = \int_{2\pi}^{0} \frac{A}{r} r d\theta = -2\pi A.$$

Note that positive Γ is defined according to the right-hand rule (positive clockwise); therefore, in the x–z plane, as in Fig. 8.10(a), the line integral must be taken in the direction opposite to that of increasing θ. Also, recall that the circulation represents the solid-body rotation times the core area ($\Gamma = 2\omega S$) as shown by Eq. (5.174). The constant A is then

$$A = -\frac{\Gamma}{2\pi}, \tag{8.49}$$

or in terms of the solid body rotation of the "core" ω,

$$A = -\frac{2\omega S}{2\pi}. \tag{8.49a}$$

and the velocity field is

$$q_r = 0, \tag{8.50a}$$

$$q_\theta = -\frac{\Gamma}{2\pi r}. \tag{8.50b}$$

As expected, the tangential velocity component decays at a rate of $1/r$, as shown in Fig. 8.10(b). The velocity potential for a vortex element at the origin can be obtained by use of the basic definition of the velocity potential, that is, by the integration of Eq. (8.50):

$$\Phi = \int q_\theta r d\theta = -\frac{\Gamma}{2\pi}\theta + C, \tag{8.51}$$

where C is an arbitrary constant that can be set to zero. Equation (8.51) indicates too that the velocity potential of a vortex is multivalued and depends on the number of revolutions around the vortex point. So when integrating around a vortex, we do find vorticity concentrated at a zero area point, but with finite circulation (see Section 5.11). However, if integrating $\vec{q} \cdot d\vec{l}$ around any closed curve in the field [not surrounding the vortex – see Eq. (5.181)] the value of the integral will be zero. Thus the vortex is a solution to the Laplace equation and results in an irrotational flow, excluding the vortex point itself. Equations (8.50) and (8.51) are for a vortex at the origin. For a vortex located at an arbitrary point (x_0, z_0) expressed in Cartesian coordinates, the formulation is

$$\Phi = -\frac{\Gamma}{2\pi} \tan^{-1} \frac{z - z_0}{x - x_0}, \tag{8.52}$$

$$u = \frac{\partial \phi}{\partial x} = \frac{\Gamma}{2\pi} \frac{z - z_0}{(x - x_0)^2 + (z - z_0)^2}, \tag{8.53}$$

$$w = \frac{\partial \phi}{\partial z} = -\frac{\Gamma}{2\pi} \frac{x - x_0}{(x - x_0)^2 + (z - z_0)^2}. \tag{8.54}$$

EXAMPLE 8.3. VELOCITY INDUCED BY A VORTEX. A vortex with a circulation of $\Gamma = 5$ m²/s is located at a point (1,1). Calculate the velocity at (0, −1) and (3, 0).

Solution: Using the velocity equations for the vortex, for the first point we get

$$u = \frac{5}{2\pi} \frac{-1 - 1}{(0 - 1)^2 + (-1 - 1)^2} = -0.318 \frac{\text{m}}{\text{s}},$$

$$w = \frac{-5}{2\pi} \frac{0 - 1}{(0 - 1)^2 + (-1 - 1)^2} = 0.159 \frac{\text{m}}{\text{s}},$$

and for the second point we get

$$u = \frac{5}{2\pi} \frac{0 - 1}{(3 - 1)^2 + (0 - 1)^2} = -0.159 \frac{\text{m}}{\text{s}},$$

$$w = \frac{-5}{2\pi} \frac{3 - 1}{(3 - 1)^2 + (0 - 1)^2} = -0.318 \frac{\text{m}}{\text{s}}.$$

EXAMPLE 8.4. A VORTEX IS A SOLUTION FOR THE LAPLACE EQUATION. Prove that the vortex of Eq. (8.51) satisfies the Laplace equation.

Solution: Let us substitute vortex equation (8.51) into Eq. (8.9):

$$\nabla^2 \Phi = \frac{\partial^2 \left(-\frac{\Gamma}{2\pi}\theta\right)}{\partial r^2} + \frac{1}{r} \frac{\partial \left(-\frac{\Gamma}{2\pi}\theta\right)}{\partial r} + \frac{1}{r^2} \frac{\partial^2 \left(-\frac{\Gamma}{2\pi}\theta\right)}{\partial \theta^2} = 0 + 0 + 0.$$

8.5.5 Advanced Topics: Solutions Based on the Green's Identity

Another method, with a much wider application range, is based on the Green's identity (see [1]). This approach is the basis for numerical methods, called panel methods, and can be used for solving the flow over complex shapes. One form of this identity, stated in Eq. (8.55), postulates that the velocity potential representing the flow over a solid body can be constructed by adding the contribution of sources and

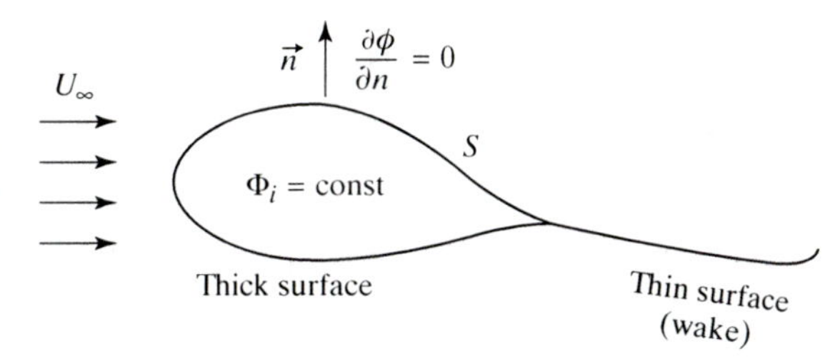

Figure 8.11. Schematic description for the use of the Green's identity to solve the flow over a 2D body.

doublets with a free stream. According to this 2D Green's formula, the potential at an arbitrary point (x, z) in the fluid can be constructed by the following summation:

$$\Phi(x, z) = \frac{1}{2\pi} \int_{S_B} \left[\sigma \ln r - \mu \frac{\partial}{\partial n}(\ln r) \right] dS + \Phi_\infty. \tag{8.55}$$

Here r, as before, represents the distance between the element and the point of interest (x, z) and S_B is the body's surface (or even its wake, as shown in Fig. 8.11).

At first it appears that this formulation is identical to the principle of superposition. However, it is also implied that the sources and doublets must be placed on the surface.

The first term in this equation, $(\sigma \ln r)$ is the source as given by Eq. (8.37); however, the doublet $\mu \frac{\partial}{\partial n}(\ln r)$ is not immediately recognizable. This is because the doublet in this formula is normal to the surface (in the direction of $\vec{n}$). However, if the derivative of ln r is taken in the x direction, then clearly the doublet potential of Subsection 8.5.3 is obtained [compare Eq. (8.38) with Eq. (8.45)].

The application of the Green's identity is shown schematically in Fig. 8.11. The correct combination of sources and doublet can be found by fulfilling the zero-normal-flow boundary condition on the surface S:

$$\frac{\partial \Phi}{\partial n} = 0 \quad \text{on } S,$$

and this velocity-based formulation is usually called the Neumann boundary condition (after Carl Neumann, German mathematician, 1832–1925). In the case of flow with forces into the vertical direction (lift) usually a zero pressure jump is forced at the sharp trailing edge where a thin wake is formed, as shown in the figure (this is called the Kutta condition). Now, if $(\partial\Phi/\partial n) = 0$ on the surface of the thick body, then the internal potential is unchanged, as shown in Fig. 8.11:

$$\Phi_i = \text{const.} \tag{8.56}$$

This is called the Dirichlet boundary condition (after the German mathematician, Johann Peter Gustav Lejeune Dirichlet, 1805–1859), which in this case is much simpler than the Neumann condition. Usually the inner potential value in Eq. (8.56) is set as zero. The application of this type of boundary condition is beyond the scope of this chapter but is described in detail in [1]. This topic will be revisited in Chapter 9, when the numerical solution of potential flows is discussed.

The concept introduced by the Green's formula in Eq. (8.55) contains the principle of superposition [as in Eq. (8.18)]. Therefore the following example serves to demonstrate the method for obtaining a solution for the potential flow by use of the principle of superposition. Although this clarifies somewhat the Green's identity

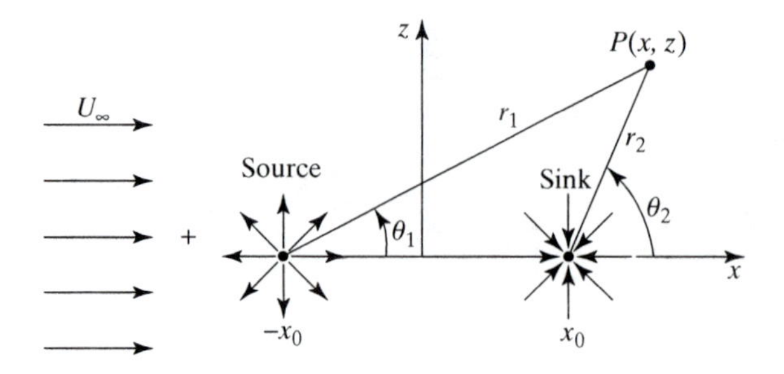

Figure 8.12. Schematic description of combining a free stream with a source and a sink.

approach, the singular elements are not placed on the solid boundary, as suggested by Eq. (8.55).

EXAMPLE 8.5. THE SUPERPOSITION OF A SOURCE, A SINK, AND A FREE STREAM. Find the velocity potential and the velocity distribution for the combination of a source, a sink, and free stream.

Solution: To demonstrate the principle of superposition, let us place a source with a strength σ at $x = -x_0$ and a sink with a strength $-\sigma$ at $x = +x_0$, both on the x axis (as shown in Fig. 8.12). The free stream with a speed U_∞ is flowing in the x direction as shown in the figure.

The velocity potential at an arbitrary point $P(x,z)$ is obtained by combining the three separate potentials:

$$\Phi(x, z) = U_\infty x + \frac{\sigma}{2\pi}\ln(r_1) - \frac{\sigma}{2\pi}\ln(r_2), \tag{8.56}$$

where $r_1 = \sqrt{(x + x_0)^2 + z^2}$, and $r_2 = \sqrt{(x - x_0)^2 + z^2}$. With these relations the velocity potential in Cartesian coordinates becomes

$$\Phi(x, z) = U_\infty x + \frac{\sigma}{2\pi}\ln\sqrt{(x + x_0)^2 + z^2} - \frac{\sigma}{2\pi}\ln\sqrt{(x - x_0)^2 + z^2}. \tag{8.56a}$$

We obtain the velocity field that is due to this potential by differentiating the velocity potential:

$$u = \frac{\partial\Phi}{\partial x} = U_\infty + \frac{\sigma}{2\pi}\frac{x + x_0}{(x + x_0)^2 + z^2} - \frac{\sigma}{2\pi}\frac{x - x_0}{(x - x_0)^2 + z^2}, \tag{8.57a}$$

$$w = \frac{\partial\Phi}{\partial z} = \frac{\sigma}{2\pi}\frac{z}{(x + x_0)^2 + z^2} - \frac{\sigma}{2\pi}\frac{z}{(x - x_0)^2 + z^2}. \tag{8.57b}$$

It appears that, ahead of the source, along the x axis, there must be a point (let us call it $x = -a$) where the velocity is zero (stagnation point). Based on these equations, the w component of the velocity along the x axis is automatically zero. To find the location of the stagnation point we equate the u component to zero:

$$u(-a, 0) = U_\infty + \frac{\sigma}{2\pi}\frac{1}{(-a + x_0)} - \frac{\sigma}{2\pi}\frac{1}{(-a - x_0)} = U_\infty - \frac{\sigma}{\pi}\frac{x_0}{(a^2 - x_0^2)} = 0,$$

and a is

$$a = \pm\sqrt{\frac{\sigma x_0}{\pi U_\infty} + x_0^2}. \tag{8.58}$$

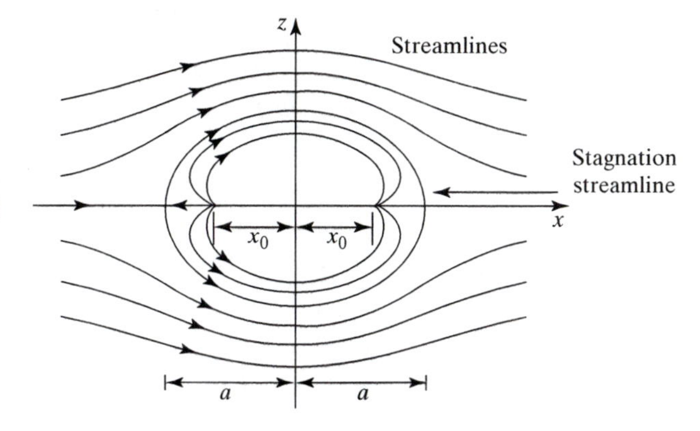

Figure 8.13. Streamlines describing the flow of combined source, sink, and free stream.

This suggests that there is a fore–aft symmetry about the vertical z axis and another stagnation point exists at $x = a$. To visualize the streamlines, we can start at a point far to the left and, based on Eq. (2.9a), march forward, using the following procedure:

$$x_{i+1} = x_i + u\Delta t,$$
$$z_{i+1} = z_i + u\Delta t. \tag{8.59}$$

Here i is a virtual time interval counter and Δt is a small time interval. The resulting streamlines are shown in Fig. 8.13, based on the results from [1].

As seen, the stagnation streamline includes a closed oval shape (called Rankine's oval, after W. J. M. Rankine, a Scottish engineer who lived between 1820 and 1872). This flow can therefore be considered to model the flow past an oval of length $2a$. For this application, the streamlines inside the oval have no physical significance. By varying the parameters σ and x_0 or a, the flow past a family of such ovals can be derived.

Also note that the velocity normal to the stagnation streamline is zero! Therefore this example demonstrates how the flow over a particular body can be obtained by the principle of superposition and by fulfilling the zero-normal-flow boundary condition. In the next section the limiting case of a circle is studied, in which the source and the sink coincide and form a doublet.

8.6 Superposition of a Doublet and a Free Stream: Flow over a Cylinder

The basic solutions developed in Section 8.4 can be combined (by use of the principle of superposition) to simulate the flow over complex shapes. The general method for doing this (for arbitrary geometries) is beyond the scope of this text, but the following examples in this section demonstrate the approach. The potential flow model assumes no flow separation, and therefore this method is powerful for attached flow cases such as the flow over airplane wings or submarine hulls. This first case, the flow over a cylinder, serves to show how the flow field over a practical shape can be obtained. However, in reality the flow over a cylinder is separated throughout most of the Reynolds number range. In spite of this model leading to an incorrect physical solution, the results create a systematic approach for developing a database for the fluid dynamic forces such as lift and drag. (In other words, if we could solve here the flow over a thick airfoil at a small angle of attack, then the calculated results would

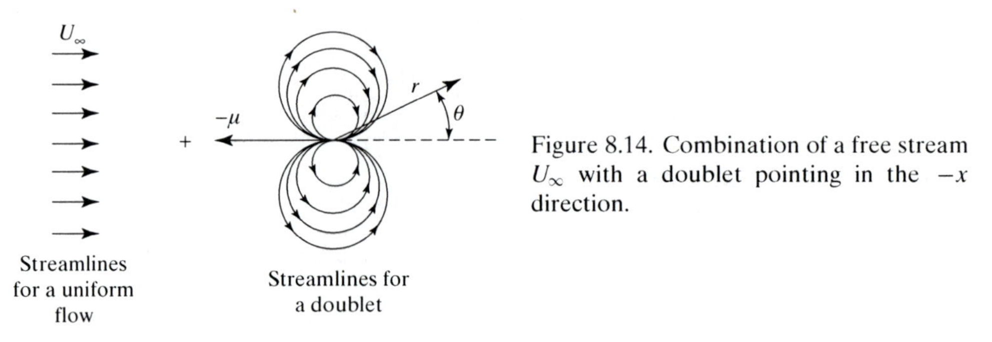

Figure 8.14. Combination of a free stream U_∞ with a doublet pointing in the $-x$ direction.

be very close to the experimental data. Unfortunately the math involved is beyond the scope of this chapter).

Let us consider the superposition of the free-stream potential of Eq. (8.21), with the potential of a doublet [Eq. (8.42)] pointing in the negative x direction, as shown in Fig. 8.14. However, it is much easier to use the r–θ coordinate system and instead of x in Eq. (8.21) we use $x = r\cos\theta$. Consequently the velocity potential of the free stream blowing into the positive x direction is

$$\Phi = U_\infty r \cos\theta. \tag{8.60}$$

Next we add a doublet at the origin but pointing in the $-x$ direction (see coordinate system in Fig. 8.14), resulting in the change of sign for the doublet potential. The combined flow has the following velocity potential:

$$\Phi = U_\infty r \cos\theta + \frac{\mu}{2\pi}\frac{\cos\theta}{r}. \tag{8.61}$$

We can obtain the velocity field of this potential by differentiating Eq. (8.61):

$$q_r = \frac{\partial\Phi}{\partial r} = \left(U_\infty - \frac{\mu}{2\pi r^2}\right)\cos\theta, \tag{8.62}$$

$$q_\theta = \frac{1}{r}\frac{\partial\Phi}{\partial\theta} = -\left(U_\infty + \frac{\mu}{2\pi r^2}\right)\sin\theta. \tag{8.63}$$

To visualize this flow we can use the property of streamlines, as stated in Eqs. (2.9) [or Eqs. (8.59)], and the result is shown in Fig. 8.15. Basically the doublet (recall the jet engine model) is blowing into the free stream. At a certain point (see P_1 in the figure) the velocity stops, and this point is called the stagnation point. It appears that a circle of radius R is dividing the streamlines of the inner flow solution from the outer one, which line is often called as the *staguation streamline*. We show later that there is no flow crossing this circle, and therefore we can consider the outer flow as the solution for the flow over a cylinder.

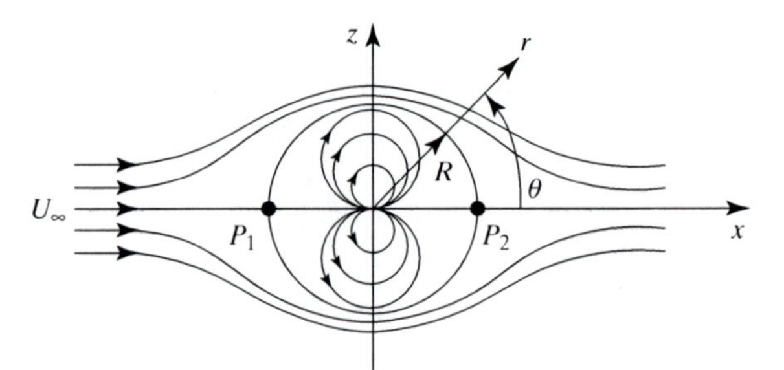

Figure 8.15. Streamlines for the combination of a free stream and a doublet (flow over a cylinder).

To verify the assumptions that there is no flow across the cylinder boundaries, let us check Eq. (8.62) for $q_r = 0$:

$$q_r = \left(U_\infty - \frac{\mu}{2\pi r^2}\right)\cos\theta = 0.$$

Suppose we wish to describe the flow over a cylinder of radius R; then we can use this equation to determine the strength μ of the doublet for this condition. Thus, substituting $r = R$ as the radius of the circle and solving for the strength of the doublet μ, we get

$$\mu = U_\infty 2\pi R^2. \tag{8.64}$$

Substituting this value of μ into the velocity potential and its derivatives [Eqs. (8.61)–(8.63)] results in the flow field around a cylinder of radius R:

$$\Phi = U_\infty \cos\theta\left(r + \frac{R^2}{r}\right), \tag{8.65}$$

$$q_r = U_\infty \cos\theta\left(1 - \frac{R^2}{r^2}\right), \tag{8.66}$$

$$q_\theta = -U_\infty \sin\theta\left(1 + \frac{R^2}{r^2}\right), \tag{8.67}$$

and the flow of interest is when $r \geq R$. The stagnation points on the circle are found by letting $q_\theta = 0$ in Eq. (8.67), and because of the $\sin\theta$ term they are located at $\theta = 0$ and $\theta = \pi$. To obtain the pressure distribution over the cylinder, the velocity components are evaluated at $r = R$:

$$q_r = 0, \quad q_\theta = -2U_\infty \sin\theta. \tag{8.68}$$

This example represents a general approach for obtaining a solution for the potential flow problem posed in Section 8.8. Instead of solving the Laplace equation directly, we combine known solutions by using the principle of superposition in a manner such that boundary condition (Eq. 8.17) is satisfied. This is exactly the condition, shown by Eq. (8.68) (e.g., $q_r = 0$), that states that the velocity normal to the surface of the cylinder is zero!

We now obtain the pressure distribution at $r = R$ by applying Bernoulli's equation on the surface of the cylinder. We take a reference point far left, where the flow is undisturbed ($p = p_\infty$ and $u = U_\infty$), and the other point is on the surface of the cylinder (where we have only q_θ):

$$p_\infty + \frac{\rho}{2}U_\infty^2 = p + \frac{\rho}{2}q_\theta^2. \tag{8.69}$$

Substituting the value of q_θ at $r = R$ yields

$$p - p_\infty = \frac{1}{2}\rho U_\infty^2(1 - 4\sin^2\theta). \tag{8.70}$$

This equation describes the pressure distribution on the surface of the cylinder. In terms of the pressure coefficient, when p_∞ is used as the reference pressure,

$$C_p = \frac{p - p_\infty}{\frac{1}{2}\rho U_\infty^2} = 1 - 4\sin^2\theta. \tag{8.71}$$

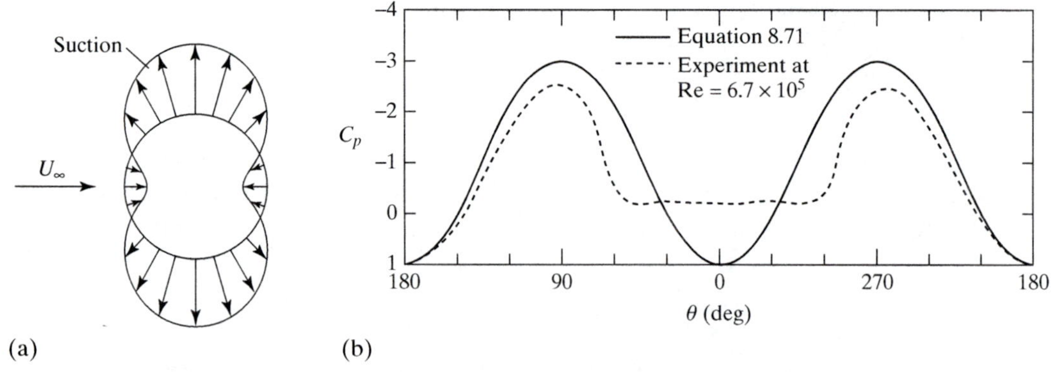

Figure 8.16. (a) Theoretical pressure distribution around a cylinder [Eq. (8.71)] and (b) comparison with experimental data. The solid curve shows the ideal-flow solution and the broken curve represents the high-Reynolds-number experimental data.

Note that by taking $p - p_\infty$ directly from Eq. (8.69) and substituting into the definition of the pressure coefficient we get

$$C_p = 1 - \frac{q^2}{U_\infty^2}. \tag{8.72}$$

The results of Eq. (8.71) are plotted schematically in Fig. 8.16(a) showing positive pressure at the front and at the back of the cylinder. As the flow accelerates around the top (or bottom) of the cylinder, the velocity increases and a large suction force results.

The same pressure distribution [of Eq. (8.71)] is also shown in Fig. 8.16(b) by the solid curve. It can be easily observed that at the stagnation points $\theta = 0$ and π (where both components of the velocity are zero: $q = 0$), the pressure coefficient is the highest at $C_p = 1$. Also, the maximum speed occurs at the top and bottom of the cylinder ($\theta = \frac{\pi}{2}, \frac{3\pi}{2}$) and the pressure coefficient there is −3. Note that there is a fore–aft symmetry as well as a top–bottom symmetry (suggesting no lift or drag).

To evaluate the components of the fluid dynamic force acting on the cylinder, the preceding pressure distribution must be integrated. Let L be the lift acting in the z direction and let D be the drag acting in the x direction. Integrating the components of the pressure force on an element of length $Rd\theta$ leads to

$$L = \int_0^{2\pi} -pRd\theta \sin\theta = \int_0^{2\pi} -(p - p_\infty)Rd\theta \sin\theta =$$
$$\frac{-1}{2}\rho U_\infty^2 \int_0^{2\pi} (1 - 4\sin^2\theta)R\sin\theta d\theta = 0. \tag{8.73}$$

$$D = \int_0^{2\pi} -pRd\theta \cos\theta = \int_0^{2\pi} -(p - p_\infty)Rd\theta \cos\theta =$$
$$\frac{-1}{2}\rho U_\infty^2 \int_0^{2\pi} (1 - 4\sin^2\theta)R\cos\theta d\theta = 0. \tag{8.74}$$

Here the pressure is replaced with the pressure difference $p - p_\infty$ term, and this has no effect on the results because the integral of a constant pressure p_∞ around a closed body is zero. As noted, because of the fore and aft symmetry the calculated

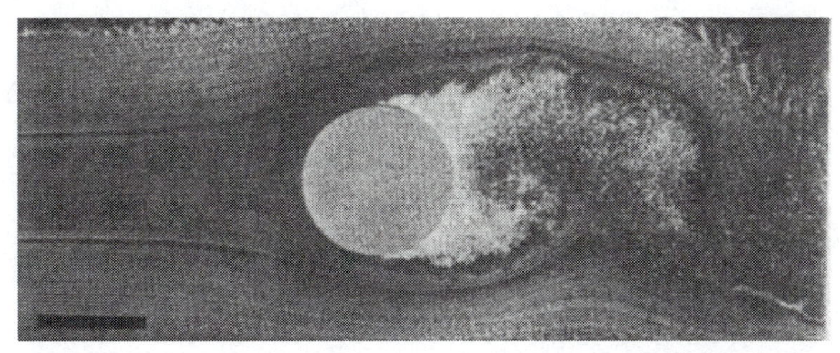

Figure 8.17. Experimental visualization of the high-Reynolds-number flow over a cylinder.

pressure loads cancel out. In reality, the flow separates (and is also unsteady) and will not follow the cylinder's rear surface, as shown in Fig. 8.17. The pressure distribution that is due to this real flow and the results of Eq. (8.71) are plotted in Fig. 8.16. This shows that, at the front section of the cylinder, where the flow is attached, the pressures are well predicted by this model. However, behind the cylinder, because of the flow separation, the pressure distribution is different. For example, near the rear stagnation point ($\theta = 0$) the experimental pressure coefficient in Fig 8.16 is negative, compared with $+1$ predicted by the ideal-flow solution. Consequently the pressure drag in an actual flow is not zero!

Note that the inviscid flow results do not account for flow separation and viscous friction near the body's surface, and therefore the calculated drag coefficient for the cylinder is *zero*. This fact disturbed the French mathematician Jean le Rond d'Alembert (1717–1783), who arrived at a similar conclusion that the drag of a closed body in 2D inviscid incompressible flow is zero (even though he realized that experimental results indicate that there is drag). Ever since those early days of fluid dynamics, this problem has been known as the *d'Alembert's paradox*.

Although the flow over a sphere was not solved here the method of solution is very similar and the resulting pressure distribution is

$$C_p = 1 - \frac{9}{4}\sin^2\theta. \tag{8.75}$$

Because of the fore–aft symmetry of the potential flow, drag is zero; the two theoretical pressure distributions (for one side – say, upper) are depicted in Fig. 8.18. The important observation here is that the maximum velocity (and the corresponding pressure coefficients) is much smaller in the 3D case, because the flow can move around the sides as well.

8.7 Fluid Mechanic Drag

This first example of the flow over a cylinder may not be the best when compared with experiments. As noted earlier, a solution of a streamlined shape (such as an airfoil) would be a much better example, but the mathematical formulation is far more complicated. Nevertheless, this solution provides the approach for calculating engineering quantities such as the pressure distribution and the force coefficients. Let us start by observing the drag of simple shapes like the cylinder and flat plate, and then extend the method to include more complex shapes (the discussion on the lift will be resumed later).

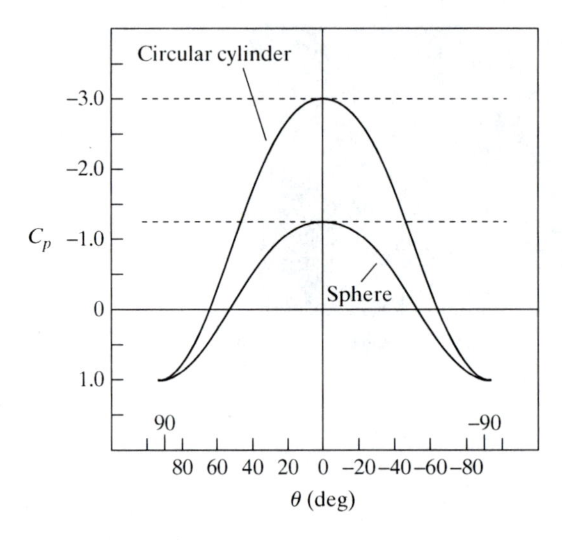

Figure 8.18. Comparison between the theoretical (ideal-flow) pressure distribution around a cylinder and a sphere.

8.7.1 The Drag of Simple Shapes

In a manner similar to plotting the friction coefficient for pipes or for the boundary layer versus the Reynolds number, we can plot the experimental drag coefficient of the cylinder versus a wide range of the Reynolds number (see Fig. 8.19). In addition, the drag coefficient of a sphere is shown, and the trends in both curves are similar. This graph seems quite complicated, and in the next paragraph an attempt is made to provide some observation-based explanations (based on the flow over a cylinder).

At a very low Reynolds number we expect laminar flow, and the drag (as the friction coefficient in the previous chapter) should decrease with increasing Reynolds number. This was validated by experiments, and flow visualizations indicated that the flow is attached up to Re < 4. This is described schematically in Fig. 8.20(a) and marked as region *a* in Fig. 8.19. For the case of the sphere in this low

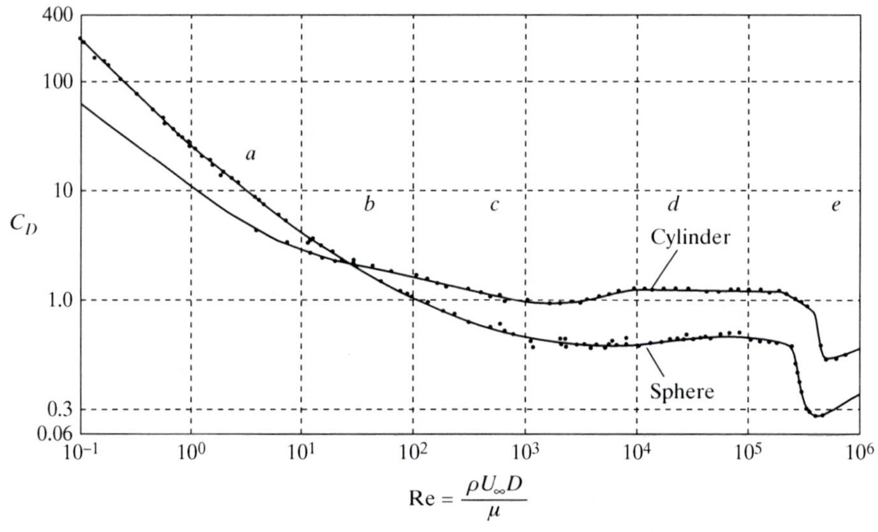

Figure 8.19. Experimental drag coefficient of a cylinder and a sphere over a wide range of Reynolds numbers.

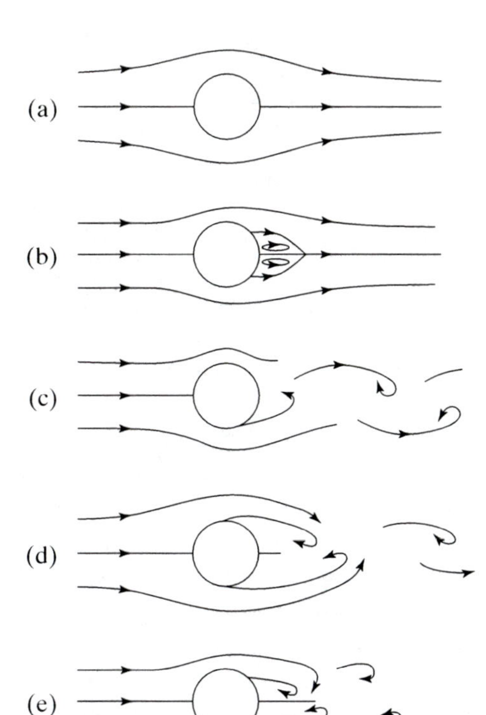

Figure 8.20. Schematic description of the various observed flow fields over a cylinder (versus Reynolds number).

Reynolds number range, Stokes (1819–1903), of Navier–Stokes fame, developed a closed-form solution, resembling the results for the friction coefficient on plates and in pipes:

$$C_D = \frac{24}{\text{Re}}. \tag{8.76}$$

This equation compares well with the experimental data in Fig. 8.19, up to a Reynolds number of about 4.

As the Reynolds number increases, the flow will separate behind the cylinder and a "stationary" separation bubble is observed [see Fig. 8.20(b)]. This condition remains up to Reynolds numbers of about 40. At the next range of Reynolds number (up to say 400) alternate vortex shedding begins [Fig. 8.20(c)] with a laminar separation bubble behind. As the Reynolds number further increases (up to 0.3×10^6), wake vortices become more turbulent, but the front is still laminar [Fig. 8.20(d)]. The separation point is actually a few degrees (up to 8) ahead of the top, which explains the larger drag (of over 1.0). The next important region is one order of magnitude larger (about $\text{Re} > 3 \times 10^6$) when the boundary layer becomes turbulent at the front and as a result the rear separation area is reduced (up to 30° back from the top). Turbulent vortex shedding continues, but the reduced area of flow separation results in much lower drag, as shown in Fig. 8.20(e) (see also the sharp drop in drag in Fig. 8.19).

This example provides valuable insight about the resistance force in a moving fluid. The skin friction originating at the boundary layer is clearly one component. However, the separated flow creates a pressure distribution that results in drag. This component of the drag is called the *pressure drag* or sometimes *form drag* and in many practical cases (e.g., a car) it is much larger than the drag that is due to skin

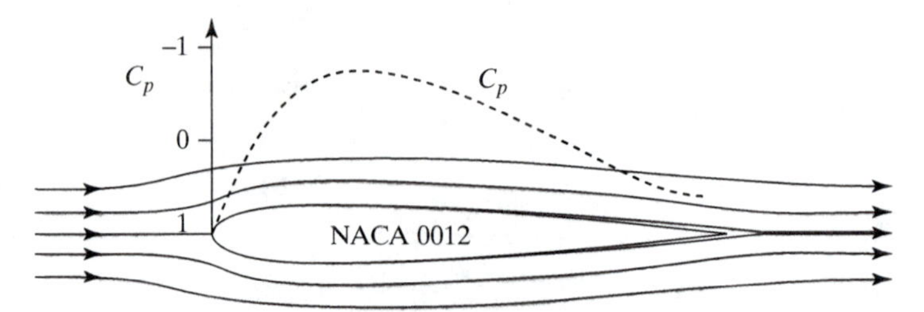

Figure 8.21. Potential flow solution of the flow over a streamlined shape and calculated pressure coefficient (which are close to experimental data). Note that the upper and lower pressure distributions are identical.

friction. Figure 8.20 clearly demonstrates that it depends on the Reynolds number and on whether the flow is laminar or turbulent. Note that this pressure drag is zero for 2D shapes in fully attached flows (as suggested by the present potential flow model – the fact that puzzled d'Alembert).

As a conclusion, a "good example" is presented in which the flow is attached. Naturally, a streamlined shape, such as shown in Fig 8.21, can qualify. The potential flow pressure distribution shown by the dashed line (calculated by adding sources and sinks to a free stream) compares well with experimental data and the drag force can be estimated by use of the skin friction data from the boundary-layer section (see Example 8.6). Consequently there is no form drag (if there is no flow separation), and the approach used in this section will be successful in predicting the pressure distribution and resulting loads.

EXAMPLE 8.6. DRAG DUE TO SKIN FRICTION ONLY. Estimate the drag coefficient at a Reynolds number of 3×10^6 on a streamlined shape, such as shown in Fig. 8.21.

Solution: In this case we assume that the flow is attached and the drag is a result of the skin friction only. We can estimate the drag coefficient from the boundary-layer data of Fig. 7.10, where we get $C_D \sim 0.004$ for turbulent flow; for laminar flow we use the Blasius formula:

$$C_D = 2\frac{0.664}{\sqrt{3 \times 10^6}} = 0.00077.$$

Based on the shape of the pressure distribution, the gradient is favorable for about 40% of the length, so we can assume that the boundary layer is laminar there, whereas it is turbulent along the rest of the surface. The total drag per side is

$$C_D = 0.4 \times 0.00077 + 0.6 \times 0.004 = 0.0027.$$

And for the two sides (2×0.0027) the drag coefficient is estimated at

$$C_D = 0.0054,$$

and this number is very close to experimental results (for this airfoil).

EXAMPLE 8.7. DRAG OF A POLE. A 15-m-tall vertical pole of diameter 0.25 m is exposed to winds of up to 30 m/s. Calculate the force on the pole and the moment at the base (assuming a 2D cylinder model). Assume air density is $\rho = 1.2$ kg/m^3 and viscosity $\mu = 1.81 \times 10^{-5}$ N s/m^2.

Solution: First we need to check the Reynolds number:

$$\mathrm{Re} = \frac{1.2 \times 30 \times 0.25}{1.81 \times 10^{-5}} = 0.497 \times 10^6.$$

In this Reynolds number range the flow is turbulent and the drag coefficient is about 0.7 (from Fig. 8.19) Note that here the drag coefficient is based on the frontal area whereas in the previous example the drag is based on the length of the streamwise surface. Nevertheless the drag numbers in this example are much larger (than in the previous one) because of the flow separations. Now we can calculate the drag force:

$$D = C_D \frac{1}{2} \rho U^2 S = 0.7 \times \frac{1}{2} \times 1.2 \times 30^2 \times 0.25 \times 15 = 1417.5\,\mathrm{N}.$$

For the moment at the base we may assume that the resultant force acts halfway to the top:

$$M = \frac{h}{2} D = \frac{15}{2} 1417.5 = 10{,}631\ \mathrm{N\,m}.$$

EXAMPLE 8.8. TERMINAL VELOCITY OF A SPHERE [RE $\sim O(1)$]. An aluminum ball of 0.5-cm diameter was dropped into a container filled with motor oil. Calculate how fast the ball will sink in the oil. Note that the density of aluminum is $\rho =$ 2700 kg/m^3 and the properties of oil are taken from Table 1.1.

Solution: This is an important example because it requires the calculation of terminal velocity (and the Reynolds number cannot be readily calculated). Let us assume that the Reynolds number is less than 4 and we can use the Stokes formula [Eq. (8.76)]. The force pulling the ball down is its weight minus the buoyancy:

$$(\rho_{\mathrm{Al}} - \rho_{\mathrm{oil}}) V g = (\rho_{\mathrm{Al}} - \rho_{\mathrm{oil}}) \frac{4}{3} R^3 g,$$

and the force acting upward is the resistance to the motion,

$$D = C_D \frac{1}{2} \rho_{\mathrm{oil}} U^2 (\pi R^2) = \frac{24}{\mathrm{Re}_D} \frac{1}{2} \rho U^2 (\pi R^2) = 6\pi U \mu.$$

During equilibrium these two forces are equal. Solving for U we get

$$U = \frac{2}{9\mu} (\rho_{\mathrm{Al}} - \rho_{\mathrm{oil}}) g R^2. \qquad (8.77)$$

Now this formula can be used to calculate the aluminum ball's sinking speed:

$$U = \frac{2}{9 \times 0.29} (2700 - 919) 9.8 \times 0.0025^2 = 0.084 \frac{\mathrm{m}}{\mathrm{s}}.$$

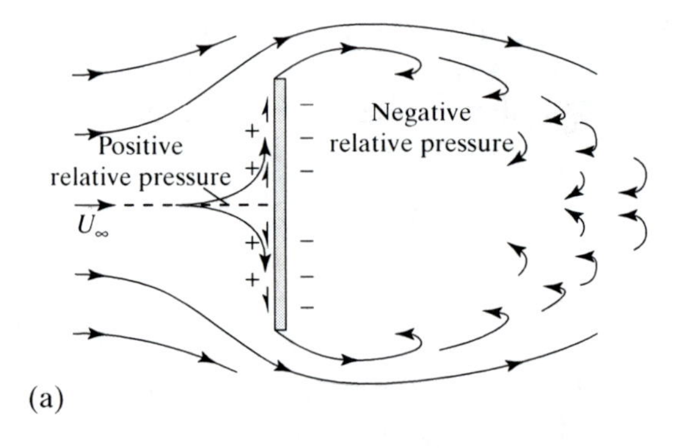

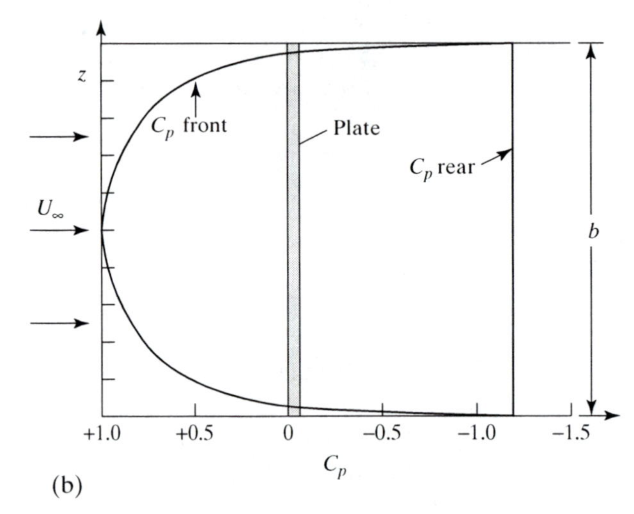

Figure 8.22. Flow normal to a flat plate and the resulting pressure distribution.

Next the Reynolds number must be calculated to validate the Stokes flow assumption,

$$\mathrm{Re} = \frac{919 \times 0.084 \times 0.005}{0.29} = 1.33,$$

and clearly the assumption was reasonable.

EXAMPLE 8.9. FLOW NORMAL TO A 2D FLAT PLATE. This is an important case because the flow separates at the two edges, as shown in Fig. 8.22a. Because the separation points are fixed, the drag coefficient is not changing significantly with Reynolds number, and $C_D \sim 1.2$ is a good estimate for $\mathrm{Re} > 10^4$ (for a width-to-height ratio of up to 5). This example also demonstrates how to extend the method to shapes other than a cylinder. Of course, tables documenting experimental results for the drag coefficient of various shapes are provided in the next section.

The measured centerline pressure distribution is shown in the lower part of the figure, and it clearly indicates that inside the aft separation bubble the pressure is almost unchanged. At the center (front) a stagnation point is present and the pressure coefficient there is about +1.0. As the fluid particles move to the sides, their velocity increases, resulting in high speed and low pressure at the edges. This low pressure prevails at the back and is the reason for the high drag.

Let us try to estimate the drag coefficient based on the pressure distribution in Fig. 8.22. The drag force is basically the average pressure difference between the front and the back of the plate multiplied by the area:

$$D = (\overline{p}_{\text{front}} - \overline{p}_{\text{rear}})S.$$

For the pressure at the front and rear we can use the average pressure coefficient:

$$\overline{p}_{\text{front}} - p_\infty = \overline{C}_p\Big|_{\text{front}} \frac{1}{2}\rho U^2 S,$$

$$\overline{p}_{\text{rear}} - p_\infty = \overline{C}_p\Big|_{\text{rear}} \frac{1}{2}\rho U^2 S,$$

and the drag is

$$D = (\overline{p}_{\text{front}} - \overline{p}_{\text{rear}})S = \left(\overline{C}_p\Big|_{\text{front}} - \overline{C}_p\Big|_{\text{rear}}\right)\frac{1}{2}\rho U^2 S,$$

or in nondimensional form

$$\overline{C}_D = \overline{C}_p|_{\text{front}} - \overline{C}_p|_{\text{rear}}.$$

Now, based on Fig. 8.22, the average quantity for $\overline{C}_p|_{\text{rear}}$ is about –1.1, and for the front it must be $\overline{C}_p|_{\text{front}} \sim +0.1$ in order for the drag coefficient to be 1.2. This is possible only if the suction at the edges of the forward-facing side is quite large, so this can counteract the high pressure at the center.

For a numerical example, assume that workers are moving a 1 m $\times$ 1 m glass plate and a wind of 15 km/h is blowing normal to it. The force on the plate is then

$$D = C_D \frac{1}{2}\rho U^2 S = 1.2\frac{1}{2}1.2\left(\frac{15}{3.6}\right)^2 1 \times 1 = 12.5 \text{ N}.$$

8.7.2 The Drag of More Complex Shapes

In the previous cases of the cylinder, the airfoil, and the flat plate, we used the pressure distribution in order to explain the resulting drag (mostly pressure drag in the case of the cylinder and the flat plate). Let us next analyze a more complex shape (as in Fig. 8.23).

The argument here is that the lessons learned from the ideal-cylinder case are applicable to more complicated shapes [as shown in Fig. 8.23(a)]. For example, at the front of the cylinder the flow faces a concave curvature and the pressure is higher. At the top, the curvature is convex, the velocity increases, and the pressure is lower (recall Bernoulli's model of molecules impinging on the surface, as discussed in Section 1.6). The same observations can be made for the automobile shape, in which the suction peaks occur where the surface is convex, as shown in Fig. 8.23(b).

When discussing the boundary-layer flow (Section 7.4), we observed that the favorable pressure gradient energizes the boundary layer and flow separation is unlikely. This is indicated in Fig. 8.23(b), in the front the pressure coefficient slopes

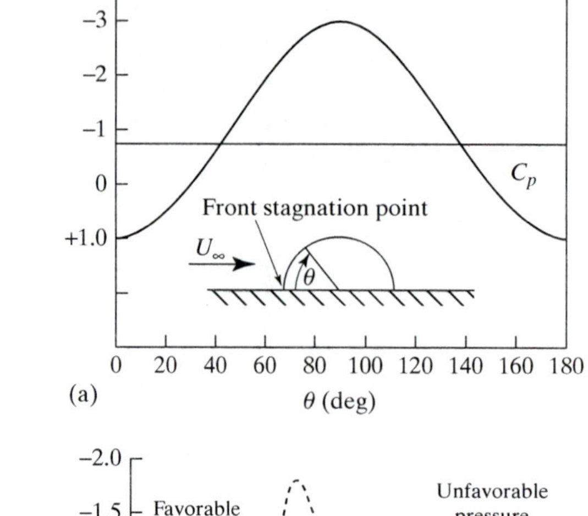

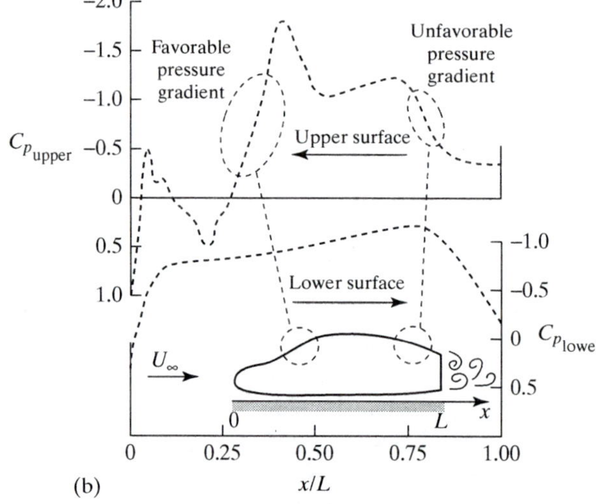

Figure 8.23. Pressure distribution (2D) on an automobile shape.

upward (hence the pressure goes from high to low = favorable). At the aft section of the cylinder and of the car, the pressure distribution is unfavorable (or adverse) and flow separation is likely. Consequently the flow separates in both cases, resulting in pressure drag (which for high-Re flows is much larger than the skin-friction drag).

To demonstrate the effect of flow separation on drag, Fig. 8.24 shows a small cylinder and a much larger airfoil shape. The high-Reynolds-number drag of both shapes is the same! The drag of the airfoil is mostly due to the skin friction in the boundary layer, whereas pressure (or form) drag is the main contributor to the drag of the cylinder.

Figure 8.25 shows the drag and lift coefficients of various and more complex configurations.

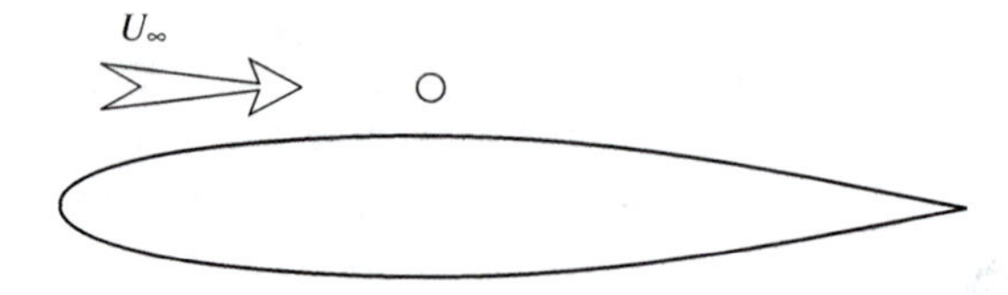

Figure 8.24. The drag of the small cylinder and that of the much larger streamlined airfoil are the same (at about $\text{Re} = 10^5$).

Shape		C_L	C_D
Circular plate		0	1.17
Circular cylinder	L/D = 0.5	0	1.15
	1		0.90
	2		0.85
	4		0.87
Rectangular plate	L/D	0	1.18
Rectangular cylinder	1	0	1.20
	5		1.30
	10		1.50
	20		1.98
Square rod 2D		0	2.00
Square rod 2D		0	1.50
Triangular cylinder 2D	60°	0	1.39
Semicircular shell 2D		0	1.20
Semicircular shell 2D		0	2.30
Hemispherical shell		0	0.39
Hemispherical shell		0	1.40
Cube		0	1.10
Cube		0	0.81
Cone–60° vertex		0	0.49
Parachute			1.20
Average man $A = 0.5 - 0.8\ \text{m}^2$	Average person		1.00–1.30
Fluttering flag $A = L{\cdot}D$	Fluttering flag, L/D = 1		0.07
	2		0.12
	3		0.15
Empire State Building			1.4
Tree	U		0.30–0.45

Shape		C_L	C_D
Low drag body of revolution		0	0.04
Low drag vehicle near the ground		0.18	0.15
Generic (older) automobile		0.32	0.43
Modern coupe		0.30	0.35
Coupe plus front splitter plate		0.20	0.35
Same + rear wing		−0.30	0.41
Prototype race car		−3.00	0.75
Motorcycle $A = 0.5 - 0.8\ \text{m}^2$		0.15 – 0.20	0.8–1.3
Bicycle			1.1
	Racing		0.88
	Drafting		0.50
	Streamlined		0.12
Trucks	Standard		0.96
	Fairing: With fairing		0.76
	Gap seal: With fairing and gap seal		0.70
Six car passenger train	Six-car passenger train		1.8

Figure 8.25. High-Reynolds-number force coefficient for various shapes (based on frontal area A apart from the flag).

Note: In most of the preceding cases the drag and lift coefficients are based on the frontal area. In certain applications the top view is used (so always verify which of the two is used).

EXAMPLE 8.10. AERODYNAMIC DRAG. My neighbor rides his bicycle at 25 km/h for 1 h. Calculate air resistance, power required, and total calories invested during 1 h.

Solution: The average frontal area of a bicycle rider is 0.36 m^2 and based on Fig 8.25, the drag coefficient is about 0.88. Using this information, we can calculate the drag force:

$$D = C_D \frac{1}{2} \rho U^2 S = 0.88 \frac{1}{2} 1.2 \left(\frac{25}{3.6}\right)^2 0.36 = 9.17\,\text{N}.$$

The power P required is simply the force times velocity,

$$P = DU, \tag{8.78}$$

and therefore

$$P = 9.17\frac{25}{3.6} = 63.65\,\mathrm{W}.$$

in terms of horsepower (hp) we divide by 745, and this is equal to 0.085 hp (and this isn't much). Also 1 cal = 4.2 J (joule) and therefore in 1 h (3600 s) the total energy spent is

$$E = P \times t = 63.65\,\mathrm{W} \times 3600\,\mathrm{s} = 229.2\,\mathrm{KJ} = 54.6\,\mathrm{K\,cal}.$$

So now he can eat his ice cream.

EXAMPLE 8.11. HYDRODYNAMIC DRAG. A submarine is cruising underwater at 20 km/h and its drag coefficient (based on frontal area) is 0.15. If its frontal area is 4 m^2 and the seawater density is 1025 kg/m^3, calculate the drag force and the power required for propelling the submarine.

Solution: The drag force on the submarine is

$$D = C_D\frac{1}{2}\rho U^2 S = 0.15\frac{1}{2}1025\left(\frac{20}{3.6}\right)^2 4 = 9491\,\mathrm{N}.$$

The power P required is simply the force times velocity:

$$P = 9491\frac{20}{3.6} = 52.7\,\mathrm{kW},$$

and this is about 71 hp.

EXAMPLE 8.12. TERMINAL VELOCITY OF A PARACHUTE. A 90-kg paratrooper jumps out of an airplane. If the chute diameter is 6 m, calculate his sinking–descent speed.

Solution: First we find that the drag coefficient of a parachute (from Fig. 8.25) is 1.2. In steady state the drag of the parachute is equal to the weight mg of the parachutist (we neglect the weight of the chute):

$$mg = C_D\frac{1}{2}\rho U^2 S.$$

Solving for U and substituting the numerical values we get

$$U = \sqrt{\frac{2mg}{C_D \rho S}} = \sqrt{\frac{2 \times 90 \times 9.8}{1.2 \times 1.2 \times \pi \times 3^2}} = 6.58\frac{\mathrm{m}}{\mathrm{s}}.$$

Based on my experience, this is too fast (4 m/s is better) and the paratrooper must lose some weight.

EXAMPLE 8.13. POWER REQUIREMENT FOR A CRUISING AUTOMOBILE. The drag coefficient of your sports car is 0.32 and its frontal area is 1.8 m^2. Although the engine output is rated at 300 hp, how much power is required for cruising at 100 km/h?

Solution: Let us first calculate the drag:

$$D = C_D \frac{1}{2} \rho U^2 S = 0.32 \frac{1}{2} 1.2 \left(\frac{100}{3.6} \right)^2 1.8 = 266.7\,\text{N};$$

the power is then simply the drag times velocity:

$$P = 266.7 \frac{100}{3.6} = 7.4\,\text{kW} = 9.9\,\text{hp}.$$

So you can probably use only one cylinder.

Now let us repeat the same exercise for a race car traveling at 300 km/h and having a frontal area of 1.4 m^2. If we assume a moderate drag coefficient of 0.75 for an open-wheel race car (and they have large drag because of the exposed wheels and wings), then the drag is

$$D = 0.75 \frac{1}{2} 1.2 \left(\frac{300}{3.6} \right)^2 1.4 = 4375\,\text{N},$$

and the required power is

$$P = 4375 \frac{300}{3.6} = 364.6\,\text{kW} = 489\,\text{hp},$$

and this car definitely needs the power.

8.8 Periodic Vortex Shedding

The experimental results for the flow over a cylinder indicated that, beyond Reynolds numbers of 40–90, a periodic vortex wake develops, as shown in Fig. 8.20. This is true in fact for other shapes (such as the flat plate in Fig. 8.22) in which such alternating vortices are visible. This phenomenon, seen on ocean currents flowing around islands or when winds cause the vibrations of telephone cables, is called the *Kármán vortex street*, after Theodore von Kármán (1881–1963), who also developed the boundary-layer integral formulation.

Figure 8.26(a) shows the alternating vortices behind the cylinder. As noted, when flow separation exists in a high-Reynolds-number flow, such a vortex street can develop; as an example, the vortices behind a large truck are shown schematically in Fig. 8.26(b). The shedding frequency is quite well defined by the Strouhal

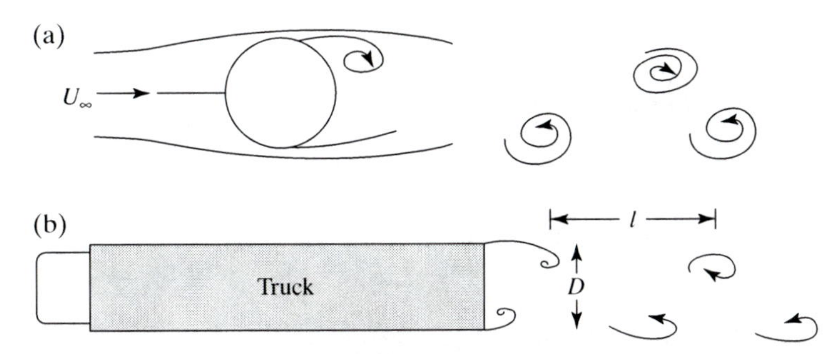

Figure 8.26. Vortex shedding behind a cylinder and a large truck.

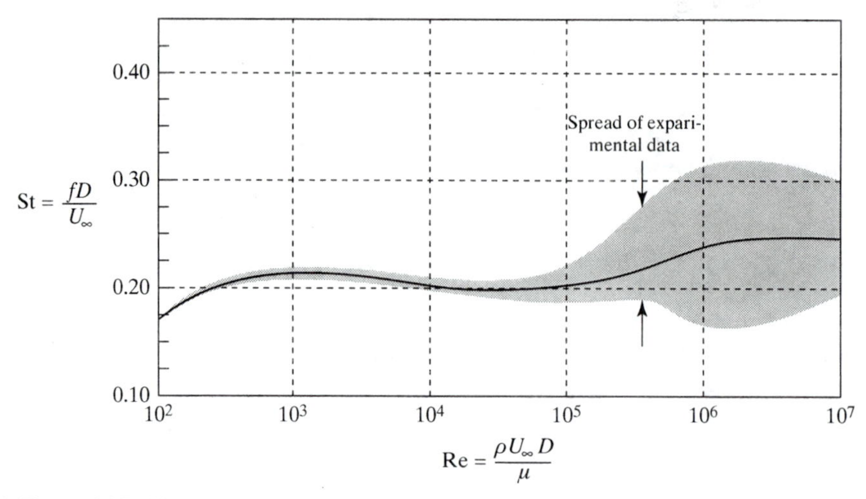

Figure 8.27. Nondimensional vortex-shedding frequency (St) versus Re for a 2D cylinder.

number (St) [Eq. (6.9)], which is very close to the value of 0.2. Figure 8.27 shows the range of observed St. versus Re, and indeed the variation is not large (the shaded area shows the range or spread of different experimental data).

The Strouhal number is defined here as

$$St = \frac{fD}{U_\infty}, \tag{8.79}$$

and f is the frequency in hertz (Hz) (cycles per second) and D is the approximate lateral spacing between the separation points (for a cylinder we take D, the diameter). The spread in experimental results usually narrows if the lateral spacing between the vortices [D in Fig. 8.26(b)] is used instead of the actual width of the body responsible for shedding the vortices. From the engineering point of view it is important to know the fluid mechanic frequencies in order not to design structures with similar frequencies (and ending with mechanical resonance). If the distance [l in Fig. 8.26(b)] between the vortices is sought, it can be calculated as

$$l = \frac{U_\infty}{f}. \tag{8.80}$$

Let us demonstrate the applicability of this simple formula by the following examples.

EXAMPLE 8.14. VORTEX-SHEDDING FREQUENCY OF A FLAGPOLE. Calculate the vortex-shedding frequency from a 0.3-m-diameter flagpole at a maximum wind speed of 35 m/s.

Solution: Let us first calculate the Reynolds number and then look up the St from Fig. 8.27:

$$Re = \frac{1.2 \times 35 \times 0.3}{1.81 \times 10^{-5}} = 0.7 \times 10^6.$$

Using this value in Fig. 8.27 we estimate St $\sim$ 0.23. We then calculate the frequency by using Eq. (8.79):

$$f = \mathrm{St}\frac{U_\infty}{D} = 0.23\frac{35}{0.3} = 27\,\mathrm{Hz}.$$

So there are 27 full cycles per second.

EXAMPLE 8.15. FLOW OSCILLATION BEHIND A LARGE TRUCK. A motorcycle travels behind a large truck and the rider feels the flow oscillations. If the truck is 2 m wide and travels at a speed of 100 km/h, calculate the shedding frequency and the spacing l between the oscillation cycles.

Solution: Let us refer to the schematics in Fig. 8.26(b). Using the width of the truck as D and calculating the Reynolds number as in the previous example, we get Re = 3.6×10^6. Let us use the same St of 0.23. The oscillation frequency is then

$$f = \mathrm{St}\frac{U_\infty}{D} = 0.23\frac{100/3.6}{2} = 3.19\,\mathrm{Hz}.$$

The distance l between the two cycles is

$$l = \frac{U_\infty}{f} = \frac{100}{3.19 \times 3.6} = 8.69\,\mathrm{m}.$$

EXAMPLE 8.16. ACOUSTIC EFFECTS OF VORTEX SHEDDING. As a musical example, calculate the "singing telephone wires" frequency in a 50-km/h crosswind. The wire diameter is 0.65 cm.

Solution: Let us calculate the Reynolds number first:

$$\mathrm{Re} = \frac{1.2 \times 50/3.6 \times 0.0065}{1.81 \times 10^{-5}} = 5985.$$

The Strouhal number from Fig. 8.27 is about 0.21, and the vortex shedding frequency is

$$f = \mathrm{St}\frac{U_\infty}{D} = 0.21\frac{50/3.6}{0.0065} = 449\,\mathrm{Hz},$$

and this is close to middle C (about 440 Hz).

8.9 The Case for Lift

We used the simple solution for the flow over a cylinder to calculate the pressure distribution and then present an explanation for the form or pressure drag. The same approach may be used for estimating the lift. This is done not only because these simple solutions are (probably) the only ones that can be presented at this introductory level, but because the lifting case contains the basic mechanism responsible for the upper–lower asymmetry in the flow field, explaining this effect.

8.9.1 A Cylinder with Circulation in a Free Stream

Up to this point the discussion was focused on the drag force; however, it is possible to create forces normal to the flow direction, and this is the topic of this subsection. For convenience we continue with the approach in which the free stream is

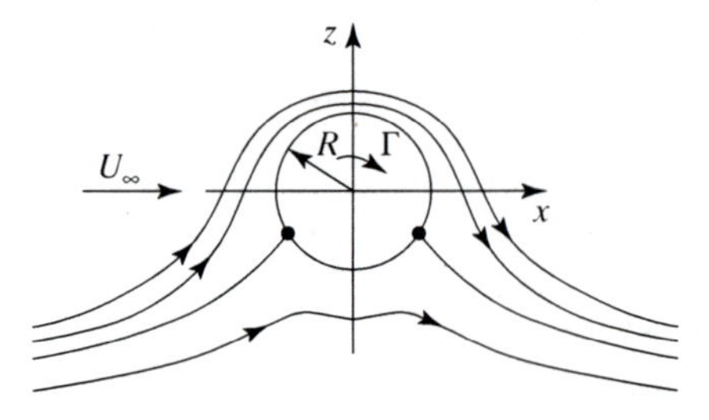

Figure 8.28. Streamlines for the flow over a cylinder with circulation.

flowing along the x axis and the normal force in the z direction is called the lift. The approach so far was to study the flow over a geometry (cylinder) that can be solved and then extrapolate (based on experimental and other data) to cases involving more complex geometries. Consequently, we can start again with the flow over a cylinder (which has upper–lower symmetry – and no lift) and search for a method to generate lift. A lifting condition can be obtained by the introduction of an asymmetry, in the form of a clockwise vortex with strength Γ situated at the origin (see Fig. 8.28). The velocity potential for this case is

$$\Phi = U_\infty \cos\theta \left(r + \frac{R^2}{r}\right) - \frac{\Gamma}{2\pi}\theta. \tag{8.81}$$

Note that we use the velocity potential of the flow over the cylinder, and the addition of the tangential vortex flow is not expected to affect the normal flow's boundary condition on the cylinder's surface at $r = R$. We can verify this by differentiating the velocity potential to get the velocity components:

$$q_r = \frac{\partial \Phi}{\partial r} = U_\infty \cos\theta \left(1 - \frac{R^2}{r^2}\right). \tag{8.82}$$

The radial component remains the same as for the cylinder without the circulation. The tangential velocity is

$$q_\theta = \frac{1}{r}\frac{\partial \Phi}{\partial \theta} = -U_\infty \sin\theta \left(1 + \frac{R^2}{r^2}\right) - \frac{\Gamma}{2\pi r}. \tag{8.83}$$

As expected, this potential still describes the flow around a cylinder because at $r = R$ the radial velocity component becomes zero. We can obtain the stagnation points by finding the tangential velocity component at $r = R$,

$$q_\theta = -2U_\infty \sin\theta - \frac{\Gamma}{2\pi R}, \tag{8.84}$$

and by solving for $q_\theta = 0$ we can see that they moved to a lower point on both sides of the cylinder:

$$\sin\theta_s = -\frac{\Gamma}{4\pi R U_\infty}. \tag{8.85}$$

These stagnation points (located at an angular position θ_s) are shown by the two dots in Fig. 8.28, and they lie on the cylinder as long as $\Gamma \leq 4\pi R U_\infty$. We find the lift and drag by using Bernoulli's equation. Substituting the tangential velocity of Eq. (8.84) yields the pressure distribution:

$$p - p_\infty = \frac{1}{2}\rho U_\infty^2 \left[1 - \left(2\sin\theta + \frac{\Gamma}{2\pi R U_\infty}\right)^2\right].$$

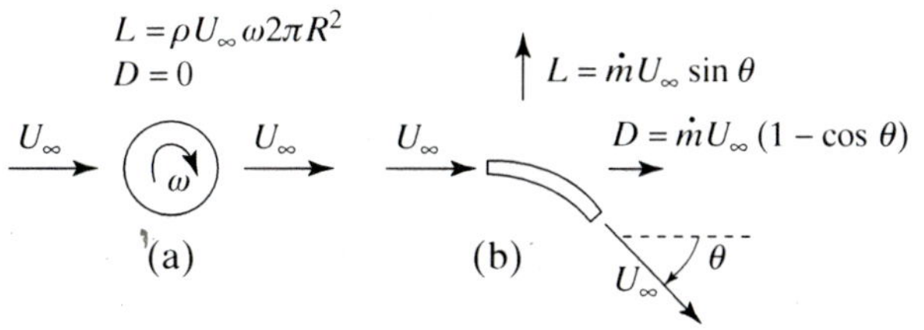

Figure 8.29. Methods of generating 2D force in a fluid: (a) by introducing angular momentum (without drag), and (b) change of linear momentum (as in the case of an impinging jet on a deflector – a case that results in drag).

Because of the fore and aft symmetry, no drag is expected from this calculation. For the lift, the tangential velocity component is substituted into the Bernoulli equation and

$$L = \int_0^{2\pi} -(p - p_\infty) R d\theta \sin\theta = -\int_0^{2\pi} \left[\frac{\rho U_\infty^2}{2} - \frac{\rho}{2} \left(2U_\infty \sin\theta + \frac{\Gamma}{2\pi R} \right)^2 \right] \sin\theta \, R d\theta.$$

But the integrals $\int_0^{2\pi} \sin\theta = \int_0^{2\pi} \sin^3\theta = 0$, and the lift integral reduces to

$$L = \frac{\rho U_\infty \Gamma}{\pi} \int_0^{2\pi} \sin^2\theta d\theta = \frac{\rho U_\infty \Gamma}{\pi} \int_0^{2\pi} \frac{1}{2}(1 - \cos 2\theta) d\theta = \rho U_\infty \Gamma. \tag{8.86}$$

This very important result states that the force in this 2D flow is directly proportional to the circulation and acts normal to the free stream. A generalization of this result was discovered independently by the German mathematician M. W. Kutta (1867–1944) in 1902 and by the Russian physicist N. E. Joukowski (1847–1921) in 1906. They observed that the lift per unit span on a lifting airfoil or cylinder is proportional to the circulation. Consequently the Kutta–Joukowski theorem is as follows: The resultant aerodynamic force in an incompressible, inviscid, irrotational flow in an unbounded fluid is of magnitude $\rho U_\infty \Gamma$ per unit width and acts in a direction normal to the free stream:

$$L = \rho U_\infty \Gamma. \tag{8.87}$$

The connection between circulation and angular rotation was established in Section 5.11. So the conclusion here is that this is a mechanism to create force in a fluid by introducing angular momentum. Consequently a fore–aft symmetry of the flow exists and no pressure drag results (if the flow is attached). Of course, this very efficient principle is used by flying birds and swimming fish (and airplanes). This remark about the efficiency can be demonstrated by considering the streamlines in Fig. 8.28 far ahead and far behind the rotating cylinder (this region is not shown in the figure). Because there is no vertical deflection of the flow, at a distance ahead and behind, the flow will be parallel to the x coordinate and there is no change in the linear momentum (hence no drag – in terms of ideal flow). This is depicted in the schematics of Fig. 8.29(a). In terms of pressure distribution the fluid particles move faster on the upper surface than on the lower one and there is a net lift force (although from a distance no change is detected between the incoming and outgoing free-stream velocity direction). On the other hand, when force is created by changing the linear momentum as in the case of an impinging jet on a deflector [see Eq. (4.8) and as shown in Fig 8.29(b)], there will be a drag force, even if the incoming and exiting velocities remain the same.

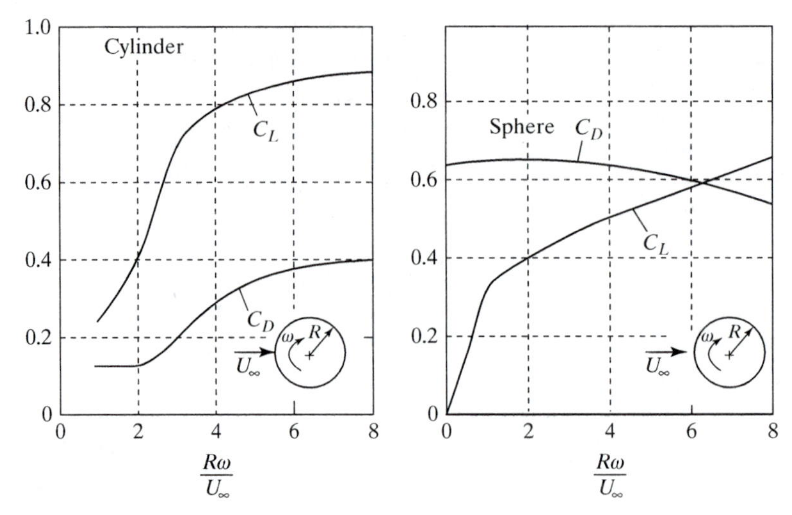

Figure 8.30. Lift and drag of rotating cylinder and sphere. Note that ω is measured in radians per second! (Re $= 0.4$–6.6×10^5, after [3, Chapter 21].)

This principle of creating lift in an attached flow is utilized by airplane wings and is discussed in the next section. However, it is interesting to examine the results for lift and drag created by both a rotating cylinder and a sphere, as depicted in Fig. 8.30. As expected, the flow over both cases is separated and therefore the drag is larger than zero and the lift is considerably less than estimated by the current model. This effect is called the *Magnus effect*, after German physicist Heinrich Gustav Magnus (1802–1870), who described this phenomenon in 1853. A similar spin effect is responsible for the curved balls in baseball or in soccer, or for the dispersion of artillery shells that is due to side winds.

To estimate the effect of flow separation on the theoretical results for an attached flow, recall Eq. (5.172), which estimates the circulation created by a rotating cylinder as

$$\Gamma = 2\pi R^2 \omega. \tag{5.172}$$

The lift calculated by Eq. (8.81) on the rotating cylinder with the attached flow is then

$$L = \rho U_\infty 2\pi R^2 \omega,$$

and the lift coefficient is

$$C_L = 2\pi \frac{R\omega}{U_\infty},$$

which is much larger than the values shown in Fig. 8.30 (so flow separation significantly reduces the lift). As noted, in addition to flow separation, surface roughness also has an effect on the lift and drag data in Fig. 8.30 (and this can be considered as a first-order estimate).

EXAMPLE 8.17. LIFT OF A ROTATING BALL. Estimate the lift and drag of a 3-cm-diameter ping-pong ball flying at 11 m/s and rotating at 7000 RPM.

Solution: First we need to calculate the lift and drag coefficients from Fig. 8.30. The nondimensional rotational parameter is

$$\frac{R\omega}{U_\infty} = \frac{0.015 \times 2\pi \times 7000/60}{11} = 1.0.$$

The lift and drag coefficients are obtained from Fig. 8.30(b) as $C_L = 0.27$ and $C_D = 0.63$. The corresponding forces are then

$$L = C_L \frac{1}{2} \rho U^2 S = 0.27 \frac{1}{2} 1.2 \times 11^2 \times \pi \times 0.015^2 = 0.014\,\text{N},$$

$$D = C_D \frac{1}{2} \rho U^2 S = 0.63 \frac{1}{2} 1.2 \times 11^2 \times \pi \times 0.015^2 = 0.032\,\text{N}.$$

If the ball moves to the left and the rotation is in the direction shown in Fig. 8.30(b), then the ball will experience lift.

8.9.2 Two-Dimensional Flat Plate at a Small Angle of Attack (in a Free Stream)

The rotating cylinder example demonstrated the concept of lift; however, because of flow separation the estimated lift didn't compare well with experimental results. A flat plate at a small incidence (or angle of attack) as shown in Fig. 8.31 is a much better example in the absence of flow separation. The solution of this problem can be obtained by use of vortices on the flat plate combined with a free stream; however, the math involved is beyond the scope of this text. The circulation for a flat plate with a chord c and at an angle of attack α (see Ref. [1, Chapter 5]) is

$$\Gamma = \pi U_\infty c \alpha. \tag{8.88}$$

The lift is calculated with Eq. (8.87),

$$L = \rho U_\infty \Gamma = \pi \rho U_\infty^2 c \alpha, \tag{8.89}$$

and the 2D lift coefficient per unit width is

$$C_L = \frac{L}{\frac{1}{2} \rho U^2 c \times 1} = 2\pi\alpha. \tag{8.90}$$

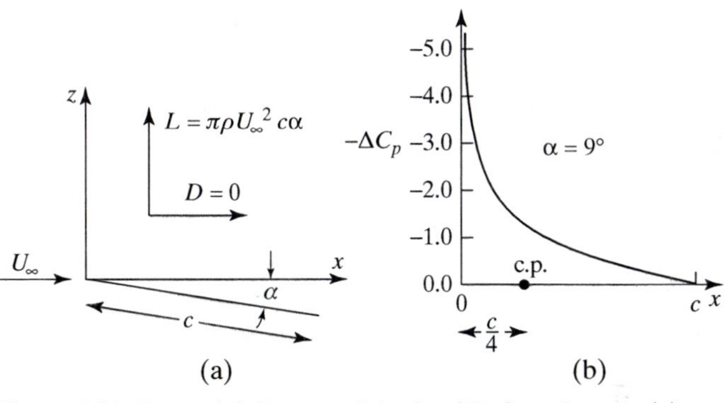

Figure 8.31. Potential flow model of a 2D flat plate at (a) an angle of attack and (b) the resulting pressure difference along the plate.

This result is amazingly close to experimental results as long as the flow is attached (up to $\alpha \sim 5°–7°$ for a thin, flat plate). The pressure difference resulting from this solution is shown in Fig. 8.31(b), and it appears that most of the lift is generated at the front. The calculated center of pressure is at the quarter chord – and this is close to experimental results as well. Also note that, in effect, the flat plate creates circulation of the magnitude given by Eq. (8.88) and the lift mechanism is similar to the lift of the rotating cylinder (however, now there is no massive flow separation). Consequently the drag C_{Di} (which is due to the pressure distribution) is zero (as in the case of the ideal flow over the cylinder)!

$$C_{Di} = 0. \tag{8.91}$$

Note that the drag that is due to skin friction (let us call it C_{d0}) is not included, and for calculating the total 2D drag it must be added:

$$C_D = C_{Di} + C_{d0}. \tag{8.92}$$

EXAMPLE 8.18. LIFT OF A 2D FLAT PLATE. The chord of a carport roof in an apartment complex is $c = 3.5$ m and its span is very wide (consider $b \sim \infty$). Calculate the lift per unit span for a 20-km/h wind blowing at $\alpha = 5°$ (straight on).

Solution: For this 2D case we use Eq. (8.90):

$$C_L = 2\pi\alpha = 2\pi\frac{5\pi}{180} = 0.548.$$

Note that α is calculated in radians. The lift per unit span is

$$L = C_L\frac{1}{2}\rho U^2 S = 0.548\frac{1}{2}1.2\left(\frac{20}{3.6}\right)^2 3.5 \times 1 = 35.53\,\text{N}.$$

8.9.3 Note about the Center of Pressure

Calculating the resultant force and its action point is very important for many engineering applications. The discussion on fluid statics (Chapter 3) demonstrated the method for the center-of-pressure calculation (albeit for simple pressure distributions only). In the case of a moving fluid over an objet, such as a baseball, a car, or an airplane, the resulting pressure distribution is complex, and the center-of-pressure calculation requires elaborate integration schemes. However, the previous examples provide some useful hints about the expected location of the center of pressure, as demonstrated in Fig. 8.32. Here the pressure distribution on the upper and lower surfaces of a flat plate are shown at the left, and the pressure distribution on one half cylinder is shown at the right.

As noted in the previous section, the center of pressure for a flat plate at an angle of attack is at 1/4 chord length from the leading edge. This location does not move when the angle of attack is changed, and aerodynamicists call it the *aerodynamic center* because the moments are not affected by the plate angle. Also note that a larger portion of the lift is due to the suction on the upper surface and not due to the high pressure on the lower surface [see Fig. 8.32(a)]!

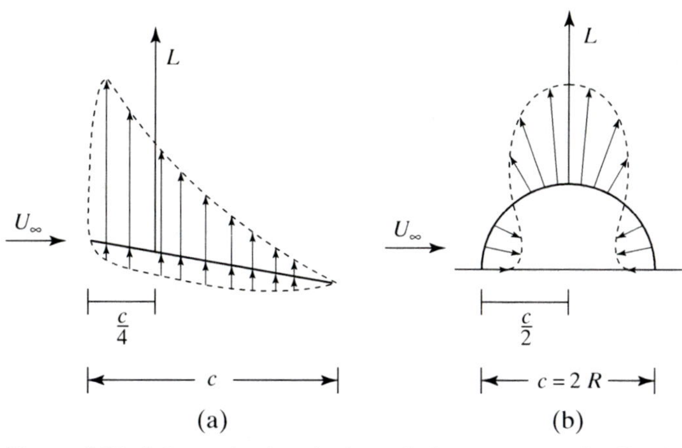

Figure 8.32. Schematic description of the pressure distribution (a) over a flat plate and (b) over the upper surface of half a cylinder in potential flow.

Figure 8.32(b) shows the pressure distribution over the upper surface of half a cylinder. The lift can be obtained by use of the integration of Eq. (8.73) between the limits: $\theta = 0, \theta = \pi$, and results in $C_L = 5/4$. Because of the fore–aft symmetry, the center of pressure is located at the center (e.g., $c/2$ in the figure). The important conclusion is that for lifting surfaces and wings the center of pressure is between the 1/4 and 1/2 chord. In the case of flat surfaces (or symmetric airfoils) it is at 1/4 chord and for cambered surfaces (or cambered airfoils) the center of pressure moves backward [but not as much as shown in Fig. 8.32 (b)].

8.10 Lifting Surfaces: Wings and Airfoils

The previous discussion about the flat plate (at a small angle of attack) established the approach for generating efficient fluid dynamic lift. Because of the importance of this topic, a short discussion on lifting surfaces (e.g., wings) is presented. Let us start with some definitions, as shown in Fig. 8.33.

A 3D wing of chord c and span b is shown at the top of the figure. The 2D cross section of a 3D wing is frequently called an *airfoil*, and a generic airfoil shape is shown by the shaded cross section in Fig. 8.33(a). Thus a 2D airfoil can be viewed as the cross section of a rectangular wing with an infinite span, and the side view of this infinitely wide wing is shown in Fig. 8.33(b) (also showing the definition of angle of attack α). The leading edge is usually rounded and the trailing edge is pointed; the letter t is used to denote the airfoil's maximum thickness. The thickness is usually measured in percentages of t/c. Figure 8.33(c) shows that an airfoil can be symmetrical or it can have a camber, and the shape of the camber line (centerline) is depicted by the dashed line.

It is interesting to look back at history, particularly at the efforts to understand the fluid mechanics of wings. Between 1914 and 1917, the group lead by Ludwig Prandtl in Göttingen, Germany, developed the so-called *lifting-line theory* to calculate the lift and drag of wings. One of the main outcomes of those studies is that for typical wings (when $b/c > 5$), the 2D airfoil-shape development can be separated from the wing planform (or top-view) shape. Following in their footsteps, we also

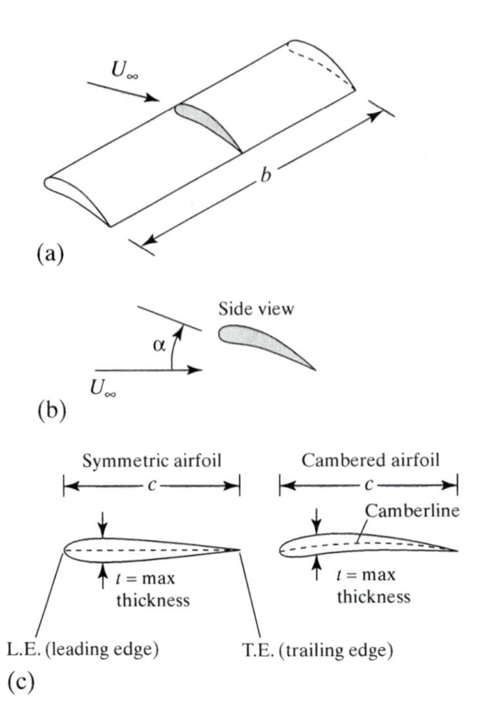

Figure 8.33. Basic definitions used to describe the shape of a wing.

discuss the flow over 2D airfoils first, and only later the effects of the 3D planform shape.

8.10.1 The Two-Dimensional Airfoil

The streamlines over a generic airfoil moving through a fluid (such as air) are presented in Fig. 8.34. As in the case of the rotating cylinder, the streamlines far ahead and behind remain parallel to the free stream. The streamline that stops under the leading edge is called the stagnation streamline because the flow stagnates (stops) at this point. The point itself is called a stagnation point. The overall effect of the airfoil on the surrounding fluid results in a faster flow above it and a slower flow under it. Because of this velocity difference the pressure above the airfoil will be lower than under it, and the resultant force (lift) will act upward. Also when comparing (from a distance) the ideal-flow streamlines in Fig. 8.34 with those for the lifting cylinder (Fig. 8.28) there is no change in the linear momentum. So from a distance the attached-flow airfoil appears as a rotating cylinder.

The shape of the pressure distribution is a direct outcome of the velocity distribution near the airfoil. For example, a fluid particle traveling along a streamline placed slightly above the stagnation streamline (in Fig. 8.34) will turn sharply to the left near the stagnation point. Because this turn is against the solid surface of the airfoil, the particle will slow down, resulting in a larger pressure near this point on the lower surface. But as it reaches the leading edge, it is forced to turn around it (but now the particle wants to move away from the surface), and therefore its acceleration increases, resulting in a very low pressure near the leading edge. A similar particle moving under the stagnation streamline experiences no major direction changes and will generally slow down near the airfoil and increase the pressure

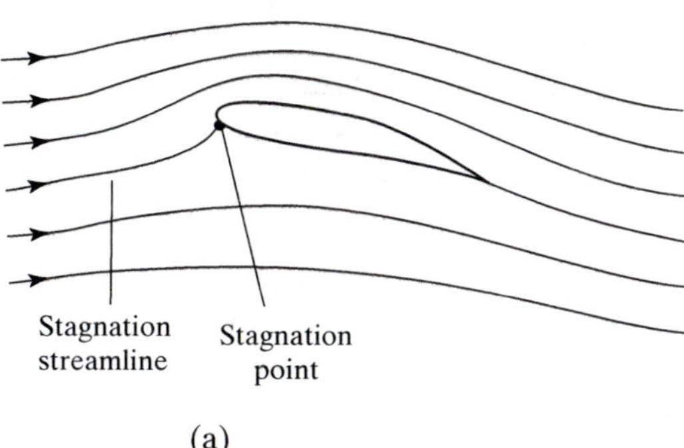

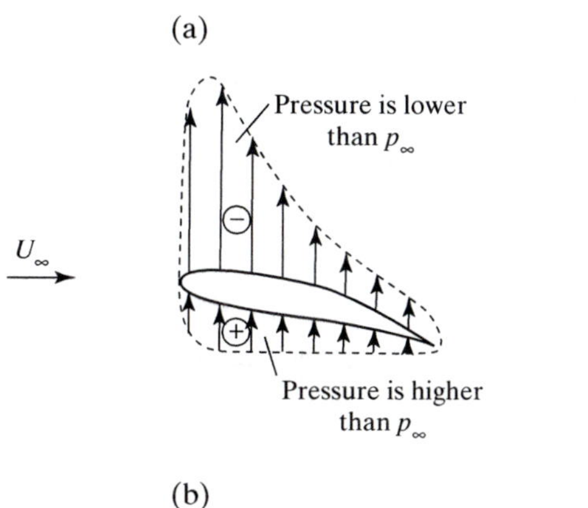

Figure 8.34. Schematic description of (a) the streamlines near an airfoil and (b) the resulting pressure distribution.

on the airfoil's lower surface. Thus the (+) sign in Fig. 8.34(b) represents the area where the pressure is higher than the free-stream static pressure, whereas the (–) sign represents the area with lower pressure. Also, in most cases the contribution of the suction side (–) to the airfoil's lift is considerably larger than that of the pressure side (+).

Next we try to demonstrate how an airfoil's geometry affects the shape of the pressure distribution. First, a typical pressure distribution on a symmetric airfoil at an angle of attack is shown in the left-hand side of Fig. 8.35. The vertical arrows show the direction of the pressure force acting on its surface. The shape of the pressure distribution on an airfoil with a cylindrical arc-shaped camber, at zero angle of attack, is shown in the center of the figure. These two generic pressure distribution shapes can be combined to generate a desirable pressure distribution, as shown on the right-hand side of the figure. Because of this observation, airfoils are frequently identified by their thickness distribution (which is a symmetric airfoil) and by an

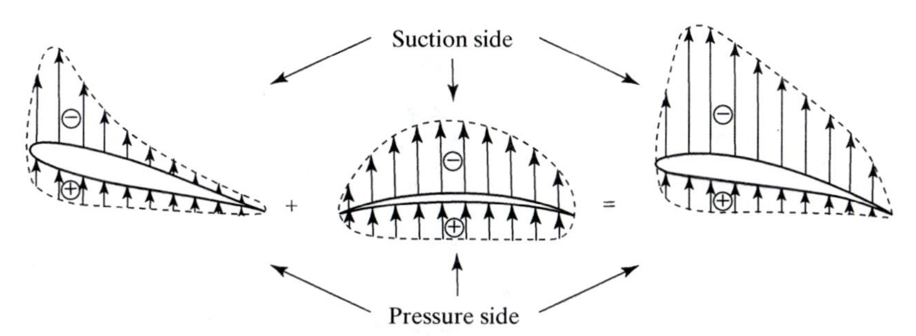

Figure 8.35. The effect of an airfoil's shape on the resulting pressure distribution.

additional centerline camber shape [the camber line shown in Fig 8.33(c)]. In conclusion, the angle of attack, the camber-line shape, and the thickness distribution determine an airfoil's pressure distribution.

The center of pressure of the symmetric airfoil is at the quarter-chord point ($x/c = 0.25$) as mentioned in the discussion about the flat plate (Subsection 8.9.3). This is verified in Fig 8.36 where the moments about the quarter chord are zero, as long as the flow is attached. On the other hand, the shape of the pressure distribution (at zero angle of attack) on the circular arc suggests that its center of pressure is located at the center (because of symmetry). Consequently it appears that by increasing the camber of an airfoil the center of pressure will shift backward. Also, the portion of the lift that is due to the symmetric airfoil, whose center of pressure is not affected by the change in angle of attack, remains at the quarter chord (a point we called earlier the aerodynamic center). This is based on the potential flow model, which permits the superposition of the symmetric and cambered airfoil's solution.

8.10.2 An Airfoil's Lift

To calculate the lift of a 2D symmetric airfoil we can start with the results for the flat plate [Eq. (8.90)]:

$$C_l = 2\pi\alpha. \tag{8.93}$$

The lowercase subscript l is usually used for the 2D case whereas the subscript L (e.g., C_L) is used for the 3D wing. Consequently, when Eq. (8.93) is used, the reference area becomes the chord c multiplied by a unit width, and the lift is also measured per unit width. Note that α is measured in radians! As an example, consider a symmetric airfoil at an angle $\alpha = 8°$; the lift coefficient is then

$$C_l = 2\pi\left(\frac{8\pi}{180}\right) = 0.877.$$

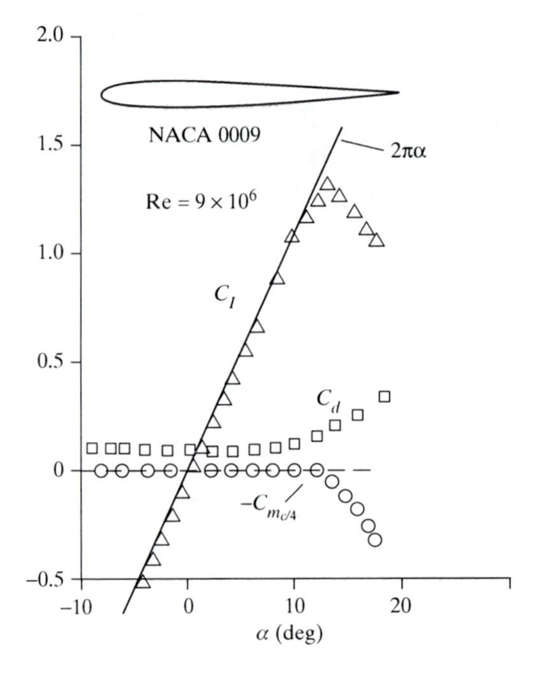

Figure 8.36. Typical data for the NACA 0009 airfoil (the last two digits, 09, mean that the airfoil is 9% thick).

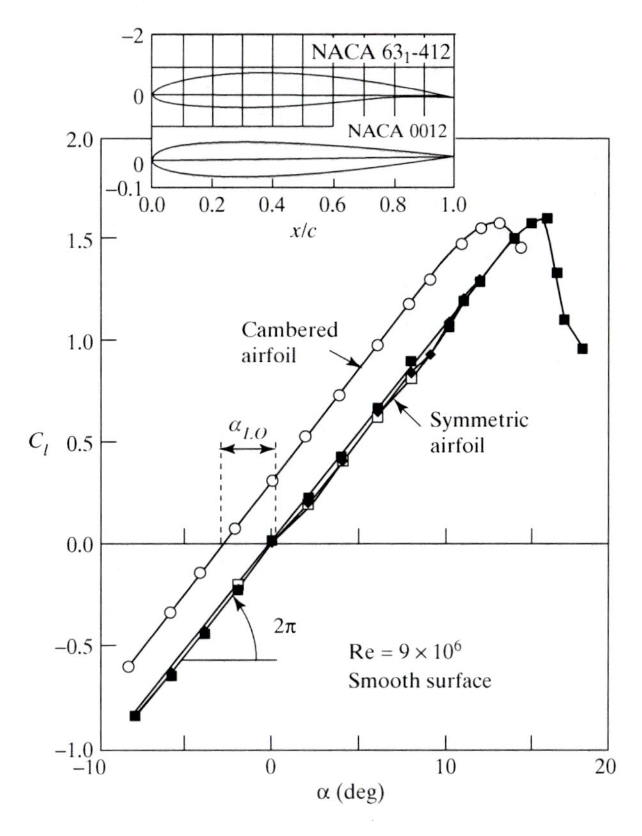

Figure 8.37. The effect of camber is like an incremental angle of attack, but it does not change the lift slope.

Figure 8.36 shows the experimental data for a typical symmetric airfoil, and the lift slope (2π) is amazingly close to the experimental results (up to stall, which in this case is beyond 10° angle of attack).

An interesting observation is that for a cambered airfoil the coefficient 2π does not change, but there is an increment in the effective angle of attack by α_{l0}. Thus the symmetric airfoil will have zero lift at $\alpha = 0$ whereas the cambered airfoil will have a lift of $2\pi\alpha_{l0}$, even at zero angle of attack. Consequently, for a cambered airfoil, Eq. (8.93) can be rewritten as

$$C_l = 2\pi(\alpha - \alpha_{l0}). \tag{8.94}$$

This is shown by the experimental data in Fig. 8.37 for two 12% thick airfoils. Note that the effect of camber can be calculated accurately by use of a combination of elementary solutions (e.g., doublets and sources).

Finally we can generalize the airfoil lift equation by using a single formula that is good for symmetric ($\alpha_{l0} = 0$) and cambered airfoils:

$$C_l = C_{l\alpha}(\alpha - \alpha_{l0}), \tag{8.95}$$

and for the 2D case the lift slope is

$$C_{l\alpha} = 2\pi. \tag{8.96}$$

8.10.3 An Airfoil's Drag

The high-Reynolds-number model developed in this chapter postulates that there is no pressure drag in ideal 2D flows (and flow separation is excluded). The experimental data in Fig 8.36 show that there is drag, and this is due to the shear stress in the boundary layer. Of course, at the higher angles of attack, flow separation results in pressure drag, which is much larger, as shown in the figure. Although the drag numbers are low, we can obtain improvements by reducing the skin friction in the boundary layer. Figure 8.38 shows two 15% thick NACA (National Advisory Committee for Aeronautics) airfoils with similar shapes. In the more recent design (NACA 64_2-415) an effort was made to keep a longer laminar region on the airfoil. This was achieved by a slight change in the shape of the upper surface so that a longer region will be exposed to a favorable pressure gradient. Because the skin friction is lower in the laminar boundary layer, visible drag reduction was achieved. This advantage is limited to a narrow range, and at higher angles of attack the gain is lost.

Note that the viscous drag shown in these two figures is the C_{d0} term in Eq. (8.92)! Here we use the lowercase d to emphasize the 2D geometry!

EXAMPLE 8.19. SIZING OF AN AIRFOIL. Frequently, experimental airfoil data, as in Fig. 8.38, are used for wing design. Suppose the design of a small airplane is aimed at low drag and therefore a lift coefficient of 0.5 is selected for cruise, using the NACA 64_2-415 airfoil. Calculate the required chord length for a lift of 50 kg_f (per unit span) at a speed of 150 km/h. Also, what is the lift-to-drag ratio at this condition?

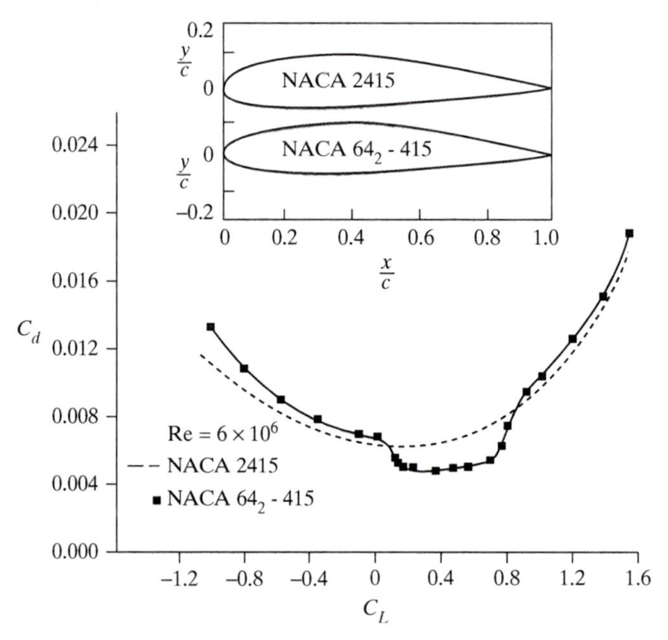

Figure 8.38. Longer laminar flow on the airfoil results in a lower drag coefficient. Of course, at larger angles of attack the boundary-layer thickness increases and the advantage disappears. (*Note*: You can view this graph as C_D versus α, because the lift coefficient in an ideal flow varies linearly with the angle of attack.)

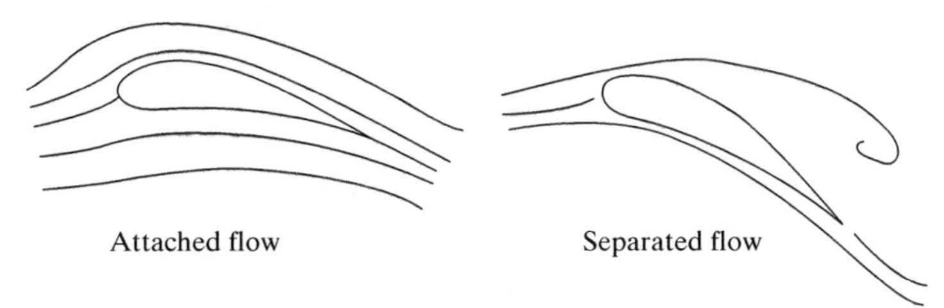

Figure 8.39. Schematic description of the streamlines near an airfoil in an attached flow and in a separated flow.

Solution: From the definition of the lift coefficient we can write

$$c = \frac{L}{0.5\rho U^2 C_l} = \frac{50 \times 9.8}{0.5 \times 1.2 \left(\frac{150}{3.6}\right)^2 0.5} = 0.94\,\mathrm{m}.$$

To estimate the drag we simply look at the figure, and it is about 0.005. Note that this represents the viscous friction drag (C_{d0}). The lift-to-drag ratio is then

$$\frac{L}{D} = \frac{0.5}{0.005} = 100,$$

and the viscous drag, per unit span, is 0.5 $\mathrm{kg_f}$.

8.10.4 An Airfoil Stall

The models developed in this Chapter were aimed at predicting the *attached* flow over streamlined shapes. Figures 8.36 and 37 show excellent agreement in the prediction of the lift, but the model fails when the flow separates. This condition is called *stall*, and airplanes or ships are not supposed to operate within this region. The streamlines for an attached airfoil are shown schematically on the left-hand side of Fig. 8.39, and they follow the shape of the airfoil. Usually when angle of attack is significantly increased, the streamlines will not follow the airfoil shapes and the flow will be separated. This will result in a dramatic loss of lift (stall) and a sudden increase in drag (see data in Fig. 8.36 for angles larger than 12°).

As the Reynolds number increases, there is a slight delay in the onset of stall; also, drooping the leading edge of the airfoil can delay the stall. The effect of stall on the pressure distribution is demonstrated in Fig. 8.40. The dashed curves shows a hypothetical shape for the pressure distribution, which would be expected without flow separation. The experimental data, however, show a sharp drop in the suction behind the separation point, an effect that is reducing lift. Also note that inside the separated flow region the pressure is not changing much.

8.10.5 The Effect of Reynolds Number

The discussion on high-Reynolds-number flows throughout the last three chapters suggests that the boundary-layer flow and flow separation are the main parameters affected by the Reynolds number (and the transition to turbulent flow). Although such flow fields are quite complex, in general, flow separation is delayed in turbulent flow because of the momentum exchange with the outer flow (but skin friction will

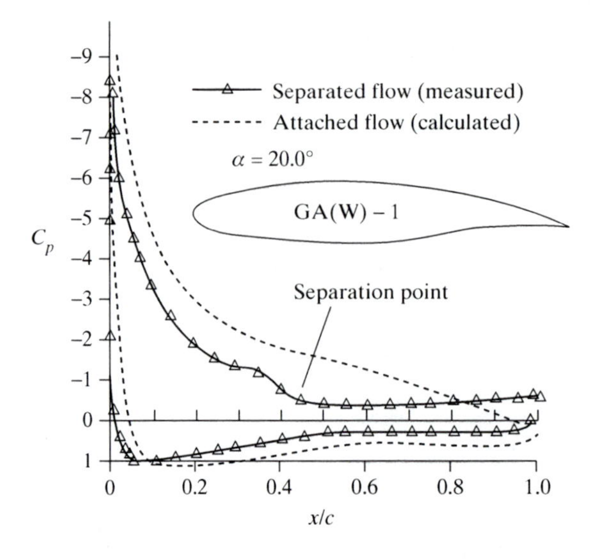

Figure 8.40. Hypothetical attached flow pressure distribution (dashed curves) on a general aviation airfoil GA(W) – 1, compared with the actual separated flow pressure distribution (triangles).

increase). On the other hand, if the flow is attached then maintaining longer laminar flow regions results in less drag. These conflicting observations indicate that optimizing a design for best L/D ratio is not trivial. Nevertheless, the effect of increasing Reynolds number can be demonstrated by testing the same airfoil at various speeds, as shown in Fig. 8.41. Again, note that when the flow is attached, the lift slope is unchanged, and the lift is the same as predicted by the outer flow model [Eq. (8.93)].

The first and most prominent effect seen on this figure is the onset of separation, which starts at smaller angles of attack for the lower Reynolds numbers. A more careful observation of this case indicates that flow separation starts at the trailing edge. By increasing the Reynolds number the trailing-edge separation is delayed and higher angles of attack can be reached without flow separation (hence the higher maximum lift). This trend continues, but near $\alpha = 16^\circ$ flow separation cannot be avoided and the lift loss is much sharper. This condition when the flow separates near the leading edge is naturally called leading-edge stall, as shown in the figure.

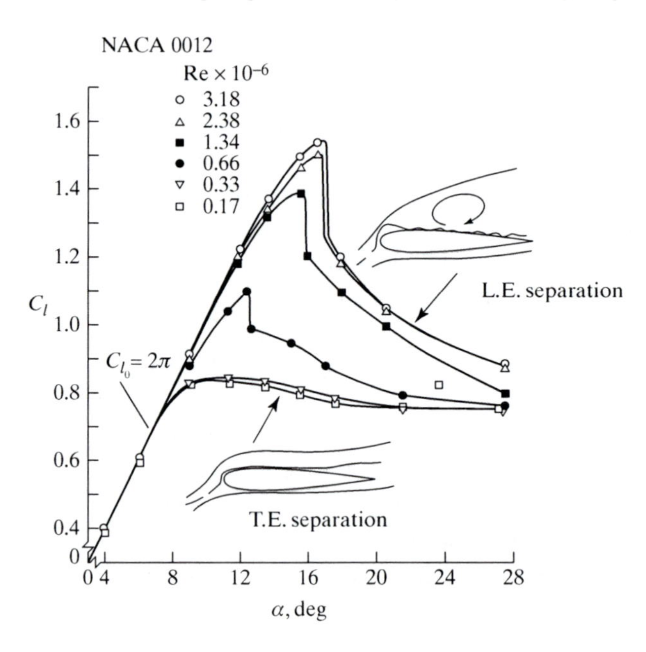

Figure 8.41. Effect of Reynolds number on the lift coefficient of a symmetric NACA 0012 airfoil (reproduced from [1, Fig. 15.16]).

8.10.6 Three-Dimensional Wings

The formulation used for the 2D airfoil can be extended to include 3D wings; however, an important wingtip effect must be taken into consideration. We can speculate that the airfoil pressure distribution (as depicted in Fig. 8.34) will be reduced if wingspan is reduced because of this wingtip effect. The lift, as before, is a result of high pressure below the wing and the low pressure above it. However, near the tips the higher pressure (below) creates a flow that escapes and rolls upward, generating a trailing vortex, as shown in Fig. 8.42. So it is clear that the lift will be less than the 2D value because of this edge effect. In other words, the pressure difference at the tip cannot be maintained and the lift there drops to zero. Therefore it is logical to assume that the pressure difference between the upper and the lower surfaces will increase with increased distance from the tip (and the same can be assumed for the local lift).

The geometry of a finite wing shape is usually identified by the 2D airfoil section (or sections), and by the planform (top view) shape. The influence of airfoil shape on the aerodynamic properties was discussed in the previous section, and the effect of planform shape is discussed briefly here. In principle, Fig. 8.42 shows that the two large tip vortices induce a downward velocity on the wing (downwash), thereby reducing its lift (and increasing the drag). Therefore it is clear that the wider the wing the less effect these tip vortices will have. This explains why sail planes have a short chord and very large span. The relative width of a 3D wing is usually identified by its *aspect ratio*, AR, which is defined as

$$\mathrm{AR} = (b^2/S). \tag{8.97}$$

Of course, for a rectangular wing,

$$\mathrm{AR} = \frac{b^2}{b\,c} = \frac{b}{c}. \tag{8.98}$$

Several generic planform shapes of planar wings are shown in Fig. 8.43. The simplest shape is the rectangular wing with a span b and a constant chord c [Fig. 8.43(a)]. The AR is then a measure of the width of the wingspan compared with its chord [Eq. (8.98)]. The wing can be swept, and in Fig. 8.43(b) the wing leading edge is

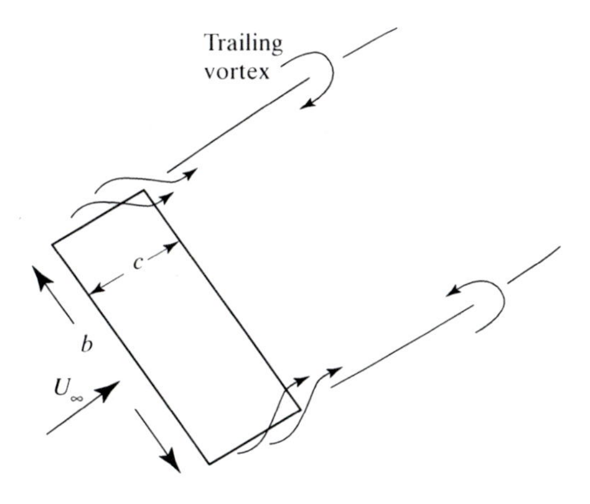

Figure 8.42. Nomenclature for the finite flat plate at an angle of attack. Note the edge effect, forming a trailing-tip vortex.

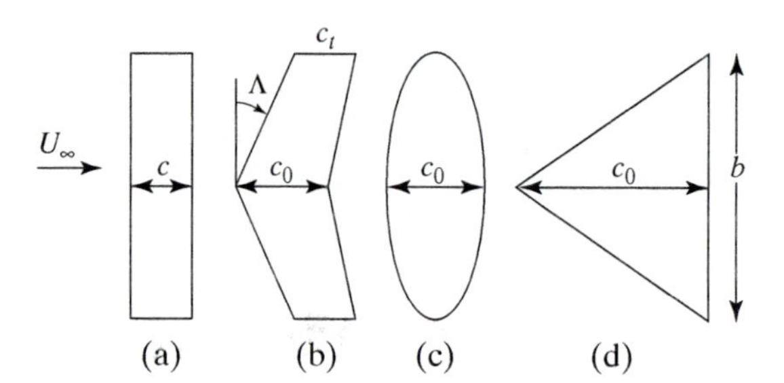

Figure 8.43. Generic planform shapes of 3D wings.

swept backward by an angle Λ. In this case the wing has a taper as well, and the tip chord c_t is smaller than the root chord c_0. The taper ratio λ simply describes the ratio between the tip and root chord lengths:

$$\lambda = (c_t / c_0). \tag{8.99}$$

The wing planform can have an elliptic shape, as shown in Fig. 8.43(c), and in this case the wing chord varies along the span, in a manner similar to an ellipse. The triangular shape of Fig. 8.43(d) is seen on many high-speed aircraft and can be viewed as a swept-aft rectangular wing with a taper ratio of zero. Any wing can be twisted so that the tip has a different angle of attack from its root chord, and it can be tilted upward at its tips (called dihedral) or downward at the tip, compared with the wing root (called anhedral). By tailoring the previously mentioned geometrical parameters, engineers can control the spanwise lift distribution on the wing. In most cases, however, the lift is the largest at the centerline and is zero at the tip (as discussed earlier).

The 3D lift and drag are calculated in a manner similar to the method used for the 2D airfoil. For the lift we may use Eq. (8.95) (but we use an uppercase L):

$$C_L = C_{L\alpha}(\alpha - \alpha_{L0}), \tag{8.100}$$

where the subscript L represents the total lift of the wing with an area S. The 3D effect (which is due to the tip vortices) that is reducing the lift slope is then included in a modified lift slope that depends on the wing AR. For most wings we can use the following simple equation:

$$C_{L\alpha} = \frac{2\pi}{1 + (1 + \delta_1)\dfrac{2}{\mathrm{AR}}}. \tag{8.101}$$

This equation was developed for elliptic wings, where $\delta_1 = 0$. In other cases δ_1 is small, and its evaluation is beyond the scope of this text (so we can approximate $\delta_1 = 0$). As discussed in the case of the flat plate, the tip vortices bend the streamlines behind the wing, creating an induced drag (which is an inviscid effect). This drag is

$$C_{Di} = \frac{1 + \delta_2}{\pi \mathrm{AR}} C_L^2. \tag{8.102}$$

Again, this equation was developed for elliptic wings where $\delta_2 = 0$. In other cases δ_2 is small, and its evaluation is again beyond the scope of this text (so, again, we can approximate $\delta_2 = 0$). The total drag of the wing then includes the airfoil-section viscous drag C_{d0} (as in Figs. 8.36 and 8.38) and the induced drag [as in Eq. (8.102) (but using an uppercase D)]:

$$C_D = C_{D0} + C_{Di}. \tag{8.103}$$

EXAMPLE 8.20. LIFT OF A 3D FLAT PLATE. The chord of a carport roof (from the example of Section 8.9) in an apartment complex is $c = 3.5$ m but its span is now $b = 10$ m. Calculate the lift of the roof for a 20-km/h wind blowing at $\alpha = 5^\circ$ (straight on).

Solution: This is now a finite plate, and because there is no camber $\alpha_{L0} = 0$. Next, we use Eq. (8.101) to approximate the 3D lift slope,

$$C_L = \frac{2\pi\alpha}{1+\dfrac{2c}{b}} = \frac{2\pi\dfrac{5\pi}{180}}{1+\dfrac{2\times 3.5}{10}} = 0.322,$$

and indeed the lift coefficient is much smaller than in the 2D case. Next we calculate the lift of the roof as

$$L = C_L\frac{1}{2}\rho U^2 S = 0.322\frac{1}{2}1.2\left(\frac{20}{3.6}\right)^2 3.5\times 10 = 209.05\,\text{N},$$

and for the average lift per unit width we divide by 10 to get 20.90 N, which is significantly less than in the previous example (where it was 35.53 N).

To calculate the induced drag coefficient we use Eq. (8.102) and AR = $10/3.5 = 2.86$,

$$C_{Di} = \frac{1}{\pi \text{AR}}C_L^2 = 0.0155,$$

and the induced drag is then

$$D_i = C_{Di}\frac{1}{2}\rho U^2 S = 7.499\,\text{N}.$$

To calculate the total drag [as in Eq. (8.103)] we can estimate C_{D0} by using the boundary-layer calculations of Chapter 7.

The approach presented here simplifies the process for estimating the lift and drag generated by lifting surfaces. The lift slope for complex shapes can be evaluated by computations, and typical results for a rectangular wing (with or without sweep) are shown in Fig. 8.44. The reduction in the lift with reduced AR [as predicted by Eq. (8.101)] is quite large, and the addition of sweep reduces the lift even further. Note the dashed line, which shows the 2D value of 2π and demonstrates the sharp loss in $C_{L\alpha}$ with reduced AR!

In general, the preceding formulation was developed for wings with AR larger than 7. For the very low range of AR < 1, a simple formula was developed by NASA scientist R. T. Jones (1910–1999):

$$C_L = \frac{\pi}{2}\text{AR}\alpha, \tag{8.104}$$

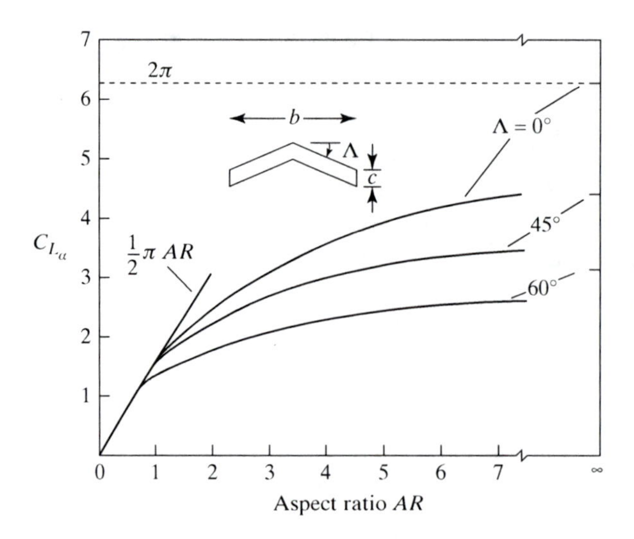

Figure 8.44. Effect of AR and sweep Λ on the lift slope of a planar rectangular wing.

and the induced drag is still calculated by Eq. (8.102). The lift slope predicted by this formula is also shown in Fig. 8.44. This formulation, however, holds for only very small angles of attack (e.g., $\alpha < 4°$).

This range of very small-AR wings is important for high-speed flight, and supersonic aircraft (such as the Concord) have such small-AR wings. At slightly higher angles of attack (e.g., $\alpha > 7°$) a leading-edge separation takes place as shown in Fig. 8.45. The side vortices then roll up and induce low pressure on the upper surface of the wing. The result (contrary to high-AR wing stall) is a significantly higher lift than predicted by Eq. (8.104). This phenomenon is called *vortex lift* and sometimes such highly swept surfaces are used ahead of larger-AR wings (called strakes).

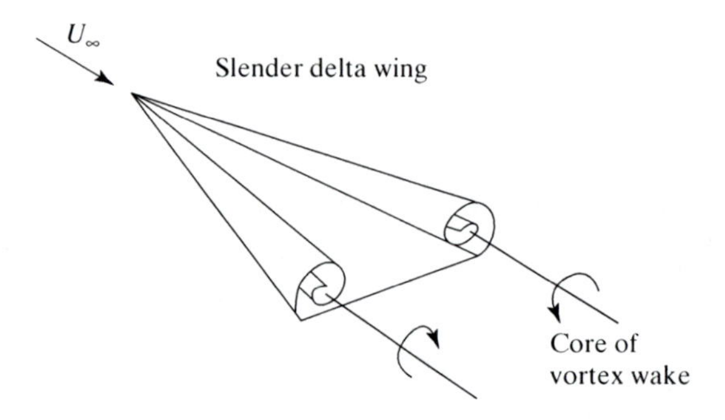

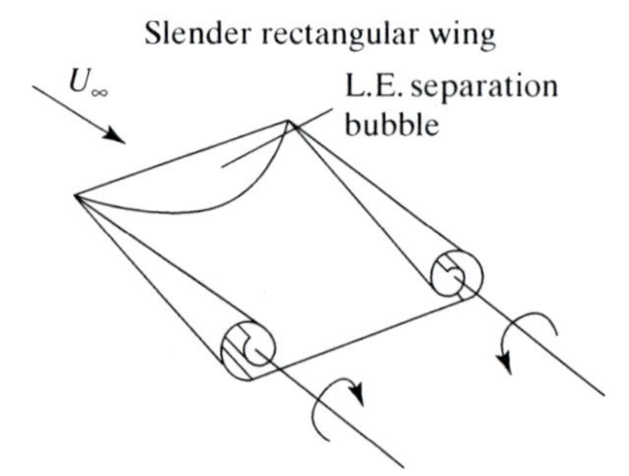

Figure 8.45. Schematic description of leading- and side-edge separation on slender wings.

For the case of these very slender wings (AR < 1) the following formulations can be used:

$$C_L = (a_1 + a_2 \mathrm{AR}) \sin\alpha, \tag{8.105}$$

$$C_D = C_L \tan\alpha, \tag{8.106}$$

and the constants are $a_1 = 0.963$ and $a_2 = 1.512$ for delta wings and $a_1 = 1.395$ and $a_2 = 1.705$ for slender rectangular wings.

The fact that such highly swept (or small-AR) surfaces can generate significant vortex lift, at very high angles of attack, was realized by aircraft designers, and most modern military airplanes have such highly swept strakes (see Fig. 8.46). For example, if such a strake is added in front of a less-swept-back wing, then the vortex originating from the strake will induce low pressures on the upper surface of the main wing, and the total gain in lift will surpass the lift of the strake alone.

This gain in lift, as shown in Fig. 8.46, begins at an angle of attack of 10° and has a significant effect up to angles of about 40°. A more careful examination of this figure reveals that the highly swept wing also stalls, but at a fairly large angle of attack. This stall, though, is somewhat different from the unswept-wing stall and is due to "vortex burst" (or breakdown). This condition is shown schematically in Fig. 8.47(a), and at a certain point the axial velocity in the vortex core is reduced and the vortex becomes unstable, its core bursts, and the induced suction on the wing disappears. So, as a result of the vortex burst, the lift of the wing is reduced and a condition similar to stall is observed (but, as noted, this takes place at a very high angle of attack).

The onset of vortex burst was investigated by many investigators, and the results for a delta wing can be summed up best by the schematic diagram in Fig. 8.48. The abscissa in this figure shows the wing AR, and the ordinate indicates the angle-of-attack range. The curve on the right-hand side indicates the boundary

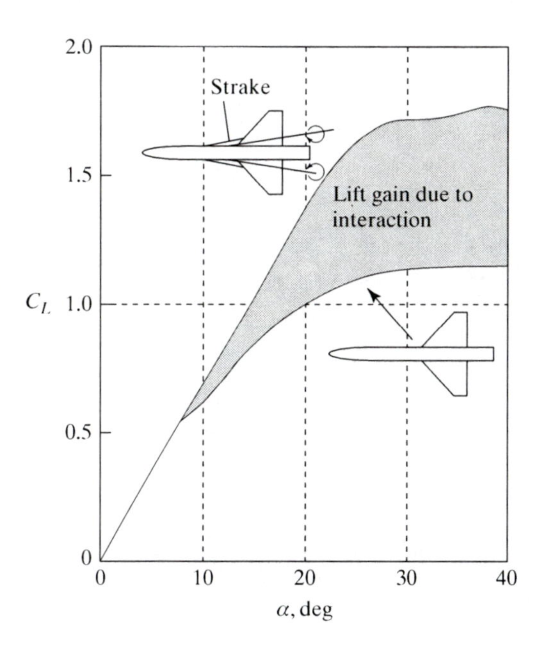

Figure 8.46. Effect of strakes on the lift of a slender wing–body configuration. (From Skow, A. M., Titiriga, A., and Moore, W. A., "Forebody/ wing vortex interactions and their influence on departure and spin resistance," published by AGARD/NATO in CP 247, High Angle of Attack Aerodynamics, 1978.)

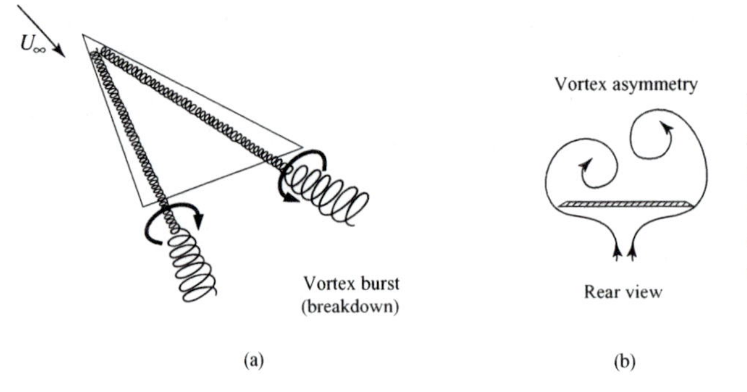

Figure 8.47. Possible high-angle-of-attack instabilities of the flow over slender wings: (a) leading-edge vortex breakdown, and (b) leading-edge vortex asymmetry (rear view).

at which the vortex burst will reach the wing's trailing edge. As an example for how to read this figure, consider a slender delta wing with, AR = 1, and let us gradually increase its angle of attack. Initially, say at $\alpha = 15°$, a vortex burst may appear, but it will be far behind the trailing edge, not affecting the lift. As the angle of attack is further increased, this vortex burst will move toward the trailing edge, and according to this figure, will reach the trailing edge near $\alpha = 35°$. Once the vortex burst moves ahead of the trailing edge, the vortex suction and resulting lift increase will be reduced, and slender-wing stall will begin. Also, based on this figure, for larger-AR wings (less leading-edge sweep) the burst will occur at lower angles of attack. On the other hand, as the wing becomes very slender, the leading-edge vortices become very strong and the burst is delayed. But for these wings another flow phenomenon, called "vortex asymmetry," is observed. This situation [shown schematically in Fig. 8.47(b)] occurs when the physical spanwise space is reduced and consequently one vortex raises above the other. The onset of this condition is depicted by the left-hand curve in Fig. 8.48. For example, if the angle of attack of an AR = 0.5 delta wing is gradually increased, then above $\alpha \approx 20°$ this vortex asymmetry will develop. If the angle of attack is increased, say up to $\alpha = 40°$, the lift will still grow and probably near $\alpha = 45°$ the vortex burst will advance beyond the trailing edge and wing

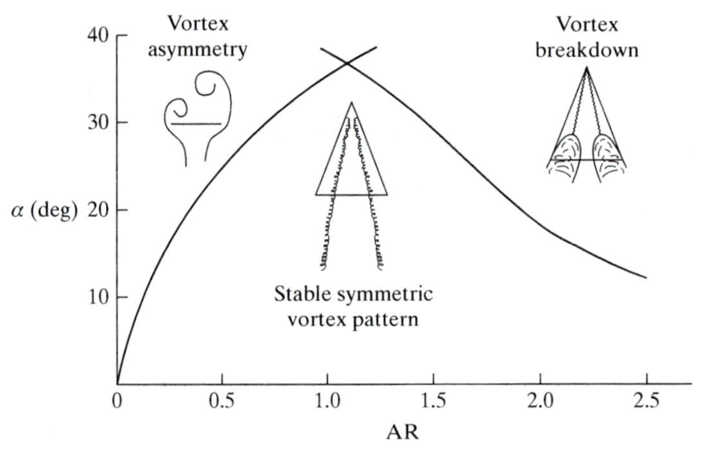

Figure 8.48. Stability boundaries of leading-edge vortices for flat delta wings in incompressible flow (adapted from [2]).

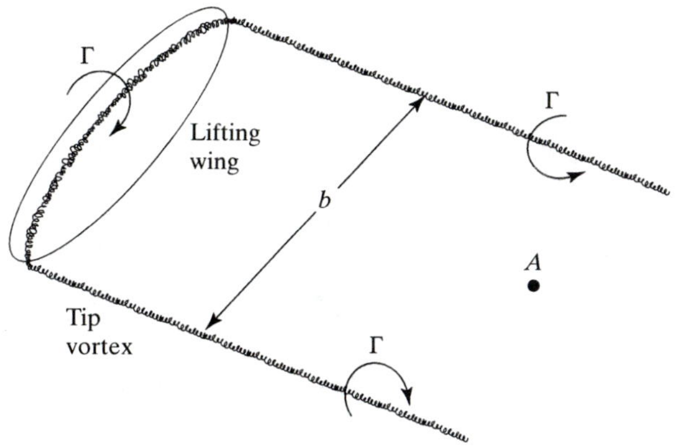

Figure 8.49. The horseshoe model for a lifting wing and its trailing vortex wake.

stall will be initiated. In general, the condition of an asymmetric vortex pattern is nondesirable because of the large rolling moments caused by this asymmetry. Furthermore, the pattern of asymmetry is sensitive to disturbances and can arbitrarily flip from side to side. The presence of a vertical fin (e.g., a rudder) between the two vortices or a central body (as in missiles) can have a stabilizing effect and delay the appearance of this vortex asymmetry.

Before concluding the discussion on 3D lifting surfaces, we discuss again the effect of the wingtip vortices and the wake. Flow visualizations, such as water-vapor condensation near the wingtips of airplanes, clearly demonstrate the existence of tip vortices. Early models, as proposed by British engineer Frederick Lanchester (1886–1946), represent the lifting wing by a horseshoe-shaped vortex, as shown in Fig. 8.49, and this model is quite useful for simple, preliminary calculations (this model was later refined by Prandtl). Recall the Kutta–Joukowski theorem of Eq. (8.87), which connects the lift to the circulation (or angular momentum). For example, the conservation of angular momentum principle suggests that the lifting vortex (representing the wing) cannot stop at the wingtip, and it is bent into the free stream and trails behind the wing, as shown (also, because the wake vortex is parallel to the free stream it creates no force).

The first effect of these trailing vortices is to create a downwash and reduce the wing's lift (as discussed earlier). But at a large distance behind the airplane the vortices continue to create a velocity field, as suggested by the figure. Between the vortices there is a strong downwash, whereas outside there is an upwash. Because of this principle, air controllers delay subsequent landings and takeoffs of airplanes to reduce the effect of the wake vortices. Probably, similar reasons encourage bird flocks to fly in a close V-shaped formation.

To demonstrate the applicability of the equations developed for the lift and drag of 3D wings, several examples are presented.

EXAMPLE 8.21. WING TRAILING VORTICES. To demonstrate the effect of an airplane trailing vortices, consider a large 100-ton airplane (such as the Boeing 767) taking off at 200 km/h. If the wingspan is $b = 47$ m (in Fig. 8.49) let us calculate the downwash far behind the airplane centerline (as in point A in Fig. 8.49).

Solution: The Kutta–Joukowski theorem of Eq. (8.87) determines the circulation per unit length, based on the lift (or the weight of the airplane). Therefore

$$\Gamma = \frac{L}{\rho U_\infty b} = \frac{100000 \times 9.8}{1.22(200/3.6)47} = 307.63\,\text{m}^2/\text{s}.$$

The resulting downwash at point A is then [based on Eq. (8.50)]

$$w = 2\frac{\Gamma}{2\pi(b/2)} = 2\frac{307.63}{\pi 23.5} = 8.33\,\text{m/s}\ (30\,\text{km/h}).$$

This could be significant for a following smaller airplane.

EXAMPLE 8.22. EFFECT OF ASPECT RATIO. Compare the lift and drag coefficients of two rectangular wings at $\alpha = 5^\circ$, both having a symmetric airfoil and chord of 0.5 m. One wing has a span of $b = 2$ m and the other has a span of $b = 4$ m.

Solution: Let us approximate the lift slope by Eq. (8.101). The aspect ratios of the two wings are

$$\text{wing 1,}\quad \text{AR} = \frac{2}{0.5} = 4; \qquad \text{wing 2,}\quad \text{AR} = \frac{4}{0.5} = 8.$$

The corresponding lift slope are, respectively,

$$C_{l\alpha} = \frac{2\pi}{1 + (2/4)} = 4.19, \qquad C_{l\alpha} = \frac{2\pi}{1 + (2/8)} = 5.02.$$

The lift coefficients are then

$$C_L = 4.19\frac{8\pi}{180} = 0.585, \qquad C_L = 5.02\frac{8\pi}{180} = 0.599.$$

The induced drags are

$$C_{Di} = \frac{1}{\pi 4}0.585^2 = 0.027, \qquad C_{Di} = \frac{1}{\pi 8}0.599^2 = 0.014,$$

and the lift-to-induced-drag ratios are

$$\frac{L}{D_i} = 21.66, \qquad \frac{L}{D_i} = 42.78.$$

Clearly the higher-AR wing is more efficient.

EXAMPLE 8.23. THE NEED FOR HIGH-LIFT DEVICES. Before landing, a 747 jet weights 250 tons. Its wingspan is 64.4 m, its wing area is 541 m^2, and its average wing chord is $c = 8.4$ m. Assuming air density at 1.1 kg/m^3 and a speed of 600 km/h, calculate the lift coefficient and estimate the lift slope.

Solution: The lift coefficient is calculated as follows:

$$C_L = \frac{L}{\frac{1}{2}\rho U^2 S} = \frac{250{,}000 \times 9.8}{\frac{1}{2}1.1\left(\frac{600}{3.6}\right)^2 541} = 0.296.$$

Let us estimate the lift slope based on the rectangular wing formula (but we know that this is a bit off because the airplane has swept wings). The aspect ratio is then

$$\mathrm{AR} = \frac{64.4^2}{541} = 7.66,$$

and the lift slope is

$$C_{L\alpha} = \frac{2\pi}{1 + \dfrac{2}{\mathrm{AR}}} = \frac{2\pi}{1 + \dfrac{2}{7.66}} = 4.98.$$

From Eq. (8.100) we can estimate the airplane angle of attack:

$$\alpha - \alpha_{L0} = \frac{C_L}{C_{L\alpha}} = 3.4^\circ.$$

In reality, the airplane fuselage at cruise is at near-zero angle relative to the free stream, and this angle represents mainly the camber effects.

Now let us repeat the same calculations for a landing speed of 220 km/h and assume that the flaps are not used. In this case the lift coefficient is

$$C_L = \frac{L}{\frac{1}{2}\rho U^2 S} = \frac{250{,}000 \times 9.8}{\frac{1}{2}1.1\left(\frac{220}{3.6}\right)^2 541} = 2.2.$$

Suppose we didn't change the wing geometry and the lift slope is unchanged; then the estimated angle of attack is

$$\alpha - \alpha_{L0} = \frac{C_L}{C_{L\alpha}} = 25.3^\circ.$$

Of course, both the lift and the required angle of attack are too high and the airplane must use flaps for landing. The flaps are the surfaces that extend at the trailing edge and significantly increase the lift coefficient (and also wing area).

This example demonstrates the need for *high-lift devices* on airplanes. Clearly, during high-speed cruise, less lift is needed (recall that lift increases with the square of speed) and wing size is designed for this condition to optimize drag. However, as this example shows, at low speed, the required lift coefficient increases dramatically, to a level that cannot be attained by simple fixed wings. The widely used solution is to change the wing geometry in order to both increase the airfoil's lift and to increase the wing area. A typical three-element airplane airfoil is shown in Fig. 8.50. By extending the rear flap the airfoil camber is increased, increasing the lift coefficient as shown by the upper curve. As the lift increases, the front stagnation point moves lower (as in the case of the circle in Subsection 8.9.1) and the leading-edge slat must be drooped to avoid stall. Note that the leading-edge element is mainly protecting against stall and thereby extending the useful range of angle of attack, but not directly increasing the lift. Also the 2D lift slope is still the same as predicted by the simple linear theory ($C_{l\alpha} = 2\pi$)!

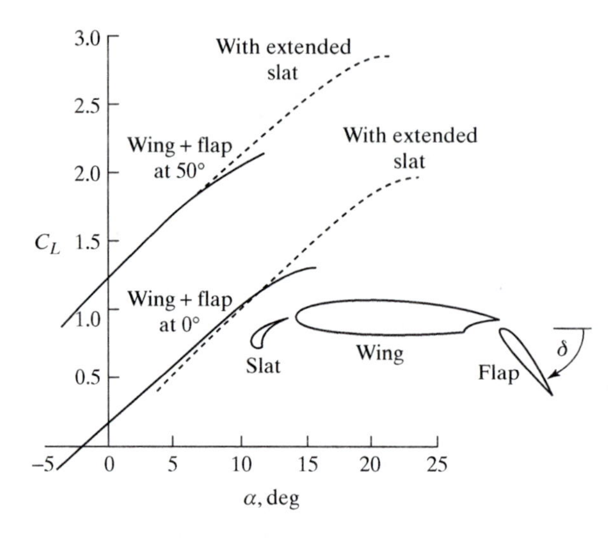

Figure 8.50. Effect of high-lift devices on the lift coefficient of a three-element airfoil (here δ represents, flap deflection).

EXAMPLE 8.24. HYDROFOIL BOAT. A high-speed hydrofoil boat is using two rectangular hydrofoils (each is 4 m wide, has a 0.5 chord, and $\alpha = 5^\circ$) to elevate it above the water level. Assuming a symmetric airfoil shape and a speed of 12 m/s, calculate the boat's weight and the power required for overcoming the induced drag.

Solution: Let us approximate the airfoil's lift slope by Eq. (8.101). The hydrofoil aspect ratio is

$$\mathrm{AR} = \frac{4}{0.5} = 8,$$

and the lift slope (approximating with the elliptic loading equation) is

$$C_{L\alpha} = \frac{2\pi}{1 + \frac{2}{8}} = \frac{8\pi}{5}.$$

The lift coefficient is then

$$C_L = \frac{8\pi}{5}\frac{5\pi}{180} = \frac{\pi^2}{22.5}.$$

The weight of the boat is calculated with the seawater density from Table 1.1 (1030 kg/m^3):

$$W = C_L \frac{1}{2}\rho U^2 S = \frac{\pi^2}{22.5}\frac{1}{2}(1030)12^2(2+2) = 130{,}120 \text{ N},$$

which is about 13.3 tons (and here we used the combined area of the two hydrofoils). To calculate the power, the induced drag is calculated:

$$C_{Di} \approx \frac{1}{\pi \mathrm{AR}} C_L^2 = \frac{1}{\pi 8}\left(\frac{\pi^2}{22.5}\right)^2 = 0.0077.$$

We assume that most of the drag is induced drag, and for the total drag we must add the contribution of the viscous effects (e.g., skin friction), as stated in Eq. (8.103). The induced drag is then

$$D_i = C_{Di}\frac{1}{2}\rho U^2 S = 0.0077\frac{1}{2}(1030)12^2(2+2) = 2271\,\text{N},$$

and the power required for overcoming the induced drag is

$$P = D_i V = 2271 \times 12 = 27{,}252\,\text{W},$$

which is 27.25 kW or about 37 hp.

EXAMPLE 8.25. LIFT OF A SLENDER WING. A supersonic delta-winged airplane is landing at a speed of 70 m/s at an angle of attack of 20°. Wing span is 15 m and wing chord is 25 m. Calculate airplane weight and its drag.

Solution: Let us first calculate the wing aspect ratio:

$$\text{AR} = \frac{b^2}{S} = \frac{b^2}{0.5bc} = \frac{2b}{c} = \frac{30}{25} = 1.2.$$

For the lift coefficient at such high angles of attack, we use Eq. (8.105):

$$C_L = (0.963 + 1.512 \times 1.2)\sin 20^\circ = 0.95.$$

The airplane weight is then

$$W = C_L\frac{1}{2}\rho U^2 S = 0.95\frac{1}{2}1.2(70^2)\frac{15 \times 25}{2} = 523{,}687.5\,\text{N},$$

which is about 53.4 tons. The drag is calculated with Eq. (8.106):

$$D = L\tan\alpha = 190{,}606.7\,\text{N},$$

which is about 19.45 tons. So lift-to-drag ratio is quite low!

8.11 Summary and Concluding Remarks

The topics discussed in this chapter have important engineering implementations. The high-Reynolds-number-flow region includes a large variety of day-to-day applications, including road, sea, and airborne vehicles. From the mathematical point of view, the governing equation can be simplified significantly. For example, we can evaluate the velocity field by solving the uncoupled continuity equation. Then we can calculate the pressures and fluid dynamic loads as a second step by using the Bernoulli equation (and not using the momentum equation – e.g., Euler). Of course all this works for attached flows. However, the method of using experimental coefficients for the lift and drag forces facilitates acceptable engineering prediction capability, even for separated and turbulent flows.

The force vector acting on a body moving through a fluid is traditionally divided into a force parallel to the free stream (drag) and into a normal component (lift). The modeling results presented here lead to a simple formulation for the drag on moving objects. For two dimensions we can write:

$$C_D = C_{D0},$$

and C_{D0} can be further split into drag that is due to skin friction and into form drag that is due to flow separations. Note the aerospace engineers use lowercase D (drag) and L (lift) for the 2D coefficients. For the 3D case a new element that is due to lift is added, the induced drag (and this is true for shapes other than wings):

$$C_D = C_{D0} + C_{Di}.$$

The lift of most objects (and certainly wings) depends on their angle of attack, and the camber effects can be included as a zero-lift angle α_{L0}. The lift slope $C_{L\alpha}$ is larger for wider objects (and the largest for the 2D case) and can be obtained by means of computations or experiments. With these assumptions, the lift can be calculated as

$$C_L = C_{L\alpha}(\alpha - \alpha_{L0}).$$

REFERENCES

[1] Katz J. and Plotkin, A., "*Low-Speed Aerodynamics*, 2nd ed., Cambridge University Press, New York, 2001.
[2] Kreyszig, E., "*Advanced Engineering Mathemathics*," 8th ed., Wiley, New York, 1999.
[3] Polhamus, E. C., "Prediction of vortex characteristics by a leading-edge suction analogy," *J. Aircraft* **8**, 193–199, 1971.
[4] Hoerner, S. F. and Borst, H. V., *Fluid Dynamic Lift*, Hoernrt Fluid Dynamics, Albuquerque, NM, 1985.

PROBLEMS

8.1. Prove that a 2D source fulfills the Laplace equation. The source equation in Cartesian coordinates is

$$\Phi(x, z) = \frac{\sigma}{2} \ln \sqrt{(x - x_0)^2 + (z - z_0)^2}.$$

8.2. Prove that a 2D doublet fulfills the Laplace equation. The doublet equation in Cartesian coordinates is

$$\Phi(x, z) = \frac{-\mu}{2\pi} \frac{x - x_0}{(x - x_0)^2 + (z - z_0)^2}.$$

8.3. The velocity components of a 2D vortex located at (x_0, z_0) are

$$u = \frac{\Gamma}{2\pi} \frac{z - z_0}{(z - z_0)^2 + (x - x_0)^2},$$

$$w = \frac{\Gamma}{2\pi} \frac{x - x_0}{(z - z_0)^2 + (x - x_0)^2}.$$

Prove that these velocity components fulfill the continuity equation.

8.4. Prove the following vector identity. Use 2D Cartesian coordinates $\vec{q} \cdot \nabla \vec{q} = -\vec{q} \times \nabla \times \vec{q} + \nabla \frac{\vec{q}^2}{2}$.

8.5. If the velocity potential of a 2D flow is given by the function

$$\Phi = U_\infty \cos\theta \left(r + \frac{R^2}{r} \right).$$

Then around the circle $r = R$.

(a) Plot the pressure coefficient for the range of $\theta = 0$ to $180°$.
(b) calculate the velocity and pressure at $\theta = 0, 90°$, and $180°$.

8.6. The velocity potential of a 2D flow is given by the function

$$\Phi = 5(x^2 - z^2).$$

(a) Derive an expression for the velocity components (u, w) in the x–z plane.
(b) Sketch the streamlines in the first quadrant (e.g., $x > 0$ and $z > 0$).

8.7. The velocity potential of a free stream is $\Phi_1 = 5x$ and for a doublet is $\Phi_2 = 5\frac{x}{x^2 + z^2}$.

(a) Write the velocity potential for the combined doublet and free stream (e.g., $\Phi = \Phi_1 + \Phi_2$).
(b) Calculate the velocity distribution (u, w) that is due to this velocity potential.
(c) Find the stagnation points along the x axis.
(d) What kind of flow field is described by Φ?

8.8. The velocity potential in the first quadrant is given by $\Phi_2 = 5(x^2 - z^2)$. Provide an expression for the velocity vector along a line $z = 2$.

8.9. A source of strength $\sigma = 10$ m^2/s is located at a point (5,1). Calculate the velocity at point (0, –2).

8.10. Calculate the velocity induced by a doublet of strength $\mu = 1$ m^3/s pointing in the $-x$ direction, at a point $x = 1$ and $z = 1$. The doublet is placed at (5, 2).

8.11. A 2D vortex with a circulation of $\Gamma = 10$ m^2/s is located at a point (0,0). Calculate the velocity at (10, 0).

8.12. Calculate the velocity components (u, v, w) that are due to a semi-infinite (but 3D) vortex line with strength $\Gamma = 1.0$ at a point (0,0,0.5).

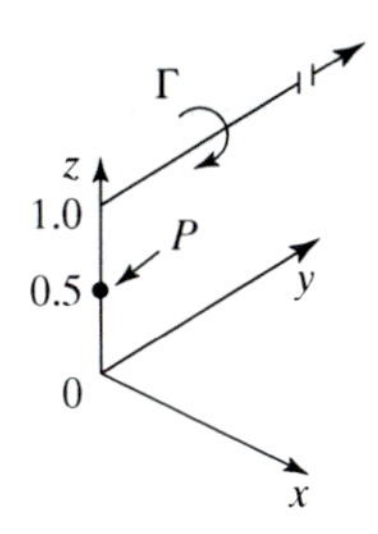

Problem 8.12.

8.13. A 2D vortex with strength $\Gamma = 1.0$ is placed at point (3,3), as shown in the figure. Find the value of the integral $\int q\, dl$ along the spiral path circling twice around the point P.

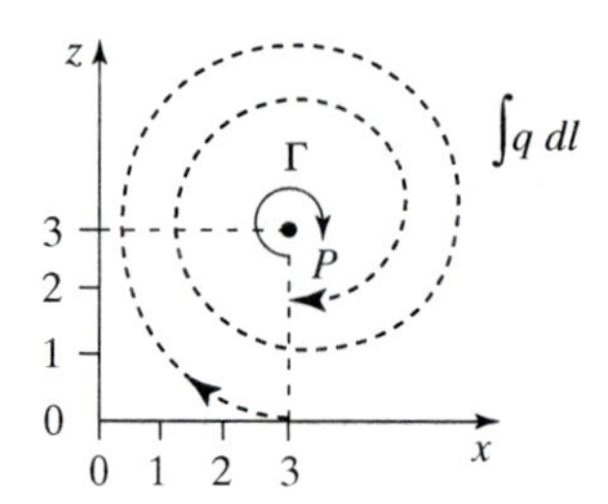

Problem 8.13.

8.14. Write the expression for the combination of a free stream of velocity U_∞ parallel to the x axis and a source of strength of K placed at (5, 0). Also calculate the velocity at a point (3, 3).

8.15. Find the location of the stagnation point (where the velocity is zero) for the flow described in the previous problem.

8.16. Calculate the value of the integral $\oint_c \vec{q} \cdot dl$ for a vortex of strength $\Gamma = 10$ placed at the origin. The path and direction for the integration is shown in the figure.

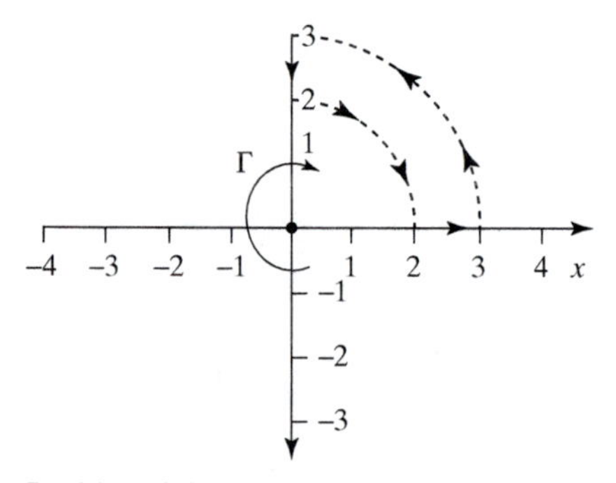

Problem 8.16.

8.17. Consider the 2D flow along a wall with a circular hump of radius R, as shown in the figure.

(a) What is the velocity potential for such a flow?
(b) Calculate the lift coefficient that is due to the upper surface of the semicircular hump.

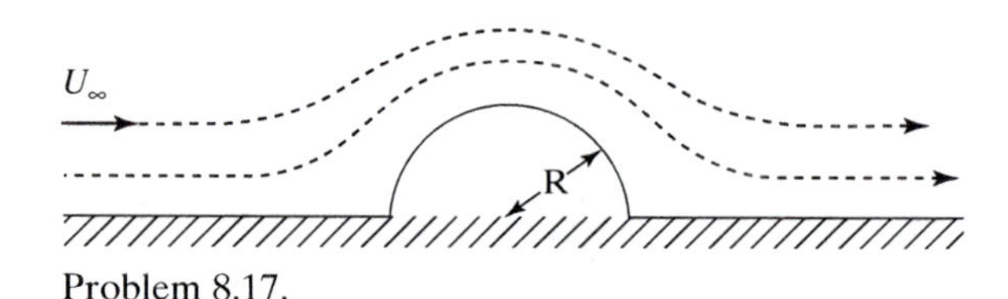

Problem 8.17.

8.18. Water flows at a velocity of 10 m/s normal to a cylinder of radius $R = 20$ cm. Using the potential flow formulas, estimate the pressure difference between the far field and points A and B.

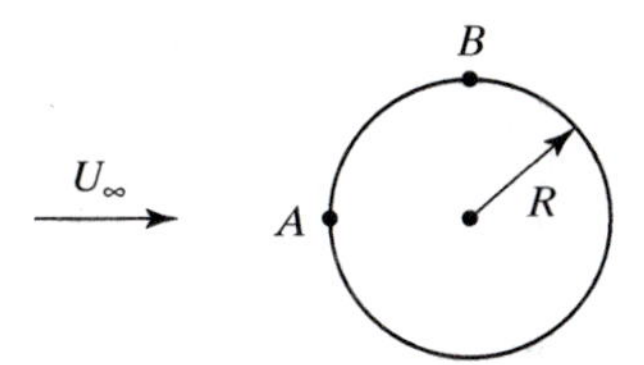

Problem 8.18.

8.19. The pressure distribution on the lifting airfoil of a hydrofoil boat is shown in the figure. Assuming the boat operates in a 40 °C warm, sweat water and the foil is about 1 m below the water surface, estimate at what speed would cavitation begin (and where). (The vapor pressure from Fig. 1.13, at 40 °C, is about 0.1 atm.)

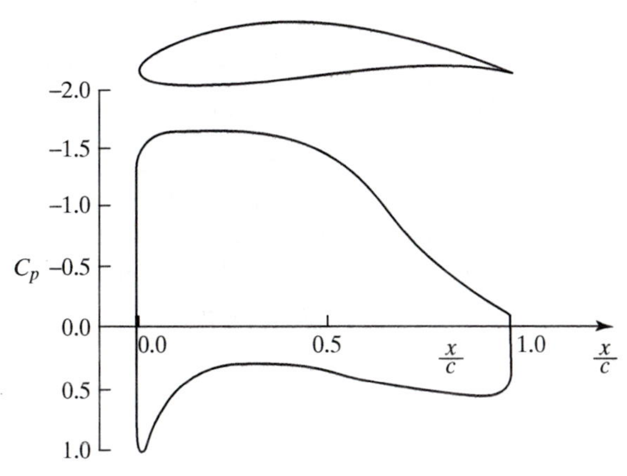

Problem 8.19.

8.20. A 40-km/h wind is blowing parallel to a storage facility, as shown in the figure. Estimate the normal force on a 0.5-m-high, 1-m-long window if the front door is left open. Assume the wind speed along the outside of the window is 1.2 U_∞ and take air density from Table 1.1.

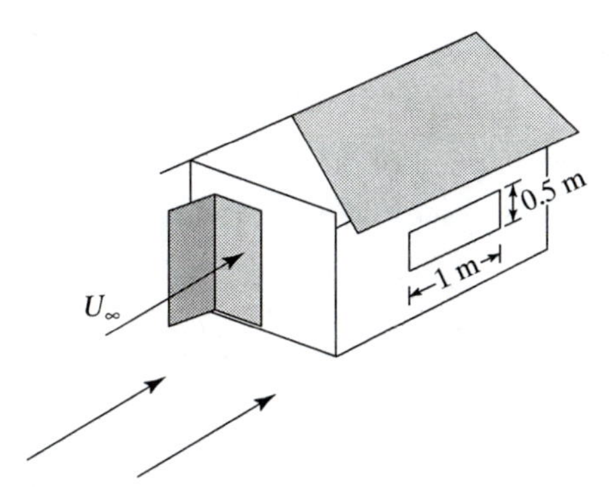

Problem 8.20.

8.21. A tall, cylindrical chimney has a 3-m diameter. Assuming wind speed is 40 m/s, calculate Re and vortex-shedding frequency. What is the distance between two subsequent vortices? ($\mu = 1.8 \times 10^{-5}$ N s/m^2, $\rho = 1.2$ kg/m^3.)

8.22. A long cylindrical antenna of 4-mm (0.004-m) diameter is mounted vertically on a vehicle moving at 100 km/h. If the Strouhal number is estimated at 0.18, then calculate the vortex-shedding frequency. Also calculate the length in centimeters between two adjacent vortices (e.g., the spacing).

8.23. Calculate the vortex-shedding frequency from a 0.5-cm-diameter vertical antenna mounted on a car traveling at 100 km/h. What is the distance between two cycles? (Use air properties from Table 1.1 and St $= 0.2$.)

8.24. The antenna of a car traveling at 90 km/h is resonating (vibrating violently) at a frequency of 500 Hz. Estimate the antenna's diameter (or which diameter to avoid).

8.25. Estimate the wind forces on a billboard 10 ft high and 30 ft wide when a 50-mph wind is blowing normal to it.

8.26. Calculate the resistance force when a 1-m-diameter submerged disk is towed behind a boat at a speed of 5 m/s. Assume the disk is perpendicular to the stream. (Use water properties from Table 1.1).

8.27. A large 1.9-m-wide, 0.5-m-high plate is carried above a truck perpendicular to the vehicle motion (like a flat plate normal to the free stream). Calculate the additional power required for carrying the plate at a speed of 72 km/h. Use the air properties from Table 1.1 and estimate the drag coefficient from Fig. 8.25.

8.28. Assume a person's drag coefficient is $C_D \sim 1.2$, frontal area is 0.55 m^2, and air density is 1.2 kg/m^3. Calculate the wind forces on the person's body when the stormy wind speed reaches 108 km/h.

8.29. A bicyclist is coasting down a hill with a slope of 8° into a head wind of 5 m/s (measured relative to the ground). The bicycle + rider mass is 80 kg and the coefficient of rolling friction is 0.02. Assuming that the frontal area is about 0.5 m^2 and the drag coefficient is about 0.5, calculate the bicycle speed (for air use $\mu = 1.8 \times 10^{-5}$ N s/m^2, $\rho = 1.2$ kg/m^3).

8.30. A 0.7-cm-diameter telephone wire is stretched between two poles, set 40 m apart. At what wind speed (blowing normal to the wire) will the wire "sing" at a frequency of 450 Hz? Also calculate the drag on the wire (use air properties from Table 1.1).

8.31. A 7-cm-diameter hockey ball is launched at a speed of 50 m/s and a spin rate of 1000 RPM. Estimate the lift and drag forces, based on Fig. 8.30, and use the air properties from Table 1.1.

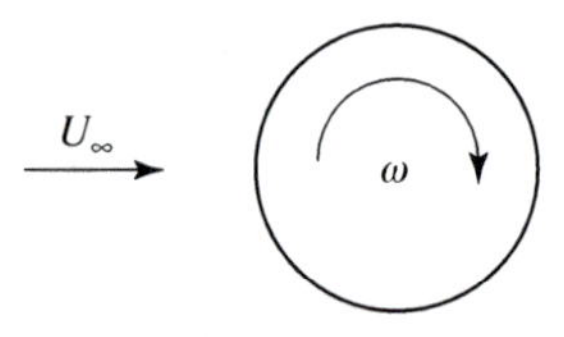

Problem 8.31.

8.32. A soccer player kicks a 22-cm-diameter ball at a speed of 30 m/s and a spin rate of 1000 RPM with the intention of creating a curved trajectory. If the ball weight is

0.45 kg, estimate the percentage of the resulting side force-to-weight ratio. (Estimate the lift force based on Fig. 8.30 and use the air properties from Table 1.1.)

8.33. The wingspan of the first successful human-powered airplane was about 30 m and the wing chord was 2.2 m. If the structural weight is about 32 kg and the pilot weight is 68 kg, calculate the lift coefficient and the induced drag coefficient. If the additional viscous drag is $C_{DO} = 0.01$, calculate the power required for a steady cruise at a speed of 16 km/h. (use the elliptic wing equations and air properties from Table 1.1).

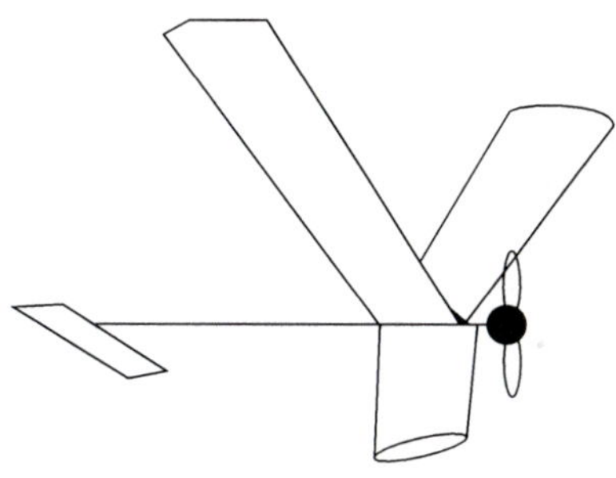

Problem 8.33.

8.34. A large cruise ship is cruising at an average speed of 40 km/h and the power supplied to the propellers is 42 MW. The sunken frontal area of a cruise ship can be estimated at 340 m^2, and the propulsion efficiency is about 0.8. Based on these numbers, estimate the hull drag coefficient.

8.35. Calculate the drag force on the ship under the conditions of Problem 8.34.

8.36. The frontal area of a 58-m-long blimp can be approximated by a circle with an 8-m radius. If the total drag coefficient is $C_D = 0.07$, estimate the power required for cruising at 60 km/h (use air properties from Table 1.1).

Problem 8.36.

8.37. The combined weight of a parachute and the jumper is 80 kg. Calculate the required chute diameter for a desirable sinking speed of 5 m/s (take air properties from Table 1.1).

8.38. An 80-kg (combined weight) skydiver jumps out of an airplane with his head initially pointing down.

(a) Calculate his terminal speed at this condition when his frontal area is about 0.22 m^2 and the drag coefficient is 1.3.

(b) After a while he changes into a prone position (looking down) and his frontal area is 0.8 m^2 with a drag coefficient of 1.2. What is his terminal speed?

(c) Finally a 7-m-diameter parachute opens; estimate his descent speed. (Take air density from Table 1.1.)

8.39. A steel ball of 3-mm diameter was dropped into a 0.5-m-deep container filled with motor oil. Assuming that the ball's sinking speed is the same from the moment it hits the oil surface, calculate how long it will take for the ball to sink to the bottom of the container. Note that the density of steel is $\rho = 7850$ kg/m^3; the properties of oil are taken from Table 1.1. Also calculate the Reynolds number.

8.40. A student is rolling down a slope of 5° with his skateboard. Neglecting the friction in the wheels and assuming he weighs 70 kg, his frontal area is 0.7 m^2, and his drag coefficient is 1.1, calculate his terminal velocity ($\rho = 1.2$ kg/m^3).

8.41. The frontal area of a small motorcycle is 0.60 m^2, its drag coefficient is 0.62, and its maximum speed is 150 km/h. Next the driver lowers his head and effectively reduces the frontal area to 0.58 m^2 and the drag coefficient to 0.60. Assuming the power of the engine is unchanged, estimate the new maximum speed ($\rho =$ 1.2 kg/m^3).

Problem 8.41.

8.42. A 50-km/h wind is blowing on a 10-m-tall vertical pole of 0.2 m diameter.

(a) calculate the Reynolds number ($\rho = 1.22$ kg/m^3, $\mu = 1.8 \times 10^{-5}$ N s/m^2).
(b) Calculate the Strouhal number, based on Fig. 8.27, and the drag coefficient from Fig. 8.19.
(c) What is the wake oscillation frequency f?
(d) What is the total drag of the pole?

8.43. Air is drawn into a wind tunnel during automobile testing, as shown in the figure. A simple manometer measures the velocity in terms of water-column height $h = 7$ cm. If the drag coefficient of the car is $C_D = 0.4$ and its frontal area is 1.5 m^2, calculate the drag force acting on the vehicle ($\rho_{\text{water}} = 1000$ kg/m^3, $\rho_{\text{air}} =$ 1.22 kg/m^3).

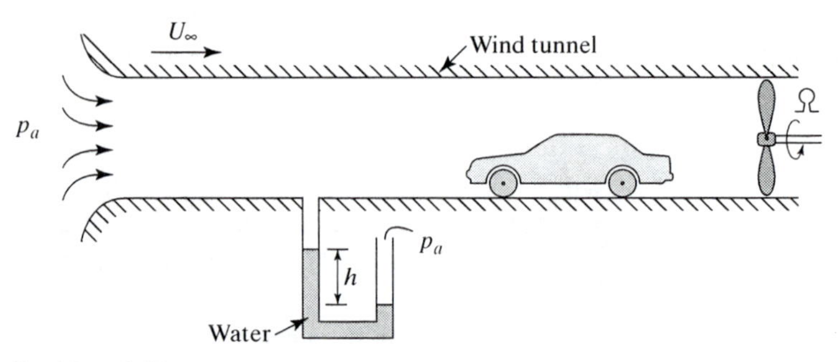

Problem 8.43.

8.44. A 0.1-m-diameter cannon ball is flying at 150 m/s and spins about a horizontal axis at 3600 RPM. Based on Fig. 8.30, estimate the vertical force on the ball. (Use $\rho = 1.1$ kg/m^3.)

8.45. An airplane has a rectangular wing with an elliptic spanwise loading. The mass of the airplane is 1200 kg, its wing area is 20 m^2, its wingspan is 14 m, and it is flying at a speed of 60 m/s. If the form drag is 0.01, calculate the total drag (e.g., calculate the induced drag) and the power required for propelling the airplane. (Use $\rho =$ 0.88 kg/m^3.)

8.46. The rectangular roof of a carport is 2 m wide and 6 m long and has an incidence angle of 5°. If a 20-m/s wind blows, as shown in the figure, calculate the lift force on the roof. Approximate the lift coefficient by using the elliptic wing formula, and use $\rho = 1.2$ kg/m^3.

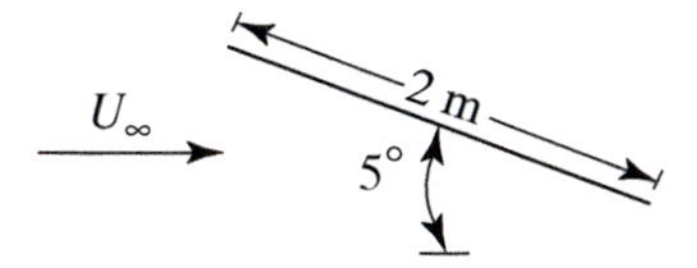

Problem 8.46.

8.47. Students built a small airplane with a rectangular wing having a span of $b = 1.5$ m and a chord of 0.4 m. They designed it to fly at a speed of 60 km/h, and the wing is at an angle of $\alpha = 1°$ relative to the free stream. If the zero-lift angle $\alpha_{L0} = -5°$ then calculate the lift coefficient (assume $\rho = 1.2$ kg/ m^3). Also, calculate the weight of the airplane.

8.48. After experimenting with the airplane in Problem 8.47, the students found that it cannot fly at angles larger than $\alpha = 10°$ (or it stalls). Calculate the lowest flight speed called "stall speed."

8.49. Let us approximate the wing of the Concord airplane by a triangle with a span of 25.6 m and a root chord of 27.6 m. Calculate the wing aspect ratio, wing area, and lift line slope $C_{L\alpha}$ (based on the equation given for slender wings). Suppose the airplane takes off at $\alpha = 17°$ and at a speed of 300 km/h – estimate its weight. (Use $\rho = 1.2$ kg/m^3.)

8.50. The space shuttle Orbiter lands at about 340 km/h and weighs about 100 tons. Approximating the wing shape with a triangle of 18-m span (the actual span with the fuselage is 23.8m) and an area of 220 m^2, calculate the angle of attack during landing. (Use the equations for slender delta wings at a high angle of attack and $\rho = 1.2$ kg/m^3.)

8.51. A 2 × 2 m glass plate is carried on top of a pickup truck. As the truck travels at 80 km/h, it hits a bump and the angle of the glass relative to the free stream is about 5°. Calculate the aerodynamic force lifting the glass ($\rho = 1.2$ kg/m^3).

8.52. John bought a 1-m-wide, 2.5-m-long door and decided to bring it home on top of his truck (he also placed it a 4° angle of attack). If his maximum speed on the highway was 100 km/h then calculate ($\rho = 1.22$ kg/m^3, $\mu = 10^{-3}$ N s/m^2) the following values:

(a) the lift and induced drag coefficients of the door (assuming no interaction with the cabin),
(b) the drag coefficient that is due to skin friction (use the Blasius formula for C_f),
(c) the total lift and drag on this surface.

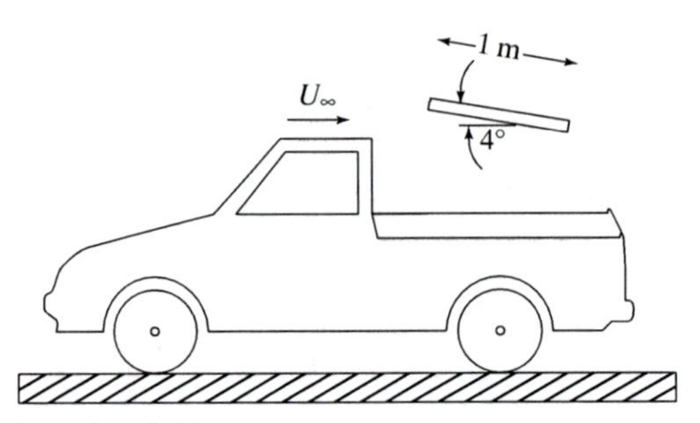

Problem 8.52.

8.53. A hydrofoil boat weights 10 tons and when fully submerged its maximum frontal area (of the submerged portion) is $S = 1.5$ m^2. When the hydrofoil is retracted, the boat's maximum speed is 20 km/h. Assuming a drag coefficient of $C_D = 0.2$ (based on S) calculate the following values.

(a) The power required for propelling the boat with the hydrofoils retracted.
(b) Suppose the boat can use two hydrofoils with a chord of 0.5 m and 2 m wide that lift the hull out of the water. Then calculate the maximum speed of the boat using the same power as in (a) (the drag is due to the hydrofoils only). Estimate the lift and drag of the hydrofoils by using Eqs. (8.101) and (8.102) (with $\delta = 0$) and assume the angle of attack is 10°.

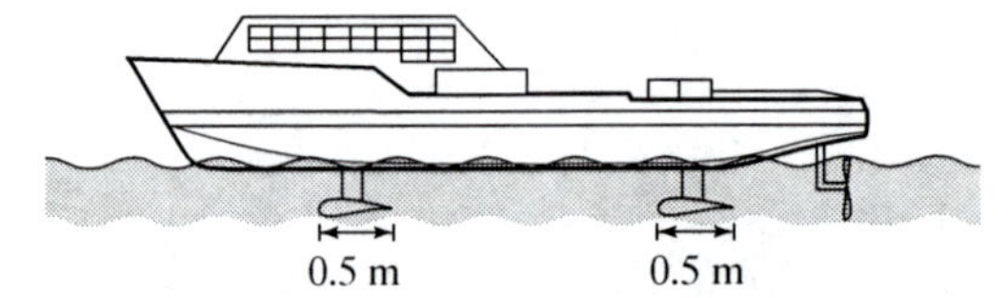

Problem 8.53.

8.54. A large bird is soaring horizontally at a speed of 25 km/h. We can approximate the wingspan at about 0.6 m and wing chord at 0.12 m. Assuming that air density is

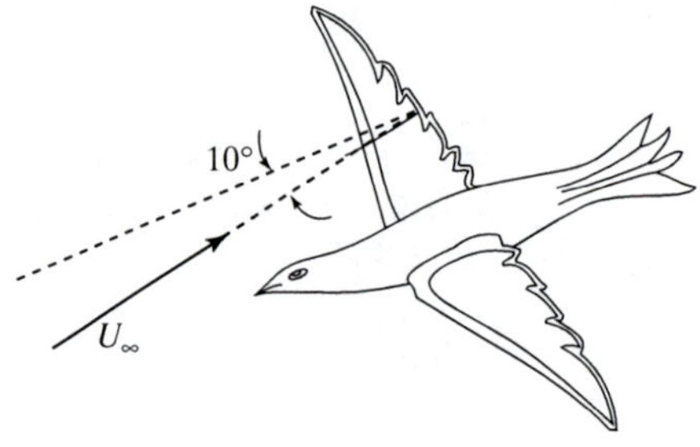

Problem 8.54.

$\rho = 1.2$ kg/m^3, angle of attack is 10°, and using the lift coefficient formulas developed for finite wings (elliptic), calculate the lift coefficient. Also calculate the bird's weight.

8.55. A 70-kg ski jumper leaves the ramp at a speed of 120 km/h and at this position, his body is at a 30° angle of attack relative to the free stream. Calculate the lift by comparing his body to a 0.6-m-wide and 1.8-m-long plate and by using the formulation of Eq. (8.105) (use air density from Table 1.1). Is the lift larger than his weight?

8.56. A small-airplane weight is 1.5 tons and it flies at sea level at a speed of 270 km/h. Its wing can be approximated by a 10-m-wide, 1.6-m-chord rectangle.

(a) Estimate the lift coefficient and power required for overcoming the induced drag (use $\delta_2 = 0$ and air density from Table 1.1).

(b) The power of the engines is about 150 kW. What percentage of available power is needed to overcome the induced drag?

8.57. It is said that some race cars develop so much downforce that they can run on an inverted (upside-down) road. If a 750-kg race car's frontal area is 1.5 m^2 and its lift coefficient is -3.5 (downforce), then how fast should it go in order to be able to drive upside down (e.g., on the ceiling)?

8.58. The pressure coefficient distribution on the upper and lower surfaces of a 2D lifting surface is approximated by a straight line, as shown in the figure. Calculate the lift coefficient and the center of pressure.

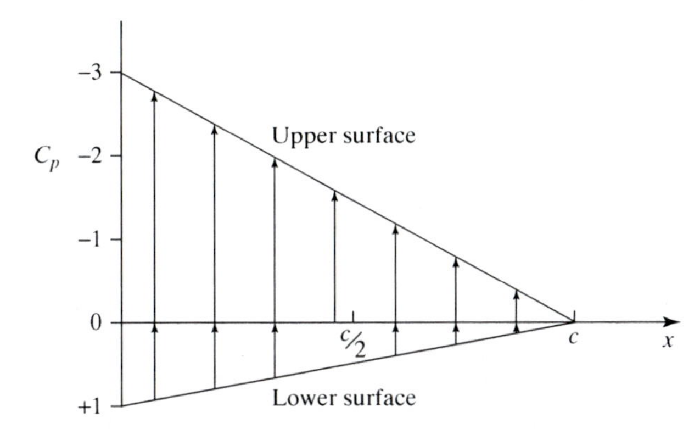

Problem 8.58.

8.59. The chord of the lifting surface shown in the previous figure is 1 m and the free-stream speed is 50 m/s. Calculate the lift of the lifting surface (assume standard air conditions).

9 Introduction to Computational Fluid Dynamics

9.1 Introduction

The fluid dynamic equations developed in Chapter 2 are complex and cannot be solved analytically for an arbitrary case. Up to this point, the classical approach was presented in which major simplifications allowed some partial solutions. In recent years, however, numerical techniques and computational power have improved significantly. This facilitated the solution of the nonlinear fluid dynamic equations, which now can be added to the growing number of practical engineering tools.

There are two major advantages to the numerical approach when one is attempting to solve the fluid dynamic equations. The first is the possibility of solving these complex equations, which cannot be solved analytically. The process begins with a numerical approximation for the fluid dynamic equations. The fluid domain is then discretized into small cells or into a grid, where the equations are applied. By specifying the equations at each cell or point, we reduce the partial differential equations to a set of algebraic relations. Thus the second major advantage of the numerical approach is the ability to replace the nonlinear partial differential equations with a set of algebraic equations, which are usually solved by iterative methods. There are a large number of methods for approximating the equations, for grid generations, and for solution methodology. This chapter attempts to explain the generic principles of the numerical approach and the process leading to the numerical solutions. A more comprehensive discussion on computational fluid dynamics (CFD) can be found in texts such as [1] or [2].

For a typical CFD solution, the computational domain must be defined (e.g., the region of interest in the flow, as shown in Fig. 9.1). In addition, the boundary or initial conditions must be specified (solid symbols) in a physically correct manner (to avoid impossible solutions). In this case the outer symbols represent points where a free-streams condition is prescribed and the inner solid symbols represent the "no-slip" boundary condition. Actual grids are significantly denser than the schematic grid shown in Fig. 9.1. For a 3D incompressible problem there are at least four unknowns (u, v, w, and p at each computational node, as shown), and therefore four equations are needed at each point or element. In general, the three directions of the momentum equation and the continuity equation can satisfy this requirement. The numerical solution then provides the values of the unknowns at

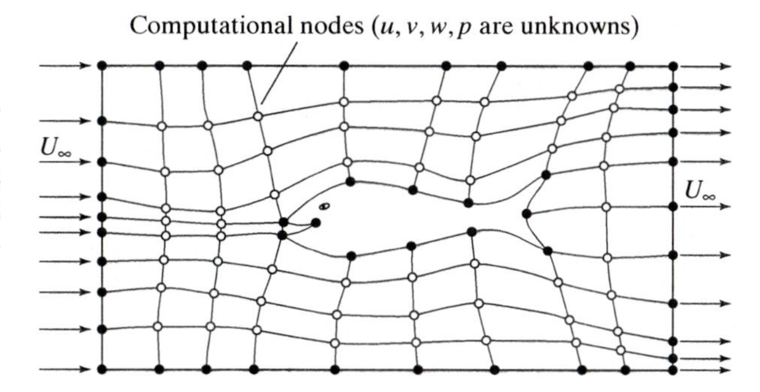

Figure 9.1. Schematic description of the grid for numerical solutions (the solid symbols represent the boundary nodes and the open circles the computational nodes).

the nodal points. Extrapolating in between the grid points or integrating the pressure field in order to obtain the forces is usually done by a postprocessor (which may introduce additional inaccuracies).

A generic process leading to the formulation of a CFD method is described next by use of the finite-difference model. This model is probably the simplest to present, but other methods (such as the finite-volume model) are more flexible for solving the flow over complex geometries. Two types of those "other methods" are discussed briefly toward the end of the chapter. Thus the objective of this chapter is to introduce the concept, particularly for those students who will not attend more advanced fluid dynamic courses (but may one day use CFD as part of a more comprehensive design package).

9.2 The Finite-Difference Formulation

There are several approaches for the numerical solution of the fluid dynamic equations, and most are based on a basic element. For example, the conservation equations can be applied to a fluid element by use of an integral or differential representation. Some methods that apply the integral form of the equation to a basic element are called finite volume. Another approach uses the finite-difference approximation for the fluid dynamic equations. As noted, for simplicity, let us follow here this latter method.

The first step of the process is to develop the numerical representation of the various terms (derivatives) in the fluid dynamic equations. Let us do this by using a 1D model as depicted in Fig. 9.2, where $u(x)$ is an analytic function along the

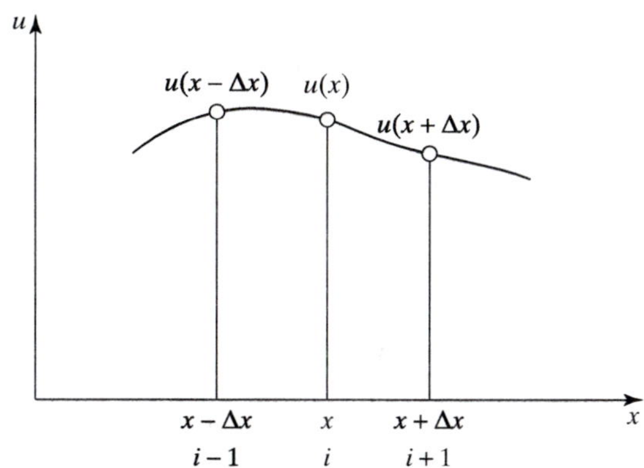

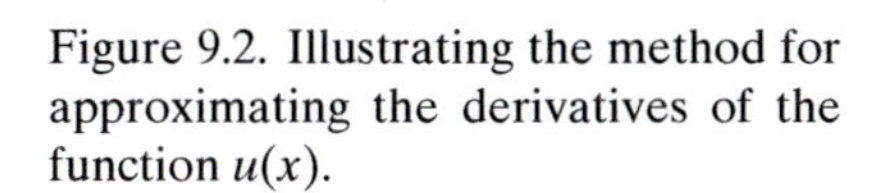
Figure 9.2. Illustrating the method for approximating the derivatives of the function $u(x)$.

coordinate x. Next we select several nodal points on the function $u(x)$ separated by the distance Δx.

The objective is then to relate the function to its derivatives, based on the function values at the nodal points. For small Δx, $u(x)$ can be expanded in a Taylor series about x:

$$u(x+\Delta x) = u(x) + \Delta x \frac{\partial u}{\partial x} + \frac{(\Delta x)^2}{2!}\frac{\partial^2 u}{\partial x^2} + \frac{(\Delta x)^3}{3!}\frac{\partial^3 u}{\partial x^3} + \cdots. \tag{9.1}$$

For example, we can find the first derivative, $\partial u/\partial x$, expressed in terms of the values at the nearby nodal points, by simple algebraic operation. Using first-order terms from Eq. (9.1) (and neglecting the higher-order terms) and solving for $\partial u/\partial x$, we get

$$\frac{\partial u}{\partial x} = \frac{u(x+\Delta x) - u(x)}{\Delta x} + O(\Delta x), \tag{9.2}$$

where we assume that Δx is small and all higher-order terms combined are of the order of Δx [so the order of the error is $O(\Delta x)$]. Note that this is equivalent to a straight-line curve fit between the two adjacent points. Equation (9.2) is called a forward-difference approximation, and by using the same method we can derive a backward-differential approximation:

$$\frac{\partial u}{\partial x} = \frac{u(x) - u(x-\Delta x)}{\Delta x} + O(\Delta x), \tag{9.3}$$

where the error is of the same magnitude as before. Also note that we are not discussing the actual "error" but the relation between the interval Δx and the error (or how fast it is reduced with reduced mesh size). We can generate a central-differential approximation by combining these two expressions [using Eq. (9.1) twice for forward and backward differences] to get

$$\frac{\partial u}{\partial x} = \frac{u(x+\Delta x) - u(x-\Delta x)}{2\Delta x} + O(\Delta x^2). \tag{9.4}$$

Note that the error is reduced for the central difference formulation (see [1]). So it is better to use this formula and retain the forward–backward formulation for the boundaries of the problem (this is usually true for evenly distributed grids where, say, Δx is constant). For the second derivative we again use a Taylor expansion, similar to Eq. (9.1):

$$u(x+2\Delta x) = u(x) + 2\Delta x \frac{\partial u}{\partial x} + \frac{(2\Delta x)^2}{2!}\frac{\partial^2 u}{\partial x^2} + \frac{(2\Delta x)^3}{3!}\frac{\partial^3 u}{\partial x^3} + \cdots. \tag{9.5}$$

By combining Eqs. (9.1) and (9.5) and solving for $\partial^2 u/\partial x^2$, we get the forward-differencing formula:

$$\frac{\partial^2 u}{\partial x^2} = \frac{u(x+2\Delta x) - 2u(x+\Delta x) + u(x)}{(\Delta)^2} + O(\Delta x). \tag{9.6}$$

Along the same lines, for first-order central and backward differencing we can write

$$\frac{\partial^2 u}{\partial x^2} = \frac{u(x+\Delta x) - 2u(x) + u(x-\Delta x)}{(\Delta x)^2} + O(\Delta x^2), \tag{9.7}$$

$$\frac{\partial^2 u}{\partial x^2} = \frac{u(x) - 2u(x-\Delta x) + u(x-2\Delta x)}{(\Delta x)^2} + O(\Delta x). \tag{9.8}$$

Note again that the central differencing is more accurate. For simplicity, let us label the first three values in Fig. (9.1) using the index i:

$$\begin{aligned} u(x+\Delta x) &\to u_{i+1}, \\ u(x) &\to u_i, \\ u(x-\Delta x) &\to u_{i-1}. \end{aligned} \tag{9.9}$$

Equations (9.4) and (9.7), the first-order central derivatives (actually have a second-order accuracy) can be expressed as

$$\frac{\partial u}{\partial x} = \frac{u_{i+1} - u_{i-1}}{2\Delta x} + O(\Delta x^2), \tag{9.10}$$

$$\frac{\partial^2 u}{\partial x^2} = \frac{u_{i+1} - 2u_i + u_{i-1}}{(\Delta x)^2} + O(\Delta x^2). \tag{9.11}$$

Higher or mixed partial derivatives can be derived with the same method. Also, these relations are called first order because higher-order terms in the Taylor expansion were neglected. For example, we can derive the central-difference representation for the second order by not neglecting the second-order terms; for central differencing they are

$$\frac{\partial u}{\partial x} = \frac{-u_{i+2} + 8u_{i+1} - 8u_{i-1} + u_{i-2}}{12\Delta x} + O(\Delta x^4), \tag{9.12}$$

$$\frac{\partial^2 u}{\partial x^2} = \frac{-u_{i+2} + 16u_{i+1} - 30u_i + 16u_{i-1} - u_{i-2}}{12(\Delta x)^2} + O(\Delta x^4), \tag{9.13}$$

and the error is of fourth order with Δx. In conclusion, the accuracy increases with reduced Δx (finer grid!) and with the higher-order approximation. Accuracy also improves when central differences are used, mainly when the grid points are evenly distributed.

9.3 Discretization and Grid Generation

One of the most important parts of a numerical solution is the grid generation. The grid can be structured or nonstructured, as shown in Fig. 9.3. Structured grids have a clear order, as shown in Fig. 9.3(a), and the discretization of the equations is simpler. However, for complex shapes, the creation of a well-structured grid becomes difficult and will require more nodal points than an unstructured grid [Fig. 9.3(b)]. These advantages of the unstructured grid are somewhat reduced by the more complex formulation of the governing equation. Usually finite-volume elements (based on the integral formulation) are used with an unstructured grid.

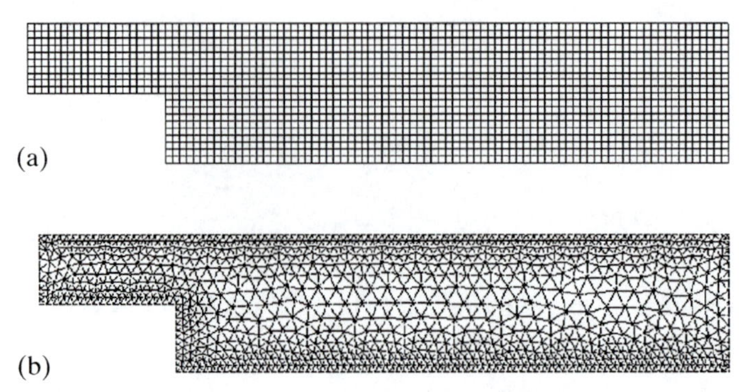

Figure 9.3. (a) Structured and (b) unstructured grids. (Courtesy of Dr. Gustaaf Jacobs, San Diego State University.)

For the numerical solution, the computational domain must be defined. A schematic description is provided in Fig. 9.4. In this case we consider the flow over a solid body, and the solution grid must extend far enough so we can consider the perturbations (which are due to the object) to be negligible. This problem in case of a subsonic flow (which is of the the elliptic equations type) requires the boundary conditions to be specified along all boundaries, as shown. Consequently, on the outer boundary we specify a constant free stream (inflow at the left and outflow at the right) and a constant pressure (see also Fig. 9.1). On the body surface we define the zero velocity (tangential and normal), as shown schematically in Fig. 9.4.

Of course, the grid generation requires a certain level of knowledge about the solution, and near areas of fast change a denser grid is required. Also note that in Eqs. (9.10)–(9.13) the error is reduced with reduced Δx; this means that, in general, the solution will improve with a finer grid. Thus the first step for generating a numerical solution is to investigate the grid. In principle, grid density should be increased until the solution appears to be independent of the grid!

9.4 The Finite-Difference Equation

At this point it is clear that the computational domain is subdivided into small elements where the equations will be specified. To demonstrate how the equations can be converted into algebraic representation at the solution cell level, let us follow the finite-difference approach (and not the finite-volume method). Therefore the first step in this process is to replace the terms in the fluid dynamic equations with their

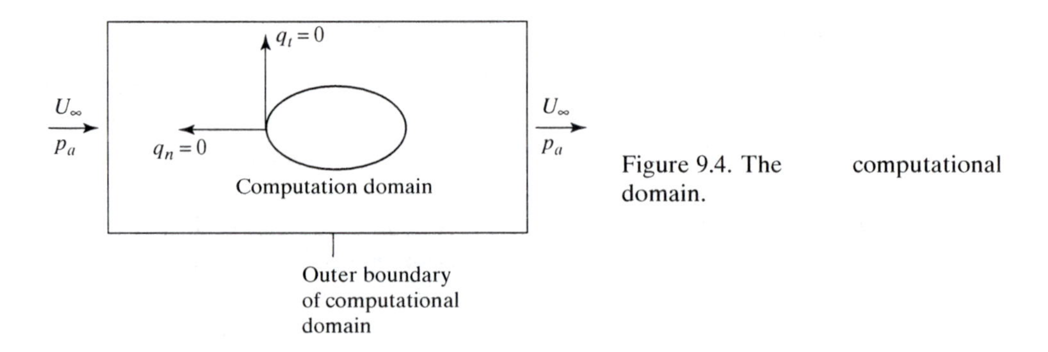

Figure 9.4. The computational domain.

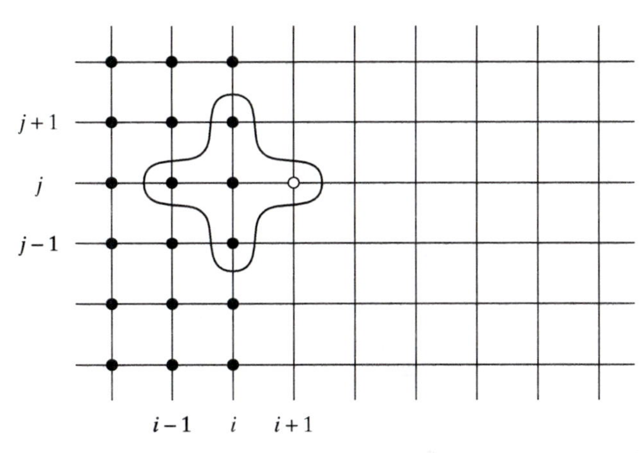

Figure 9.5. Typical explicit iteration scheme; the unknown is the open circle.

finite-difference equivalent, in effect creating a set of algebraic equations. This is a major simplification and allows the solutions of problems that cannot be solved analytically. For simplicity let us consider the 2D incompressible continuity equation [Eq. (8.16)]:

$$\frac{\partial^2 \Phi}{\partial x^2} + \frac{\partial^2 \Phi}{\partial z^2} = 0. \tag{9.14}$$

Replacing the terms with their finite- (central-) difference equivalent results in

$$\frac{\Phi_{i+1,j} - 2\Phi_{i,j} + \Phi_{i-1,j}}{(\Delta x)^2} + \frac{\Phi_{i.j+1} - 2\Phi_{i,j} + \Phi_{i,j-1}}{(\Delta z)^2} + O((\Delta x)^2, (\Delta z)^2) = 0. \tag{9.15}$$

This equation must be specified for all the grid points in Fig. 9.1 or Fig. 9.3. Note that near the boundaries either the forward or backward difference must be used for one of the terms in Eq. (9.14). One approach for the solution of similar equations is the explicit formulation in which only one unknown per equation remains. For example, an iteration scheme in the x direction [based on Eq. (9.15)] can be proposed such that

$$\Phi^{n+1}_{i+1,j} = \left[2\Phi u_{i,j} - \Phi_{i-1,j} - \left(\frac{\Delta x}{\Delta z}\right)^2 \Phi_{i.j+1} - 2\Phi u_{i,j} + \Phi_{i,j-1}\right]^n,$$

$$\text{error} \approx O((\Delta x)^2, (\Delta z)^2) \tag{9.16}$$

where n represents the iteration counter.

The solution must start from the boundaries where the values at the nodal points are known (boundary conditions). The finite-difference formulation near the boundaries may use a different scheme, depending on the type of boundary condition (e.g., if Φ is given then this is a Dirichlet condition, and if $\partial\Phi/\partial x$ is given then this is called the Neumann condition). The process is described schematically in Fig. 9.5, where the solid dots represent known values. By use of the information from the neighboring point, the solution can march forward (to the right) as depicted by the small open circle. This example serves only to demonstrate the explicit approach and will *not* work without using some of the tricks of numerical analysis. The major advantage of the explicit formulation is its simplicity, but

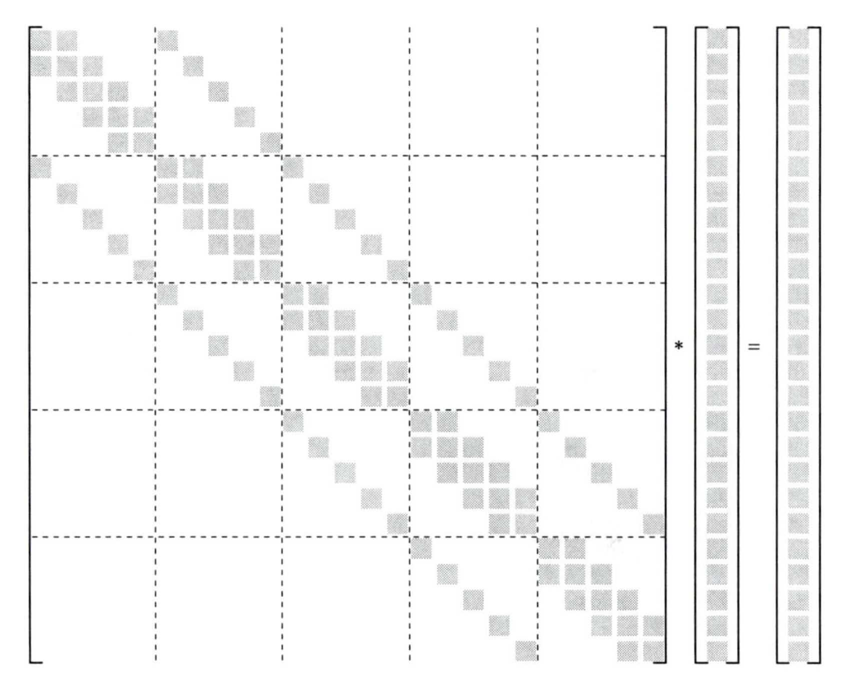

Figure 9.6. A typical set of linear algebraic equations resulting from the discretization of the fluid dynamic equations.

the solutions are usually less stable and more denser grids are required (and usually used for time-dependent problems).

An alternative approach, called the implicit formulation, has more than one unknown per node, as expressed in Eq. (9.15). Then, when this relation is specified for each of the N unknown grid points, N equations of the following form will result [based on Eq. (9.15)]):

$$\Phi_{i,j} = \frac{1}{\left[2+2\left(\frac{\Delta x}{\Delta z}\right)^2\right]}\left[\Phi u_{i+1,j} + \Phi_{i-1,j} + \left(\frac{\Delta x}{\Delta z}\right)^2 (\Phi_{i,j+1} + \Phi_{i,j-1})\right],$$

$$\text{error} \approx O[(\Delta x)^2, (\Delta z)^2] \qquad (9.17)$$

Of course, in this case too, the boundary conditions are used near the boundaries of the computational domain. In practice, the number of such equations can be of the order of several thousands or even millions (depending on the number of grid points and the computational power). The final algebraic problem has the form of a matrix in which the unknowns are centered near the diagonal, as shown in Fig. 9.6.

In principle, all unknowns are solved at once (with the matrix inversion), and therefore the implicit method is usually more stable than the explicit one and also requires a less dense grid. On the other hand, because all equations are solved at once, the programming effort is much larger and the required computation time is longer.

Because of the large size of the matrix, iterative solvers are used, but, as shown, the matrix is usually diagonally dominant and otherwise mostly contains zeros, which simplifies the solution.

Note that in this very simple example [of Eq. (9.14)] only one unknown per point is solved. To demonstrate the approach used with more unknowns, let us consider the x component of the steady-state, 2D, incompressible Navier–Stokes equation [from Chapter 5, Eq. (5a)]:

$$u\frac{\partial u}{\partial x} + w\frac{\partial u}{\partial z} = \frac{-1}{\rho}\frac{\partial p}{\partial x} + \frac{\mu}{\rho}\left(\frac{\partial^2 u}{\partial x^2} + \frac{\partial^2 u}{\partial z^2}\right). \tag{9.17}$$

Replacing the partial derivatives with their corresponding central-difference representation, we get the following finite-difference equation:

$$u_{i,j}\frac{u_{i+1,j} - u_{i-1,j}}{2\Delta x} + w_{i,j}\frac{u_{i,j+1} - u_{i,j-1}}{2\Delta z}$$
$$= \frac{-1}{\rho}\frac{p_{i+1,j} - p_{i-1,j}}{2\Delta x} + \frac{\mu}{\rho}\left[\frac{u_{i+1,j} - 2u_{i,j} + u_{i-1,j}}{(\Delta x)^2} + \frac{u_{i,j+1} - 2u_{i,j} + u_{i,j-1}}{(\Delta z)^2}\right]$$
$$+ O((\Delta x)^2, (\Delta z)^2). \tag{9.18}$$

This shows that at each point there are three unknowns (u, w, p) and therefore three equations are needed. Of course, there are the continuity and the momentums equations in the z direction that must be used. In practice, several algebraic substitutions are used to simplify the resulting equations (or the solution matrix).

9.5 The Solution: Convergence and Stability

In the previous sections the principle of the numerical approach was presented briefly by use of the finite-difference formulation. The grid generation and overseeing the solution process are usually left for the end user of a CFD program. Therefore, first, the grid must be evaluated in terms of density, particularly in regions of rapid change. The objective is for the solution to be independent of the grid, a process that usually increases the number of grid points. Once the grid geometry is set, the unknown quantities at each nodal point are solved with an iterative scheme. The solution process may involve large matrices and various iterative methods. The applicability of the method often depends on the actual fluid dynamic problem. At the end, after a certain numbers of iterations, a solution is expected. For example, if we attempt to calculate the drag force on a sphere, then we can define an error as the difference between a known value and the calculated one:

$$\text{error} = \frac{D_{\text{calculated}} - D_{\text{measured}}}{D_{\text{measured}}}. \tag{9.19}$$

If there are no experimental data, then an average of the recent calculations can be used for D_{measured} and the difference in the numerator can be replaced with the change during the last iteration. Most computer programs will display the residue of the solution, which is the normalized difference between the previous and the current iteration. The residual principle can be applied to the equations; then, for example, the residues of the continuity and the momentum equations are provided. A desirable convergence of the solution is shown by curve 1 in Fig. 9.7. In this case the numerical solution gradually approaches a value that can be defined as the "solution." For example, if the drag coefficient of an object is estimated, then the error bar could be set at a few percent.

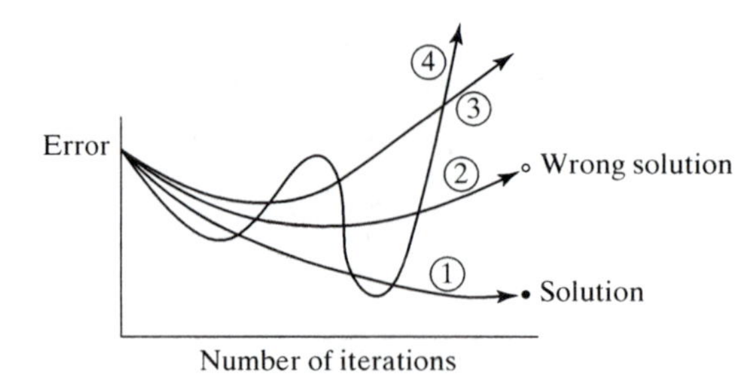

Figure 9.7. Schematic description of convergence and instability of numerical solutions.

In the case of nonlinear equations, the solution may end with a false solution, as shown by curve 2. This may be a result of wrong formulations, wrong boundary conditions, or even a bad grid. In the worst-case scenario the results will diverge with an increasing number of iterations, and this process could be gradual (curve 3) or completely unstable and oscillatory as in curve 4. This suggests that we must have an idea about the expected results and should be able to recognize convergence and stability issues. This is why the programmer must be familiar with the analytical models presented in the previous chapters.

Even if a solution is obtained we must remember that this is an approximate approach and there are several sources for an error the most common of which are subsequently listed:

1. Errors that are due to the model (e.g., the equations used do not contain all required physics). Typical examples include the modeling of boundary-layer transition, turbulence modeling, or vortex flow (and vortex breakdown).
2. Errors that are due to discretization, such as shown by Eqs. (9.1)–(9.13). These equations clearly indicate that the solution improves with a finer grid (smaller Δx in our 1D example).
3. Errors that are due to computer accuracy, such as roundoff or truncation errors.
4. Errors that are due to numerical convergence, as depicted in Fig. 9.7.
5. Errors in postprocessing – integration of shear and pressure, etc.

In spite of this list of errors, computational tools are quite accurate when the proper model is used (e.g., laminar flow model when the flow is expected be to laminar, or turbulence modeling when the flow is expected to be turbulent, etc.). Typical areas of weakness are cases with flow separation or in which transition from laminar to turbulent flow takes place.

9.6 The Finite-Volume Method

The main elements of a generic numerical solution were demonstrated in the previous sections. The discussion was based on the differential form of the Navier–Stokes equations, leading to the finite-difference discretization scheme. It is possible, however, to use the integral form of the equations [see Eqs. (2.20) and (2.24)] and their discretization is obtained by the so-called *finite-volume* approach. So for the finite-difference method the differentials are approximated, whereas for the finite-volume methods the integrals are approximated. The finite-volume methods are gaining popularity, mainly because of the inherent flexibility of using nonuniform grids and because the integral formulation conserves mass and momentum. Such grids are

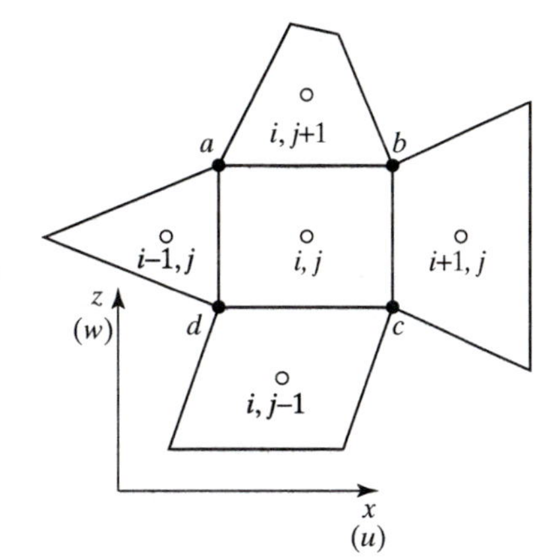

Figure 9.8. Schematic description of a 2D finite-volume cell.

necessary when one is modeling the flow over complex shapes, so that grid smoothness is not a major concern. As noted, by solution of the integral form, properties such as mass or momentum are conserved compared with other methods (in which numerical errors can introduce excess mass, etc.). Therefore finite-volume methods are considered to be more suitable for the solution of fluid mechanics problems. Because of the significance of the finite-volume-based methods, a brief description, highlighting the element of the model, is presented next.

As noted, the finite-volume model is based on the integral form of the fluid mechanic equations, which were applied to a control volume (see Section 2.6). To demonstrate the discretization process, based on this model, consider the incompressible, steady-state continuity equation from Eq. (2.25):

$$\int_{\text{c.s.}} \rho(\vec{q} \cdot \vec{n}) dS = 0. \tag{9.20}$$

As in all numerical solutions, the flow field is subdivided into small cells, or control volumes, and this equation is then applied to that finite-volume element. In practice, the grid can consist of different shape elements and usually tetrahedral, triangular (in two dimensions), or any polygon shape can be used. For simplicity, however, a 2D rectangular element is used here, as shown at the center of Fig. 9.8.

Of course, the element could be trapezoidal or triangular, but then the formulation becomes more complex. Each of these cells will have a finite volume or a finite area in the 2D case. In one approach, the *cell-centered model*, a central point (marked with the open circle in the figure) is assigned to each cell. The indexes i, j are used to refer to the cell, and the average variables u, w, etc., for the whole cell are assigned at this central point. Also, for this simple case it is assumed that, along each of the cell surfaces, the properties are constant, and this will be clarified later. Next the integral of Eq. (9.20) is applied to the control volume (cell), but because the changes are small it is approximated by the following summation:

$$\int_{\text{cell}} \rho(\vec{q} \cdot \vec{n}) dS \approx \sum_{\text{sides}} \rho(\vec{q} \cdot \vec{n}) S = 0. \tag{9.21}$$

This is a vector expression and for a complex cell shape must be applied accordingly. For the present simple 2D presentation, a perfectly aligned (with the

x–z coordinates) i–j cell is used. The four corners of the cells are marked by a, b, c, and d, respectively, and constant velocity is assumed along the cell boundary. For example u_{ad} represents the velocity in the x direction along the line a–d. For this perfect cell, therefore, the dot product can be separated into the x and the z directions, as follows (also, a constant density is assumed):

$$\sum_{\text{sides}} (\vec{q} \cdot \vec{n}) S = -u_{ad} \cdot \Delta z + u_{bc} \cdot \Delta z - w_{dc} \cdot \Delta x + w_{ab} \cdot \Delta x = 0. \tag{9.22}$$

However, this relation for the cell must be related to the properties in the centroid. This can be approximated (first order) as the average between the adjacent cells (even if their shapes are not exactly the same as the shape of the i–j cell):

$$\begin{aligned} u_{ad} &= \frac{1}{2}(u_{i-1,j} + u_{i,j}), \\ u_{bc} &= \frac{1}{2}(u_{i,j} + u_{i+1,j}), \\ w_{dc} &= \frac{1}{2}(w_{i,j} + w_{i,j-1}), \\ w_{ab} &= \frac{1}{2}(w_{i,j+1} + w_{i,j}). \end{aligned} \tag{9.23}$$

Substituting these into Eq. (9.22) results in

$$\frac{(u_{i+1,j} + u_{i-1,j})}{2\Delta x} + \frac{(w_{i,j+1} + w_{i,j-1})}{2\Delta z} = 0. \tag{9.24}$$

This is the finite-volume representation of the incompressible continuity equation at cell i–j. It looks similar to the finite-difference representation, but the difference is that the variables (e.g., u and w) represent an average for the cell and not the value of the same variable at a nodal or grid point (as in the case of finite differences). This representation is more flexible when unstructured grids are used. Also, this is a first-order representation, and for higher-order and 3D tetrahedral elements more complex formulations are used.

This short description serves only to demonstrate the application of the integral approach in CFD. Once the governing equations are converted to the algebraic form [as in Eq. (9.24)], the solution methods are similar to those described for the finite-difference approach. The applicability of the method, using the finite-volume model, is demonstrated in the next section.

9.7 Example: Viscous Flow over a Cylinder

As an example, let us solve the flow over a 2D cylinder. This type of flow was discussed in Section 8.6, and experimental results for the drag coefficients are provided in Section 8.7. By selecting a Reynolds number of 100, we can use the laminar momentum equations, but because of the vortex shedding behind the cylinder, we use an unsteady solver. For simplicity, a structured grid, shown in Fig. 9.9, is selected with about 50,000 cells. This number is considered *low resolution*, particularly behind the cylinder, in the wake, where asymmetric vortex shedding is expected. The boundary conditions are simple inflow (free stream) from the left

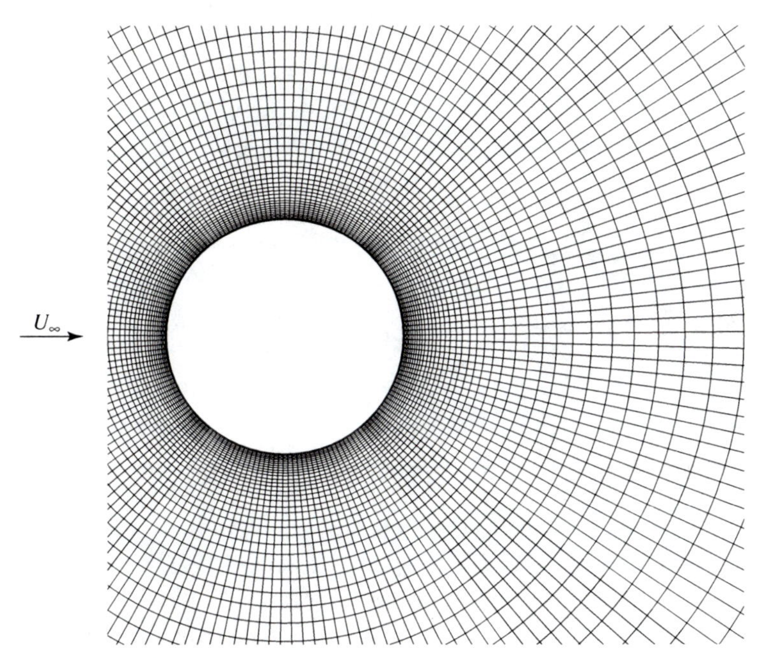

Figure 9.9. Structured grid used for calculating the laminar flow over a cylinder (Re = 100). Only a small portion of the grid is visible!

and exit from the right, and constant pressure at the outer boundaries was also specified.

For the solution, a first-order finite-volume method is used, with 200 iterations per time step, and the solution's convergence history is shown in Fig. 9.10. The time frame shown in this figure represents a total of about 270,000 iterations. The drag coefficient stabilized first at a value close to $C_d = 1.1$, which is close to the experimental values shown in Fig. 8.19. The residuals of the continuity equations are also shown, indicating a stable convergence to the solution presented here.

The calculated streamlines are presented in Fig. 9.11. The computations clearly capture the asymmetric vortex shedding, and the Strouhal number is close to the values indicated in Fig. 8.30. Because of the radial grid that lacked sufficient resolution in the wake, the finer details of the vortex streets were not captured. However,

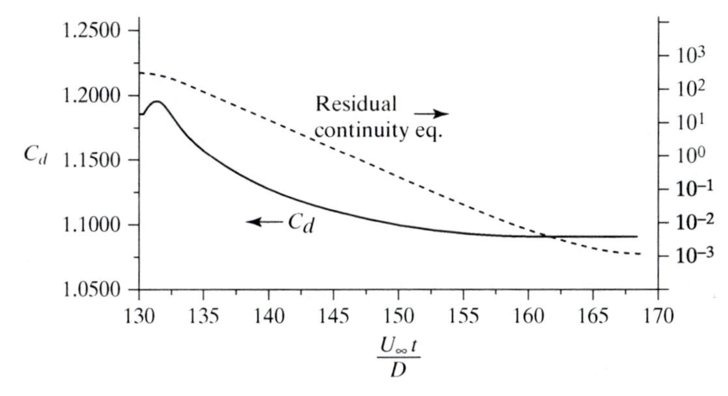

Figure 9.10. Reduction in the residuals (and the error) with iteration number. The data in the figure is obtained after a restart after a nondimensional time of 130.

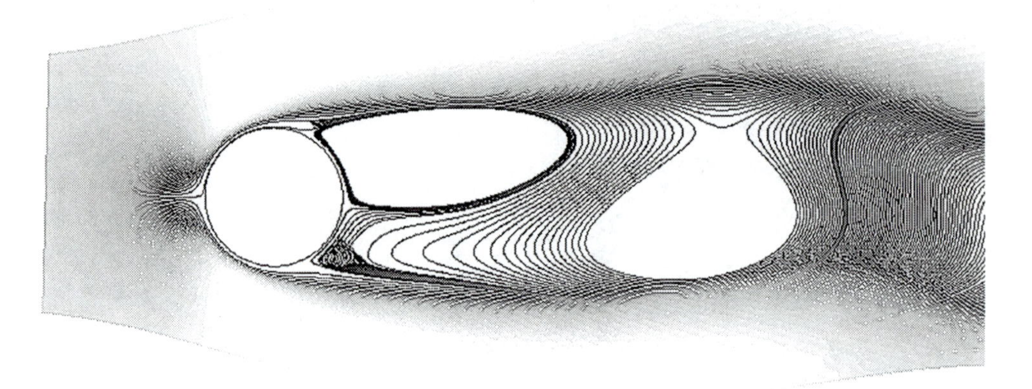

Figure 9.11. Display of the streamlines obtained from the numerical solution.

near the cylinder the flow compares well with experimental results. For example, the separation line is ahead of the top, as discussed in reference to Fig. 8.20.

The computed pressure distribution around the cylinder is plotted in Fig. 9.12 with the same format that was used in Fig. 8.16. The ideal-flow "exact" solution of Eq. (8.71) is shown by the solid curve, and the dashed curve describes the time-averaged pressure coefficient. The $\theta = 180°$ position represents the front centerline and $\theta = 0°$ is at the back. Because of the aft-flow separation, the pressure does not recover to the values of the ideal flow and remains near $C_p \sim -1.1$. Of course, the high pressure at the front and the low pressure at the back are the sources of the calculated drag force. Also, comparing the computations with the high-Reynolds-number experimental results of Fig. 8.16 shows earlier flow separation here, resulting in a higher suction behind the cylinder and in a higher drag coefficient.

This example served to demonstrate the type of expected results from CFD computations. The presented results could be improved, by refining the grid behind the cylinder. Also, in this type of grid, the cell size increases with the distance from the cylinder. To improve resolution in the wake, fairly dense and constant-size cells should be used in the separated wake area behind the cylinder. For more details on the separation point, a denser grid near the cylinder surface could be used. In spite of all the preceding inaccuracies, the major features and even the magnitude of the forces was captured reasonably well with this simple (low-resolution) solution.

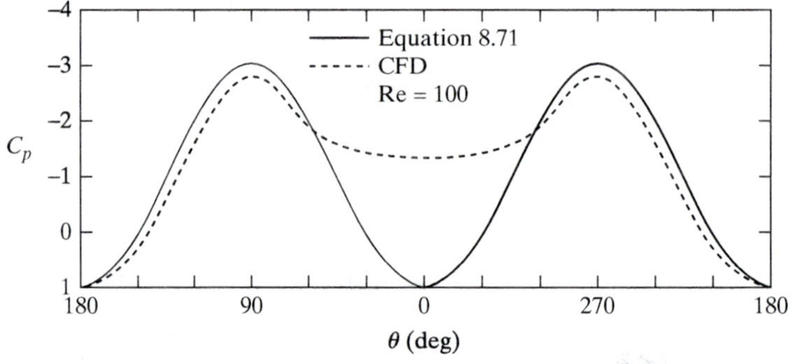

Figure 9.12. Comparison between the average pressure coefficient distribution predicted by CFD with the ideal flow results of Eq. (9.71).

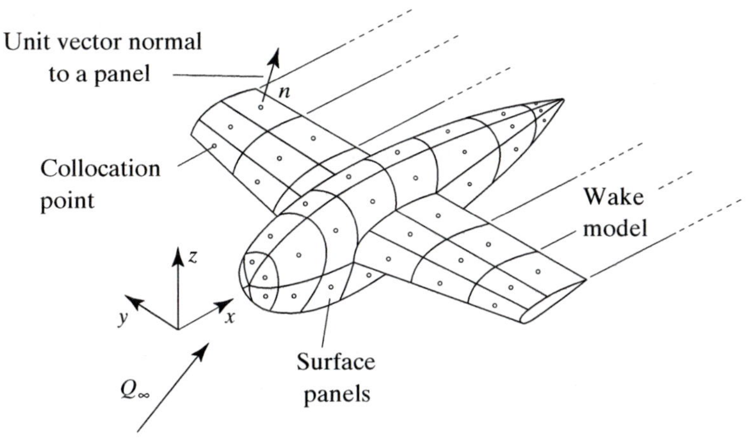

Figure 9.13. Schematic description of the potential flow over a closed body.

9.8 Potential Flow Solvers: Panel Methods

The computational methods presented so far are used for solving the viscous Navier–Stokes equations. In Chapter 6, however, it was shown that, for high-Reynolds-number attached flows, the viscous terms can be neglected outside of a thin boundary layer. This assumption led to the potential flow model, presented in Chapter 8, which allows a simple and efficient numerical solution for 3D external flows over complex shapes. These methods, called *panel methods*, are based on using the surface singularity distributions (see Section 8.5). The advantage, compared with the finite-difference (or finite-volume) approach, is that the unknown elements are distributed on the surface and not in the whole fluid volume, thereby significantly reducing computational effort. Another advantage is that the velocity field is obtained by solution of the continuity equation only (decoupled from the momentum equation), and instead of three velocity components at each point (e.g., u, v, w) only one unknown, namely the velocity potential, is sought. In conclusion, these methods are very efficient numerically but applicable only to inviscid, attached flows.

The theoretical background of such panel methods is shown schematically in Fig. 9.13. A coordinate system is attached to the body, whose surface is subdivided to panel elements, as shown. The steady-state free-stream magnitude and direction, expressed in this frame of reference, is $Q_\infty = (U_\infty, V_\infty, W_\infty)$ and the continuity equation is

$$\nabla^2\Phi = 0, \tag{9.25}$$

where Φ is the velocity potential in the body's frame of reference. The boundary conditions [see Eq. (8.17)] require that the normal component of velocity on the solid boundaries of the body be zero:

$$\nabla\Phi \cdot \vec{n} = 0, \tag{9.26a}$$

where $\vec{n}$ is an outward normal vector to the surface (as shown for one panel in Fig. 9.9). An alternative form of this boundary condition, called the Dirichlet condition, was introduced in Eq. (8.56). This alternative boundary condition requires that the potential inside a closed body be constant. By setting the constant to zero, this boundary condition becomes

$$\Phi_i = 0. \tag{9.26b}$$

Another form of the boundary condition is possible by setting the perturbation potential to zero inside the body (since the total potential $= \Phi + \Phi_\infty$):

$$\Phi_i = \Phi_\infty. \tag{9.26c}$$

The solution of Eqs. 9.26 provides the velocity field through the whole fluid domain. The pressures and corresponding fluid dynamic loads are then calculated separately by the use of the Bernoulli equation (recall that the continuity and momentum equations are not coupled for ideal flow).

There are a large number of numerical methods for solving the preceding problem. They can be based on source, doublet, or vortex distributions (or a combination of all of the above) and can use various forms of the boundary conditions or shapes of the panel elements. The solution methodology is usually very similar, and here only one variant is presented. Because the Green's theorem [Eq. (8.55)] postulates that the solution of the problem consists of sources and doublets distributed on the surface, the problem reduces to finding a singularity distribution that will satisfy Eqs. (9.26) (because the sources and doublets are already a solution of the Laplace equation). Once this distribution is found, the velocity $\vec{q}$ at each point in the field is known and the corresponding pressure p can be calculated from steady-state Bernoulli equation (8.15). With the preceding stipulations the velocity potential, based on the Green's theorem, can be constructed as a sum of source σ and doublet μ distributions placed on the surface S: The three-dimensional version of Eq. (8.55) is then

$$\Phi = \frac{-1}{4\pi}\int_S \left[\sigma\frac{1}{r} - \mu\frac{\partial}{\partial n}\left(\frac{1}{r}\right)\right] dS + \Phi_\infty. \tag{9.27}$$

We can find the values of the unknowns (σ, μ) by applying the boundary conditions, and r is the distance between the element at (x_0, y_0, z_0) and an arbitrary point (x, y, z) that is placed sequentially on the other panels. The free-stream potential, based on Eq. (8.23), is

$$\Phi_\infty = U_\infty x + V_\infty y + W_\infty z. \tag{9.28}$$

The unknowns are then the strengths of the sources and doublets assigned to each panel. However, if the source and doublet strengths on each surface panel are known, then the potential at any point can be calculated by Eq. (9.27). By applying the boundary condition of Eq. (9.26c) inside the closed body and setting the inside potential as $\Phi_i = \Phi_\infty$, we find that Eq. (9.27) becomes

$$\frac{-1}{4\pi}\int_S \left[\sigma\frac{1}{r} - \mu\frac{\partial}{\partial n}\left(\frac{1}{r}\right)\right] dS = 0. \tag{9.29}$$

In this case, however, the source distribution is known (see [1, p. 209] of Chapter 8),

$$\sigma = \vec{n}\cdot\vec{Q}_\infty, \tag{9.30}$$

and only the unknown doublet distribution remains to be solved. The solution is then based on discretizing Eq. (9.29) and the methodology is described next.

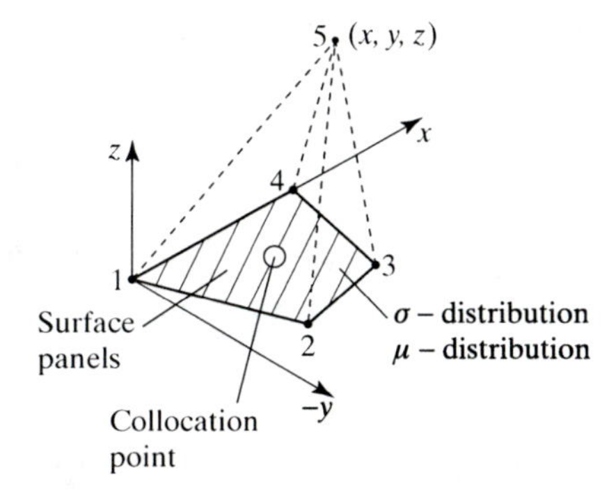

Figure 9.14. Schematic description of a generic panel element.

The numerical solution begins with a surface grid of N elements (called panels), as shown in Fig. 9.13. The surface can be made of flat rectilinear panels or more complex shapes (called higher-order panels). Similarly, for each panel the singularity distributions (source and doublets) are specified. If a constant strength source σ or doublet μ for a particular panel is assumed, then this is usually called a first-order method. Higher-order approximations are usually based on a polynomial distribution of the source or doublet strength within the panel. Thus one of the most basic components of the method is the panel element, which is shown schematically in Fig. 9.14. At the centroid of the panel, a collocation point is placed where the boundary conditions will be applied (see also small circles in Fig. 9.13). For example, when solving Eq. (9.29), instead of performing the integral over the whole surface S, we calculate this integral for a generic panel element, as shown in the figure.

The results for a generic panel with a constant source distribution and an index k will have the form (see [1, p. 214] in Chapter 8)

$$\frac{-1}{4\pi}\int_{1,2,3,4}\left(\frac{1}{r}\right)dS\bigg|_k \equiv B_k, \tag{9.31}$$

where the constant B_k depends on the panel's four corner points and on the field point (5) where the potential that is due to this panel is evaluated. Similarly, the influence of a constant doublet distribution at point 5 (C_k) can be calculated by the integral

$$\frac{1}{4\pi}\int_{1,2,3,4}\frac{\partial}{\partial n}\left(\frac{1}{r}\right)dS\bigg|_k \equiv C_k. \tag{9.32}$$

The most important feature of these influence coefficients (e.g., B_k, C_k) is that this calculation is based on the geometry only (e.g., the location of the preceding five points) and we can execute it without knowing the strength of the singularity elements. As an example let us use the boundary condition of Eq. (9.29), which is based on an unknown doublet distribution. The influence coefficients C_k are precalculated and do not depend on the strength μ. Next, for each collocation point (in Fig. 9.13) the potentials due to all elements must add to zero inside the body (so the collocation point is assumed to be on the inner surface of the body). Consequently, when Eq. (9.29) is specified at one collocation point with the index, j, it will have the form

$$\sum_{k=1}^{N} B_{jk}\sigma_k + C_{jk}\mu_k = 0. \tag{9.33}$$

This equation basically states that the potential at the collocation point of panel k is the sum of the potentials of all the surface panels [note that the source strength

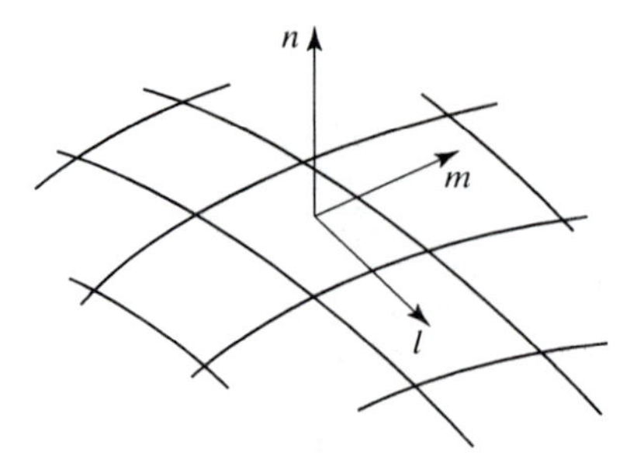

Figure 9.15. Nomenclature for calculating the local (perturbation) tangential velocity. Note that the normal velocity is zero [see boundary condition (9.26a)].

is known and given by Eq. (9.30)]. This equation is then applied at all N collocation points with the N unknowns μ_k, thus reducing the integral in Eq. (9.29) into a set of linear algebraic equations (matrix of the order of N). By solving the N equations, we calculate the unknown doublets for each panel, and therefore the potential Φ is known everywhere. In reality, vortex wakes must be modeled too, and this is explained in [1, Section 9.3] of Chapter 8. Let us reiterate that this method is based on the principle of superposition. Instead of solving the continuity equation, only the strengths of the elementary solutions (sources and doublets) is obtained by using the boundary conditions.

Once Eq. (9.33) is solved, the unknown singularity values are obtained (μ_k in this case), and the local velocity components can be evaluated. Also note that the doublet in Eq. (9.33) represents the potential jump between the outside and the inside of the body (where $\Phi_i = \Phi_\infty$). Therefore the perturbation velocity components are the derivatives of the velocity potential [see Eq. (8.4)], and in terms of the panel local coordinates (l, m, n) the two tangential velocity components are

$$q_l = -\frac{\partial \mu}{\partial l}, \qquad q_m = -\frac{\partial \mu}{\partial m}. \tag{9.34}$$

The differentiation is done numerically by use of the values on the neighbor panels, as shown in Fig 9.15. Once we calculate the velocity (and the free-stream velocity is added), we can calculate the surface pressure and resulting lift and drag by using Bernoulli equation (8.15) or in terms of the pressure coefficient [Eq. (8.72)]:

$$C_p = 1 - \frac{q^2}{U_\infty^2}. \tag{9.35}$$

The matrix representing Eq. (9.33) is usually diagonally dominant, because the influence of the panel on itself is the largest. Also, the number of unknowns is significantly less than in the finite-difference or finite-volume methods, and therefore numerical convergence and stability issues are almost nonexistent. The main sources of errors when calculating the velocity field are then due to insufficient or incorrect grid distribution. Additional errors can occur during the pressure integration over complex shapes. Of course, the main limitation of the method is the significantly simplified physics (e.g., neglecting viscosity as discussed in Chapter 6).

9.9 Summary

Computational fluid mechanics has changed the face of the discipline. Instead of using considerably simplified models involving complex mathematics, CFD provides an easily accessible tool to solve a large variety of fluid mechanics problems. At the beginning of the third millennium, however, the method is still not fully developed and is still controlled by available computer power and the abundance of various methods. It is expected that in the future artificial intelligence can control processes such as numerical modeling and grid generation and optimize solution algorithms on the fly (and therefore less programmer interference will be required).

In the meantime, (until CFD is perfected) there is no replacement for studying classical fluid dynamics. The potential user of such computational methods must be prepared with the knowledge of the classical models and also understand the hierarchy of approximations in CFD, which can be summarized as follows:

1. The actual, true physics may be more complex than the most advanced models.
2. The equations solved by CFD are approximations to the physics mentioned in 1. For example, in the Navier–Stokes equations, we assume a Newtonian fluid.
3. Many computer models solve a simpler model (such as laminar, incompressible, or inviscid flow, and for high-Reynolds-number flow there are various models for turbulence).
4. At the next level, where the equations are discretized (finite difference, finite volume, or element, etc.), additional inaccuracies are introduced.
5. Grid generation, particularly when inappropriately done, can introduce further errors.
6. The numerical solution of a large number of equations can create another form of errors (such as roundoff errors, convergence, stability, etc.).
7. Finally, once a solution is obtained, the postprocessing can further reduce accuracy (e.g., errors that are due to integrating pressures and shear stresses).

At each level of the preceding hierarchy, some portion of the true physics is compromised. Being aware of this and understanding the implication is necessary for obtaining good results. At the time of writing this text, there are still many unresolved issues such as boundary-layer transition, turbulence modeling, and the preservation of vorticity. In spite of the shortcomings just listed, CFD is a powerful tool in the hands of a knowledgeable fluid dynamicist.

REFERENCES

[1] Hoffman, K. A. and Chiang, S. T., *Computational Fluid Dynamics for Engineers*, Vols. I and II, Engineering Education System, Wichita, KS, 1993.

[2] Ferziger, J. H. and Peric, M., *Computational Methods for Fluid Mechanics*, 2nd ed., Springer-Verlag, New York, 1999.

PROBLEMS

9.1. Develop a finite-difference representation for

$$\frac{\partial g}{\partial x} = k\frac{\partial^2 g}{\partial y^2}.$$

Use forward differencing for the x derivative and central differencing for the y derivatives.

9.2. Modify the proposed finite-difference scheme for the boundary at $y = 0$.

9.3. Develop a finite-difference representation for

$$\frac{\partial f}{\partial t} = k\left(\frac{\partial^2 f}{\partial x^2} + \frac{\partial^2 f}{\partial z^2}\right).$$

Use forward differencing for the time derivative and central differencing for the spatial terms.

9.4. Calculate the first derivative of the function $f(x) = \sin(\pi x)$ at $x = 0.25$:

(a) using a forward-difference approximation with $\Delta x = 0.01$,
(b) using a central-difference approximation with $\Delta x = 0.01$,
(c) using the exact value.

9.5. Calculate the first derivative of the function $f(x) = \cos(\pi x)$ at $x = 0.3$ for three different increments of $\Delta x = 0.02, 0.01$, and 0.005, and compare with the exact result (use forward-difference scheme).

9.6. The oscillation of a mass M suspended on a spring K is described by the equation $M\frac{d^2x}{dt^2} + Kx = 0$, where the spring length x is a function of time $x(t)$.

(a) Derive a finite-difference representation of this problem.
(b) Describe the first step using the numerical values for $t = 0$: $\Delta t = 0.01$, $M = 1$, and $K = 1$. Assume initial conditions: $x = 0.1$, $dx/dt = 0.1$.

For the following problems the students must have access to computational tools (either locally or from the Internet).

9.7. Use a laminar CFD solver to calculate the flow over a 2D cylinder and compare the results for the drag coefficient with the experimental data in Fig. 8.19. Study two cases: (a) Re $= 1$ and (b) Re $= 5$. Also plot the surface pressure distribution.

9.8. Use a laminar CFD solver to calculate the flow over a 2D cylinder at a Reynolds number of 40. Repeat the computations three times and increase grid density twice (second computation) and four times (third computation). Compare the results for the drag coefficient with the experimental data in Fig. 8.19 and discuss the effect of grid density on the solution.

9.9. Calculate the viscous laminar flow on a sphere at a Reynolds number of 1, and compare your results for the drag coefficient with the Stokes formula [Eq. (8.76)].

9.10. Calculate the 2D flow normal to a flat plate (as in Fig. 8.22), and compare the drag results with the value of $C_D = 1.17$ from Fig. 8.25. Assume Re $= 100$.

9.11. Use a time-dependent CFD model to calculate the flow over a cylinder at Re $= 60$. Compare the drag coefficient and the Strouhal number with the experimental data presented in Chapter 8.

9.12. Use a 3D panel code to calculate the lift and (induced) drag of a rectangular wing at an angle of attack of 5°. Assume aspect ratios of 5, 7, and 10, and compare with calculated results based on Eqs. (8.101) and (8.102).

9.13. Calculate the laminar flow inside a long cylindrical pipe for Re $= 50$, based on the pipe diameter. Assume uniform velocity profile at the entrance to the pipe. How long does it take along the pipe (in terms of diameters) before the velocity profile becomes parabolic, as in Eq. (5.79)?

10 Elements of Inviscid Compressible Flow

10.1 Introduction

The fluid dynamic models presented so far, for both liquids and gases, were based on negligible fluid compressibility. However, there are certain cases, mostly related to gas flow, in which compressibility cannot be neglected. In these situations, the internal energy (and temperature) changes are not negligible, requiring a more careful observation of the fluid properties. For example, let us assume that the cylinder in Fig. 10.1 is filled with an ideal gas. By compressing the cylinder from point 1 to point 2, more molecules will hit the wall and both pressure and temperature will increase. The relation among the various ideal-gas properties was given by Eq. (1.11) as

$$\frac{p}{\rho} = RT. \tag{10.1}$$

The density, however, is the inverse of the volume per unit mass $\rho = 1/v$, and the relation between the two points in Fig. 10.1 is

$$\frac{p_1 v_1}{T_1} = R = \frac{p_2 v_2}{T_2}. \tag{10.2}$$

Next, the process involving the motion of the piston from point 1 to point 2 must be addressed. For example, if the gas temperature is kept constant during the compression by cooling the walls we get

$$p_1 v_1 = p_2 v_2, \tag{10.3}$$

which indicates that, by reducing the volume, the pressure will increase by the same ratio. If the process is adiabatic (no heat transfer) then based on thermodynamics (see [1] in Chapter 1), the process is

$$\frac{p}{\rho^\gamma} = \text{const.} = C_1, \tag{10.4}$$

and γ is the specific heat ratio defined by Eq. (1.19). In terms of the volume per unit mass Eq. (10.4) yields:

$$p v^\gamma = \text{const.} = C_1. \tag{10.5}$$

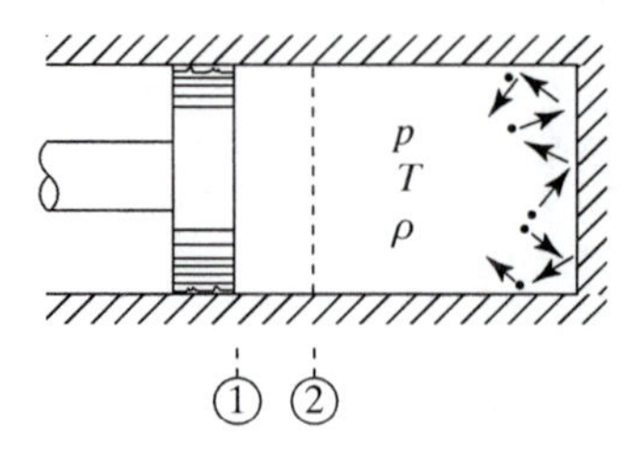

Figure 10.1. Gases inside a sealed cylinder.

Using the ideal-gas relation from Eq. (10.1) to replace the pressure with the temperature, we get

$$Tv^{\gamma-1} = \text{const.} = C_2 \tag{10.6}$$

or by replacing v with p [from Eq. (10.5)], we get

$$Tp^{\frac{\gamma-1}{\gamma}} = \text{const.} = C_2. \tag{10.7}$$

If one of the fluid properties will change through an adiabatic process (no heat exchange), then the preceding formulas allow the calculation of the other fluid properties.

The following topics are based on the simple 1D inviscid flow concept, and the main objective is to demonstrate the elementary effects of compressibility on the fluid flow.

10.2 Propagation of a Weak Compression Wave (the Speed of Sound)

Small compression perturbations, such as sound, move in a fluid at a finite speed approximated by Eq. (1.30). To evaluate this *speed of sound*, let us consider a cylinder filled with stationary gas, as shown in Fig. 10.2. Because of an infinitesimal movement Δu of the piston (much as in a loudspeaker's membrane) a weak compression front forms that travels forward at a speed of a. It is assumed that the compression process is isentropic (adiabatic and reversible), so there are no losses in this process. Also, it appears that the speed on the left-hand side of the sound wave is $a - \Delta u$ whereas on the right-hand side it is a.

By using the 1D continuity equation, Eq. (2.26), we find that the flow rate on the left-hand side becomes $(\rho + \Delta\rho)(a - \Delta u)A$, where A is the cylinder cross-section area. This flow rate is equal to the flow on the right-hand side of the pressure jump $\rho a A$ and, from the 1D continuity equation we can write

$$(\rho + \Delta\rho)(a - \Delta u)A = \rho a A. \tag{10.8}$$

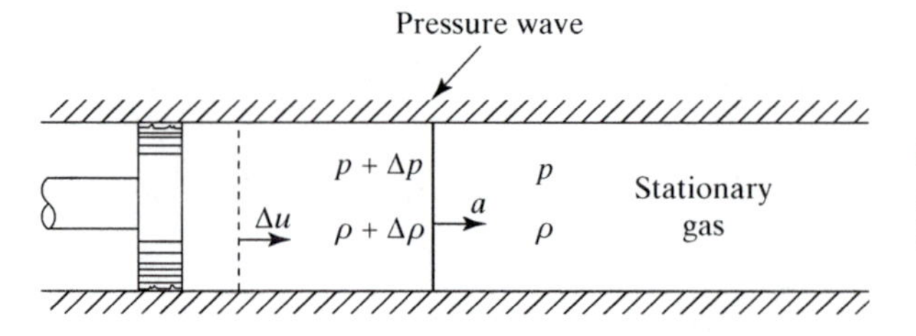

Figure 10.2. A weak compression wave traveling at a speed a in a stationary fluid.

Solving for Δu and after neglecting the smallest term, we approximate Δu as

$$\Delta u = a\frac{\Delta \rho}{\rho}. \tag{10.9}$$

Assuming no friction or body force, we can write the 1D momentum equation [Eq. (4.6)] for the two sides of the pressure wave:

$$0 = \dot{m}a - \dot{m}(a - \Delta u) + pA - (p + \Delta p)A. \tag{10.10}$$

Because the mass flow rate is the same on both sides of the pressure wave we can write $\dot{m} = \rho a A$, and momentum equation (10.10) reduces to

$$0 = \rho a A \Delta u - \Delta p A$$

or

$$a = \frac{\Delta p}{\rho \Delta u}. \tag{10.11}$$

Substituting Δu from Eq. (10.9) yields

$$a^2 = \frac{\Delta p}{\Delta \rho}.$$

We can conclude that for an isentropic process the speed of sound is

$$a = \sqrt{\frac{\partial p}{\partial \rho}}, \tag{10.12}$$

which is the same result presented in Eq. (1.30). This indicates that the speed of sound is a result of the compressibility. If in the previous chapters we often used the incompressible flow assumption, this meant that the speed of sound there is infinite! For an ideal gas undergoing an isentropic process, the relation between pressure and density is given by Eq. (10.4). We then obtain the derivative appearing in Eq. (10.12), by deriving Eq. (10.4):

$$\frac{\partial p}{\partial \rho} = \gamma C_1 \rho^{\gamma - 1} = \gamma \frac{p}{\rho^{\gamma}} \cdot \rho^{\gamma - 1} = \gamma \frac{p}{\rho}.$$

However, for an ideal gas, $(p/\rho) = RT$, and therefore

$$\frac{\partial p}{\partial \rho} = \gamma RT.$$

Substituting this into Eq. (10.12) results in

$$a = \sqrt{\gamma RT}, \tag{10.13}$$

confirming that the speed of sound in an ideal gas is a function of temperature only!

EXAMPLE 10.1. THE SPEED OF SOUND IN AIR. Calculate the speed of sound in air at a temperature of 300 K.

Solution: Substituting the value of $\gamma = 1.4$ for air and R from Eq. (1.13) into Eq. (10.13) results in

$$a = \sqrt{1.4 \times 286.6 \times 300} = 346.95 \text{ m/s}.$$

Because the speed of sound is finite, the ratio between the local velocity q in the fluid to the speed of sound a can signal the importance of compressibility effects. This ratio, called the Mach number, after Ernst Mach (Austrian physicist, 1838–1916) was already defined by Eq. (6.15):

$$M = \frac{q}{a}. \tag{10.14}$$

For example, if $M < 0.5$ everywhere in the flow, it can be considered incompressible. In general, flows in which $0 < M < 0.8$ are called subsonic.

$$0 < M < 0.8, \; subsonic.$$

However, above Mach numbers of $M = 0.5$, the compressibility effect may become noticeable. If the local speed in the fluid exceeds the speed of sound, then the flow is called transonic. For example, an airplane can fly at $M = 0.85$, but the flow over the wing can accelerate beyond $M = 1$. Consequently this region is defined as transonic:

$$0.8 < M < 1.2, \; transonic.$$

Because the local Mach number in transonic flow can be above 1, shock waves may be present. When the flow is significantly faster, then this flow regime is called supersonic:

$$1.2 < M < 3.0, \; supersonic.$$

Some very fast airplanes do operate within this flow range. For much faster flows, a serious temperature increase may take place and this flow regime is called hypersonic. Some researchers define this range as being above $M = 3.0$ and some define it as being above $M = 5.0$ So, let us call the range $3.0 < M < 0.5$ high supersonic and

$$5.0 < M, \; hypersonic.$$

There is another interesting aspect to the fact that the velocity of sound is finite. For example, let us consider an airplane flying at a subsonic speed, as depicted schematically in Fig. 10.3(a). We can start our observation at point 1 and assume that after 1 s the airplane will be at point 2 and after another second at point 3, and so on. The perturbation is generated by the motion or the engine noise propagates at the speed of sound, and after 4 s, the four circles represent the regions where this information

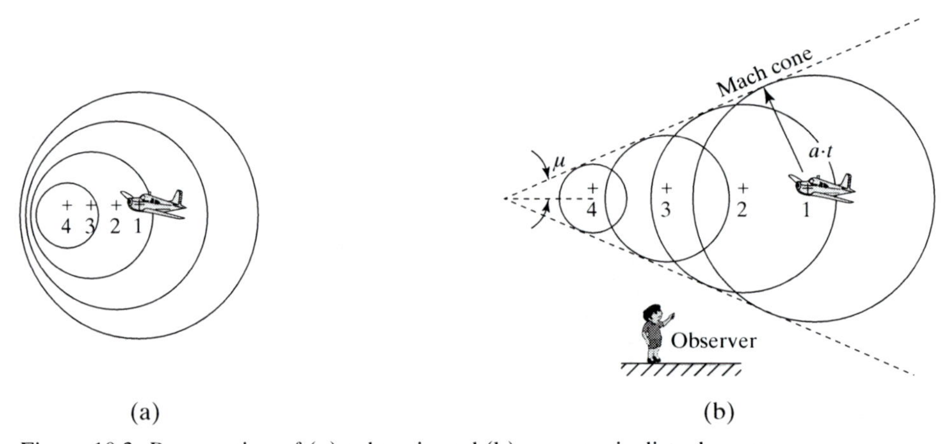

Figure 10.3. Propagation of (a) subsonic and (b) supersonic disturbances.

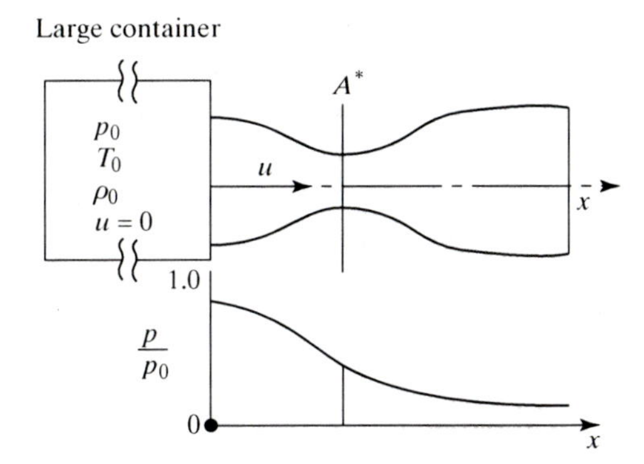

Figure 10.4. Model for 1D isentropic compressible flow.

(or noise) is heard. Of course the noise generated at point 1 has the longest time to spread and is represented by the largest circle. Now if the airplane flies at subsonic speeds, as depicted by Fig. 10.3(a), then the airplane stays inside these "information" circles – meaning that the noise it generates propagates faster than the flight speed (so the pilot can hear the engine jet noise). It is also clear that the faster the airplane moves, the more the left-hand side of each circle is shifted to the left; at $M = 1$ they will coincide so that no information is passed forward of the flight direction.

As the airplane moves faster than the speed of sound ($M > 1$) the picture described in Fig. 10.3(b) evolves. In fact, the airplane moves outside the circle representing the region where the disturbance created by it can be heard (so the pilot cannot hear the engine jet noise). The circles shown in the figure (actually spheres in three dimensions) create a cone inside which the sound generated by the airplane can be heard, but no disturbance can be felt outside this cone, called the *Mach cone*. An observer on the ground may look and see the airplane passing by but cannot hear its noise. It may take a short while before the edge of the cone reaches him, so that he can hear the engine noise (but by then the airplane is long gone).

The half-angle of the Mach cone is called the *Mach angle* μ, and we can calculate it by observing that the distance between the apex and point 1 is $U_\infty \Delta t$, where U_∞ is the airplane speed and Δt is the flight time between these two points. During the same time, the sound traveling to the edge of the cone is $a\Delta t$. Consequently, from the geometry, we can write

$$\sin \mu = \frac{a\Delta t}{U_\infty \Delta t} = \frac{1}{M}. \tag{10.15}$$

The angle μ is shown in Fig. 10.3(a) (and can be defined for $M \geq 1$).

10.3 One-Dimensional Isentropic Compressible Flow

The simplest case involving fluid compressibility is the 1D ideal flow. To visualize such a flow, consider the 1D flow in the x direction, as shown in Fig. 10.4. At the left there is a large container where the velocity is zero. In this tank, the gas properties will not change with time because of its large volume compared with the volume of the flow escaping on the right-hand side. The fluid properties inside the tank are called stagnation conditions because the velocity there is zero and denoted by the subscript 0_0. At the other end of the axisymmetric circular tube, the pressure is very low, causing the flow to move to the right. Flow velocity will change mainly because of the cross-section-area variations, as we shall see later. The objective now is to evaluate the changes in fluid properties and the resulting velocity.

For the present model we assume that the fluid inside the tank is an ideal gas and the process is isentropic (adiabatic and reversible). Therefore the schematics in Fig. 10.4 can be reversed to simulate an inlet, capturing high-speed flow and slowing it down to zero velocity inside a large container. Because no energy is lost we can first observe the changes in the enthalpy h of the gas (see [1, p. 85] of Chapter 1):

$$dh = c_p dT. \tag{10.16}$$

The change between two section along the tube is then

$$h_2 = h_1 + \int_{T_1}^{T_2} c_p dT.$$

The total energy of the fluid at any section is constant and also includes the kinetic energy; per unit mass at an arbitrary point 1, it becomes

$$h_1 + (u_1^2/2).$$

If we write this relation for a point inside the stagnation tank where the velocity is zero and for an arbitrary point inside the tube, then

$$h_0 = c_p T_0 = c_p T + (u^2/2), \tag{10.17}$$

and here we assume a constant c_p. Rearranging this equation provides the relation between the velocity and the temperature in the expansion tube:

$$\frac{T_0}{T} = 1 + \frac{u^2}{2c_p T}. \tag{10.18}$$

However, we can rewrite the velocity u by using the speed of sound and the Match number,

$$u = Ma = M\sqrt{\gamma RT}, \tag{10.19}$$

and therefore

$$\frac{T_0}{T} = 1 + \frac{M^2 \gamma RT}{2c_p T}.$$

Next, we rearrange Eq. (1.20), connecting the gas constant R with the heat capacity,

$$R = c_p \frac{\gamma - 1}{\gamma}, \tag{10.20}$$

and after substituting this into the temperature ratio we get

$$\frac{T_0}{T} = 1 + \frac{\gamma - 1}{2} M^2. \tag{10.21}$$

Note that the x coordinate in Fig. 10.4 can be replaced with a corresponding Mach number coordinate. So if we know the local Mach number, the temperature ratio is given by Eq. (10.21). To calculate the pressure at each point along the tube, we use Eq. (10.7) for the isentropic process,

$$\frac{p_0}{p} = \left(\frac{T_0}{T}\right)^{\frac{\gamma}{\gamma-1}} = \left(1 + \frac{\gamma - 1}{2} M^2\right)^{\frac{\gamma}{\gamma-1}}, \tag{10.22}$$

and the variation in density is obtained by use of the ideal-gas equation:

$$\frac{\rho_0}{\rho} = \frac{p_0}{p}\frac{T}{T_0} = \left(1 + \frac{\gamma - 1}{2}M^2\right)^{\frac{1}{\gamma-1}}. \tag{10.23}$$

To calculate the mass flow, we use the continuity equation,

$$\dot{m} = \rho u A,$$

where A is the cross-section area, and the local velocity is given by Eq. (10.19):

$$u = M\sqrt{\gamma R T}. \tag{10.19}$$

We obtain the velocity, in terms of the stagnation conditions, by using Eq. (10.21), to replace T:

$$u = M\sqrt{\frac{\gamma R T_0}{1 + \frac{\gamma-1}{2}M^2}}. \tag{10.24}$$

EXAMPLE 10.2. TEMPERATURE INCREASE IN SUPERSONIC SPEEDS. Assume that, for structural reasons, the temperature at any point on an airplane cannot be above 850 K. Calculate the maximum flight speed of such an airplane operating at an ambient temperature of 270 K.

Solution: At any stagnation point on a moving object, such as the wing leading edges, the temperature will reach the stagnation temperature. Therefore, by using Eq. (10.21) ($\gamma = 1.4$), we get

$$M = \sqrt{\frac{2}{(\gamma-1)}\left(\frac{T_0}{T} - 1\right)} = \sqrt{\frac{2}{(1.4-1)}\left(\frac{850}{270} - 1\right)} = 3.28,$$

so at this speed portions of the wing, if made of aluminum, may melt. Along the same line we can explore another speed "limit" and use this model for reverse flow (recall that the flow is isentropic). In this case we model the inlet of a high-speed airplane, and the flow is slowed down by such a diffuser to subsonic speeds as it reaches the combustion chamber. Suppose the temperature of the combustion products inside a jet engine is 2500 K. At what Mach number will the air heat up to this temperature behind the diffuser, and therefore fuel cannot be burned (in a subsonic combustion chamber) to generate thrust?

Again, using the same equation, we get

$$M = \sqrt{\frac{2}{(1.4-1)}\left(\frac{2500}{270} - 1\right)} = 6.42.$$

EXAMPLE 10.3. PRESSURE AND TEMPERATURE RATIO AT $M = 1$. A large tank contains compressed air at 300 K and at 4 atm (1 atm = 101,300 N/m^3). Assuming the flow escapes through a small circular tube (as shown in Fig. 10.4), calculate the pressure and temperature at a section where $M = 1$.

Solution: Assuming $\gamma = 1.4$ for air and using Eqs. (10.22) and (10.23), we get

$$\frac{T_0}{T} = 1 + \frac{\gamma - 1}{2}M^2 = 1.2,$$

$$\frac{p_0}{p} = \left(1 + \frac{\gamma - 1}{2}M^2\right)^{\frac{\gamma}{\gamma-1}} = 1.893.$$

The pressure at this section therefore is $4/1.893 = 2.11$ atm and the temperature is 250 K. Note that a pressure ratio that is slightly less than 2 is sufficient to accelerate the flow to sonic velocities.

10.3.1 Critical Conditions

The condition at which $M = 1$ is called the critical condition, and usually an asterisk is used to note the properties in this section. Based on Eq. (10.21), the temperature ratio is then

$$\frac{T_0}{T^*} = \frac{\gamma + 1}{2}, \tag{10.25}$$

the pressure ratio, based on Eq. (10.22) is

$$\frac{p_0}{p^*} = \left(\frac{\gamma + 1}{2}\right)^{\frac{\gamma}{\gamma-1}}, \tag{10.26}$$

and the density ratio, based on Eq. (10.23), is

$$\frac{\rho_0}{\rho^*} = \left(\frac{\gamma + 1}{2}\right)^{\frac{1}{\gamma-1}}. \tag{10.27}$$

The local velocity is calculated with the temperature ratio from Eq. (10.24):

$$u^* = a^* = \sqrt{\gamma RT} = \sqrt{\frac{2\gamma RT_0}{\gamma + 1}}. \tag{10.28}$$

Up to this point the cross-section area or the shape of the tube in Fig. 10.4 has not been discussed. To evaluate the area changes, we observe the continuity equation:

$$\dot{m} = \rho u A.$$

This relation holds at any section of the tube, and by substituting the velocity from Eq. (10.24) and the density from Eq. (10.23), we get

$$\frac{\dot{m}}{A} = \rho u = \frac{\rho_0}{\left(1 + \frac{\gamma - 1}{2}M^2\right)^{\frac{1}{\gamma-1}}} M \sqrt{\frac{\gamma RT_0}{1 + \frac{\gamma - 1}{2}M^2}} = \frac{\rho_0 M \sqrt{\gamma RT_0}}{\left(1 + \frac{\gamma - 1}{2}M^2\right)^{\frac{\gamma+1}{2(\gamma-1)}}}. \tag{10.29}$$

It is much easier to evaluate this ratio at the critical section, where $M = 1$ and $A = A^*$:

$$\frac{\dot{m}}{A^*} = \frac{\rho_0 \sqrt{\gamma RT_0}}{\left(\frac{\gamma + 1}{2}\right)^{\frac{\gamma+1}{2(\gamma-1)}}}. \tag{10.29a}$$

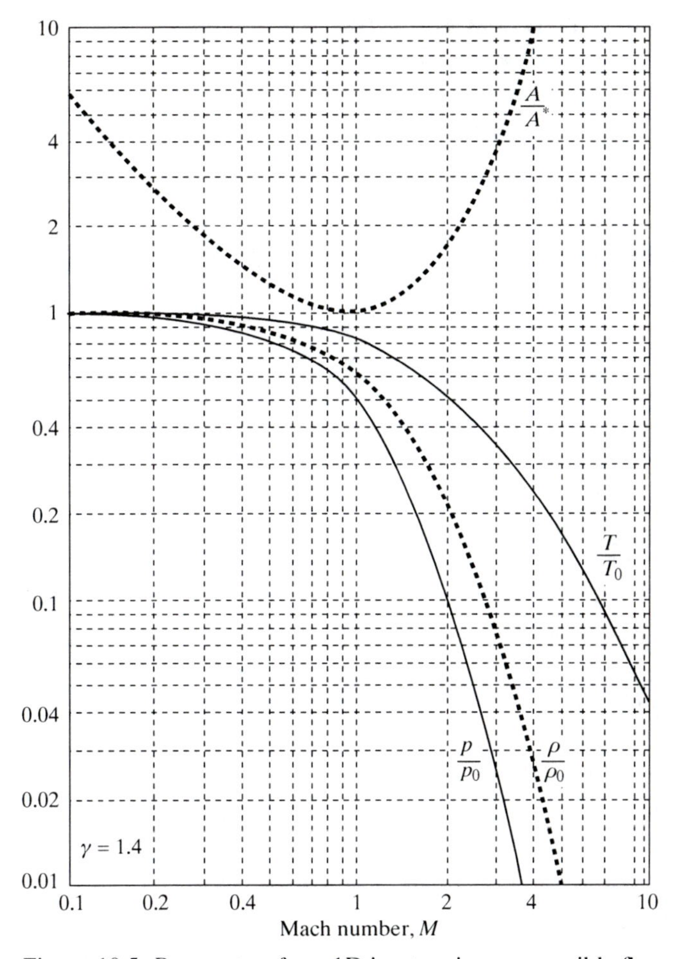

Figure 10.5. Parameters for a 1D isentropic compressible flow of an ideal gas.

We obtain the area ratio by dividing Eq. (10.29a) by Eq. (10.29):

$$\frac{A}{A^*} = \frac{\dot{m}}{A^*}\frac{A}{\dot{m}} = \frac{1}{M}\left(\frac{1+\dfrac{\gamma-1}{2}M^2}{\dfrac{\gamma+1}{2}}\right)^{\frac{\gamma+1}{2(\gamma-1)}}$$

and after some algebra inside the parentheses we get

$$\frac{A}{A^*} = \frac{1}{M}\left\{\frac{1}{\gamma+1}[2+(\gamma-1)M^2]\right\}^{\frac{\gamma+1}{2(\gamma-1)}}. \tag{10.30}$$

This is a very interesting result, and this area variation and the pressure, density, and temperature are plotted in Fig. 10.5 versus the Mach number. These values for the isentropic 1D flow are also presented in a table form in Appendix B. Also note that we can replace the x coordinate with the Mach number coordinate in Fig. 10.4, and clearly, with an increasing Mach number, the pressure, density, and temperature will be reduced (as expected). However, the cross-section area has a converging–diverging shape, with the smallest cross section being at $M = 1$.

The smallest area, A^*, is the throat area, and based on Eq. (10.29a), it fixes the mass flow rate. Using the ideal-gas relation of Eq. (10.1) to replace the density in Eq. (10.29a), with $\frac{p_0}{RT_0}$ we get

$$\dot{m} = p_0 A^* \sqrt{\frac{\gamma}{RT_0}} \left(\frac{2}{\gamma+1}\right)^{\frac{\gamma+1}{2(\gamma-1)}}, \tag{10.31}$$

and this equation clearly indicates that, for a given set of stagnation conditions, the mass flow rate depends on the throat area.

The terms containing γ are sometimes combined to simplify the preceding formula:

$$\dot{m} = \frac{p_0 A^*}{\sqrt{RT_0}} \Gamma, \tag{10.31a}$$

and the definition of Γ is then

$$\Gamma = \sqrt{\gamma} \left(\frac{2}{\gamma+1}\right)^{\frac{\gamma+1}{2(\gamma-1)}}. \tag{10.32}$$

Equation (10.31a), in its second form, shows more clearly the linear dependence between stagnation pressure and mass flow rate (and the same can be stated about the throat area). To simplify the calculations involving Γ, we construct Table 10.1.

Returning to Figs. 10.4, and 10.5, it is obvious that a subsonic nozzle has a converging shape. However, to accelerate the flow to a supersonic condition, a diverging section is required. Therefore in a supersonic nozzle the cross-section area increases, with an increased Mach number.

Because the assumption so far is that the flow is isentropic (e.g., also reversible), the picture in Fig. 10.4 can be reversed such that the flow is flowing into the container. In this case the supersonic converging section is a supersonic diffuser, whereas the diverging section (behind the throat) is a diverging subsonic diffuser! In practice, however, a supersonic converging–diverging diffuser is prone to internal shock waves and seldom used.

10.3.2 Practical Examples for One-Dimensional Compressible Flow

Compressed gases are used in numerous engineering disciplines, and the preceding formulation is applicable for several applications. For example, Eq. (10.26) shows

Table 10.1. *Numerical values for the γ and Γ terms*

γ	Γ
1.10	0.6284
1.15	0.6386
1.20	0.6485
1.25	0.6581
1.30	0.6674
1.35	0.6761
1.40	0.6847

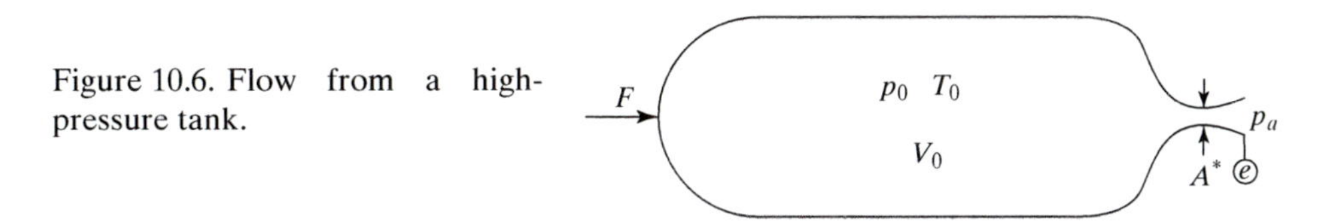

Figure 10.6. Flow from a high-pressure tank.

that compressed air can reach sonic velocities because of pressure ratios near 2! This indicates that compressible flow is present in a large number of gas flows, as demonstrated in the following examples.

EXAMPLE 10.4. FLOW FROM A HIGH-PRESSURE TANK. Consider a high-pressure tank, such as used in the welding industry, or a solid-propellant rocket engine immediately after burnout. The conditions of an ideal gas inside the tank are known as p_0, T_0, and the volume V_0 (and the flow is choked, $p_0/p_a \gg 2$). At this point the flow exits through a small nozzle because of the high internal pressure. Assuming frictionless flow, calculate the momentary pressure in the tank, the mass flow rate (versus time), and the force F (shown in Fig. 10.6) required for holding the tank in place.

Solution: With the preceding assumptions, the exit mass flow rate can be estimated by Eq. (10.31). However, as the flow leaves the tank, the pressure will be gradually reduced. The mass inside the tank is $\rho_0 V_0$ and the conservation of mass (the continuity equation) requires that the reduction of mass inside the tank be equal to the flow $\dot{m}_{\text{out}}$ leaving through the nozzle:

$$\frac{d}{dt}(\rho_0 V_0) = -\dot{m}_{\text{out}}.$$

Replacing the density with the ideal-gas relation and $\dot{m}_{\text{out}}$ with Eq. (10.31), we get

$$\frac{d}{dt}\left(\frac{p_0}{RT_0}V_0\right) = -p_0 A^* \sqrt{\frac{\gamma}{RT_0}}\left(\frac{2}{\gamma+1}\right)^{\frac{\gamma+1}{2(\gamma-1)}}.$$

After rearranging the terms and separation of the variables we get

$$\frac{dp_0}{p_0} = -\frac{A^*}{V_0}\sqrt{\gamma RT_0}\left(\frac{2}{\gamma+1}\right)^{\frac{\gamma+1}{2(\gamma-1)}} dt.$$

The integration between the initial pressure $p_0(0)$ and the momentary pressure $p_0(t)$ yields

$$\ln\frac{p_0(t)}{p_0(0)} = -\frac{A^*}{V_0}\sqrt{\gamma RT_0}\left(\frac{2}{\gamma+1}\right)^{\frac{\gamma+1}{2(\gamma-1)}} t. \tag{10.33}$$

This equation clearly describes an exponential decay of the pressure inside the tank. We can simplify this equation by using the definition of Γ from Eq. (10.32):

$$\ln\frac{p_0(t)}{p_0(0)} = -\left(\frac{A^*\Gamma}{V_0}\sqrt{RT_0}\right) t. \tag{10.33a}$$

Once the momentary pressure variation is solved, the momentary flow rate and axial force can be calculated (although this requires extensive algebra). First,

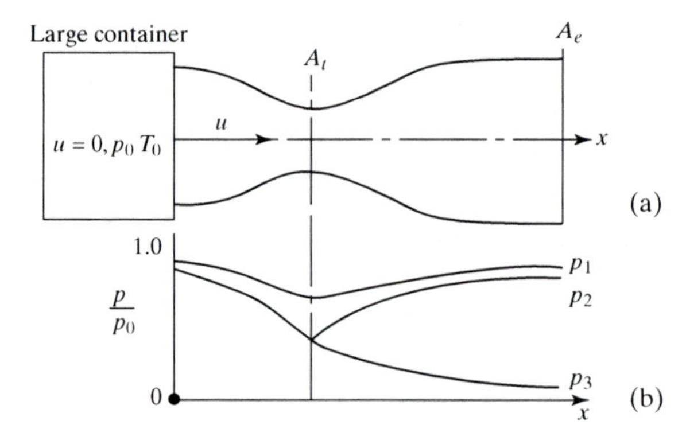

(a) (b) Figure 10.7. Flow in a converging–diverging nozzle.

we can estimate the change in stagnation temperature T_0 by using the ideal-gas relation (the volume V_0 and the momentary mass inside the tank determine the density). Next, we can calculate the nozzle mass flow rate $\dot{m}$ by using Eq. (10.31a). Finally, we evaluate the axial force on the tank by using Eq. (2.29):

$$F = \dot{m}u_e + (p_e - p_a)A_e,$$

and we can calculate the exit conditions (u_e and P_e) by using Eqs. (10.22) and (10.24.)

EXAMPLE 10.5. FLOW THROUGH A CONVERGING–DIVERGING NOZZLE. The stagnation conditions in a large container, shown in Fig. 10.7, are $p_0 = 4$ atm and $T_0 = 300$ K. The flow is exhausted through a circular nozzle having a throat area of $A_t = 5\ \text{cm}^2$ and exit area $A_e = 10\ \text{cm}^2$. Assuming that the exit pressure (or back pressure) can be controlled, calculate the exit Mach numbers.

Solution: Case 1. Suppose the back pressure at the exit is set at $p_1 = 3.8$ atm; then the exit Mach number can be calculated from Eq. (10.22) or from the tables in Appendix B. For the pressure ratio,

$$\frac{p_0}{p} = \frac{4}{3.8} = 0.95.$$

From Appendix B we get $M_1 = 0.26$ and $(A_e/A^*) = 2.32$ therefore A^* for this case is

$$A^* = \frac{10}{2.32} = 4.31\ \text{cm}^2.$$

This area is smaller than the throat area A_t, and we can calculate the area ratio at the throat

$$\frac{A_t}{A*} = \frac{5}{4.31} = 1.16.$$

Therefore the Mach number at the throat is $M_t = 0.64$ [this can be calculated with Eq. (10.31) as well, but the algebra is quite complex].

So when the pressure ratio p/p_0 at both ends of the tube is larger than the one shown by the line marked p_2 in Fig. 10.7, the flow is entirely subsonic. Initially the flow accelerates into the throat because of the converging shape and the pressure is reduced. However, the diverging section appears as a diffuser to this subsonic flow, and in this section the speed is reduced and the pressure increases to match the exit pressure p_1.

Case 2. This is the limiting case in which the back pressure p_2 is lowered to a level at which a sonic speed is reached at the throat. Therefore the throat area is the same as the critical area (e.g., $A^* = A_t$).

The area ratio at the exit is therefore $(A_e/A_t) = (A_e/A^*) = 2.00$ from Appendix B. For this ratio at the exit we get $M_2 = 0.3$ and $p_2/p_0 = 0.939$.

The exit back pressure for this case is therefore

$$p_2 = 0.939 p_0 = 3.75 \text{ atm}.$$

So when the pressure at the exit is lowered a bit, the velocity at the throat becomes sonic. The diverging section of the tube is still a diffuser and the flow velocity is slowing down toward the exit.

Case 3. The next question is this: How low should the pressure p_3 at the exit be in order to generaste supersonic speeds at the exit (so that the diverging section becomes a supersonic nozzle). Again, instead of using the equation, we use the supersonic side of the table in Appendix B. The area ratio at the exit is: $(A_e/A_t) = (A_e/A^*) = 2.00$ from Appendix B. From the supersonic part, we get $M_2 = 2.2$ and $p_3/p_0 = 0.094$.

The exit back pressure for this case is therefore

$$p_2 = 0.094 p_0 = 0.376 \text{ atm}.$$

So the pressure ratio $\frac{P_3}{P_0}$ at the two ends of the tube must be significantly smaller to create a supersonic nozzle. It is also clear from this example that, if the exit pressure drops below 0.376 atm, the flow inside the tube will not change. As a matter of fact, the same flow exit conditions are expected in vacuum as well.

The more difficult question, though, is what happens if the back pressure is in between case 2 and case 3 (or between p_2 and p_3)? According to our model so far, only the previously described conditions are possible. Consequently, if the back pressure is higher (say 1 atm), then an adjustment in the form of a shock wave can take place. This type of discontinuity (shock wave) causing a sharp pressure change is discussed next.

10.4 Normal Shock Waves

The ideal-flow model developed in the previous section indicates that the local Mach number in a nozzle, as seen in Fig. 10.4, depends mainly on the area ratios. Once the nozzle geometry is fixed, then the pressure ratio is also determined by the ideal-flow equations (see Fig. 10.5). However, if at the nozzle exit the pressure is higher, then usually a shock wave pattern will correct the pressure differences. Such a scenario is shown schematically in Fig. 10.8(a), where a supersonic flow velocity is reduced to subsonic by means of a shock wave. Another possibility of creating normal shock waves is when a blunt object or an inlet flies at supersonic speeds [Fig. 10.8(b)]. In this case the velocity near the inlet must slow down to subsonic velocities, or even to zero in the case of a blunt object. This sudden reduction in flow speed usually results in a normal shock wave, and the main objective of this section is to determine the changes in velocity and fluid properties across the shock wave. Note that this

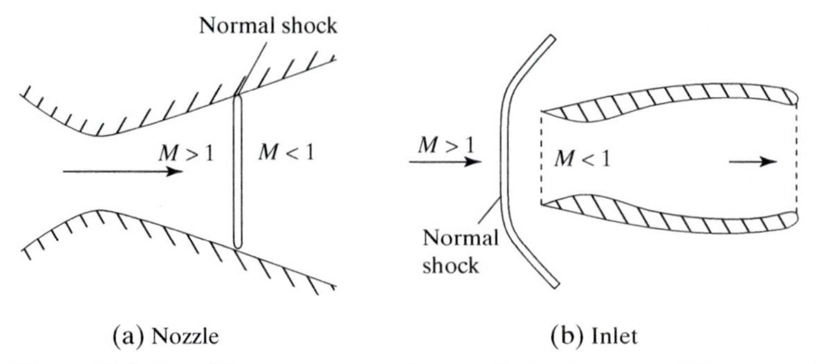

Figure 10.8. Possible occurrence of normal shock waves: (a) converging–diverging nozzle, (b) ahead of a nacelle or a blunt object.

is a *nonreversible* phenomenon and a subsonic flow *cannot* become supersonic by crossing a shock wave!

A simple 1D model for a normal shock wave is depicted in Fig. 10.9. Here the flow is captured between two parallel stream lines, and the normal shock represents a thin discontinuity, normal to the streamlines. Again, note that this is a strong discontinuity and the changes are not isentropic or reversible (also meaning that stagnation pressure losses are expected). Consequently the simple isentropic model used for a weak pressure disturbance in Section 10.2 cannot be used. On the left-hand side of the model in Fig. 10.9, a supersonic flow is entering the control volume (which includes both sides of the shock wave) and the subsonic flow leaves at the right-hand side.

The fluid properties and incoming Mach number M_1 are known on the left-hand side of the shock wave. It is also assumed that the fluid is an ideal gas. The proposed model should provide the fluid properties and the Mach number on the other side of the shock wave. The first equation used for this model is the steady-state continuity equation,

$$\rho_1 u_1 A_1 = \rho_2 u_2 A_2,$$

but the area A is the same on both sides of the discontinuity and we can simply write

$$\rho_1 u_1 = \rho_2 u_2. \tag{10.34}$$

Similarly, the 1D momentum equation reduces to

$$p_1 + \rho_1 u_1^2 = p_2 + \rho_2 u_2^2. \tag{10.35}$$

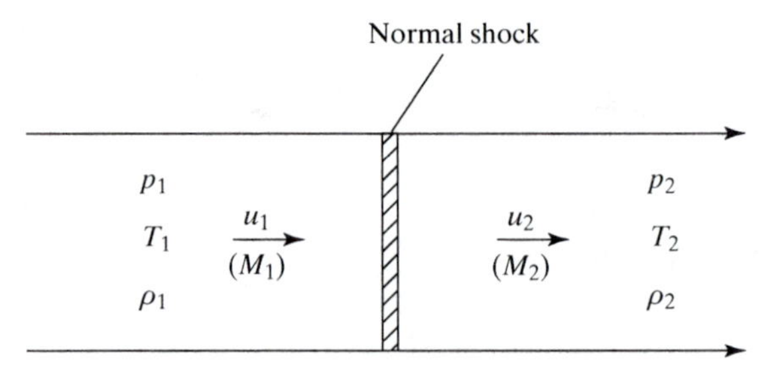

Figure 10.9. 1D model for a normal shock wave.

Also there is no heat exchange (adiabatic flow) and the stagnation enthalpy is unchanged (e.g., conservation of energy). Because we approximated the enthalpy of an ideal gas by Eq. (10.17), we can write

$$T_{01} = T_{02}. \tag{10.36}$$

With the aid of these equations, we calculate the changes across the shock wave, knowing the conditions ahead of the shock. For example, the temperature ratio can be formulated, based on the Mach numbers on both sides [and using the isentropic relations from Eq. (10.21)]:

$$\frac{T_2}{T_1} = \frac{\dfrac{T_{01}}{T_1}}{\dfrac{T_{02}}{T_2}} = \frac{1 + \dfrac{\gamma - 1}{2} M_1^2}{1 + \dfrac{\gamma - 1}{2} M_2^2}. \tag{10.37}$$

and we shall calculate M_2 later. Of course, according to Eq. (10.36), $T_{02} = T_{01}$. If the Mach number is known, then the velocity ratio is easily calculated:

$$\frac{u_2}{u_1} = \frac{M_2\sqrt{\gamma R T_2}}{M_1\sqrt{\gamma R T_1}} = \frac{M_2}{M_1}\sqrt{\frac{T_2}{T_1}}. \tag{10.38}$$

The density ratio is calculated with the continuity equation (Eq. 10.34):

$$\frac{\rho_2}{\rho_1} = \frac{u_1}{u_2} = \frac{M_1}{M_2}\sqrt{\frac{T_1}{T_2}}$$

Using the temperature ratio from Eq. (10.37), we get

$$\frac{\rho_2}{\rho_1} = \frac{M_1}{M_2}\sqrt{\frac{1 + \dfrac{\gamma - 1}{2} M_2^2}{1 + \dfrac{\gamma - 1}{2} M_1^2}}. \tag{10.39}$$

We then obtain the pressure ratio from momentum equation (10.35), and by using the ideal-gas relation $\rho = (p/RT)$, we get

$$p_1 + \frac{p_1}{RT_1} u_1^2 = p_2 + \frac{p_2}{RT_2} u_2^2.$$

Now recalling that the term

$$\frac{u^2}{RT} = \frac{M^2 \gamma RT}{RT} = M^2 \gamma$$

and substituting this into the momentum relation, we get

$$p_1 \left(1 + M_1^2 \gamma\right) = p_2 \left(1 + M_2^2 \gamma\right).$$

The static pressure ratio is then

$$\frac{p_2}{p_1} = \frac{\left(1 + M_1^2 \gamma\right)}{\left(1 + M_2^2 \gamma\right)}. \tag{10.40}$$

The last missing unknown is the Mach number behind the shock, M_2. For this we can use the ideal-gas relation, namely,

$$\frac{T_2}{T_1} = \frac{\dfrac{p_2}{\rho_2 R}}{\dfrac{p_1}{\rho_1 R}} = \frac{p_2 \rho_1}{p_1 \rho_2}.$$

Next we substitute the pressure and density ratios from Eqs. (10.39) and (10.40):

$$\frac{T_2}{T_1} = \frac{p_2 \rho_1}{p_1 \rho_2} = \frac{\left(1 + M_1^2 \gamma\right) M_2}{\left(1 + M_2^2 \gamma\right) M_1} \sqrt{\frac{1 + \dfrac{\gamma - 1}{2} M_1^2}{1 + \dfrac{\gamma - 1}{2} M_2^2}}.$$

Substituting the temperature ratio from Eq. (10.37) into the left-hand side we get an equation relating M_1 to M_2:

$$\frac{1 + \dfrac{\gamma - 1}{2} M_1^2}{1 + \dfrac{\gamma - 1}{2} M_2^2} = \frac{\left(1 + M_1^2 \gamma\right) M_2}{\left(1 + M_2^2 \gamma\right) M_1} \sqrt{\frac{1 + \dfrac{\gamma - 1}{2} M_1^2}{1 + \dfrac{\gamma - 1}{2} M_2^2}}$$

or

$$\sqrt{\frac{1 + \dfrac{\gamma - 1}{2} M_1^2}{1 + \dfrac{\gamma - 1}{2} M_2^2}} = \frac{\left(1 + M_1^2 \gamma\right) M_2}{\left(1 + M_2^2 \gamma\right) M_1}.$$

By squaring both sides of the equation we get a quadratic equation for M_2^2 that provides two solutions. The first is $M_2^2 = M_1^2$, which is trivial, and the second solution is

$$M_2^2 = \frac{2 + (\gamma - 1) M_1^2}{\gamma \left(2 M_1^2 - 1\right) + 1}. \tag{10.41}$$

Last, the stagnation pressure loss is evaluated. This can be done again by use of the isentropic relation [Eq. (10.22)] on both sides of the shock wave:

$$\frac{p_{02}}{p_{01}} = \frac{\dfrac{p_{02}}{p_2}}{\dfrac{p_{01}}{p_1}} \frac{p_1}{p_2} = \frac{\left(1 + \dfrac{\gamma - 1}{2} M_2^2\right)^{\frac{\gamma}{\gamma - 1}}}{\left(1 + \dfrac{\gamma - 1}{2} M_1^2\right)^{\frac{\gamma}{\gamma - 1}}} \frac{\left(1 + M_2^2 \gamma\right)}{\left(1 + M_1^2 \gamma\right)} \tag{10.42}$$

and the static pressure ratio is taken from Eq. (10.39).

At this point, all quantities at the other side of the shock wave are calculated. However, it is more practical to rearrange the equations in terms of the incoming

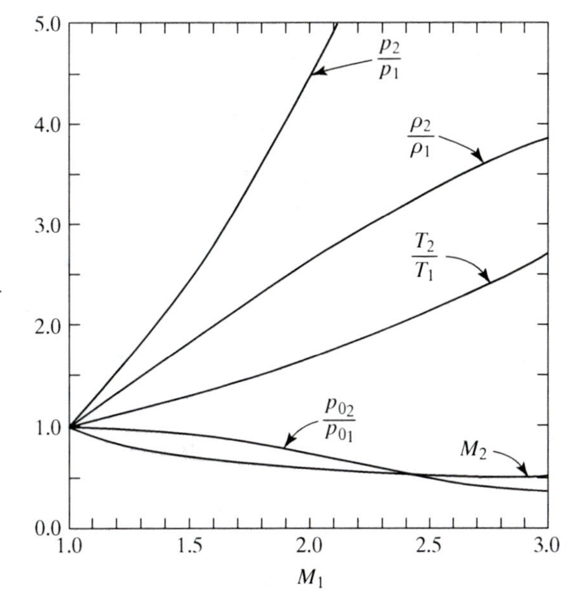

Figure 10.10. Changes in flow properties across a normal shock wave.

Mach number M_1. The algebra involved is quite elaborate and therefore only the final results are summarized:

$$\frac{p_{02}}{p_{01}} = \frac{\left(\dfrac{\frac{\gamma+1}{2}M_1^2}{1+\frac{\gamma-1}{2}M_1^2}\right)^{\frac{\gamma}{\gamma-1}}}{\left(\dfrac{2\gamma}{\gamma+1}M_1^2 - \dfrac{\gamma-1}{\gamma+1}\right)^{\frac{1}{\gamma-1}}}. \tag{10.43}$$

$$\frac{T_2}{T_1} = \frac{\left[2+(\gamma-1)\,M_1^2\right]\left(2\gamma M_1^2-\gamma+1\right)}{(\gamma+1)^2\,M_1^2}, \tag{10.44}$$

$$\frac{p_2}{p_1} = \frac{2\gamma}{\gamma+1}M_1^2 - \frac{\gamma-1}{\gamma+1}, \tag{10.45}$$

$$\frac{\rho_2}{\rho_1} = \frac{u_1}{u_2} = \frac{(\gamma+1)\,M_1^2}{2+(\gamma-1)\,M_1^2}, \tag{10.46}$$

and M_2 is given by Eq. (10.41). These equations are tabulated in Appendix C as a function of the incoming Mach number M_1. The results are also plotted in Fig. 10.10.

The data plotted in Fig. 10.10 clearly indicate the large changes across the shock wave. For example, the static pressure jump is the highest, but the static temperature increase is also significant. Consequently the density jump is also very high. These large increments result in a sharp loss in velocity and the Mach number behind the shock wave is always subsonic. As a matter of fact, the faster the incoming flow M_1, the weaker the exiting Mach number M_2. The loss in stagnation pressure is increasing with M_1 and clearly indicates the loss of usable energy (or the irreversibility of the flow).

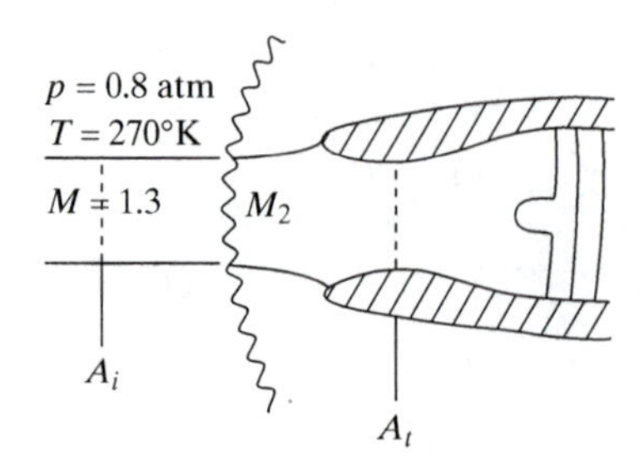

Figure 10.11. Schematic description of an inlet operating slightly above the speed of sound.

10.5 Some Applications of the One-Dimensional Model

There are several practical cases in which such 1D models provide feasible preliminary engineering data. When the pressure gradient is favorable, as in the case of nozzles, these tools are considerably better compared with the adverse pressure cases found in diffusers. Also, in most cases oblique shocks are present (which are beyond the scope of this text) but if the normal shock model is used, preliminary designs of diffusers and nozzles is still possible. As an example for the successful use of this model the following examples are presented.

10.5.1 Normal Shock Wave ahead of a Circular Inlet

Airplanes with a circular inlet (like the first supersonic fighters of the 1950s) were capable of flying at low supersonic speeds. The inlet is basically a subsonic diffuser and at speeds above Mach 1, a normal shock forms ahead of the inlet. Because the losses below $M = 1.3$ are quite small, such an approach was workable, particularly when supersonic speeds were used only in extreme distressed conditions.

For a numerical example, consider an airplane with a circular inlet (as shown in Fig. 10.11) that is flying at $M = 1.3$ and the smallest cross-section area in the inlet is $A_t = 0.15\ \mathrm{m}^2$. As discussed earlier, a normal shock wave is present ahead of the inlet. The most important question then relates to the stagnation pressure loss that is due to the shock wave. Also it is important to calculate the Mach number behind the shock, the capture area A_i, and the mass flow rate.

The condition depicted in Fig. 10.11 is called a detached shock wave and the Mach number at the smallest area A_t is equal to 1. The Mach number behind the normal shock is easily calculated by Eq. (10.40) or from the tables in Appendix C ($\gamma = 1.4$):

$$M_2^2 = \frac{2 + (\gamma - 1)\, M_1^2}{\gamma\,(2M_1^2 - 1) + 1} = 0.618$$

or $M_2 = 0.786$.

At the same time the stagnation pressure ratio is given in the same Appendix C or can be calculated from Eq. (10.43) as

$$\frac{p_{02}}{p_{01}} = 0.979.$$

So the stagnation pressure loss is not too high (e.g., about 2%). Because the flow behind the shock wave is subsonic, by assuming ideal flow, we can calculate the area ratio from Appendix B or from Eq. (10.31):

$$\frac{A_i}{A_t} = \frac{A}{A^*} = \frac{1}{M_2}\left\{\frac{1}{\gamma+1}\left[2+(\gamma-1)M_2^2\right]\right\}^{\frac{\gamma+1}{2(\gamma-1)}} = 1.05.$$

Note that the area A_i ahead and behind the shock is unchanged and therefore the capture area is

$$A_i = 1.05 \times 0.15 = 0.1575 \text{ m}^2.$$

We now calculate the mass flow rate by using the information ahead of the shock wave [and the velocity from Eq. (10.19)]. Let us assume that the static temperature is $T = 270$ K and the pressure is $p = 0.6$ atm:

$$\begin{aligned}\dot{m} &= \rho u A_i = \rho M\sqrt{\gamma RT}A_i = \frac{p}{RT}M\sqrt{\gamma RT}A_i = pMA_i\sqrt{\frac{\gamma}{RT}}\\ &= 0.6\times 10^5 \times 1.3 \times 0.1575\sqrt{\frac{1.4}{286.6\times 270}}\\ &= 52.25 \text{ kg/s},\end{aligned}$$

and here we assume that 1 atm = 10^5 N/m^2.

This example clearly shows that this model allows the sizing of inlets, even in the presence of a normal shock.

10.5.2 The Converging–Diverging Nozzle (de Laval Nozzle)

Rockets and high-pressure jet engines use converging–diverging nozzles to generate supersonic exit velocity and high thrust. Such designs are often called in the literature the *de Laval nozzle*, after Karl Gustaf Patrik de Laval (1845–1913), the Swedish engineer who improved steam turbine performance by using converging–diverging nozzles. This concept was already discussed in relation to the 1D compressible isentropic flow (see Fig. 10.7), and for the present analysis, the converging–diverging nozzle was redrawn in Fig. 10.12. In this case, however, back-pressure conditions for a wider range are considered. Let us assume that the flow is entering the nozzle with given stagnation pressure p_0 and temperature T_0 at the left, and by gradually lowering the back pressure, the flow rate through the nozzle can be increased. Of course if $p_1 = p_0$ there will be no flow, but by lowering p_1 a bit, the flow will accelerate toward the throat and then slow down (because of the area change) toward the exit. Note that the flow in this case is entirely subsonic (e.g., when the exit pressure is p_1). The two diagrams below the nozzle show the variations of the pressure (with the area change) and the velocity, in terms of the Mach number, along the nozzle. The next interesting condition is when the back pressure at the exit is lowered to p_2 and the velocity at the throat reaches sonic speed, but behind the throat the flow will slow down and stay subsonic all the way to the exit as discussed in reference to Fig. 10.7. Also, based on Eq. (10.26), the pressure ratio at the throat (where now $M = 1$) is $p^*/p_0 = 0.528$, as shown (for $\gamma = 1.4$). Up to this point, the 1D isentropic

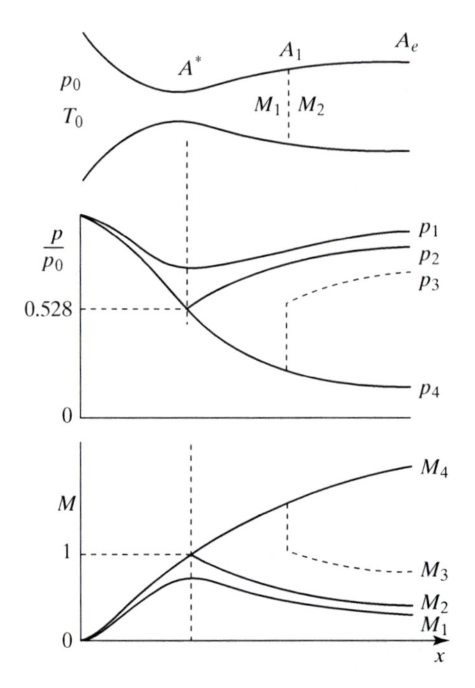

Figure 10.12. The de Laval nozzle and corresponding pressure and Mach number variations along the nozzle.

flow model is applicable and local velocities (or Mach numbers) can be calculated by this model.

To create supersonic flow in the diverging section, a significantly lower back pressure is needed, as shown schematically by the pressure p_4 in the figure (and M_4 is the exit mach number). If the pressure outside the nozzle is lower than p_4 then the flow inside the nozzle will not change and an external expansion will adjust for the pressure outside. This condition is called *underexpanded*, meaning that larger exit area could have been used for the nozzle. Also note that, if the pressure outside is less than p_2, all flow parameters in the nozzle remain unchanged!

Of particular interest is the case in which the pressure at the exit is between p_2 and p_4. In this case a fully developed supersonic flow cannot exist and a shock wave will adjust the pressures, as shown in the figure. To match an exit pressure of p_3, the flow behind the shock wave becomes subsonic and the diverging section actually becomes a diffuser (because, owing to the increased cross-section area, the pressure will increase toward the exit). This condition is called an *overexpanded* nozzle and is highly undesirable for all sorts of propulsion application (because thrust will be reduced).

EXAMPLE 10.6. THRUST OF A SMALL CONTROL ROCKET. The stagnation conditions inside the combustion chamber of a small satellite control rocket are $T_0 = 2100$ K and $p_0 = 3.5$ atm. The nozzle throat area is $A_t = 4$ cm^2 and the exit-to-throat-area ratio is $A_e/A_t = 4.0$. Calculate thrust of this module in space and on Earth where $p_a = 1$ atm (for simplicity use the properties of air).

Solution: To calculate the thrust in space the nozzle exit velocity must be calculated. From the isentropic flow relations (or by using the tables from Appendix

B) and scanning the area ratio column we get for $(A_e/A_t) = (A/A^*) = 4$,

$$\textit{isentropic flow} \Rightarrow \begin{cases} M_e = 2.94 \\ p_e/p_0 = 0.030. \\ T_e/T_0 = 0.366 \end{cases}$$

This condition corresponds to the supersonic curve (p_4) in Fig. 10.12 and the exit pressure is $p_e = 0.030 \times 3.5 = 0.105$ atm. Also note that in space the back pressure is zero and therefore this is an underexpanded condition. To calculate the thrust we use Eq. (2.29);

$$F_x = \rho_e u_e^2 A_e + (p_e - p_a) A_e.$$

The exit velocity can be calculated as

$$u_e = M\sqrt{\gamma RT} = 2.94\sqrt{1.4 \times 286.6(2100 \times 0.366)} = 1632.7 \text{ m/s},$$

and the exit density is then (using the ideal-gas relation)

$$\rho_e = \frac{p_e}{RT_e} = \frac{0.105 \times 10^5}{286.6 \times 2100 \times 0.366} = 0.04766 \text{ kg/m}^3,$$

and here we assume that 1 atm $= 10^5$ N/m^2. The thrust in space can now be calculated:

$$F_x = 0.0476 \times 1632.7^2 \times 16 \times 10^{-4} + (0.105 \times 10^5 - 0) \times 16 \times 10^{-4} = 219.82 \text{ N}.$$

Next the thrust of the same unit at sea level is investigated. The previous calculations showed that, to maintain supersonic flow, the exit pressure should be $p_e = 0.105$ atm. Because sea-level pressure is assumed to be 1 atm, then clearly an overexpanded condition exists (as marked by the p_3 line in Fig. 10.12). To calculate this conditions, it is best to use an iterative method in conjunction with the tables in Appendices B and C. Because the exit Mach number for the ideal supersonic case is $M_e = 2.94$ the normal shock occurs at a lower Mach number. The iterative process begins with guessing the location of the shock (in terms of M_1) and then calculating the pressure at the exit. If the resulting exit pressure is too high, a stronger shock is guessed next. Usually two to three iterations are required, and the final result can be obtained by simple extrapolation. For simplicity let us present only the final (actually third) iteration and assume that the normal shock is in the section where $M_1 = 2.9$. Based on this assumption, the normal shock tables in Appendix C provide the values behind the shock wave:

$$M_1 = 2.9, \quad \textit{normal shock} \Rightarrow \begin{cases} M_2 = 0.481 \\ p_{02}/p_{01} = 0.358 \end{cases}$$

Next, the location of section A_1 must be identified. From the isentropic tables for the supersonic flow (in Appendix B) scanning down the Mach number column, we get

$$M_1 = 2.9, \quad \textit{isentropic flow} \Rightarrow \frac{A_1}{A^*} = 3.85.$$

Knowing the nozzle geometry, we can calculate A_1 as $A_1 = 3.85A^*$. Next, the condition at the same cross section but in the subsonic side is explored by use of the subsonic tables (in Appendix B) scanning down the Mach number column,

$$M_2 = 0.481, \qquad \textit{isentropic flow} \Rightarrow \frac{A_1}{A^{**}} = 1.38.$$

Here A^{**} is the (imaginary) throat area for the flow behind the shock wave. Next the pressure at the exit must be calculated to see if it matches the $p_e = 1$ atm condition. Knowing that the flow behind the shock wave is subsonic, we estimate the exit Mach number by first evaluating the area ratio and then scanning down the area ratio column in Appendix B to Match A_e/A^{**}:

$$\frac{A_e}{A^{**}} = \frac{A_1}{A^{**}}\frac{A^*}{A_1}\frac{Ae}{A^*} = 1.433; \quad \textit{isentropic flow} \Rightarrow \begin{cases} M_e = 0.45 \\ p_e/p_{02} = 0.870. \\ T_e/T_0 = 0.961 \end{cases}$$

With the area ratio information the exit conditions can be calculated, using the ratios shown to the right of the curly brace. Foremost, the exit pressure is

$$p_e = p_0 \frac{p_{02}}{p_0}\frac{p_e}{p_{02}} = 3.5 \times 0.358 \times 0.870 = 1.09 \text{ atm}.$$

This is very close to the target $p_e = 1$ atm (taking into account the accuracy of the numbers in the Appendices B and C) and we conclude that this is an acceptable iteration for the solution. In reality, after two or three iterations, a curve fit can be used for a more accurate solution.

To calculate the thrust at sea level we need to calculate the exit conditions. The exit temperature (note that the stagnation temperature does not change across the shock wave) is

$$T_e = 0.961 T_0 = 2018 \text{ K}.$$

The exit velocity is then

$$u_e = M\sqrt{\gamma RT} = 0.45\sqrt{1.4 \times 286.6(2100 \times 0.961)} = 404.9 \text{ m/s},$$

and the exit density is

$$\rho_e = \frac{p_e}{RT_e} = \frac{1.09 \times 10^5}{286.6 \times 2100 \times 0.961} = 0.18845 \text{ kg/m}^3.$$

The thrust at sea level is then

$$F_x = 0.1884 \times 404.9^2 \times 16 \times 10^{-4} + (1.09 \times 10^5 - 1.00 \times 10^5)16 \times 10^{-4} = 63.83 \text{ N},$$

and this is significantly less than the previously calculated thrust in space! The conclusion is that, when possible, either matched or underexpanded designs should be used.

10.5.3 The Supersonic Wind Tunnel

A quite popular supersonic wind-tunnel design is based on the blow-down concept, shown in Fig. 10.13. In this case, high-pressure gas is introduced into a large stagnation chamber, from where the flow is accelerated along a converging–diverging

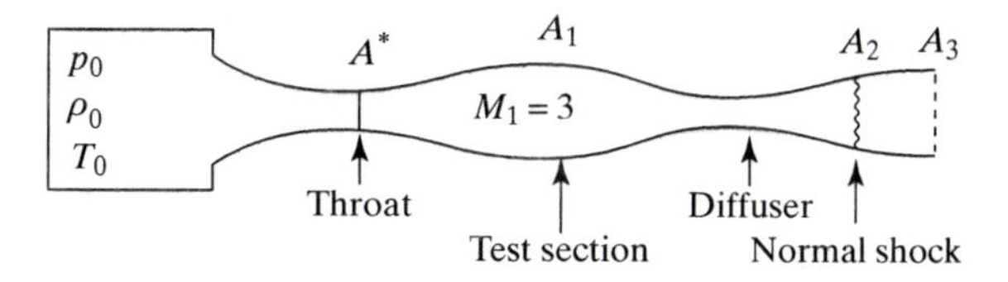

Figure 10.13. Schematic description of a (blow-down) supersonic wind tunnel.

nozzle. Next a nearly constant-area test section follows (see A_1) where the flow is supersonic and various models can be tested. Behind the test section a converging–diverging diffuser is placed, which is supposed to slow down the flow and create an exit pressure, matching the outside condition. If properly designed, no shock wave will be present in the diffuser. However, a slightly off design conditions can create shock waves in the diffuser, as shown. Also note that such wind tunnels operate for only a short time until the pressure in the compressed air reservoir is depleted.

For a numerical example, let us investigate the parameters of such a wind tunnel with a test section speed of $M_1 = 3$. Note that the subscript 1 is used for the test section, subscript 2 for the normal shock section, and subscript 3 for the exit section. The throat-to-test-section cross-section area can be obtained by scrolling down the Mach number column in the supersonic portion of the isentropic flow model (in Appendix B):

$$M_1 = 3.0, \quad \textit{isentropic flow} \Rightarrow \begin{cases} \dfrac{A_1}{A^*} = 4.23 \\ p_1/p_{02} = 0.027. \\ T_1/T_0 = 0.357 \end{cases}$$

Based on these numbers, the temperature drops over 60% and the pressure is only about 3% from the full stagnation pressure value. The throat at the diffuser, in principle, is the same as the throat in the nozzle, but because of boundary-layer and other losses it is slightly larger. Suppose the exit area A_3 is designed to be at atmospheric pressure, and for practical reasons it has the same size as the test section. Because we fixed the exit area ratio by using the subsonic section of the isentropic flow tables in Appendix B, we get

$$\frac{A_3}{A^*} = 4.23, \quad \textit{isentropic flow} \Rightarrow \begin{cases} M_3 = 0.14 \\ p_3/p_0 = 0.986. \\ T_1/T_0 = 0.9961 \end{cases}$$

Based on this, if p_3 (at exit section 3) is atmospheric, only a small amount for pressure is required for driving this wind tunnel. In reality, pressure ratios of larger than 2 are needed to establish sonic speed in the nozzle [Eq. (10.26)] and to overcome friction. As a numerical example, let us estimate the conditions in this wind tunnel for a stagnation pressure of $p_0 = 2$ atm and a temperature of $T_0 = 300\,K$. The pressure in the test section is calculated based on the pressure ratio obtained earlier,

$$p_1 = 0.027 \times 2 = 0.054 \text{ atm},$$

and for an ideal flow across the wind tunnel with a subsonic diffuser the pressure at the exit would be

$$p_3 = 0.986 p_0 = 1.972 \text{ atm}.$$

Of course this is too high, and based on the de Laval nozzle example, it is clear that a shock wave must adjust for the excess pressure. In this case the flow remains supersonic through the diffuser up to section A_2, as shown. Based on Appendix C, a stagnation pressure loss of near 50% is possible only for Mach numbers near 2.65. Therefore the first guess assumes a shock where $M_2 = 2.65$, and based on Appendix C and scrolling down the Mach number column, we find

$$M_2 = 2.65, \quad \textit{normal shock} \Rightarrow \begin{cases} M_{2a} = 0.4996 \\ p_{02}/p_{01} = 0.4416 \end{cases},$$

where M_{2a} is the Mach number in the subsonic side of the normal shock wave at section 2. Next, the location of section A_2 must be identified. Based on the isentropic tables for the supersonic flow (for the left-hand side of the shock) we get (Appendix B):

$$M_2 = 2.65, \quad \textit{isentropic flow} \Rightarrow \frac{A_2}{A^*} = 3.04.$$

Knowing the throat area, we can calculate A_2 as $A_2 = 3.04A^*$. Next, we explore the condition at the same cross section, but on the subsonic side, by using the subsonic tables in Appendix B:

$$M_{2a} = 0.499, \quad \textit{isentropic flow} \Rightarrow \frac{A_2}{A^{**}} = 1.34.$$

Here again A^{**} is the (imaginary) throat area for the flow behind the shock wave. Next the pressure at the exit must be calculated to see if it matches the $p_e = 1$ atm condition. Knowing that the flow behind the shock wave is subsonic, we estimate the exit Mach number by first evaluating the area ratio there, and with this ratio scrolling down the area ratio column in Appendix B, we find:

$$\frac{A_3}{A^{**}} = \frac{A_2}{A^{**}} \frac{A^*}{A_2} \frac{A_3}{A^*} = 1.866, \quad \textit{isentropic flow} \Rightarrow \begin{cases} M_e = 0.33 \\ p_e/p_{02} = 0.9274 \end{cases}.$$

With this information the exit conditions can be calculated. Therefore the exit pressure is

$$p_e = p_0 \frac{p_{02}}{p_0} \frac{p_3}{p_{02}} = 2 \times 0.4416 \times 0.9274 = 0.819 \text{ atm}.$$

This pressure is too low and a weaker shock (with less losses) can increase the exit pressure. Therefore, for the next iteration, a weaker shock (at A_2) is selected. Assuming that the shock there is at $M_2 = 2.44$ and from Appendix C, we get

$$M_2 = 2.44, \quad \textit{normal shock} \Rightarrow \begin{cases} M_2 = 0.5189 \\ p_{02}/p_{01} = 0.5234 \end{cases}.$$

Next the location of the section A_2 must be identified. From the isentropic tables for the supersonic flow (Appendix B), we get

$$M_2 = 2.44, \quad \textit{isentropic flow} \Rightarrow \frac{A_2}{A^*} = 2.493.$$

Knowing the nozzle geometry, we can calculate A_2 as $A_2 = 2.493A^*$. Next, we explore the condition at the same cross section, but on the subsonic side, by using the subsonic tables (Appendix B):

$$M_{2a} = 0.5189, \quad \textit{isentropic flow} \Rightarrow (A_2/A^{**}) = 1.30.$$

Here again A^{**} is the (imaginary) throat area for the flow behind the shock wave. Next the pressure at the exit must be calculated to see if it matches the $p_e = 1$ atm condition. Knowing that the flow behind the shock wave is subsonic, we estimate the exit Mach number by first evaluating the area ratio:

$$\frac{A_3}{A^{**}} = \frac{A_2}{A^{**}}\frac{A^*}{A_2}\frac{A_3}{A^*} = 2.210, \quad \textit{isentropic flow} \Rightarrow \begin{cases} M_e = 0.27 \\ p_e/p_{02} = 0.951 \end{cases}.$$

With this information the exit conditions can be calculated. The exit pressure for this iteration is

$$p_e = p_0 \frac{p_{02}}{p_0}\frac{p_3}{p_{02}} = 2 \times 0.523 \times 0.951 = 0.99 \text{ atm},$$

which is quite close, and we assume that this last iteration represents the condition along this supersonic wind tunnel.

10.6 Effect of Compressibility on External Flows

The 1D ideal-flow models discussed so far in this chapter have demonstrated the effect of compressibility, but mainly internal flow examples were presented. The external compressible flows that are due to speeds near and above Mach 1 are also important, but their analytical treatment is more complex. For completeness, however, a brief discussion follows that shows some basic features of high-Mach-number external flows. As an example, the trends in the drag coefficient of a sphere and of a slender missile configuration versus Mach number are depicted in Fig. 10.14. The change in angle of attack for the case of the missile configuration does not affect the basic variations, which show a large increase near the transonic region.

To gain some insight, particularly into the high subsonic side of the diagram, let us return to the 1920s. As mentioned in Section 8.10, the pioneering group of Prandtl in Göttingen, Germany, was busy laying the foundations for analytical aerodynamics. By this time, airplane propeller tip speeds had begun to reach sonic velocity, experiencing performance losses, as expected from the data in Fig. 10.14. In an effort to understand the phenomenon and to extend the potential flow theory (Section 8.5) into the compressible flow regime, the Prantl-Glauert correction was developed. (Ludwig Prantl, 1875–1953 and Hermann Glauert, a British scientist, 1892–1934.) This correction provided an early explanation for the increase in fluid dynamic forces and is valid for the range of $M < 0.8$ (for small angles of attack and for slender configurations, as in the missile sketched in Fig. 10.14).

The method is based on the ideal-flow model of Section 8.4, but now the high-speed free stream is aligned with the x axis (M is the Mach number of the free stream), and the velocity components (perturbations) in the other directions are

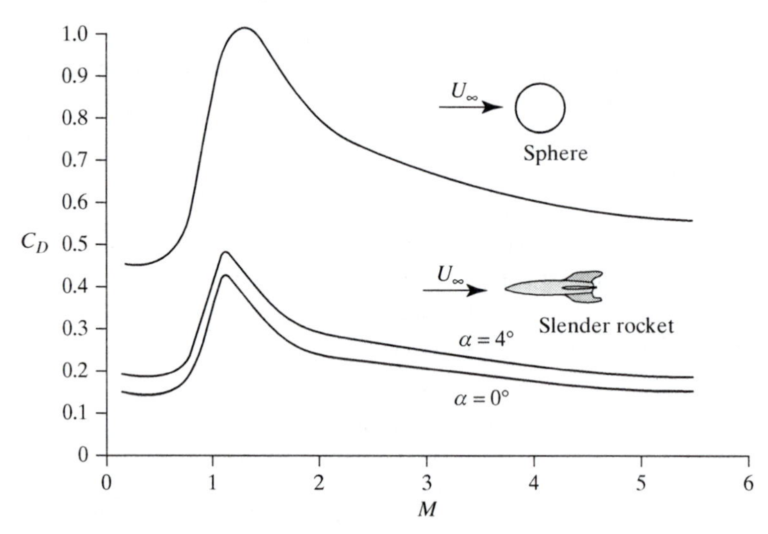

Figure 10.14. Effect of compressibility on the drag coefficient of a sphere and a slender rocket (Re > 10^6).

much smaller. After neglecting the smaller terms, we find that the continuity equation for the perturbations becomes (see [1, Chapter 11)]:

$$(1 - M^2)\frac{\partial u}{\partial x} + \frac{\partial v}{\partial y} + \frac{\partial w}{\partial z} = 0. \tag{10.47}$$

For an irrotational flow a velocity potential Φ can be defined [see Eqs. (8.4)] such that

$$u = \frac{\partial \Phi}{\partial x}, \qquad v = \frac{\partial \Phi}{\partial y}, \qquad w = \frac{\partial \Phi}{\partial z}. \tag{10.48}$$

Substituting these into the continuity equation results in the following expression:

$$(1 - M^2)\frac{\partial^2 \Phi}{\partial x^2} + \frac{\partial^2 \Phi}{\partial y^2} + \frac{\partial^2 \Phi}{\partial z^2} = 0. \tag{10.49}$$

The idea behind the Prandtl–Glauert correction is to define a new set of spatial variables in which the x coordinate is being stretched as the Mach number increases;

$$x' = \frac{x}{\sqrt{1 - M^2}}, \quad y' = y, \quad z' = z. \tag{10.50}$$

By use of these variables, the $(1 - M^2)$ term cancels, and the continuity equation returns to the incompressible flow format:

$$\frac{\partial^2 \Phi}{\partial x'^2} + \frac{\partial^2 \Phi}{\partial y'^2} + \frac{\partial^2 \Phi}{\partial z'^2} = 0. \tag{10.51}$$

The outcome of this analysis is that now we can use the results of incompressible ideal (potential) flow and simply correct the results by using the Prandtl–Glauert correction. For example, if we find the pressure coefficient by using the methods of Chapter 8 (e.g., at $M = 0$), the pressure coefficient at higher Mach numbers is

$$C_p = \frac{C_p(M = 0)}{\sqrt{1 - M^2}}. \tag{10.52}$$

The same correction applies to the integral of the pressures, namely, the lift and the drag coefficients:

$$C_L = \frac{C_L(M = 0)}{\sqrt{1 - M^2}}, \tag{10.53}$$

$$C_D = \frac{C_D(M = 0)}{\sqrt{1 - M^2}}. \tag{10.54}$$

This correction explains the initial increase in the forces as the Mach number increases (see Fig. 10.14) and can be used as an initial approximation as long as shock waves are not present.

EXAMPLE 10.7. EFFECT OF COMPRESSIBILITY ON THE LIFT OF A FLAT PLATE. Consider the case of the flat plate at $M = 0$, as presented in Subsection 8.9.2. The lift coefficient, based on Eq. (8.90), is

$$C_L = 2\pi\alpha.$$

This 2D formula is valid in the subsonic flow range, but for $M = 0.7$ the Prandtl–Glauer correction yields

$$C_L = \frac{2\pi\alpha}{\sqrt{1 - M^2}} = 1.40 \times 2\pi\alpha.$$

This is a significant increase and follows the trends shown in Fig. 10.14.

At this point let us take a closer look at Eqs. (10.52)–(10.54), mainly because at $M = 1$ the lift and drag approach infinity! It was already stated that this correction is not valid close to $M = 1$; however, in the 1930s and even later this was interpreted as the *sound barrier*. Also, if the molecules in a gas move at a velocity close to the speed of sound than they cannot escape and will accumulate near the nose of an airplane flying near or above the speed of sound, effectively creating a "brick wall." This line of thought suggested that supersonic flight is impossible.

Let us move a few years forward, knowing that supersonic flight is a daily event, and try to understand the increase in the drag coefficients near the sonic speed. For the discussion, consider the flow over an airfoil, as depicted in Fig. 10.15. Because the flow accelerates near the airfoil, sonic (and above sonic) speeds can be reached

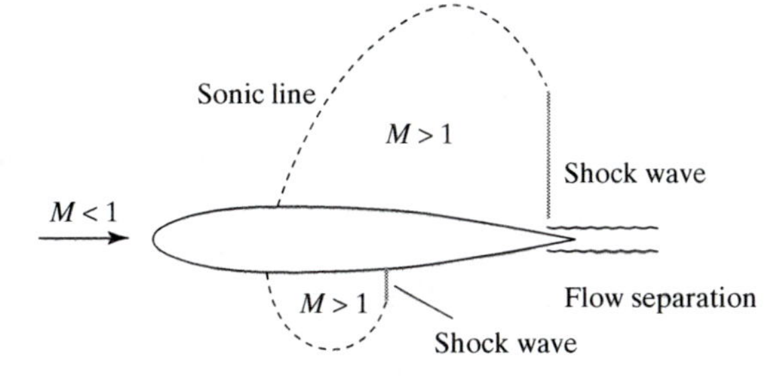

Figure 10.15. Schematic description of the supersonic Mach contour (dashed curve), at speeds slightly above the critical Mach number on a transonic airfoil.

even at subsonic flight speeds. The flight speed where local supersonic flow begins on an airfoil is called the critical Mach number and can be as low as $M = 0.7$ or as high as $M = 0.9$. This region is often called "transonic," as noted earlier, characterized by the sharp increase in drag. It is clear that the speed is higher around a thick airfoil than a thin one, or on those having a sharp suction peak near the leading edge (see Fig. 8.34). Therefore the critical Mach number can be delayed by use of thinner airfoils and by elimination of the sharp suction peaks (these airfoils are sometimes called critical or transonic airfoils). In the case shown in Fig. 10.15, supersonic conditions exist on both the upper and lower surfaces, resulting in shock waves. As the free-stream Mach number increases toward $M = 1$, the shock wave strength increases (more losses) and also flow separation may result behind the shock wave, significantly increasing the drag. Once the flight Mach number increases above $M = 1$, oblique shock waves form and the drag coefficient is reduced (actually following the trend shown by the Prandtl–Glauert correction but using $\sqrt{M^2 - 1}$)). In this higher supersonic range (e.g., $M > 1.2$) the lift coefficient of a flat plate, at small angles of attack, can be estimated by the following formula ([1, Chapter 12]):

$$C_L = \frac{4\alpha}{\sqrt{M^2 - 1}}. \tag{10.55}$$

Also, at supersonic speeds the pressure distribution acts normal to the plate and the drag is the projection into the flight direction (e.g., lift times $\tan\alpha$). Assuming small α, and $\tan\alpha \approx \alpha$, we get:

$$C_D = \frac{4\alpha^2}{\sqrt{M^2 - 1}}. \tag{10.56}$$

This is in contrast to the subsonic case in which the ideal 2D pressure drag (see Subsection 8.9.2) is zero!

As a summary of the preceding discussion, it is obvious that the lift and mainly the drag will increase significantly near transonic conditions. The supersonic drag will be larger than the subsonic value, making commercial supersonic flight considerably more expensive. Also, supersonic airfoils will be thinner than subsonic designs and supersonic wings will be swept. Although this was not discussed, by sweeping the leading edge more than the Mach cone, the flow normal to the airfoil may appear as subsonic, reducing shock wave effects. Therefore supersonic airplanes use highly swept slender wings, and some of the simple formulas [e.g., Eq. (8.104)] may still be applicable.

10.7 Concluding Remarks

This chapter served to demonstrate certain effects of compressibility. The first effect discussed is the speed of sound, which indicates that information, such as a weak pressure perturbation, is traveling at a finite speed in a fluid. If a vehicle speed exceeds the speed of sound, perturbations such as the sound cannot reach certain areas in the fluid.

The 1D isentropic flow model developed provided information on temperature and pressure changes in fast-moving flows. Also, a converging subsonic nozzle may turn into a diffuser in a supersonic flow. The 1D shock wave model, which is non-reversible, introduced the concept of strong discontinuities in high-speed flows and the resulting stagnation pressure losses.

Finally, the effect of compressibility in external flows was addressed and the sharp increase in drag near the transonic region was discussed.

REFERENCE

[1] Anderson, J. D. Jr., *Fundamentals of Aerodynamics*, 3rd ed., McGraw-Hill, New York, 2001.

PROBLEMS

10.1. A commercial airplane is designed to cruise at a Mach number of $M = 0.8$. Calculate its actual speed at sea level ($p = 1$ atm, $T = 300$ K), at an altitude of 5 km ($p = 0.53$ atm, $T = 255$ K), and at an altitude of 10 km ($p = 0.26$ atm, $T = 223$ K).

10.2. Airplanes have many aluminum parts on them, including most of the skin. If pure aluminum melts at 933 K, then calculate the speed (and Mach number) at which this temperature is reached near the stagnation points on the airplane. Use the atmospheric conditions at an altitude of 5 km, where $p = 0.53$ atm and $T = 255$ K.

10.3. Some of the fastest airplanes can cruise at a Mach number of 3. If such an airplane flies at an altitude of 10 km ($p = 0.26$ atm, $T = 223$ K) then calculate the stagnation temperature along the wing leading edges.

10.4. A reentry vehicle gliding down from its orbit reaches an altitude of 30 km at a Mach number of 5. If the ambient temperature at this altitude is $T = 226$ K, then calculate the stagnation temperature (near the nose and the leading edges).

10.5. A supersonic airplane flies horizontally at an altitude of 7 km and at a Mach number of 2. A bystander looks up and sees the airplane exactly above him. How long does it take (in seconds) for the bystander to hear the airplane's noise and how far away is the airplane at this moment? Also calculate the Mach angle μ (assume that the average speed of sound is 320 m/s).

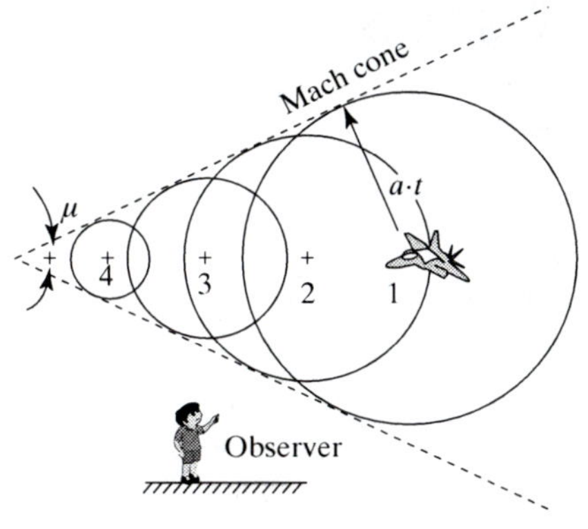

Problem 10.5.

10.6. An airplanes flies at an altitude of 5 km where the atmospheric conditions are $p = 0.53$ atm and $T = 255$ K. By the time the sonic shock reaches an observer's ears on the ground, the airplane reached a distance of 10 km to the left. Calculate the Mach number, Mach angle, and airplane speed (assume that the speed of sound is not changing much as it travels down from an altitude of 5 km).

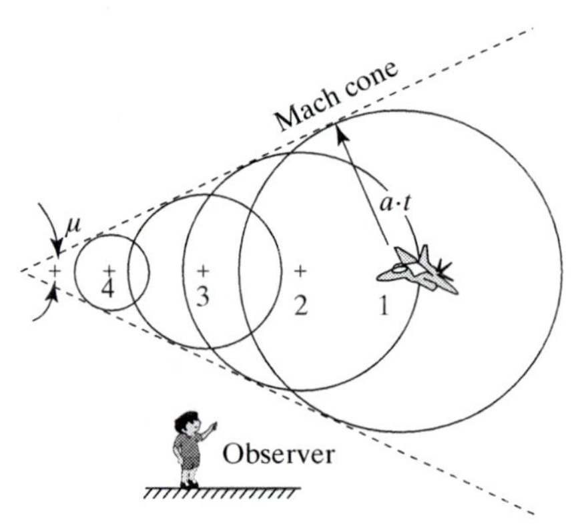

Problem 10.6.

10.7. The performance of a nozzle with an area ratio of $A_e/A_t = 6.5$ is investigated. The stagnation chamber conditions are $p_0 = 50$ atm, $\rho_0 = 55$ kg/m^3, and $T_0 = 300$ K. If the throat area is $A_t = 2$ cm^2 and the ambient pressure is $p_a = 1$ atm, then calculate the nozzle exit velocity and the thrust of this unit ($\gamma = 1.4$, $R = 287$ m^2/s^2 K).

10.8. A 50-L container (as shown in Fig. 10.6) is filled with air at a pressure of 40 atm and a temperature of 300 K. Suddenly a small nozzle with a throat area of 1 cm^2 is opened, allowing the air to flow outside (where the pressure is 1 atm). Calculate the variation of the pressure with time inside the tank and the axial force it creates ($\gamma = 1.4$, $R = 287$ m^2/s^2 K).

10.9. A 50-L container (as shown in Fig. 10.6) is filled with air at a pressure of 50 atm and at a temperature of 300 K. Suddenly a small nozzle with a throat area of 2 cm^2 is opened, allowing the air to flow outside (where the pressure is 1 atm). How long it will take for the pressure in the tank to reach 25 atm ($\gamma = 1.4$, $R = 287$ m^2/s^2 K).

10.10. An airplane flies at $M = 1.42$ and a normal shock wave is formed ahead of the engine inlet lip. Calculate the capture-area-to-throat ratio A_a/A_t, the stagnation pressure loss, and the Mach number behind the shock wave.

10.11. Compare the static thrust generated by two nozzle geometries. The stagnation conditions ahead of the nozzle are $p_0 = 25$ atm, $T_0 = 2000$ K, and at the exit $p_a = 1$ atm. The first design is a converging nozzle, with $A_t = 0.05$ m^2, and the second is a converging–diverging nozzle, having the same A_t, and $A_e/A_t = 2.5$.

10.12. Calculate the thrust generated by compressed air blowing through a converging nozzle into the atmosphere where $p_a = 1$ atm. The stagnation conditions ahead of the nozzle are $p_0 = 30$ atm and $T_0 = 300$ K, and the converging nozzle throat area is $A^* = 3$ cm^2 ($\gamma = 1.4$, $R = 287$ m^2/s^2 K).

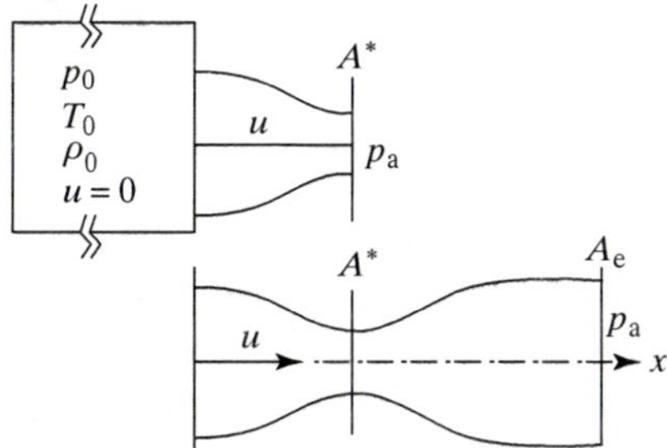

Problem 10.12.

10.13. Calculate the thrust generated by the previous compressed air nozzle, but now a diverging section is added, as shown. Exit and stagnation conditions are, as before, $p_a = 1$ atm, $p_0 = 30$ atm, and $T_0 = 300$ K. The converging nozzle area is A$^* = 3$ cm^2 and the diverging nozzle area ratio is $A_e/A^* = 3$ ($\gamma = 1.4$, $R = 287$ m^2/s^2 K).

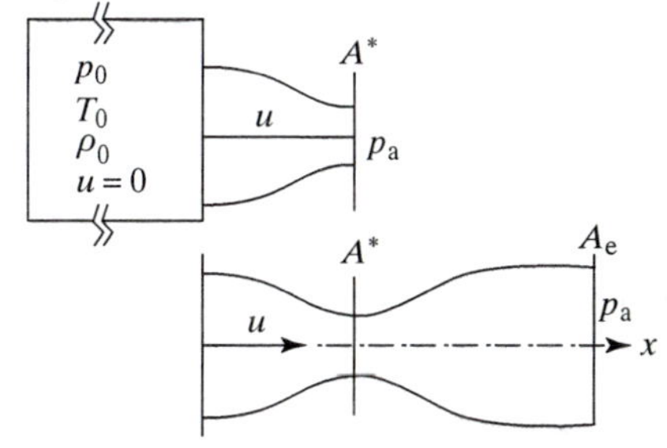

Problem 10.13.

10.14. A thrust chamber is tested statically by use of a high-pressure air supply at $p_0 = 50$ atm ($T_0 = 300$ K, $\rho_0 = 50$ kg/m^3). Calculate the thrust if $A_e/A_t = 5.5$ and $A_t = 0.0005$ m^2, $\gamma = 1.4$, $R = 286.78$ m/s K, $p_a = 0.9$ atm.

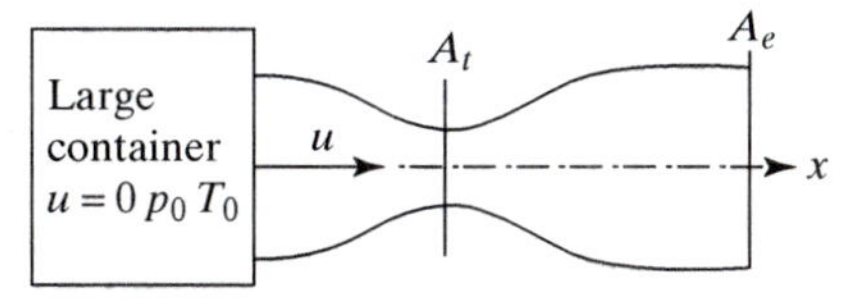

Problem 10.14.

10.15. The stagnation conditions inside a large container are $p_0 = 1.5$ atm, $\rho_0 = 1.82$ kg/m^3, $T_0 = 300$ K, and the ambient pressure is $p_a = 1$ atm. If the nozzle throat area is $A^* = 3$ cm^2 and the diverging nozzle area ratio is $A_e/A^* = 2$, calculate the exit Mach number and the thrust of this nozzle.

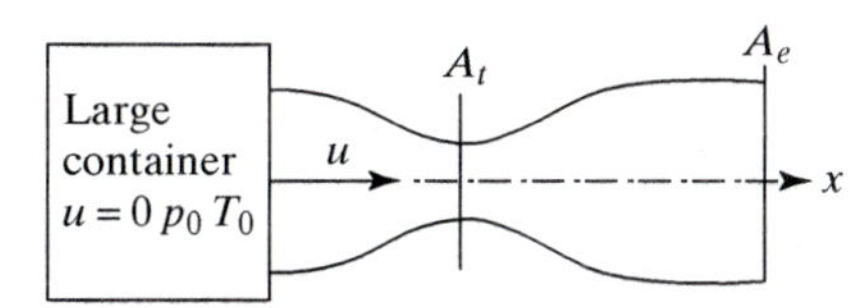

Problem 10.15.

10.16. Repeat the previous problem but now the stagnation conditions inside a large container are $p_0 = 15$ atm, $\rho_0 = 18.30$ kg/m^3, $T_0 = 300$ K, and the ambient pressure is $p_a = 1$ atm. If the nozzle throat area is $A^* = 3$ cm^2 and the diverging nozzle area ratio is $A_e/A^* = 2$, calculate the exit Mach number and the thrust of this nozzle.

10.17. The thrust of a rocket engine is a result of the pressure distribution on the inner and outer surfaces of the thrust chamber, as shown in the figure. It is logical to assume that the thrust is close to the open area times the internal pressure (e.g., thrust $= p_0 \times A_t$). In fact, the thrust coefficient C_T is defined as thrust $= C_T \times p_0 \times A_t$. Consider a case in which $p_0 = 50$ atm, $T_0 = 2000$ K, $p_a = 1$ atm, $A_t = 0.0004$ m^2, and throat-to-exit-area ratio is $A_t/A_e = 0.5$ (assuming $\gamma = 1.4$). Calculate the thrust and the thrust coefficient.

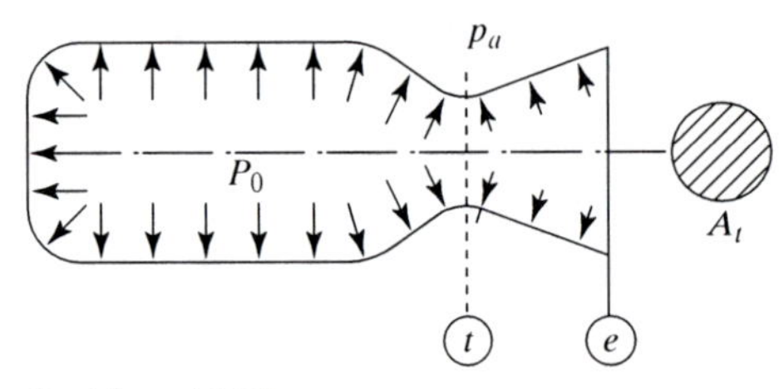

Problem 10.17.

10.18. Calculate the thrust and the thrust coefficient for the previous problem, but now the nozzle is converging only (so the exit area is equal to 0.0004 m^2). Is the thrust coefficient still larger than 1.0?

10.19. The pressure inside the thrust chamber of a rocket engine is 100 atm and the combustion temperature is $T_0 = 2000$ K. The nozzle-throat-to-exit-area ratio is $A^*/A_e = 0.1342$ and the exit area is $A_e = 0.02$ m^2. Assuming $\gamma = 1.4$ estimate the thrust at sea level where $p_a = 1$ atm and the thrust in space.

10.20. An airplane is cruising at $M_a = 0.8$ where the outside pressure is 0.7 atm and its inlet geometry is shown schematically. The inlet area ratio is $A_2/A_1 = 1.3$ and it is slowing down the airspeed to $M_2 = 0.4$ at station 2.

(a) Calculate the pressure at station 2.
(b) What is the Mach number at station 1?
(c) Calculate the area ratio A_2/A_a.

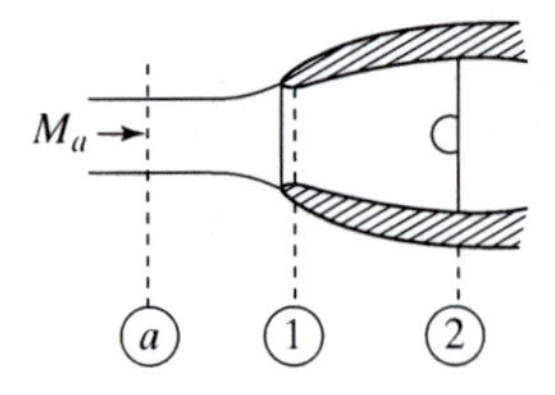

Problem 10.20.

10.21. An airplane is cruising at $M_a = 0.9$, where the outside pressure is 0.8 atm and its inlet geometry is shown schematically. The inlet area ratio is $A_2/A_1 = 1.3$ and it is slowing down the airspeed to $M_2 = 0.4$ at station 2.

(a) Calculate the pressure at station 2.
(b) What is the Mach number at station 1?
(c) Calculate the area ratio A_2/A_a.

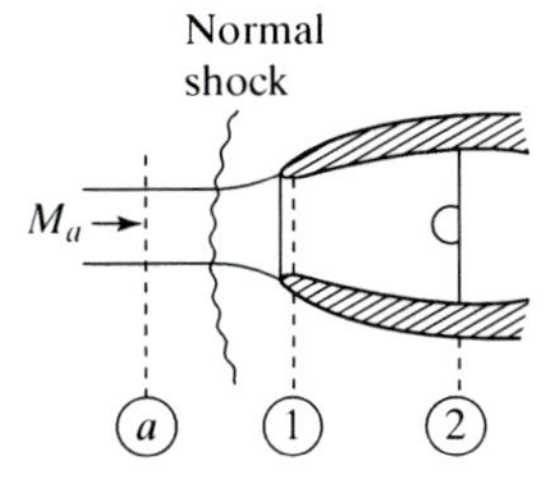

Problem 10.22.

10.22. An airplane is cruising at $M_a = 1.42$ and a normal shock is formed ahead of the inlet. If the inlet area ratio is $A_2/A_1 = 1.24$, calculate the following values:

(a) The area ratio A_1/A_a.
(b) The stagnation pressure loss.
(c) The Mach number M_2 at station 2.

10.23. An airplane is cruising at $M_a = 1.6$ and a detached normal shock is formed ahead of the inlet. Ambient conditions are $p_a = 0.25$ atm and $T = 230$ K. If the inlet's smallest cross-section area (at station 1) is 300 cm^2 calculate the mass flow rate for this case ($\gamma = 1.4$). Estimate the capture area A_a in the figure for Problem 10.22.

10.24. In the early days of jet airplanes, a converging diverging inlet was proposed for airplanes flying faster than the speed of sound. Consider the case shown in the figure in which the flight speed is $M = 1.42$ and the area ratio is $A_i/A_t = 1.24$. Assuming a normal shock wave will form ahead of the inlet, calculate the stagnation pressure loss, the area ratio A_a/A_t, and the inlet Mach number M_i.

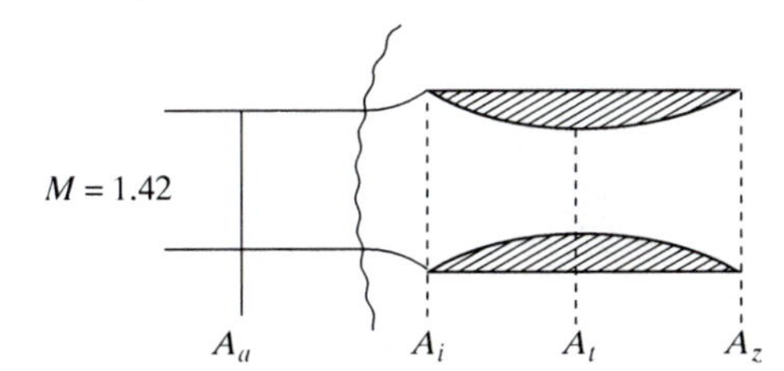

Problem 10.24.

10.25. At what speed the detached shock in the figure for Problem 10.24 will reach the inlet lip (just before the shock will be swallowed)?

10.26. The capture area ratio of the cooling air inlet on a supersonic jet (see previous figure) is $A_a/A_t = 1.2$. Calculate the flight Mach number and the stagnation pressure loss that is due to the shock wave.

10.27. A supersonic airplane is cruising at a Mach number of 3 at an altitude of 10 km ($p = 0.26$ atm, $T = 223$ K) and a normal shock is present ahead of its nose. Calculate the stagnation temperature and pressure behind the shock wave.

10.28. An airplane flies at a supersonic speed and a normal shock is present at the lip of the converging diverging diffuser. If the Mach number ahead of the compressor is $M_2 = 0.4$, then find the area ratio A_t/A_2. Also, if $A_1 = 0.75\,A_2$, calculate the flight Mach number and the stagnation pressure ratio ahead of the compressor p_2/p_{0a}.

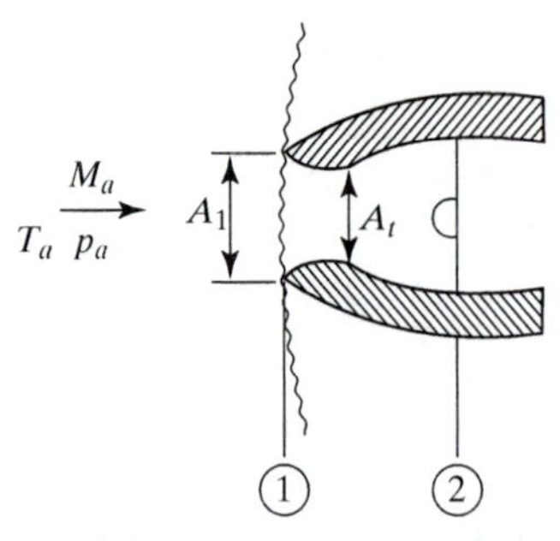

Problem 10.28.

10.29. The area ratios for the diffuser described in the previous problem are $A_t/A_1 = 0.85$ and $A_t/A_2 = 0.63$. Calculate the flight Mach number and the stagnation pressure ratio ahead of the compressor p_2/p_{0a}.

10.30. Calculate the lift (per unit width) of a 1-m-long flat plate at a 1° angle of attack and at a Mach number of 0.8 ($\rho = 1.22$, $T = 300$ K, and $\gamma = 1.4$).

10.31. Calculate the lift and drag (per unit width) of a 1-m-long flat plate at a 1° angle of attack and at a Mach number of 1.8 ($\rho = 1.22$, $T = 300$ K, and $\gamma = 1.4$).

11 Fluid Machinery

11.1 Introduction

The need to channel water flow and other fluids must have originated in the early civilizations, and one of the better-known inventions, the Archimedes screw, dates back to the third century B.C.E. The Archimedes screw, or screwpump, shown in Fig. 11.1, was used to transfer water from lower reservoirs into higher irrigation ditches.

Over the years, many inventions focused on developing various machines either for pumping fluids or using fluid energy to drive other machinery (e.g., turbines). These machines may be classified as positive-displacement or continuous-flow machines. Some mechanical solutions using these two types of hardware are shown schematically in Fig. 11.2.

For example, the most basic configuration is a piston sliding inside a cylinder, as shown in Fig. 11.2(a). If this schematic is considered a pump, then while the piston is moving to the left it is pushing the fluid out of the cylinder through an open valve. The pumping operation of fresh fluid can continue when the piston is moving backward, closing this (exhaust) valve and opening the intake valve, creating a reciprocating cycle. This type of machine is called a positive-displacement machine because a fixed volume of fluid is captured in the cylinder and then transferred across the pump. Another example is the rotating-gear pump, shown in Fig. 11.2(b). This is also a positive-displacement machine, because there are fixed volumes of fluid between the outer wall and the gears. Figures 11.2(c) and 11.2(d) show two types of continuous-flow machines; the first is an axial fan, and the second is a centrifugal compressor. In this type of machinery, the fluid is not contained inside an enclosed volume and, for example, pressure cannot be maintained if the machine is not running (whereas a cylinder can hold a compressed fluid). Also note that any of these machine principles can be used as a pump or as an engine or turbine.

The difference in the performance between these two types of mechanical solutions is demonstrated schematically in Fig. 11.3. Assume that the positive-displacement pump of Fig. 11.2(a) is delivering an incompressible fluid through a pipe, which has a valve on it. When the valve is partially closed, the pressure increases, but the volume of the flow (per cycle) remains constant. This is shown schematically by the vertical line in Fig. 11.3. In continuous-flow pumps, however,

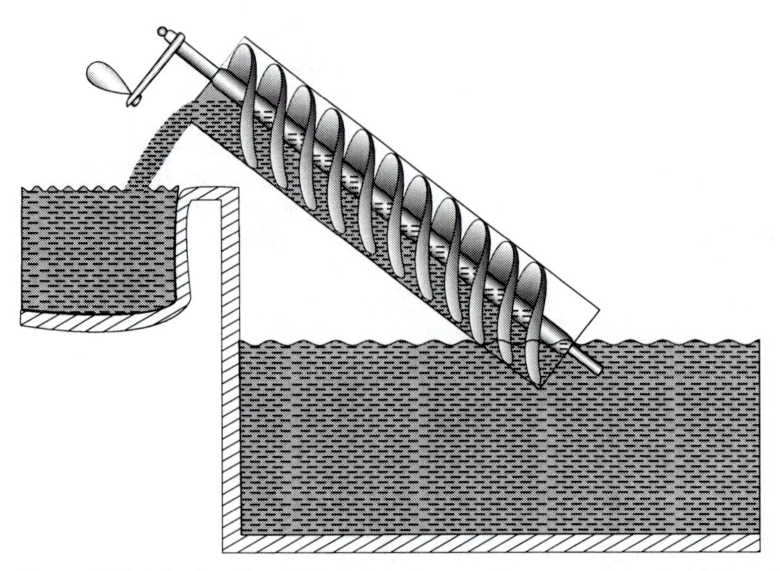

Figure 11.1. The Archimedes screwpump used to pump water to higher elevations.

when a rotor moves the fluid, the rotor can rotate even when the valve stops the flow entirely. In such pumps usually the mass flow rate will increase when the restrictor valve is opened, resulting in the other type of curves shown in Fig. 11.3. In the case of an axial pump, as shown in Fig. 11.2(c), the pressure decreases with increasing mass flow rate, whereas with the centrifugal pump design of (Fig. 11.2(d) the

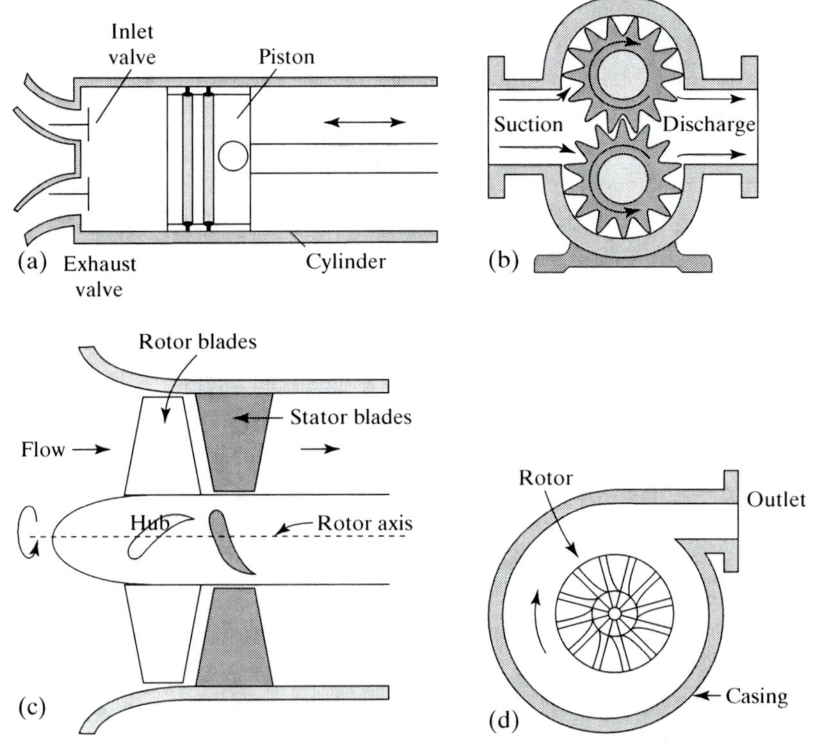

Figure 11.2. Various types of fluid machinery: (a), (b) Positive displacement; and (c), (d) continuous flow.

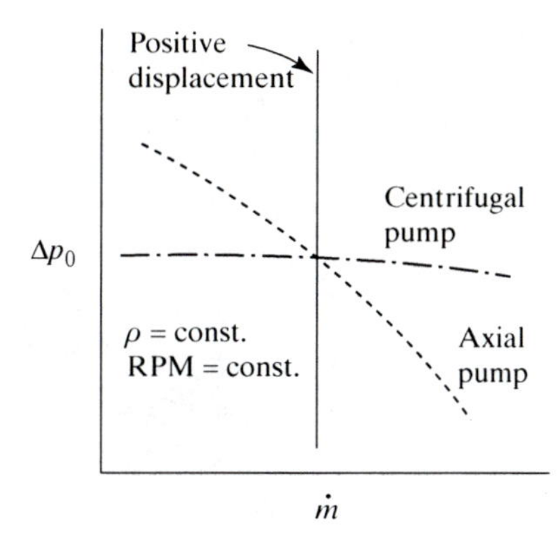

Figure 11.3. Schematic description of the performance difference between positive-displacement and continuous-flow pumps.

pressure changes are much smaller. Actually, the centrifugal pump curve depends on the impeller shape, as explained in Subsection 11.4.2. These types of continuous-flow machines require deeper understanding of the fluid dynamic design and are discussed next.

The objective of this chapter is to provide an introductory level formulation to estimate the power requirements, the pressure ratios, and some of the fluid mechanic principles needed to design such a machine. Consequently ideal-flow models with constant properties are used in the following formulations. This need to estimate fluid pumping power was addressed briefly in Chapter 5 when the power requirements for an elementary pump were discussed. Equation (5.109) was derived for an incompressible fluid pump for which the work W is simply the pressure difference times the volume of the liquid:

$$W = \Delta p \; V. \tag{5.109}$$

If the conditions inside a moving cylinder are changing, then this can be expressed in terms of the integral:

$$W = \int_{\text{cycle}} p dV. \tag{11.1}$$

From the thermodynamic point of view, the work in general is related to the change in stagnation enthalpy, h_0, which includes the internal energy as well. Consequently we can write for an incompressible fluid (and recall that $m/\rho = V$)

$$W = m\Delta h_0 \approx m\frac{\Delta p_0}{\rho} = V\Delta p_0, \tag{11.2}$$

which is the same result obtained from mechanical considerations leading to Eq. (5.109). In the case of a compressible fluid, the change in enthalpy can be estimated by $\Delta h_0 \approx c_p \Delta T_0$, and the work of the machine is then

$$W = m\Delta h_0 \approx m \; c_p \Delta T_0, \tag{11.3}$$

where c_p is the heat capacity, as defined in Eq. (1.17) (we also assume that c_p is constant). Note that the fluid stagnation–enthalpy includes the energy of the moving fluid; hence, for an incompressible fluid having a velocity q, the stagnation pressure is

$$\frac{p_0}{\rho} = \frac{p}{\rho} + \frac{q^2}{2}, \tag{11.4}$$

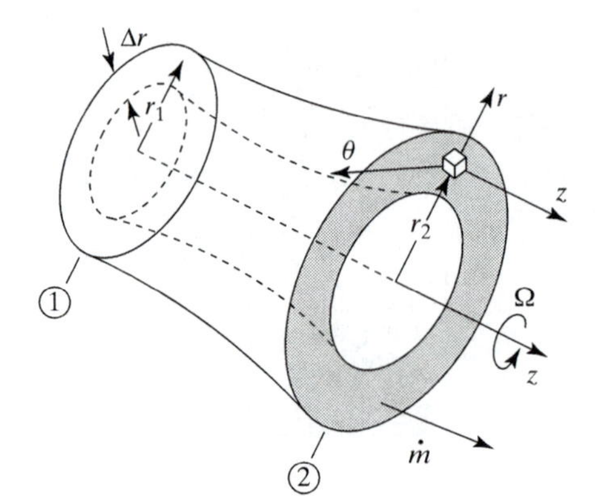

Figure 11.4. An axisymmetric control volume for calculating the torque and power of rotating turbomachinery.

and for a compressible fluid the stagnation temperature is

$$c_p T_0 = c_p T + \frac{q^2}{2}. \tag{11.5}$$

As noted earlier, this chapter focuses on the continuous-flow design often called *turbomachinery*. The basic approach is developed first for an axial compressor and then extended to centrifugal compressors and axial turbines. To estimate machine performance, the relation between the fluid flow and power requirement must be established. For a successful design the effects of component geometry on performance must be clarified as well. (Note that propellers and wind turbines are not discussed here because of the difficulty in modeling blade-tip effects – similar 3D wingtip effects were discussed briefly in Subsection 8.10.5).

The first-order approach presented in the following sections is based on simplifications such as the average radius and the steady flow assumption. More accurate models are beyond the scope of this text and require complex numerical solutions. However, in spite of the simplicity, the relation between turbomachinery geometry and its performance is established, providing a satisfactory preliminary design tool.

11.2 Work of a Continuous-Flow Machine

The first task is to develop a relation between the work of a rotating shaft and fluid motion. Let us consider the axisymmetric control volume shown in Fig. 11.4. The fluid enters at ring-shaped section 1 and leaves at section 2. Note that the flow is continuous across the control volume and there are no changes with time. However, a particle entering at one side will experience accelerations because of the changes between the two stations (1 and 2). The cylindrical coordinate system is placed on the rotating axis (z) and the r–θ coordinates are as shown in the figure. A particle of mass Δm can enter the control volume at section 1 and leave at section 2 (one particle is shown leaving at section 2), and its velocity components $\vec{c}$ are also depicted in the figure.

This velocity vector is measured in an inertial frame of reference, which is not rotating with the shaft:

$$\vec{c} = (c_r, c_\theta, c_z). \tag{11.6}$$

To calculate the tangential force (in the θ direction) we can use the momentum principle,

$$\Delta F_\theta = \frac{d}{dt}(\Delta m\, c_\theta),$$

and the torque Tq required for accelerating this particle is

$$\Delta Tq = r\,\Delta F_\theta = r\frac{d}{dt}(\Delta m\, c_\theta).$$

The unit normal vectors $\vec{n}$ of the inlet–exit areas (of the control volume) point in the $\pm z$ direction, and based on Eq. (2.19) the fluid flow rate entering or leaving the control volume is

$$\frac{d}{dt}\Delta m \equiv \Delta \dot{m} = \rho(\vec{q}\cdot\vec{n})dS = \rho c_z dS.$$

As noted, the fluid exchange takes place only through the two ring-shaped surfaces at stations 1 and 2 and if Δr is small c_θ is constant at each station. To calculate the total torque applied to the fluid in the control volume, the contributions of the mass elements Δm must be added by integrating over the control surface (c.s.):

$$Tq = \int_{\text{cs}} \Delta Tq\, dS = \int_{\text{cs}} r\, c_\theta\, \rho c_z dS = \int_{s2} (r\, c_\theta)\rho c_z dS - \int_{s1} (r\, c_\theta)\rho c_z dS. \tag{11.7}$$

Next let us introduce the mean-radius approximation. This is equivalent to stating that the ring-shaped inlet and exit surfaces are very thin and the radial variations in the velocity of the fluid elements are negligible. Consequently all particles at station 1 are considered as entering as $r = r_1$ and having a tangential velocity of $c_{\theta 1}$. Assuming the same for the exit (station 2) we get

$$Tq = r_2\, c_{\theta 2}\int_{s2} \rho c_z dS - r_1\, c_{\theta 1}\int_{s1} \rho c_z dS = \dot{m}(r_2\, c_{\theta 2} - r_1\, c_{\theta 1}), \tag{11.8}$$

and of course the mass flow rate entering and leaving the control volume is the same:

$$\int_{s1} \rho c_z dS = \int_{s2} \rho c_z dS = \dot{m}.$$

For cases with significant radial variations, we can use Eq. (11.8) by creating sublayers inside the control volume (see later Subection 11.3.3). The power P required is simply the product of torque times the rotation speed.

$$P = Tq\cdot\Omega = \dot{m}\Omega(r_2\, c_{\theta 2} - r_1\, c_{\theta 1}). \tag{11.9}$$

It is convenient to define a tip velocity U such that

$$U \equiv r\Omega, \tag{11.10}$$

and with this definition the power equation becomes

$$P = \dot{m}(U_2\, c_{\theta 2} - U_1\, c_{\theta 1}) \tag{11.11}$$

and the work per unit mass flow w_c is

$$w_c = (P/\dot{m}) = (U_2\, c_{\theta 2} - U_1\, c_{\theta 1}). \tag{11.12}$$

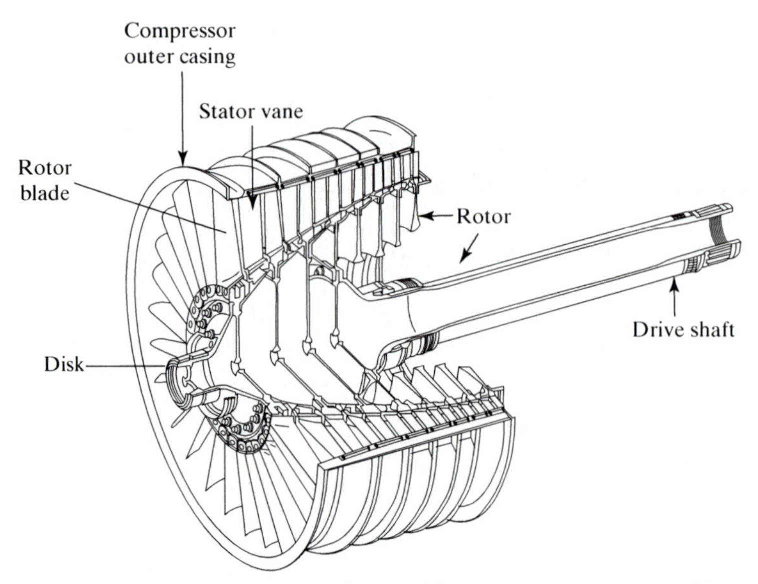

Figure 11.5. Schematics of a typical multistage compressor assembly.

Equations (11.8)–(11.12) state that the torque and the work are related to the change in angular momentum. Because the focus of this book is on fluid mechanics, the next step is to relate these quantities to the internal geometry of the turbomachinery.

11.3 Axial Compressors and Pumps (The Mean-Radius Model)

The objective of this section is to formulate a preliminary design approach for an axial compressor or pump geometry, along with the ability to estimate the compression ratio and the required power. This method can be extended later to study the performance of centrifugal compressors and axial turbines. Because the discussion is now focused on continuous-flow machinery, the fluid mechanics of the inner components (e.g., rotor and stator blades) must be clarified. The cross section of a typical multistage axial compressor is shown in Fig. 11.5. It consists of a rotating shaft with airfoil-shaped rotor blades attached to a central hub. In between each row of rotors there is a row of airfoil-shaped stator vanes attached to the outer casing – which are stationary. Naturally the flow is highly unsteady, mainly because of the multiple blades in each rotor and stator row. In the following model, however, a steady flow model is assumed in which, for example, the tangential velocity c_θ is not changing tangentially (when $z = \text{const.}$).

A schematic cross section of this multistage compressor is shown in Fig. 11.6. For example, the entrance to one of the rotor stages is marked as station 1 in Fig. 11.6, and the exit, as station 2 (which is also the entrance to the stator behind it). A typical axial compressor stage consists of a combination of a rotor and stator. Thus the rotor–stator assembly between sections 1 and 3 can be considered a typical stage. Similarly, the stage between sections 3 and 5 (and so on) can also be considered a typical stage.

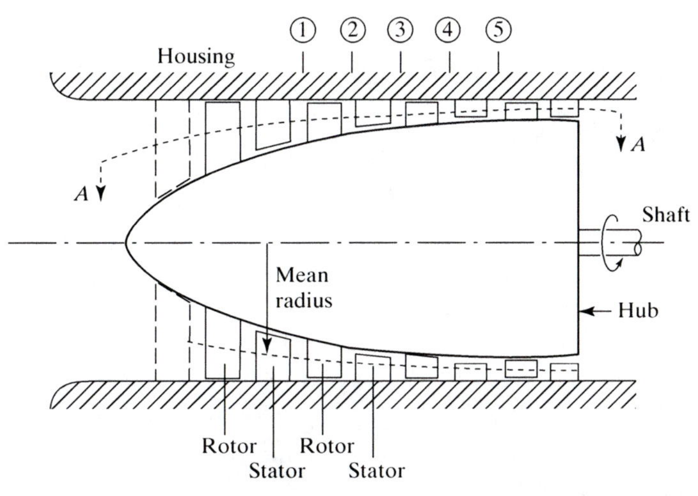

Figure 11.6. Cross section of a typical axial compressor assembly. Note section AA cuts at the average radius at each stage.

To relate to the "mean-radius" approach of the previous section, let us create a cut (through section AA) looking down from the top, as shown in Fig. 11.6. This view is described schematically in Fig. 11.7, showing stations 1–5, and now the airfoil shape of both rotor and stator blades is visible. Because the cut is at a constant radius, the 2D $z - \theta$ coordinate system is also shown in this figure. Our viewpoint is stationary and therefore the rotor blades appear to move at a velocity of $U = r\Omega$ into the tangential direction. For proper operation, the flow direction must be reasonably aligned with the airfoil-shaped blades of both the rotor and stator, but this is not clearly visible at this point. Consequently the velocity vectors at each station must be identified.

For example, we can define a velocity vector $\vec{w}$ in a frame of reference attached to the rotor blade; its velocity components are

$$\vec{w} = (w_r, w_\theta, w_z). \tag{11.13}$$

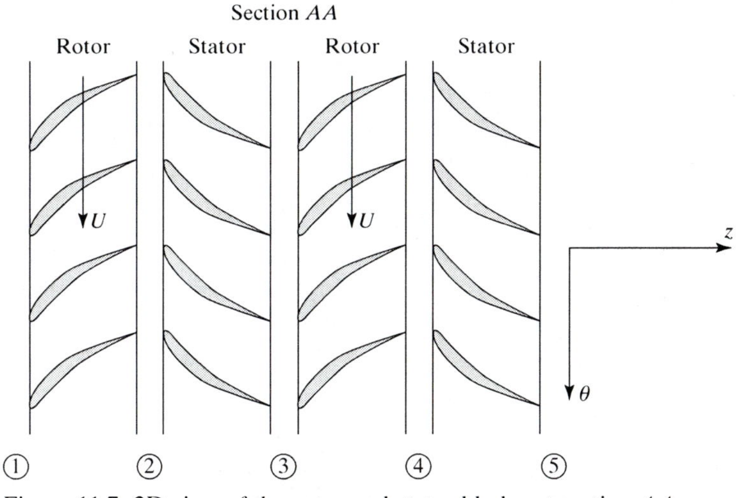

Figure 11.7. 2D view of the rotor and stator blades at section AA.

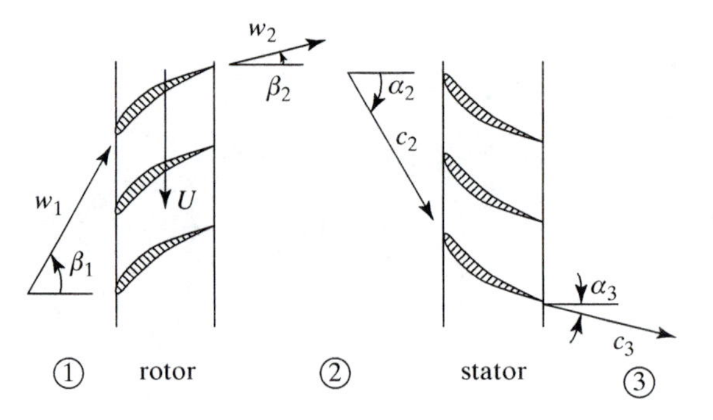

Figure 11.8. Velocity vectors entering and leaving the rotor–stator blades.

Most textbooks on turbomachinery use the symbol $\vec{w}$ for the rotor velocity (not to be confused with the velocity in the z direction, as used in this book). Consequently this nomenclature is used in this chapter only!

Returning to the geometry in Fig. 11.7 and assuming attached flow over the rotor and stator blades, we can sketch the schematic diagram of Fig. 11.8.

It is expected that the velocity vector w_1 ahead of the rotor blade will have a moderate angle of attack and the exit velocity w_2 is parallel to the trailing-edge bisector for an attached flow (as shown in Fig. 11.8). The same is assumed for the stationary stator, but now the velocity $\vec{c}$ is measured at the inertial frame (because the stator is not rotating). Note the definition of the inlet and exit angles α and β, as shown in Fig. 11.8. Next, let us think about an experiment in which we seed the flow with a visible tracer so that when we are looking at section AA (Fig. 11.7) the velocity vector $\vec{c}$ (in the inertial frame) becomes visible. This is described in Fig. 11.9, where the velocity c_1 leaves parallel to the trailing edge of the previous stator. Observe the 2D coordinate system used (because $r =$ const.), where z points into the horizontal direction and the tangential coordinate points down. Once the fluid passed through the rotor rows the particles must have increased their tangential velocity, and their exit velocity c_2 is pointing more in the tangential direction. To view this velocity from the frame of reference attached to the rotor (as shown on the left-hand side of Fig. 11.8), the velocity diagram shown in Fig. 11.9 is constructed. This is based on the kinematic transformation between the velocity $\vec{w}$ measured in

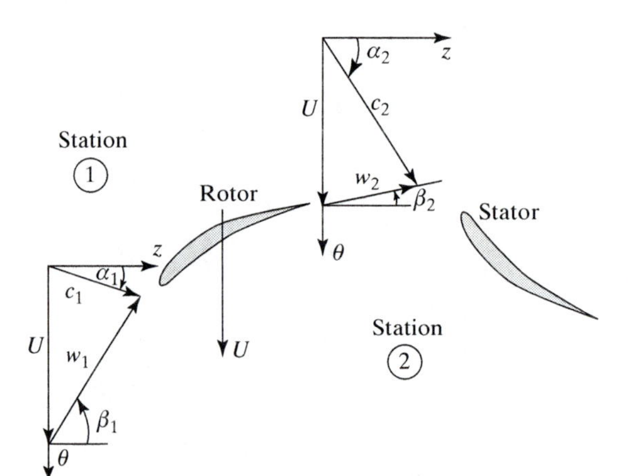

Figure 11.9. Velocity triangles ahead and behind the rotor.

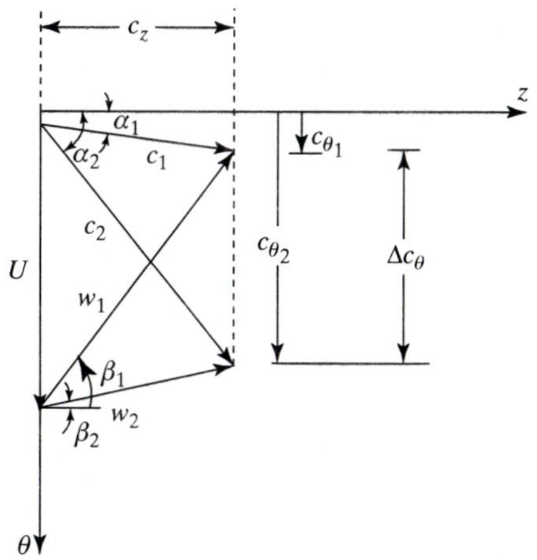

Figure 11.10. The average-radius velocity triangle for one rotor stage (between stations 1 and 2).

a moving frame of reference, and the velocity $\vec{c}$ measured in the stationary (inertial) frame of reference:

$$\vec{c} = \vec{q}_{\text{origin}} + \vec{w}.$$

Therefore $\vec{q}_{\text{origin}}$ is the velocity of the origin of the frame of reference where $\vec{w}$ is measured (and $\vec{w}$ is the relative velocity in this frame). The velocity of the origin is

$$\vec{q}_{\text{origin}} = (0, r\Omega, 0).$$

But the tip speed was already defined in Eq. (11.10), and based on this, the origin's velocity vector is defined as

$$\vec{U} = (0, r\Omega, 0) \tag{11.14}$$

and the transformation of the velocities between the two frames of reference becomes

$$\vec{c} = \vec{U} + \vec{w}. \tag{11.15}$$

This formula allows the construction of the velocity diagrams in the moving rotor frame of reference and the observation of the proper flow angles β, as shown in Figs. 11.8 and 11.9. Consequently, when a compressor rotor is designed, the blade orientation can be properly aligned to ensure desirable performance (and to avoid blade stall). Note that the direction of U points down but, relative to the rotating blade, its direction is reversed.

11.3.1 Velocity Triangles

Based on Eq. (11.15), the rotor velocity triangle shown in Fig. 11.10 can be drawn. In this case both the incoming (station 1) and exiting (station 2) velocity diagrams are superimposed on a z–θ coordinate system. It is assumed that the axial velocity is not changing much so c_z is the same for incoming and exiting flows. Note that, to calculate the power in Eq. (11.11), the change in tangential velocity is required. Therefore $c_{\theta 1}$ and $c_{\theta 2}$ are shown in the figure, basically allowing the calculation of the power.

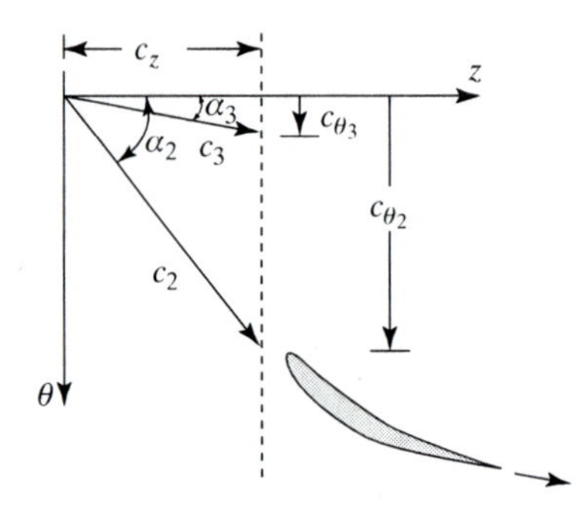

Figure 11.11. Stator velocity vectors of the incoming (station 2) and exiting flows (at station 3). Note that c_3 is parallel to the stator trailing edge.

The torque on the rotor stage is then calculated by use of Eq. (11.8), assuming no change in the radius across the rotor blade (for axial design only),

$$Tq = \dot{m}r(c_{\theta 2} - c_{\theta 1}) = \dot{m}r\,\Delta c_\theta, \tag{11.16}$$

and the power required for driving the rotor is calculated with Eq. (11.11),

$$P = \dot{m}U(c_{\theta 2} - c_{\theta 1}) = \dot{m}U\Delta c_\theta. \tag{11.17}$$

A similar representation of the velocity vectors for the stator (between stations 2 and 3) is shown in Fig. 11.11. Because the stator is not moving, both vectors are viewed in the stationary frame of reference. The tangential velocity components are shown in the figure and no change in the axial velocity c_z is assumed here as well.

The torque on the stator can be calculated with Eq. (11.8), and it is similar to the formulation for the rotor,

$$Tq = \dot{m}r(c_{\theta 3} - c_{\theta 2}), \tag{11.18}$$

and because the stator is not rotating, $U = 0$. Based on Eq. (11.11), the power is therefore zero!

$$P = 0. \tag{11.19}$$

EXAMPLE 11.1. SIMPLE AXIAL FAN. A simple cooling fan with symmetric airfoil-shaped blades is pumping air for a cooling system, as shown in Fig. 11.12. The rotor with an average radius of 0.3 m rotates at 3000 RPM and the axial velocity is 61 m/s. Draw the velocity diagram and calculate the flow angle β_1 ahead of the rotor blade.

Solution: First let us calculate the blade velocity U:

$$U = 2\pi r\Omega = 2\pi 0.3\frac{3000}{60} = 94.25\frac{\text{m}}{\text{s}}.$$

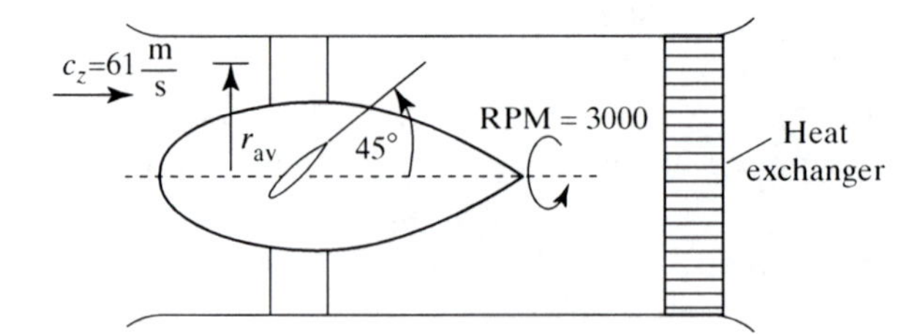

Figure 11.12. A single-stage fan rotating in a cooling duct.

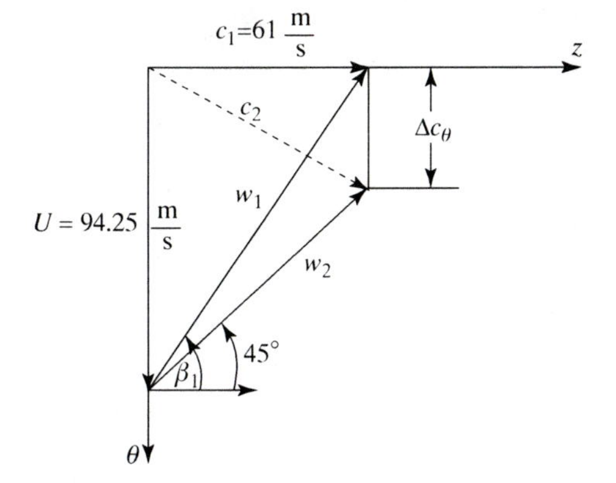

Figure 11.13. Velocity triangles for the cooling fan shown in Fig. 11.12.

Because there is no turning vane ahead of the rotor, the incoming flow is assumed to have no swirl (e.g., $\alpha_1 = 0$). This means that $\vec{c}_1 = (0, 0, c_z)$. With this information, the velocity diagram can be constructed (see Fig. 11.13). First, the vector $U = 94.25$ m/s and $c_1 = c_z = 61$ m/s are drawn. And by closing the triangle, we find the vector w_1. The incoming relative angle is then

$$\beta_1 = \tan^{-1} \frac{U}{c_z} = 57°.$$

Because the blade is oriented at 45°, the blade angle of attack is 17°, which is a bit high but workable if there is a dense cascade (called high solidity). Next, assuming the flow is attached, it leaves the blade parallel to the trailing edge at $\beta_2 = 45°$ and w_2 can be drawn, as shown. It is possible to calculate the change in the tangential velocity because $w_{\theta 2} = 61$ ($= c_z$, because of the 45° angle) and $w_{\theta 1} = U = 94.25$ m/s. Thus the change in tangential velocity is

$$\Delta c_\theta = c_{\theta 1} - c_{\theta 2} = w_{\theta 1} - w_{\theta 2} = 94.25 - 61 = 33.25 \text{ m/s}.$$

11.3.2 Power and Compression-Ratio Calculations

Based on the first law of thermodynamics and assuming an adiabatic system (without heat transfer), the work done on the fluid is equal to the change in the fluid enthalpy. This was already stated by Eqs. (11.2) and (11.3) for incompressible and compressible flows, respectively. The power P is then the time derivative of the work W:

$$P = \frac{d}{dt} W = \dot{m} \Delta h_0. \tag{11.20}$$

At this point the stage compression efficiency η_c is defined as the ratio between the enthalpy change in an isentropic compression Δh_{0s} over the actual enthalpy change Δh_0 (note that isentropic means adiabatic and reversible, so no losses). Also the stage efficiency represents the combined losses across the rotor and stator (for one stage):

$$\eta_c = \frac{\Delta h_{0s}}{\Delta h_0}. \tag{11.21}$$

Consequently the power per unit mass invested in the rotor is calculated with Eq. (11.17):

$$(P/\dot{m}) = \Delta h_0 = U \Delta c_\theta. \tag{11.22}$$

The ideal increase in enthalpy (and in the compression ratio) is less because of the losses, such as friction. From Eq. (11.2) for the ideal compression of an incompressible fluid we get

$$\Delta h_{0s} = \frac{\Delta p_0}{\rho}. \tag{11.23}$$

Combining Eqs. (11.21)–(11.23) provides the pressure rise for the incompressible case (pump):

$$\frac{\Delta p_0}{\rho} = \eta_c U \Delta c_\theta. \tag{11.24}$$

Note that Eq. (11.24) is based on the enthalpy increase in the rotor only. However, because the work in the stator is zero, this represents the change in the whole stage (rotor and stator). Consequently Eq. (11.24) shows the stagnation pressure rise in the whole stage! For a compressible fluid, the change in enthalpy, based on Eq. (11.3), is

$$\Delta h_0 = c_p \Delta T_0, \tag{11.25}$$

and by using Eq. (11.22) we get

$$c_p \Delta T_0 = c_p (T_{02} - T_{01}) = U \Delta c_\theta. \tag{11.26}$$

To calculate the pressure rise we assume an isentropic process for which the relation between the temperature and pressure change is

$$\frac{p_{02}}{p_{01}} = \left(\frac{T_{02s}}{T_{01}} \right)^{\frac{\gamma}{\gamma-1}}. \tag{11.27}$$

Here T_{02s} is the isentropic value of the temperature for the compression to p_{02} and $\gamma = c_p/c_v$, as defined in Eq. (1.19). However, because of the losses in the system, such as friction, the process efficiency is defined as in Eq. (11.21)

$$\eta_c = \frac{\Delta h_{0s}}{\Delta h_0} = \frac{c_p (T_{02s} - T_{01})}{c_p (T_{02} - T_{01})}, \tag{11.28}$$

and of course T_{02} is larger than T_{02s}. By combining Eqs. (11.27) and (11.28), we calculate the compression ratio as

$$\frac{p_{02}}{p_{01}} = \left(\frac{T_{02s}}{T_{01}} \right)^{\frac{\gamma}{\gamma-1}} = \left(1 + \frac{T_{02s} - T_{01}}{T_{01}} \right)^{\frac{\gamma}{\gamma-1}} = \left(1 + \eta_c \frac{T_{02} - T_{01}}{T_{01}} \right)^{\frac{\gamma}{\gamma-1}},$$

and here the actual stagnation temperature rise is exchanged with the isentropic ratio, as stated in Eq. (11.28). The actual temperature ratio is given by Eq. (11.26) as

$$\frac{(T_{02} - T_{01})}{T_{01}} = \frac{U \Delta c_\theta}{c_p T_{01}}.$$

Substituting this into Eq. (11.29) we get the pressure rise for the compressible case (for gases):

$$\frac{p_{02}}{p_{01}} = \left(1 + \eta_c \frac{U \Delta c_\theta}{c_p T_{01}}\right)^{\frac{\gamma}{\gamma-1}}. \tag{11.29}$$

Equations (11.24) and (11.29) calculate the stagnation pressure rise across the single-stage rotor. Again, the stator is not moving, and its power (and work) is zero! Because the stage efficiency accounts for the losses in both the rotor and the stator, there is no change in the stagnation quantities in the stator. Consequently, the stagnation pressure rise in the rotor is the same as for the whole stage (between station 1 and station 3, in Fig. 11.7 or 11.8). In summary, the pressure rise for an incompressible pump stage is then [Eq. (11.24)]

$$\frac{p_{03} - p_{01}}{\rho} = \eta_c U \Delta c_\theta, \tag{11.30}$$

and for a compressible fluid

$$\frac{p_{03}}{p_{01}} = \left(1 + \eta_c \frac{U \Delta c_\theta}{c_p T_{01}}\right)^{\frac{\gamma}{\gamma-1}}. \tag{11.31}$$

Quite often engineers need to estimate the power required for a specified flow rate and pressure ratio. For the incompressible case this is obtained from Eqs. (11.30) and (11.22),

$$P = \dot{m} \frac{p_{03} - p_{01}}{\eta_c \rho}, \tag{11.32}$$

and for the compressible case, from Eqs. (11.31) and (11.22),

$$P = \dot{m} \frac{c_p T_{01}}{\eta_c} \left[\left(\frac{p_{03}}{p_{01}}\right)^{\frac{\gamma-1}{\gamma}} - 1\right]. \tag{11.33}$$

EXAMPLE 11.2. PRESSURE RISE IN AN AXIAL COMPRESSOR. The axial velocity across the third stage of an axial compressor is $c_z = 120$ m/s, the average radius is $r_{av} = 0.5$ m, and the stagnation temperature ahead of the rotor is $T_{01} = 300$ K (see Fig. 11.14). If the compressor rotates at 4000 RPM, the stator angle from the previous stage is $\alpha_1 = 22°$, and the flow leaves the rotor at an angle

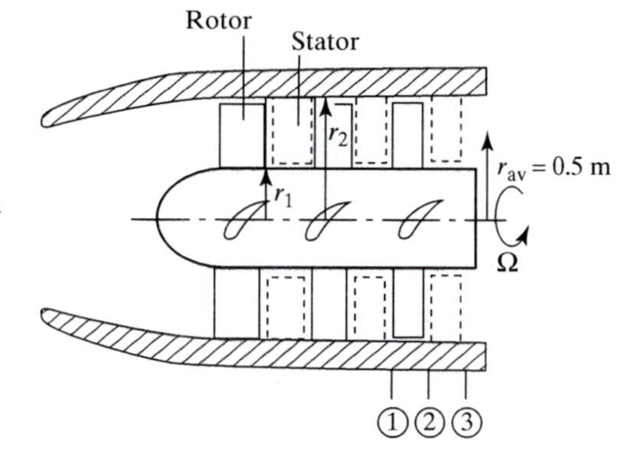

Figure 11.14. Schematic description of a typical compressor stage for example 11.2.

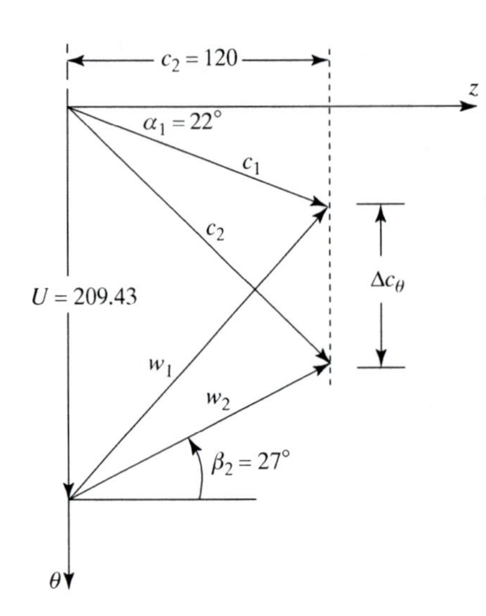

Figure 11.15. Velocity triangles for Example 11.2.

$\beta_2 = 27°$, draw the velocity diagram and calculate the stagnation pressure rise p_{03}/p_{01} across this stage [$\eta_{\text{stage}} = 0.98$, $c_p = 0.24$ kcal/(kg °C), $\gamma = 1.4$].

Solution: Let us start with calculating the tip speed U:

$$U = 2\pi r\Omega = 2\pi 0.5\frac{4000}{60} = 209.43\frac{\text{m}}{\text{s}}.$$

Now we can start constructing the velocity diagram as shown in Fig. 11.15. Because c_z is known, by drawing a line pointing down at $\alpha_1 = 22°$, we define the vector c_1. By adding the vector U, we complete the inflow (station 1) velocity triangle (by drawing w_1). We form the exit-velocity triangle by drawing a line at $\beta_2 = 27°$, ending at $c_z = 120$ (see the vector w_2). This also defines c_2.

Now that the velocity triangles are complete, the pressure rise can be calculated. First let us calculate Δc_θ based on simple trigonometrical relations in the diagram:

$$\Delta c_\theta = U - c_z(\tan\alpha_1 + \tan\beta_2) = 99.80 \text{ m/s},$$

and the compression ratio is then

$$\frac{p_{03}}{p_{01}} = \left(1 + \eta_c\frac{U\Delta c_\theta}{c_p T_{01}}\right)^{\frac{\gamma}{\gamma-1}} = \left(1 + 0.98\frac{209.43 \times 99.80}{0.24 \times 4200 \times 300}\right)^{3.5} = 1.26.$$

Note that 1 kcal = 4200 J.

11.3.3 Radial Variations

The "average-radius" assumption used so far ignores several important variations in compressor performance along the blade at different radial positions. We could partially address this by subdividing the compressor inner volume into several radial layers (e.g., hub, average, and tip as shown in Fig. 11.6) and analyzing each separately with the average-radius approach. The radial variations in a typical compressor stage can be separated into the following categories:

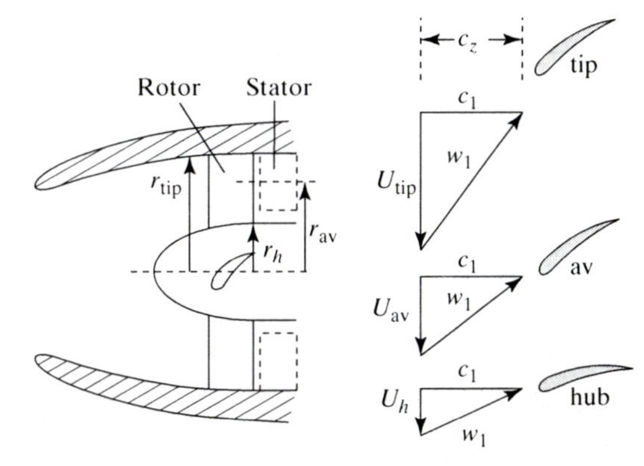

Figure 11.16. Effect of radial variations on rotor blade twist.

1. Effects that are due to radial variations in the axial velocity. So far our assumption has been that the axial velocity c_z is constant along the blade; however, it may change if the blade-tip-to-hub ratio is large. Also, near the compressor walls (hub and outer casing) the boundary layer slows down the axial flow and compression is reduced.
2. Static pressure varies with the radius because of the fluid tangential rotation. This is simply the centrifugal acceleration term resulting in increased pressure with increasing radius (e.g., $\frac{dp}{dr} \approx \rho \frac{c_\theta^2}{r}$, as in solid-body rotation).
3. Effect of tangential velocity. This is explained best by observation of the stagnation–enthalpy increase in Eq. (11.22):

$$\Delta h_0 = U \Delta c_\theta. \tag{11.22}$$

   Clearly the tangential velocity will increase with r ($U = r\Omega$) and usually also Δc_θ will increase, resulting in a radial increase in the compression ratio.
4. The increase in blade speed U: This was mentioned in the previous paragraph. However, the increase in radius also results in blade twist. This can be explained by dividing the rotor into three layers (hub, average radius, and tip) as shown in Fig. 11.16. This also demonstrates the approach of using three sublayers (and assuming average radius in each) to better model a compressor stage with significant radial variations.

To explain this last argument, the incoming-velocity triangles for the rotor blade are sketched at three radial positions (namely, at the hub, at the tip, and at the average radius). Assuming constant axial velocity c_z and no initial swirl (e.g., no turning vanes ahead of rotor and therefore $\alpha_1 = 0$), then only the blade velocity U will increase with r ($U = r\Omega$). The effect of this change in U on the incoming flow velocity triangles is shown on the right-hand side of the figure. This variation not only increases the velocity w_1 facing the blade but also changes the angle β_1. This angle change dictates a blade twist and even increased camber toward the hub, as shown in the figure (recall that a too large angle will stall the airfoil). Of course, propellers and cooling fans will have blade twist for exactly the same reason.

In multistage high-compression-ratio compressors, the designers reduce the radial variations by simply increasing the hub diameter (if possible) as shown in Fig. 11.6. In conclusion, the effect of radial variations cannot be eliminated entirely,

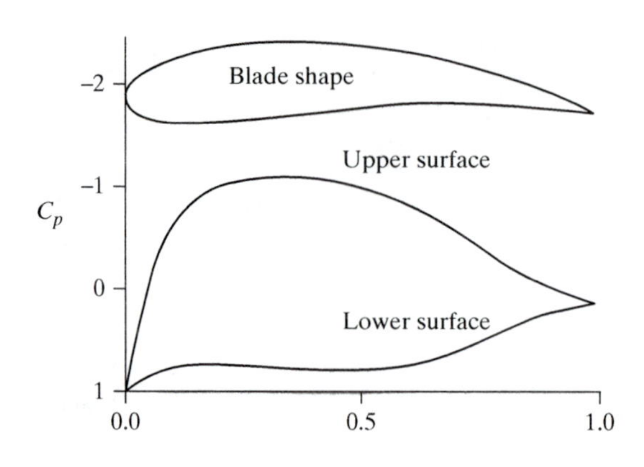

Figure 11.17. Typical pressure distribution on a rotor blade.

and one of the most common compromises is the "free-vortex" design, in which the product rc_θ is kept constant along the rotor blade exit.

11.3.4 Pressure-Rise Limitations

The previous sections established the relation between the velocity vectors and the rotor–stator blade components, leading to estimated pressure-rise calculations. It is clear that the airfoil-shaped blade performance is similar to the fluid mechanics of airfoil, as discussed in Section 8.10. Also, the discussion here is limited to subsonic flows without shock waves. Because of the close spacing between the rotor (or stator) blades, instead of a single airfoil, a cascade of airfoil is tested. In spite of the close proximity between the blades and the pulsating flow effects as the rotor blades rotate relative to the stator blades, the average pressure distribution resembles the attached airfoil case as shown in Fig. 11.17. The airfoil shapes usually have more camber and because of the denser cascade (called solidity) the flow is attached for a wider range of angles of attack than for a similar but isolated airfoil (see Section 8.10).

The next question is about how to relate the airfoil angle of attack (as discussed in Section 8.10) to the performance of a cascade. For example, a cascade of airfoils is depicted in Fig. 11.18(a), where the blade mounting angle, λ, called a stagger angle, is shown. Clearly the angle of attack is $\beta_1 - \lambda$! However, instead of using

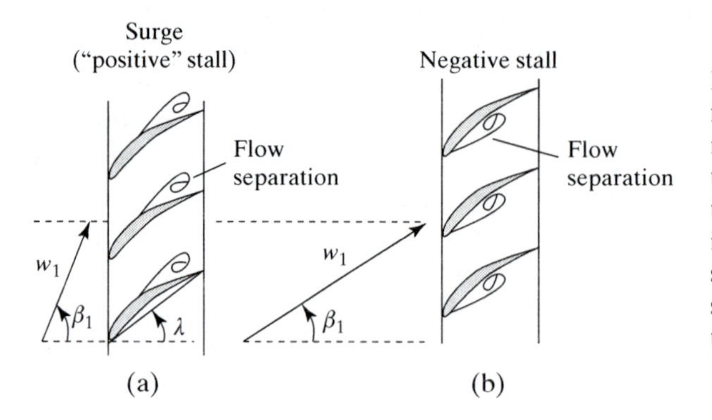

Figure 11.18. Effect of incoming flow angle on cascade performance. If the angle is too high the flow separates and (a) positive stall is observed. However, if the incoming flow angle is too small, separation on the lower surface is possible, or (b) negative stall.

a lift coefficient as a measure of the loading on an isolated airfoil, a pressure-rise coefficient C_P is defined for the cascade of airfoils:

$$C_P = \frac{p_2 - p_1}{\frac{1}{2}\rho w_1^2}. \tag{11.34}$$

Now return to Fig. 11.18(a), where the incoming flow angle of attack appears to be too high, causing partial stall. Of course, for proper performance the angle of attack must be reduced, and in cascade terminology this is approximated as

$$C_P < C_P|_{\max}. \tag{11.35}$$

For most airfoil shapes a safe empirical assumption is that

$$C_P|_{\max} = 0.6;$$

however, highly cambered and optimized airfoils can generate larger pressure-rise coefficient values. Now, recall that the discussion on cascades includes stators and rotors as well (so this condition applies to both). If we assume a small compression ratio, or "almost incompressible flow," then by applying the Bernoulli equation between stations 1 and 2 we get

$$C_P = \frac{p_2 - p_1}{\frac{1}{2}\rho w_1^2} \approx \frac{\frac{1}{2}\rho\left(w_1^2 - w_2^2\right)}{\frac{1}{2}\rho w_1^2} = 1 - \frac{w_2^2}{w_1^2}, \tag{11.36}$$

or by applying the same consideration to the stator stage (between stations 2 and 3) we get

$$C_P = \frac{p_3 - p_2}{\frac{1}{2}\rho c_2^2} \approx 1 - \frac{c_3^2}{c_2^2}. \tag{11.37}$$

This can be rewritten in terms of the cascade angle. For example, based on trigonometrical relations, the rotor angles are (see Fig. 11.10)

$$\cos\beta_1 = \frac{c_z}{w_1}, \qquad \cos\beta_2 = \frac{c_z}{w_2}. \tag{11.38}$$

By substituting this into Eq. (11.36) we can calculate the pressure-rise coefficient based on rotor angles only:

$$C_P = 1 - \frac{w_2^2}{w_1^2} = 1 - \frac{\cos\beta_1^2}{\cos\beta_2^2}. \tag{11.39}$$

This equation, when combined with the limitation of Eq. (11.35), limits the angular change across the cascade. As noted, a similar argument is also valid for the stator row.

Turbomachinery performance may be limited by other parameters. For example, if the airfoil upper surface pressure (as shown in Fig. 11.17) drops below the vapor pressure in a pump, cavitation will result (bubbles will be created), even if the flow is attached. In the case of compressible gases, at very high speeds sonic shocks may reduce performance. Consequently a limiting Mach number M_1 is defined by use of the local sonic speed a_1:

$$M_1 = (w_1/a_1) < 0.85. \tag{11.40}$$

This provides the limit for subsonic compressor performance, although some transonic compressors operate at speeds closer to the speed of sound. By combining this with the pressure-rise limit [Eq. (11.35)], an approximate maximum compression ratio can be calculated. First, let us rearrange the definition of the pressure-rise coefficient [Eq. (11.34)]:

$$p_2 - p_1 = C_P \frac{1}{2} \rho w_1^2.$$

Next, dividing by p_1, we get

$$\frac{p_2}{p_1} = 1 + \frac{C_P}{2} \frac{\rho}{p_1} w_1^2 = 1 + \frac{C_P}{2} \frac{w_1^2}{RT_1} = 1 + \frac{C_P}{2} \frac{\gamma w_1^2}{\gamma RT_1} = 1 + \frac{\gamma C_P}{2} M_1^2. \quad (11.41)$$

Here we first use the ideal-gas assumption and then we use the definition of the speed of sound from Eq. (1.33):

$$a_1 = \sqrt{\gamma RT_1}. \quad (1.33)$$

For a numerical example, consider an airplane multistage compressor, compressing air, for which $\gamma = 1.4$. Using the limits for M_1 from Eq. (11.40) and for the pressure-rise coefficient from Eq. (11.35a) we get

$$\frac{p_2}{p_1} = 1 + \frac{\gamma C_P}{2} M_1^2 = 1 + \frac{1.4 \times 0.6}{2} 0.85^2 = 1.3.$$

Assuming similar high loading in the stator, the maximum compression ratio of a single-stage axial compressor is

$$\frac{p_3}{p_1} = \frac{p_2}{p_1} \times \frac{p_3}{p_2} = 1.3 \times 1.3 = 1.69.$$

This condition represents a case in which both stator and rotor are on the verge of stall, and for maximum stage efficiency, compression ratio is much lower (e.g., 1.1–1.4).

11.3.5 Performance Envelope of Compressors and Pumps

From the fluid mechanics point of view, continuous-flow machines can be designed to operate near an optimum performance point. However, quite often, actual operation dictates conditions (e.g., in terms of axial velocity or rotation speed) that are far from ideal. To understand the effect of these off-design conditions, let us discuss the effect of changing the axial velocity.

Figure 11.18a can help in explaining the effect of changing the axial velocity c_z on the incoming flow angle β_1. For simplicity, let us consider a single stage compressor as shown in Fig. 11.19a. Assuming a constant rotation speed (which makes $U = \text{const.}$), then β_1 depends on the horizontal component of the velocity triangle, namely the axial velocity (or mass flow rate). If c_z is too small, the angle β_1 is large and the cascade may stall, as shown at the left-hand side of Fig. 11.19(b). At a higher flow rate (than design) the angle β_1 is much smaller, and this may result in negative stall, as shown in Fig. 11.18(b). This could also be explained by the hypothetical experiment shown in Fig. 11.19(a). Here a single-stage axial compressor pumps air into a duct, at the end of which there is a large valve. Assuming a constant rotation

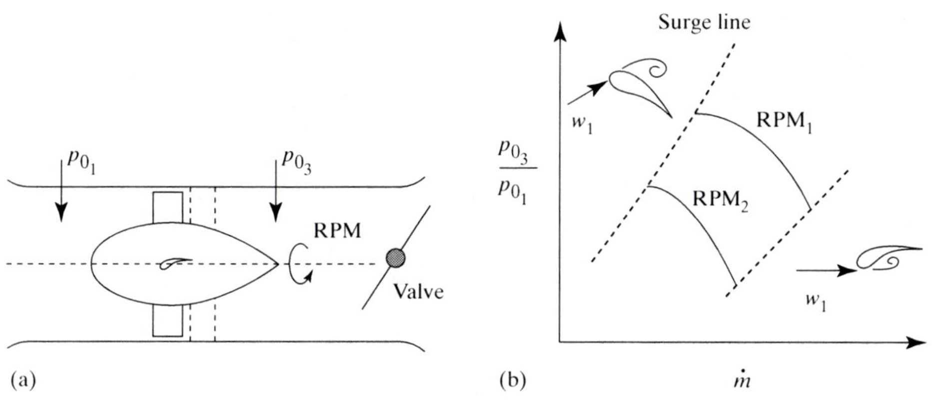

Figure 11.19. Estimated performance curves of a single-stage axial compressor.

speed, the mass flow rate across this system can be controlled by closing or opening this valve. For example, if the valve is slightly closed, then the axial flow is reduced and the pressure will increase because β_1 will increase as shown in Fig. 11.18(a). At a certain point, while the valve is closing, the axial velocity is too slow and the rotor will stall. In practice, when several blades are stalled, then the axial velocity will increase across the rest of the blades on the same rotor disk (reattaching the flow). This is a quite complex fluid mechanic phenomenon called *rotating stall.* As the flow rate is further reduced, a strong vibration may develop and a surge line can be defined. When compressor or pump performance is plotted versus mass flow rate, as in Fig. 11.19(b), then this surge line can be identified by the dashed line, as shown. In general we can conclude that the pressure ratio will increase with reduced mass flow rate. Now if the valve behind the compressor is opened, then the axial velocity will increase, β_1 will decrease, and the compression ratio will decrease. If the valve is fully opened and there is no resistance, negative stall may be present (it depends on the design point of the system – or on the blade stagger angle). Compressor operation in this region (sometimes called the chocked region) is not recommended and the boundaries are shown schematically by the dashed line in Fig. 11.19(b). Based on the preceding considerations, Fig. 11.19(b) can be viewed as a generic (continuous-flow) compressor–pump map. The same behavior is observed at other rotation speeds and, for example, RPM$_1$ is larger than RPM$_2$ in Fig. 11.19(b).

EXAMPLE 11.3. EFFECT OF RADIAL VARIATIONS. To demonstrate the effects of radial variations and the importance of pressure-rise limitations, consider the first fan stage of a turbofan engine (Fig. 11.20). For simplicity let us assume

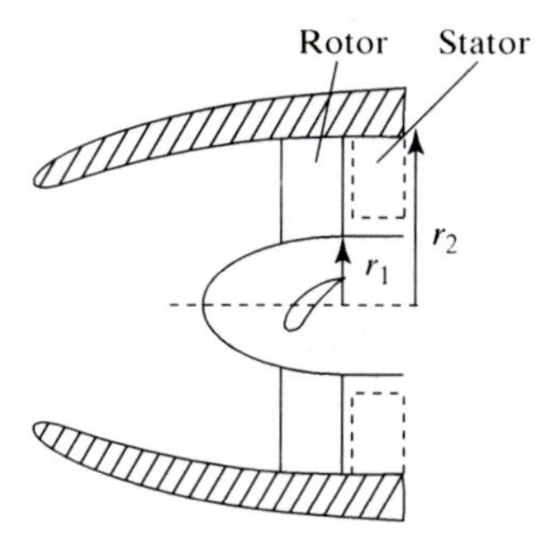

Figure 11.20. A generic single-stage compressor stage used for Example 11.3.

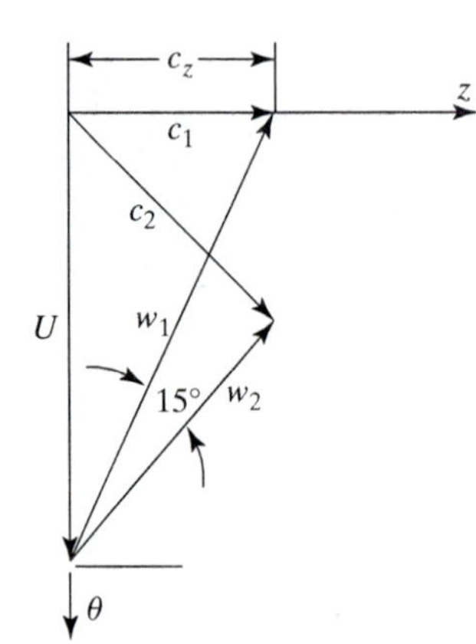

Figure 11.21. Generic velocity diagram for the first stage in Example 11.3.

that the rotor blade airfoil shape is the same along the blade and it turns the flow by $\beta_1 - \beta_2 = 15°$. The tip radius is $r_2 = 0.5$ m, the hub radius is $r_1 = 0.2$ m, rotation speed is 5500 RPM, and the axial velocity of $c_z = 120$ m/s. Calculate the pressure-rise coefficient at the hub and at the tip and the corresponding compression ratios (assume the air temperature is $T = 350$ K, $\gamma = 1.4$, and the stage efficiency is $\eta_c = 0.98$).

Solution: To use Eq. (11.39), the velocity triangles must be constructed. The blade velocity U vectors at these two radial locations are

$$U_{\text{hub}} = 2\pi r_2 \Omega = 2\pi 0.2 \frac{5500}{60} = 115.19 \frac{\text{m}}{\text{s}},$$

$$U_{\text{tip}} = 2\pi r_2 \Omega = 2\pi 0.5 \frac{5500}{60} = 287.98 \frac{\text{m}}{\text{s}}.$$

The velocity triangles for the hub and tip can now be constructed because there is no turning blade ahead of the rotor and therefore $\alpha_1 = 0$. This is shown schematically in Fig. 11.21. By connecting the U vector with the c_1 vector, we identify the incoming velocity w_1 and its direction β_1. We obtain the velocity vector w_2 by simply reducing the angle β_1 by 15°, as shown (assuming constant c_z).

Let us check if the tip velocity is close to sonic (because the velocities are always the highest at the tip). The tip velocity is then

$$w_1 = \sqrt{U^2 + c_z^2} = 311.95 \frac{\text{m}}{\text{s}}.$$

We calculate the speed of sound by using Eq. (1.33):

$$a_1 = \sqrt{\gamma RT} = \sqrt{1.4 \times 286.6 \times 350} = 374.75 \frac{\text{m}}{\text{s}},$$

where R was taken from Eq. (1.13). The local Mach number is then

$$M_1 = \frac{w_1}{a_1} = 0.83,$$

which is slightly less than the limit stated in Eq. (11.40). Once the velocity triangles are established, the pressure-rise coefficients can be calculated. Let us start at the hub:

$$\beta_1 = \tan^{-1}\frac{U_{\text{hub}}}{c_z} = 43.83^\circ, \qquad \beta_2 = \beta_1 - 15^\circ = 28.83^\circ.$$

The pressure-rise coefficient is

$$C_P = 1 - \frac{\cos\beta_1^2}{\cos\beta_2^2} = 0.32,$$

and this is far from stall. To calculate the compression ratio, the tangential velocity change is calculated based on the velocity triangle geometry:

$$\Delta c_\theta = U - c_z \tan\beta_2 = 49.1\ 3 \text{ m/s},$$

and the compression ratio is then

$$\frac{p_{03}}{p_{01}} = \left(1 + \eta_c \frac{U\Delta c_\theta}{c_p T_{01}}\right)^{\frac{\gamma}{\gamma-1}} = \left(1 + 0.98\frac{115.19 \times 49.13}{0.24 \times 4200 \times 350}\right)^{3.5} = 1.056.$$

Next, we repeat these calculations for the rotor tip. The blade angles there are calculated as follows:

$$\beta_1 = \tan^{-1}\frac{U_{\text{tip}}}{c_z} = 67.38^\circ, \qquad \beta_2 = \beta_1 - 15^\circ = 52.38^\circ,$$

and the pressure-rise coefficient at the tip is

$$C_P = 1 - \frac{\cos\beta_1^2}{\cos\beta_2^2} = 0.60,$$

which is on the verge of stall. Next we calculate ΔC_θ:

$$\Delta c_\theta = U - c_z \tan\beta_2 = 132.27 \text{ m/s}.$$

The compression ratio is then

$$\frac{p_{03}}{p_{01}} = \left(1 + \eta_c \frac{U\Delta c_\theta}{c_p T_{01}}\right)^{\frac{\gamma}{\gamma-1}} = \left(1 + 0.98\frac{287.98 \times 132.27}{0.24 \times 4200 \times 350}\right)^{3.5} = 1.42.$$

Note that, even if the same airfoil shape is used, the tip compression ratio is significantly higher. Usually a higher camber and a higher blade incidence at the hub can increase the compression ratio there.

EXAMPLE 11.4. EFFECT OF CHANGING THE AXIAL VELOCITY. Large changes in axial velocity are encountered during the starting process of multistage compressors of jet engines and on high-speed military airplanes that operate over a wide range of flight speeds. For example, the compressor's compression ratio of a typical airplane engine is over 30, and, as a result, the density toward the rear stages is very high (compared with the first stage). When the engine is started, however, the density is almost the same across all stages and even if the axial velocity at the aft stages is larger than "design," the mass flow rate is significantly less! This dramatic reduction in mass flow rate results in similar reduction of the axial velocity in the early compressor stages. Consequently, the rotor inflow

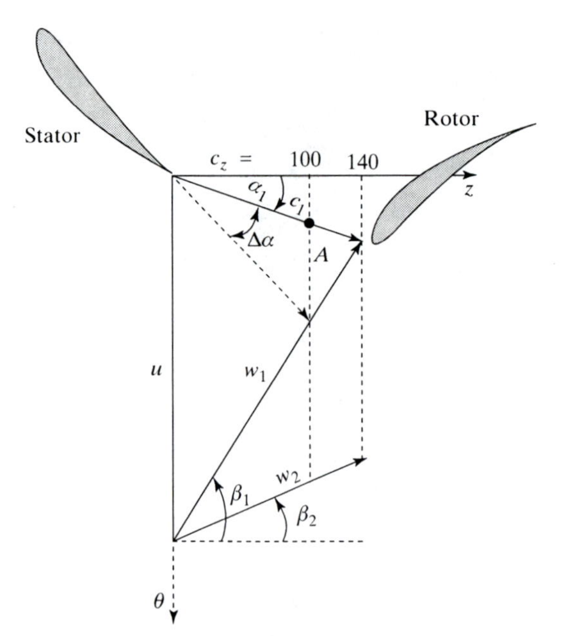

Figure 11.22. Effect of axial velocity change on rotor performance (velocity diagram for example 11.4).

angle changes accordingly, bringing it closer to stall condition. A possible remedy for large axial velocity changes is to use rotating stators to reduce the flow angle ahead of the rotor. This approach is used on military aircraft engines that must perform well for a wide range of axial velocities in the compressor (because of large thrust and flight speed changes). As an example, consider the flow conditions in the second stage of a typical jet engine:

Assume that the average radius at this stage is $r = 0.5$ m, the axial velocity is $c_z = 140$ m/s, and the rotor turns at 4000 RPM. The stagnation temperature ahead of the rotor is $T_0 = 300$ K, the incoming velocity angle is $\alpha_1 = 20°$, and the pressure rise coefficient for this rotor stage is $C_p = 0.45$.

First, let us calculate the stagnation pressure rise p_{03}/p_{01} across this stage (using the values $\eta_c = 0.98$, $c_p = 0.24$, $\gamma = 1.4$).

Second, assume that during the starting process the axial velocity is reduced to $c_z = 100$ m/s but the rotation stays at the operational level of 4000 RPM. How many degrees should the stator blade (ahead of this rotor) be rotated so that the rotor blade angle β_1 will not change?

Solution: First, we need to generate the velocity diagram for the rotor under nominal operational conditions. The rotation speed is then

$$U = r\Omega = 0.5 \times 2\pi \frac{4000}{60} = 209.43 \text{ m/s}.$$

Because the axial velocity and α_1 are known, the velocity triangle for the rotor inlet can be drawn (see Fig. 11.22), and the angle β_1 can be calculated:

$$\tan \beta_1 = \frac{U - c_z \tan \alpha_1}{c_z} = \frac{209.43 - 140 \tan 20°}{140} = 1.131 \Rightarrow \beta_1 = 48.5°.$$

Because the pressure-rise coefficient is given, the angle β_2 can be calculated too:

$$C_P = 1 - \frac{\cos\beta_1^2}{\cos\beta_2^2} = 0.45 = 1 - \frac{\cos 20^2}{\cos\beta_2^2} \Rightarrow \beta_2 = 26.7^\circ.$$

With the aid of this angle the vector w_2 can be drawn and the rotor velocity diagram is complete. To calculate the compression ratio we first calculate the change in the tangential velocity,

$$\Delta c_\theta = U - c_z(\tan\alpha_1 + \tan\beta_2) = 88.12 \text{ m/s}$$

and the compression ratio is

$$\frac{p_{03}}{p_{01}} = \left(1 + \eta_c \frac{U \Delta c_\theta}{c_p T_{01}}\right)^{\frac{\gamma}{\gamma-1}} = \left(1 + 0.98 \frac{209.43 \times 88.12}{0.24 \times 4200 \times 300}\right)^{3.5} = 1.22.$$

For the second part of this problem, the axial velocity is reduced to 100 m/s, as shown in the velocity diagram. If no action is taken, then the w_1 vector tip will be at point A on the diagram; a significant increase in the angle of attack. To avoid such high angles of attack on the rotor blade, the angle β_1 must be kept unchanged (at 48.5°). This is accomplished by turning the stator ahead of this stage to adjust the incoming anlge α_1, as shown by the dashed line in the figure. Based on the new triangle geometry, the incoming flow angle is calculated:

$$\tan\alpha_1 = \frac{U - c_z \tan\beta_1}{c_z} = \frac{209.43 - 100 \tan 48.5^\circ}{100} = 0.964 \Rightarrow \alpha_1 = 43.9^\circ.$$

Therefore *the stator ahead of* this stage must be rotated, as shown in the figure by

$$\Delta\alpha = 43.9^\circ - 20^\circ = 23.9^\circ.$$

To calculate the compression ratio for this second case we calculate the change in the tangential velocity

$$\Delta c_\theta = U - c_z(\tan\alpha_1 + \tan\beta_2) = 62.90 \text{ m/s},$$

and the compression ratio is

$$\frac{p_{03}}{p_{01}} = \left(1 + \eta_c \frac{U \Delta c_\theta}{c_p T_{01}}\right)^{\frac{\gamma}{\gamma-1}} = \left(1 + 0.98 \frac{209.43 \times 62.90}{0.24 \times 4200 \times 300}\right)^{3.5} = 1.15.$$

And by adjusting the stator angle α_1, rotor stall is avoided and compression ratio stays close to the design condition.

11.3.6 Degree of Reaction

Many axial compressors consist of multiple stages and from the fluid dynamic point of view (also from the manufacturing point of view), similar rotor and stator blades are desirable (although they may not result in the most efficient design). In Subsection 11.3.2, using the thermodynamic point of view, it was concluded that there is no stagnation pressure rise in the stator. Therefore, the first impression is that stator design is less important. The rotor and the stator in a compressor, in fact, are similar fluid dynamic components, and therefore it is useful to define the static pressure rise in the rotor versus the whole stage. This is done by the *degree of reaction*, R, which is the ratio of enthalpy increase in the rotor versus the whole stage. For simplicity, by assuming "almost incompressible flow" in one stage, we can estimate the static enthalpy change as:

$$h_2 - h_1 \approx \frac{1}{\rho}(p_2 - p_1),$$

and the total stagnation enthalpy change in the whole stage is

$$h_{03} - h_{01} \approx \frac{1}{\rho}(p_{03} - p_{01}).$$

Thus the degree of reaction is defined as

$$R = \frac{p_2 - p_1}{p_{03} - p_{01}}. \tag{11.42}$$

Using the incompressible Bernoulli relation for the numerator,

$$p_2 - p_1 = \frac{\rho}{2}\left(w_1^2 - w_2^2\right),$$

and Eq. (11.24) for the denominator

$$\Delta h_0 \approx \frac{p_{03} - p_{01}}{\rho} = U(c_{\theta 2} - c_{\theta 1}),$$

we get a relation, using rotor parameters only!

$$R = \frac{w_1^2 - w_2^2}{2U(c_{\theta 2} - c_{\theta 1})}. \tag{11.43}$$

Because the velocity vector in the z–θ coordinate system has two components, we can write

$$R = \frac{w_{1z}^2 + w_{1\theta}^2 - w_{2z}^2 - w_{2\theta}^2}{2U(c_{\theta 2} - c_{\theta 1})} = \frac{w_{1\theta}^2 - w_{2\theta}^2}{2U(c_{\theta 2} - c_{\theta 1})} = \frac{(w_{\theta 1} - w_{\theta 2})(w_{\theta 1} + w_{\theta 2})}{2U(c_{\theta 2} - c_{\theta 1})}$$

$$= \frac{(w_{\theta 1} + w_{\theta 2})}{2U}, \tag{11.44}$$

and here the tangential velocity difference $(w_{\theta 1} - w_{\theta 2}) = (c_{\theta 2} - c_{\theta 1})$ is the same. We can obtain a more informative expression by observing the velocity triangles (e.g., in Fig. 11.15). For the two tangential components of the rotor velocity,

$$w_{\theta 1} = U - c_z \tan\alpha_1,$$
$$w_{\theta 2} = c_z \tan\beta_2.$$

Substituting these relations into Eq. (11.44) yields

$$R = \frac{1}{2} - \frac{c_z}{2U}(\tan\alpha_1 - \tan\beta_2). \tag{11.45}$$

This equation clearly shows that for a 50% degree-of-reaction stage (e.g., $R = \frac{1}{2}$), in which the static pressure rise is the same for the rotor and the stator, the velocity diagram is symmetric ($\alpha_1 = \beta_2$) and both rotor and stator airfoils see similar incoming and exiting flow angles (see Fig. 11.23)! This condition is desirable in the inner

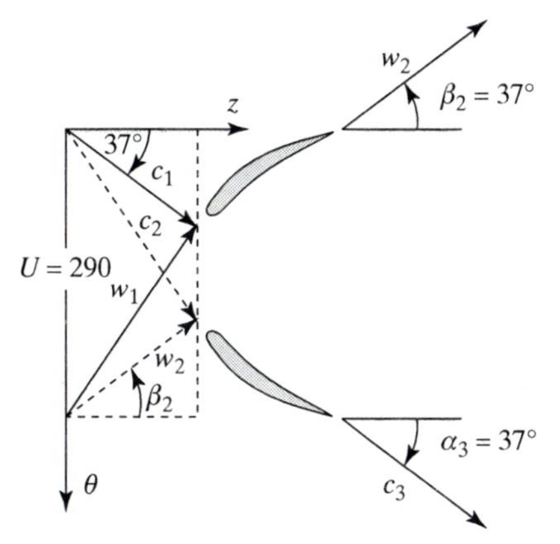

Figure 11.23. Velocity diagram for a 50% degree-of-reaction compressor stage for Example 11.5.

stages of a multistage compressor because the aerodynamics of both rotor and stator blades is the same. The aerodynamic symmetry, in terms of the velocity triangle, is demonstrated by the symmetry (about an imaginary horizontal line), shown in Fig. 11.23.

EXAMPLE 11.5. 50% DEGREE-OF-REACTION COMPRESSOR. The fourth stage of an axial compressor is designed to have a degree of reaction of $R = 0.5$. The velocity leaving the stator ahead of this stage is $c_1 = 150$ m/s and the rotor blade velocity (at the average radius) is $U = 290$ m/s. The angle behind the rotor is $\beta_2 = 37°$ and the air temperature is $T_{01} = 280$ K (also $\gamma = 1.4$ and stage efficiency is $\eta_c = 0.85$). Draw a velocity diagram and calculate the compression ratio.

Solution: For a 50% degree-of-reaction compressor, according to Eq. (11.45) $\alpha_1 = \beta_2$. Consequently the velocity triangle for the incoming flow can be drawn (e.g., $U = 290$ m/s, $c_1 = 150$ m/s and $\alpha_1 = 37°$). Because $\beta_2 = 37°$ the exit-velocity triangle can be drawn as well, and clearly there is a symmetry about a horizontal centerline. The rotor and stator blades are drawn as well, and clearly, from the fluid dynamic point of view, the rotor and stator airfoil shapes are the same (assuming that the stator at station 3 is the same as in station 1, $\alpha_1 = \alpha_3$).

Once the velocity diagram is established the change in tangential velocity can be calculated by means of trigonometric relations:

$$c_{\theta 1} = c_1 \sin\alpha_1 = 150 \sin 37° = 90.27 \text{ m/s},$$
$$\Delta c_\theta = U - 2c_1 \sin\alpha_1 = 109.46 \text{ m/s}.$$

To calculate the pressure-rise coefficient, β_1 must be calculated:

$$\tan\beta_1 = \frac{U - c_{\theta 1}}{c_z} = \frac{U - c_{\theta 1}}{c_1 \cos 37°} = 1.67 \Rightarrow \beta_1 = 59.04°.$$

The pressure-rise coefficient is then

$$C_P = 1 - \frac{\cos\beta_1^2}{\cos\beta_2^2} = 0.584.$$

This is high loading but probably will not stall. The compression ratio is then

$$\frac{p_{03}}{p_{01}} = \left(1 + \eta_c \frac{U \Delta c_\theta}{c_p T_{01}}\right)^{\frac{\gamma}{\gamma-1}} = \left(1 + 0.85 \frac{290 \times 109.46}{0.24 \times 4200 \times 280}\right)^{3.5} = 1.38.$$

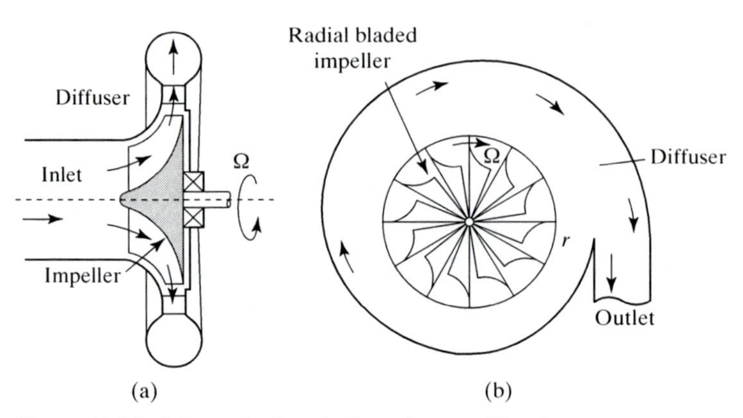

Figure 11.24. Schematic description of a centrifugal compressor–pump.

11.4 The Centrifugal Compressor (or Pump)

Centrifugal compressors and pumps are less sensitive to design parameters such as the rotor blade angle of an axial compressor. They are capable of significantly higher compression ratios than a single-stage axial compressor or pump and are usually less expensive to manufacture. Consequently they are used in numerous applications, spanning domestic, automotive, or agricultural applications. In jet engines they are used less because of the large frontal area and difficult multistaging. A centrifugal design usually consists of a rotor (often called an impeller) with radial vanes that is rotating inside an outer casing (see Fig. 11.24).

The fluid enters at the center inlet, and then it is captured by the rotating radial vanes, accelerating it toward the outer collector. This collector (called a diffuser) slows down the flow and increases the pressure toward the exit. The schematics in Fig. 11.24 show only one exit, but multiple exits are possible. Also note that, at the inlet to the rotor, the radial vanes are turned “into” the flow to reduce local flow separation. This is clearly shown in Fig. 11.25 where an impeller was photographed outside the casing.

Figure 11.25. Typical geometry of a centrifugal pump impeller. Note how the leading edge of the blade is turned toward the incoming flow direction.

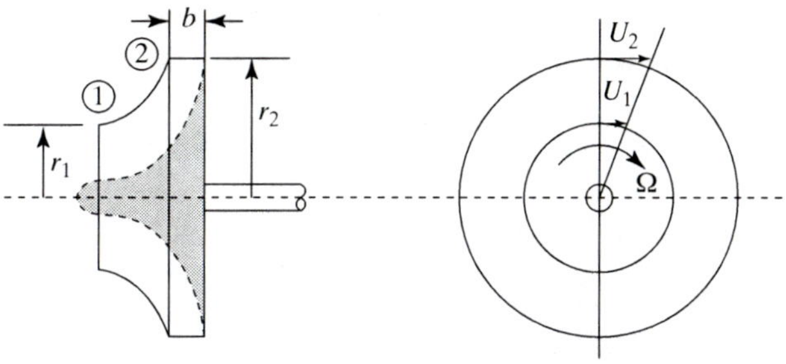

Figure 11.26. Schematic description of the rotor (impeller).

In spite of the different geometry, the model developed for the axial turbomachinery is applicable here as well. A typical single-stage compressor therefore consists of a rotor and a stator and with the geometry shown in Fig. 11.24 the entrance can be identified as rotor station 1, the exit from the rotor is station 2, and the exit from the diffuser is station 3. These stations are described with more detail in the following subsections.

11.4.1 Torque, Power, and Pressure Rise

Because the outer housing (casing) is stationary, only the rotor is turning and therefore the torque, power, and work per unit mass flow can be readily calculated (based on rotor parameters only). For example, the torque required for turning the impeller is calculated using Eq. (11.8):

$$Tq = \dot{m}(r_2\, c_{\theta 2} - r_1\, c_{\theta 1}). \tag{11.46}$$

The subscript 1 relates to the rotor inlet, as shown in Fig. 11.26 and similarly station 2 is the rotor exit. Note that at the inlet we used the maximum radius instead of the average radius – but this is only a small inaccuracy, as we shall see later.

Once the torque is estimated, the power (P) is given by Eq. (11.9) as

$$P = Tq\Omega = \dot{m}\Omega(r_2\, c_{\theta 2} - r_1\, c_{\theta 1}). \tag{11.47}$$

Again, the impeller tip speeds at the inlet and exit are defined (as shown in Fig. 11.26):

$$\begin{aligned} U_1 &= r_1\Omega, \\ U_2 &= r_2\Omega. \end{aligned} \tag{11.48}$$

With these definitions the power equation becomes

$$P = \dot{m}(U_2\, c_{\theta 2} - U_1\, c_{\theta 1}) \tag{11.49}$$

and the work per unit mass flow w_c is

$$w_c = \frac{P}{\dot{m}} = (U_2\, c_{\theta 2} - U_1\, c_{\theta 1}). \tag{11.50}$$

The calculation of the pressure rise is similar to the axial compressor–pump case. The work per unit mass flow, according to Eq. (11.2), is equal to the increase in stagnation enthalpy:

$$w_c = \Delta h_0. \tag{11.51}$$

Combined with the definition of stage efficiency η_c [Eq. (11.23)], the pressure rise for the incompressible case [see Eq. (11.26)] is

$$\frac{\Delta p_0}{\rho} = \eta_c(U_2\, c_{\theta 2} - U_1\, c_{\theta 1}) \tag{11.52}$$

and for a compressible fluid [as in Eq. (11.31)] is

$$\frac{p_{02}}{p_{01}} = \left(1 + \eta_c \frac{T_{02} - T_{01}}{T_{01}}\right)^{\frac{\gamma}{\gamma-1}}. \tag{11.53}$$

The stagnation temperature change is calculated by Eq. (11.28):

$$c_p(T_{02} - T_{01}) = (U_2\, c_{\theta 2} - U_1\, c_{\theta 1}). \tag{11.54}$$

The pressure rise is therefore related to the change in the angular velocities. For example, Eq. (11.54) can be rearranged such that

$$\frac{T_{02} - T_{01}}{T_{01}} = \frac{(U_2\, c_{\theta 2} - U_1\, c_{\theta 1})}{c_p T_{01}} = \frac{U_2^2}{c_p T_{01}}\left[\frac{c_{\theta 2}}{U_2} - \left(\frac{U_1}{U_2}\right)^2 \frac{c_{\theta 1}}{U_1}\right]$$

$$= \frac{U_2^2}{c_p T_{01}}\left[\frac{c_{\theta 2}}{U_2} - \left(\frac{r_1}{r_2}\right)^2 \frac{c_{\theta 1}}{U_1}\right]. \tag{11.55}$$

To calculate the pressure rise for the compressible flow case, Eq. (11.55) must be substituted into Eq. (11.53) (although it is simpler to evaluate these two equations separately):

$$\frac{p_{02}}{p_{01}} = \left\{1 + \eta_c \frac{U_2^2}{c_p T_{01}}\left[\frac{c_{\theta 2}}{U_2} - \left(\frac{r_1}{r_2}\right)^2 \frac{c_{\theta 1}}{U_1}\right]\right\}^{\frac{\gamma}{\gamma-1}}. \tag{11.53a}$$

When the same algebra is applied to Eq. (11.52), the pressure rise in a pump (incompressible) becomes

$$\frac{\Delta p_0}{\rho} = \eta_c U_2^2 \left[\frac{c_{\theta 2}}{U_2} - \left(\frac{r_1}{r_2}\right)^2 \frac{c_{\theta 1}}{U_1}\right]. \tag{11.52a}$$

Now if there are no turning vanes ahead of the rotor inlet then $c_{\theta 1} \approx 0$. In addition $(r_1/r_2)^2 \ll 1$, and therefore an approximate relation for the temperature rise is

$$\frac{T_{02} - T_{01}}{T_{01}} = \frac{U_2^2}{c_p T_{01}} \frac{c_{\theta 2}}{U_2}$$

in case of straight vanes *only*; then $(c_{\theta 2}/U_2) \approx 1$ and the pressure rise for a compressible fluid is

$$\frac{p_{02}}{p_{01}} \approx \left(1 + \eta_c \frac{U_2^2}{c_p T_{01}}\right)^{\frac{\gamma}{\gamma-1}} \tag{11.56}$$

and for an incompressible fluid is

$$\frac{\Delta p_0}{\rho} = \eta_c U_2^2. \tag{11.57}$$

This indicates that a centrifugal compressor–pump performance can be easily estimated based on rotation speed and rotor exit radius (and not on inlet radius).

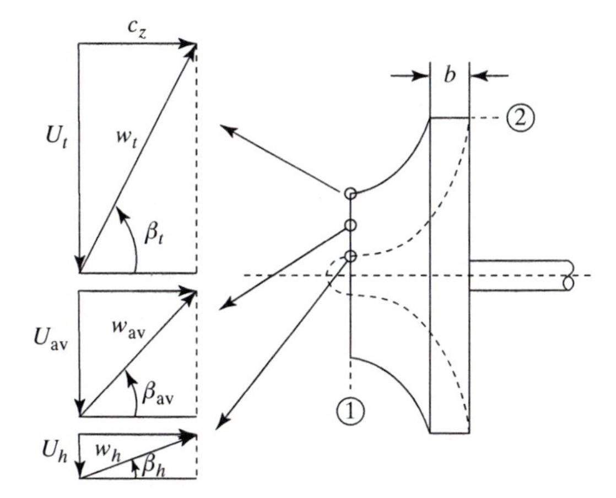

Figure 11.27. Rotor (impeller) inlet geometry and the reason for the leading-edge twist shown in Fig. 11.24.

This is why, in many applications, portions of the impeller inlet are eliminated. Some air conditioning centrifugal fans have only the exit tip vanes for the same reason.

11.4.2 Impeller Geometry

The formulation for the pressure rise in the centrifugal compressor is readily transferred from the axial compressor model. The rotor geometry, however, is somewhat different and the inlet (station 1) and exit (station 2) velocity triangles cannot be superimposed (as in the case of the axial design). In spite of Eqs. (11.56) and (11.57) implying that the compression ratio is not much affected by the inlet condition, a poor inlet can cause flow separations and reduce pump–compressor efficiency. The velocity vectors at the inlet can be drawn exactly as was done for the axial compressor case (as in Fig. 11.9). To investigate the radial variations, the velocity triangles are shown schematically in three radial locations along the rotor (impeller) leading edge (Fig. 11.27). It is assumed that there are no turning vanes ahead of this stage (no swirl) and $c_{\theta 1} = 0$; therefore $c_1 = c_z$, as shown. The blade velocity U, however, is changing along the blade, resulting in an increase in the flow angle β_1 relative to the rotor blade. In order not to stall the blade leading edge, it must be turned into the flow, and more turning is required with increasing radius. This is the reason for the increasing twist (with increasing radial position) at the impeller inlet, as shown in the impeller's front view in Fig. 11.25.

Next, a limit on the rotor leading-edge droop is addressed. This could be viewed as the lower limit on the mass flow rate, for a given RPM, prior to leading-edge stall (or what is the maximum value of β_1 before stall). To clarify this, the inlet area of the impeller is drawn separately in Fig. 11.28. For a fixed radial position, the inlet segment of the rotor can be viewed as an axial compressor with the inlet at station 1 and the exit section is immediately behind, as the blade becomes straight (let us call this station $*$, as shown in Fig. 11.28). The incoming velocity triangle is exactly the same as drawn in Fig. 11.27. At the exit (station $*$), however, the flow must leave parallel to the blades, which are basically pointing straight backward (as shown). Consequently the exit velocity has only an axial component and $\beta_* = 0$.

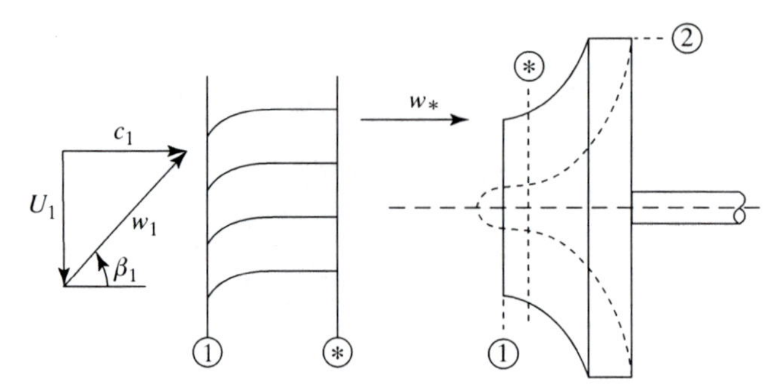

Figure 11.28. Rotor (impeller) inlet geometry. Note that the section between station 1 and station $*$ was stretched at the left-hand side (for illustration purpose only).

Assuming no major changes in the cross-section area, the axial velocity will not change much, and therefore

$$c_1 = w_{z1} = w_*,$$

or

$$w_* = w_1 \cos\beta_1.$$

where w_* is the exit-velocity vector. To estimate the maximum turning angle β_1 we can use the limits applied to the axial flow machines. The pressure coefficient [Eq. (11.39)] written for this case is

$$C_P = 1 - \frac{w_*^2}{w_1^2} = 1 - \left(\frac{w_1 \cos\beta_1}{w_1}\right)^2 = 1 - \cos^2\beta_1. \tag{11.58}$$

Using the empirical limit for the pressure-rise coefficient (e.g., 0.6), as given by Eq. (11.35a), provides an estimate for the maximum turning angle:

$$C_P = 1 - \cos^2\beta_1 = 0.6,$$
$$\beta_1 = 50.8^\circ. \tag{11.59}$$

This allows the calculation of the lowest axial velocity (or mass flow rate) for a given rotation speed of the impeller. In the case of a compressor, the inlet tip velocity should be subsonic to avoid shock waves and flow separation, and we can write, similar to Eq. (11.40),

$$M_1 = \frac{w_1}{a_1} < 0.9. \tag{11.60}$$

For pumps the vapor pressure must be checked at this point to avoid cavitation.

As noted earlier, the rotor exit velocity cannot be drawn in the z–θ plane because the flow is turned entirely in the radial direction. Consequently the exit-velocity triangles are plotted in the r–θ plane, as shown in Fig. 11.29.

This view clarifies the effect of the flow exit angle β_2. For example, considering three impellers that have the same dimensions, apart from the shape of the blades (near the exit), the resulting velocity triangles are as shown in Fig. 11.29. The exit

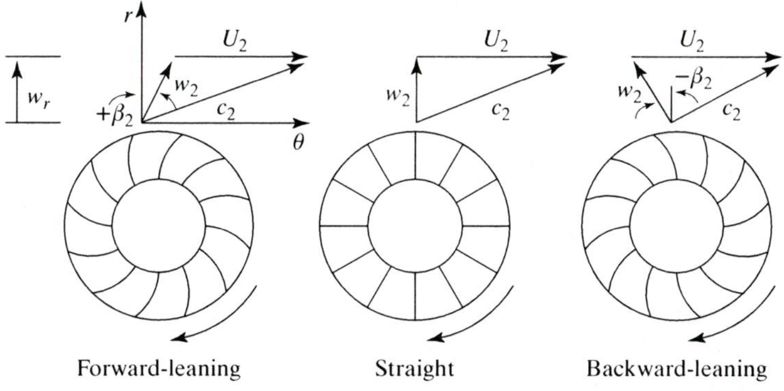

Figure 11.29. Effect of rotor blade exit orientation on a velocity triangle.

radial velocity w_{r2} depends on the blade height b, as shown in Fig. 11.27, and from a simple continuity consideration it can be calculated as

$$w_{r2} = \frac{\dot{m}}{\rho_2 2\pi r_2 b}. \tag{11.61}$$

If the three rotors in Fig. 11.29 have the same radial velocity, then (assuming the flow leaves parallel to the vane's trailing edge) the forward-leaning blades will produce the highest tangential velocity component and the highest compression ratio [see Eqs. (11.52)–(11.54)]. Quite often, from the manufacturing point of view, the blades are straight, as shown in the middle. The backward-leaning design is also quite popular because of quieter operation (but with a lower compression ratio). If the exit angle β_2 is defined as positive for the forward-leaning blades, then, based on the velocity triangle, the tangential velocity component is

$$c_{\theta 2} = U_2 + w_{r2} \tan \beta_2. \tag{11.62}$$

Substituting the radial velocity from Eq. (11.61), we get

$$c_{\theta 2} = U_2 + \frac{\dot{m}}{\rho_2 2\pi r_2 b} \tan \beta_2. \tag{11.63}$$

Substituting this into Eq. (11.52a) for the incompressible case, we get

$$\frac{\Delta p_0}{\rho} = \eta_c U_2^2 \left[1 + \frac{\dot{m} \tan \beta_2}{\rho_2 2\pi r_2 b U_2} - \left(\frac{r_1}{r_2} \right)^2 \frac{c_{\theta 1}}{U_1} \right], \tag{11.64}$$

and for a compressible fluid [using Eq. (11.53)], we get

$$\frac{p_{02}}{p_{01}} = \left\{ 1 + \eta_c \frac{U_2^2}{c_p T_{01}} \left[1 + \frac{\dot{m} \tan \beta_2}{\rho_2 2\pi r_2 b U_2} - \left(\frac{r_1}{r_2} \right)^2 \frac{c_{\theta 1}}{U_1} \right] \right\}^{\frac{\gamma}{\gamma - 1}}. \tag{11.64}$$

As noted earlier, the last term, $(\frac{r_1}{r_2})^2 \frac{c_{\theta 1}}{U_1}$, is negligible, and therefore increasing the mass flow rate will increase the pressure rise for the forward-leaning blades and the opposite for the backward-leaning blades. For straight blades, the compression ratio will not be affected by a change in the mass flow rate [however, inlet flow angles cannot be larger than 50°, as determined by Eq. (11.59)].

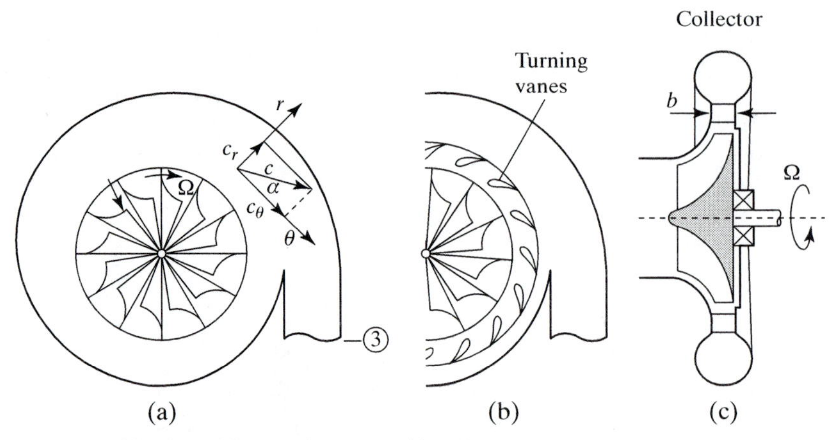

Figure 11.30. The diffuser of the centrifugal pump–compressor.

11.4.3 The Diffuser

The fluid leaving the rotor at station 2 is slowed down by the diffuser, increasing the pressure at the exit (station 3). The velocity vector $\vec{c}$ is shown schematically in Fig. 11.30(a) and it has two components, one in the radial direction and one in the tangential directions. Because the diffuser (collector) is stationary, there is no velocity triangle. The radial velocity at station 2 is the same as the exit velocity of the rotor,

$$c_{r2} = w_{r2}, \tag{11.65}$$

and the tangential velocity is given by Eq. (11.62),

$$c_{\theta 2} = U_2 + w_{r2} \tan \beta_2. \tag{11.62}$$

There are two major types of diffusers; those without internal turning vanes, as shown in Fig. 11.30(a), or those with guide (or turning) vanes, as shown in Fig. 11.30(b). Turning vanes can reduce the size of the diffuser, but they also add weight and marginally reduce efficiency (more friction). Vaneless diffusers are preferred (if size is not an issue) because they allow a wider operation range than a diffuser with turning vanes. Figure 11.30(c) shows the side view of a typical centrifugal pump, and the width of the collector [(b) in the figure] is increasing to further slow down the radial flow. In the case of a vaneless design and where b = const., the exit velocity can be approximated as follows. For the radial velocity component, consider the continuity equation:

$$\rho c_r 2\pi r b = \text{const.}$$

Applying this between station 2 and exit station 3,

$$\rho c_r 2\pi r b|_2 = \rho c_r 2\pi r b|_3 ,$$

or assuming incompressible flow (and b = const.), we get

$$\frac{c_{r3}}{c_{r2}} = \frac{r_2}{r_3}. \tag{11.66}$$

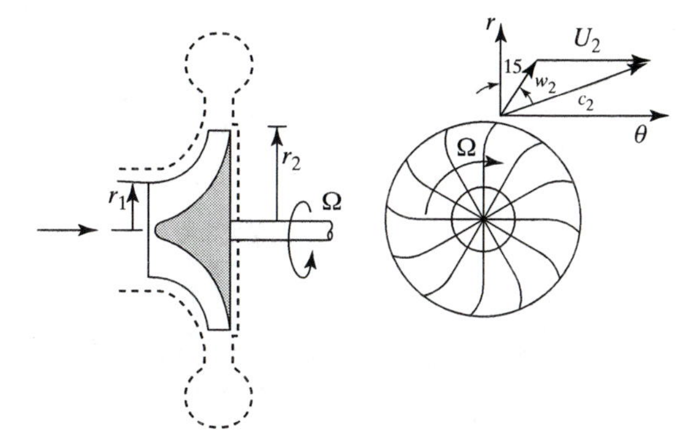

Figure 11.31. The centrifugal compressor for Examples 11.5 and 11.6.

Similarly, for the tangential velocity component of the vaneless diffuser, the conservation of angular momentum can be applied:

$$c_\theta r = \text{const.}$$

Applying this between station 2 and station 3, we get

$$\frac{c_{\theta 3}}{c_{\theta 2}} = \frac{r_2}{r_3}. \tag{11.67}$$

Because both velocity components are being reduced inversely with the radius, the angle α between these components is the same at any station in the diffuser, as shown in Fig. 11.30(a). In conclusion, we can write for this case that

$$\frac{c_3}{c_2} = \frac{r_2}{r_3}. \tag{11.68}$$

For example, to reduce the impeller exit velocity by two, the radius of the exit must be twice as large as the impeller diameter. In airborne applications, such as turbojets, a reduction in diffuser size is desirable and internal turning vanes are used.

EXAMPLE 11.6. CENTRIFUGAL COMPRESSOR. Estimate the compression ratio of a centrifugal compressor rotating at 5000 RPM (see Fig. 11.31). The fluid is air, the impeller dimensions are $r_1 = 0.1$ m and $r_2 = 0.4$ m, and the tangential velocity at the impeller inlet is negligible. At the exit, the blade is leaning forward at $\beta = 15^\circ$ and we assume that $w_{r2} = U_2/2$. Also, $\eta_c = 0.90$, $c_p = 0.24$, $\gamma = 1.4$, and $T_{01} = 300$ K.

Solution: The compression ratio can be calculated with Eqs. (11.53) and (11.55), but first the exit-velocity diagram must be constructed (as shown in Fig. 11.31). The blade tip velocity at station 2 is

$$U_2 = 2\pi 0.4 \frac{5000}{60} = 209.44 \text{ m/s}.$$

The exit radial velocity is then

$$w_{r2} = U_2/2 = 104.72 \text{ m/s}.$$

The exit tangential velocity is calculated based on the velocity triangle at the exit

$$c_{\theta 2} = w_{r2} \tan 15 + U_2 = 237.50 \text{ m/s}.$$

The stagnation temperature rise is calculated with Eq. (11.55),

$$\frac{T_{02}-T_{01}}{T_{01}}=\frac{(U_2\,c_{\theta 2}-U_1\,c_{\theta 1})}{c_p T_{01}}=\frac{(209.44\times 237.50-0)}{0.24\times 4200\times 300}=0.1645,$$

and the pressure rise, based on Eq. (11.53), is

$$\frac{p_{02}}{p_{01}}=\left(1+\eta_c\frac{T_{02}-T_{01}}{T_{01}}\right)^{\frac{\gamma}{\gamma-1}}=(1+0.9\times 1.645)^{3.5}=1.62.$$

EXAMPLE 11.7. CENTRIFUGAL PUMP. Estimate the compression ratio of a centrifugal pump rotating at 1000 RPM. The fluid is water, the impeller dimensions are $r_1=0.05$ m and $r_2=0.10$ m, and the exit blade is straight ($\beta=0$). Also, inlet swirl is negligible and $\eta_c=0.90$.

Solution: To calculate the pressure rise in the pump, Eq. (11.52) is used. Note that, for straight blades, at the exit $U_2=c_{\theta 2}=2\pi r_2\Omega$,

$$\frac{\Delta p_0}{\rho}=\eta_c(U_2\,c_{\theta 2}-0)=0.90\left(2\pi\times 0.1\times\frac{1000}{60}\right)^2=174.53\left(\frac{\text{m}}{\text{s}}\right)^2$$

and the pressure rise is

$$\Delta p_0=1000\times 174.53=174{,}532.3\left(\frac{\text{N}}{\text{m}^2}\right),\quad\text{or}\ \sim 1.78\,\text{atm}.$$

Next, let us estimate the lowest mass flow rate, on the verge of inlet stall [e.g., when $\beta_1\approx 50^\circ$, based on Eq. (11.59)]. The tip speed at the inlet is then

$$U_1=2\pi\,0.05\frac{1000}{60}=5.24\ \text{m/s}.$$

The axial velocity is then calculated, based on the inlet velocity triangle at the tip (see Fig. 11.28),

$$c_1=U_1\tan(90-\beta_1)=5.24\times\tan 40=4.39\ \text{m/s},$$

and the mass flow rate is

$$\dot{m}=\rho c_1 S=1000\times 4.39\times\pi\times 0.05^2=34.50\ \text{kg/s}.$$

The minimum power at this RPM is then calculated with Eq. (11.49),

$$P=\dot{m}(U_2\,c_{\theta 2}-0)=34.50\left(2\pi\,0.1\frac{1000}{60}\right)^2=3783.35\ \text{W},$$

which is about 5.14 hp.

This example shows again that a centrifugal compressor performance depends mainly on the impeller exit radius.

11.4.4 Concluding Remarks: Axial versus Centrifugal Design

After this brief survey of the design principles of continuous-flow pumps–compressors, our next question is this: Which type of machinery is preferred for a particular application? When we compare the centrifugal and the axial designs, which deliver the same mass flow rate and pressure ratio, there are no major differences in efficiencies. This is probably true when both designs are optimized for a specific operation point. In general, however, the axial design tends to achieve higher efficiencies. In terms of weight, there is also no major difference; however, the shape quite frequently dictates the applications. Usually the axial design is longer and its smaller frontal area makes it more attractive for aircraft compressors. On the other hand, the compression ratio of centrifugal design is significantly higher (per stage). If we consider the compression of air, then the maximum compression ratio of a single-stage axial compressor was estimated in the previous section at about 1.7, but more realistically it is closer to 1.2. At the same time, a centrifugal compressor stage can generate a compression ratio of 4! This is why most automotive or hydraulic pumps have a centrifugal design. Another consideration is multi-staging, particularly for very high compression ratios (as in jet engines). Stacking up rows of compressor stages, one behind the other, is natural for the axial design but quite awkward for the centrifugal machine.

11.5 Axial Turbines

Turbines are machines that extract power from a moving fluid and were used for centuries, mainly in form of a large bladed wheel rotated by falling water. In modern times turbines are used for power generation, in jet engines, or even to drive the turbochargers in passenger cars. A turbine design can be of the axial or centrifugal or even mixed geometry, and the discussion is similar to the approach used for the axial and centrifugal compressors. Also, because the pressure drop across a turbine is favorable (from high to low) it is less sensitive to flow separation compared with a similar compressor or pump. Therefore turbine efficiencies are usually higher than for an equivalent compressor. In many applications in which the turbine is driven by combustion products (such as gas turbines of jet engines) blade cooling is also a very important aspect of the fluid dynamic considerations. Because the axial and centrifugal designs were discussed in detail, only the axial turbine is discussed here, and the formulation can be easily extended to treat centrifugal designs as well (as was done for the compressors).

A schematic description of a single-stage axial turbine is shown in Fig. 11.32. Although the approach is the same as used for the axial compressor in Section 11.3, there are some minor differences:

1. A typical stage consists of rotor and stator blade rows; however, the stator (called now the nozzle) is ahead of the rotor. Consequently station 2 is the exit from the nozzle and the rotor is between stations 2 and 3.

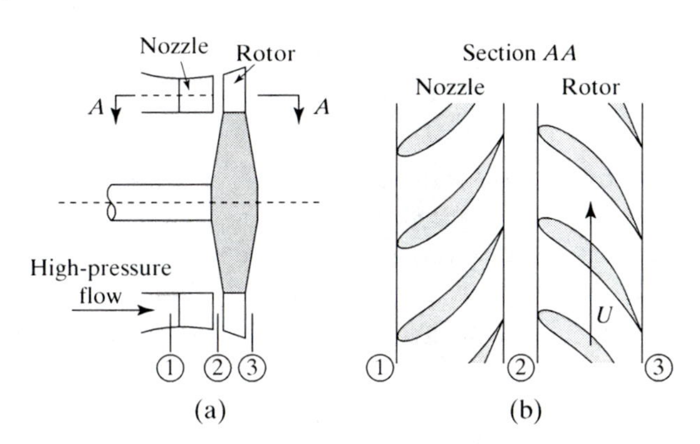

Figure 11.32. Schematic description of (a) a single-stage turbine and (b) the blade geometry as seen from section cut AA.

2. Because the fluid works on the rotor (a change in sign from the thermodynamic point of view), the direction of rotation is reversed to highlight this (also the direction of the coordinate θ is reversed).

Blade shape and orientations are shown through cut AA in Fig. 11.32(b). This is a 2D representation of the average-radius model, and radial variations can be treated following the approach presented in Subsection 11.3.3. There are also multistage turbines, and then a stator (nozzle) row will follow station 3 in Fig. 11.32.

11.5.1 Torque, Power, and Pressure Drop

The torque on the row of stators or rotors can be calculated by Eq. (11.8). When applied between station 2 and station 3 (rotor),

$$Tq = \dot{m}(r_2\, c_{\theta 2} - r_3\, c_{\theta 3}). \tag{11.69}$$

The power for the rotor is then calculated by Eq. (11.9) (or by simply multiplying the torque by the rotation speed):

$$P = Tq\,\Omega = \dot{m}\Omega(r_2\, c_{\theta 2} - r_3\, c_{\theta 3}). \tag{11.70}$$

Similarly for the nozzle (stator row) we can write

$$Tq = \dot{m}(r_1\, c_{\theta 1} - r_2\, c_{\theta 2}). \tag{11.69a}$$

and the torque acting on the nozzle is in the direction opposite to the torque on the rotor. Also, because the nozzle is stationary, it generates no power:

$$P = 0. \tag{11.70a}$$

At this point, if we limit the discussion to axial turbines and introduce the average-radius assumption, then the rotor blade speed is defined as before:

$$U = r\Omega. \tag{11.71}$$

With this definition, the equation for the power generated becomes

$$P = \dot{m}U(c_{\theta 2} - c_{\theta 3}), \tag{11.72}$$

and the work per unit mass w_t is

$$w_t = \frac{P}{\dot{m}} = U(c_{\theta 2} - c_{\theta 3}). \tag{11.73}$$

The calculation of the pressure drop would be similar to the axial compressor–pump case. However, the stage efficiency η_t now represents the ratio between the actual work of the turbine divided by the work invested. In terms of the enthalpy change this becomes

$$\eta_t = \frac{\Delta h_0}{\Delta h_{0s}}. \tag{11.74}$$

Note that the losses through both the nozzle and rotor rows are included in this formulation. The actual work invested, per unit mass, is

$$\Delta h_0 = \frac{\Delta p_0}{\rho}. \tag{11.75}$$

The work generated by the turbine w_t is less (because of the losses) and therefore

$$w_t = \eta_t \frac{\Delta p_0}{\rho}.$$

Substituting w_t from Eq. (11.73) shows that, for the incompressible case, the stagnation pressure drop (for a given power) is increased with reduced efficiency!

$$\frac{\Delta p_0}{\rho} = \frac{U}{\eta_t}(c_{\theta 2} - c_{\theta 3}). \tag{11.76}$$

In the case of a turbine it is more useful to rewrite the power generated as a function of the actual pressure drop. We do this by combining Eqs. (11.76) and (11.73):

$$P = \dot{m}\eta_t \frac{\Delta p_0}{\rho} = \dot{m}U(c_{\theta 2} - c_{\theta 3}). \tag{11.78}$$

For the compressible case we can rewrite Eq. (11.74) in terms of the stagnation temperature as

$$\eta_t = \frac{\Delta h_0}{\Delta h_{0s}} = \frac{c_p(T_{01} - T_{03})}{c_p(T_{01} - T_{03s})}. \tag{11.79}$$

Note that the enthalpy change is calculated across the whole stage, including both rotor and stator. The actual temperature change (for calculating Δh_0) is equal to the work per unit mass, in Eq. (11.72):

$$\Delta h_0 = c_p(T_{01} - T_{03}) = w_t = U(c_{\theta 2} - c_{\theta 3}). \tag{11.80}$$

Consequently we can write

$$\frac{\Delta T_0}{T_{01}} = \frac{T_{01} - T_{03}}{T_{01}} = \frac{U(c_{\theta 2} - c_{\theta 3})}{c_p T_{01}}. \tag{11.81}$$

As noted earlier, it is more useful to relate the work to the pressure drop across the turbine. This can be accomplished by use of the isentropic (adiabatic and reversible) process as defined by Eq. (11.27):

$$\frac{p_{03}}{p_{01}} = \left(\frac{T_{03s}}{T_{01}}\right)^{\frac{\gamma}{\gamma-1}}. \tag{11.82}$$

The ideal temperature change can be replaced with the actual values, based on Eq. (11.79):

$$\frac{p_{03}}{p_{01}} = \left(\frac{T_{01} - T_{01} + T_{03s}}{T_{01}}\right)^{\frac{\gamma}{\gamma-1}} = \left(1 - \frac{T_{01} - T_{03s}}{T_{01}}\right)^{\frac{\gamma}{\gamma-1}} = \left(1 - \frac{T_{01} - T_{03}}{\eta_t T_{01}}\right)^{\frac{\gamma}{\gamma-1}}. \tag{11.83}$$

Substituting the temperature change from Eq. (11.81), we find that the pressure drop across the turbine becomes

$$\frac{p_{03}}{p_{01}} = \left[1 - \frac{U(c_{\theta 2} - c_{\theta 3})}{\eta_t c_p T_{01}}\right]^{\frac{\gamma}{\gamma-1}}. \tag{11.84}$$

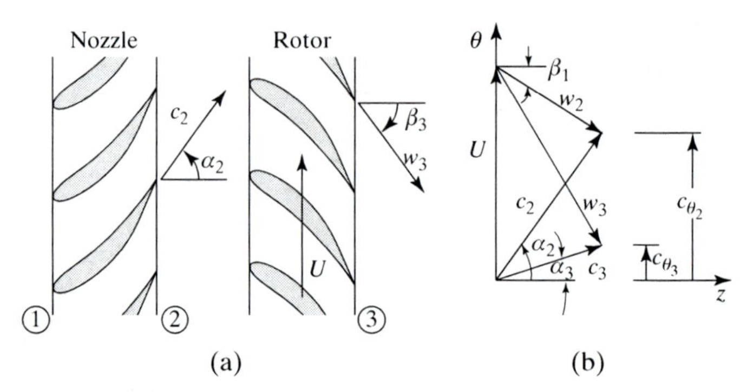

Figure 11.33. Schematic description of an axial turbine nozzle and rotor and resulting rotor velocity triangle.

In the case of the compressible flow turbine it is also of interest to estimate the power generated for a given pressure drop across the whole stage (nozzle and rotor). We do this by combining Eqs. (11.72) and (11.84):

$$P = \dot{m} c_p T_{01} \eta_t \left[1 - \left(\frac{p_{03}}{p_{01}} \right)^{\frac{\gamma-1}{\gamma}} \right]. \tag{11.85}$$

11.5.2 Axial Turbine Geometry and Velocity Triangles

To calculate the pressure drop or the power generated by the turbine, the various velocity components must be identified. Furthermore, drawing of the velocity diagrams allows the proper alignment of the turbine blades relative to the local velocity components. This is needed from the blade airfoil section point of view (to avoid flow separations). A typical axial turbine velocity diagram is shown in Fig 11.33. The stationary nozzle is accelerating the high-pressure fluid between stations 1 and 2 and the exit velocity c_2 is shown in the figure (and the positive direction for measuring α). The velocity triangles for the rotor are shown in Fig. 11.33(b). Note the direction of the U vector, and, of course, in the rotor frame of reference the direction is reversed. By adding the blade velocity to the incoming velocity c_2, we obtain the relative rotor inlet velocity w_2. This clarifies the orientations of the rotor blades and the rotor exit-velocity w_3 direction (which for attached flow should be parallel to the rotor trailing-edge bisector). Once the incoming- and exiting-velocity triangles are drawn [as in Fig. 11.33(b)], assuming no change in c_z between stations 2 and 3, the tangential velocity components are identified. These components ($c_{\theta 2}, c_{\theta 3}$) then can be used in the pressure-drop and power calculations developed in the previous section.

EXAMPLE 11.8. AXIAL TURBINE. A turbine wheel shown in Fig. 11.34(a) is operated by a water jet with a cross section of 3 cm^2 and a flow rate of 15 kg/s. The nozzle angle is $\alpha_2 = 68.5°$ and the blade inflow and outflow angles are the same ($\beta_2 = -\beta_3$). The average radius of the turbine blades is $r = 0.3$m. The turbine operates a pulley, which is lifting a mass of 100 kg [see Fig. 11.34(a)], and the rotation rate is stabilized at 1000 RPM. Calculate the blade angle β_2 and how fast the weight is being lifted.

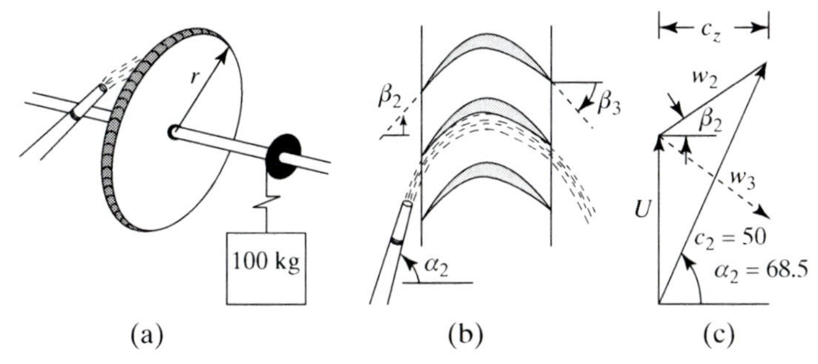

Figure 11.34. An axial turbine operated by a water jet, raising a mass of 100 kg.

Solution: The rotor velocity triangle can be established by the calculation of U and c_2. The blade-tip speed is then

$$U = r\Omega = 0.3\, 2\pi \frac{1000}{60} = 31.41 \text{ m/s}.$$

The nozzle velocity is calculated with the continuity equation (the water density is 1000 kg/m^3):

$$c_2 = \frac{\dot{m}}{\rho S} = \frac{15}{1000 \times 0.0003} = 50.00 \text{ m/s}.$$

Knowing the magnitude of two vectors; the tip speed U, the jet speed c_2 and the nozzle angle α_2, we can draw the velocity triangle [see Fig. 11.34(c)]. The angle β_2 is now easily calculated based on the triangle trigonometry; but first the axial velocity is calculated:

$$c_z = c_2 \cos\alpha_2 = 18.32 \text{ m/s}.$$

The rotor inlet angle is

$$\tan^{-1}\beta_2 = \frac{c_2 \sin\alpha_2 - U}{c_z} = 0.824,$$

$$\beta_2 = 39.5^\circ.$$

and this angle is shown in Fig. 11.34(b). The power of the turbine is calculated with Eq. (11.72):

$$P = \dot{m}U(c_{\theta 2} - c_{\theta 3}) = \dot{m}U(2c_z \tan\beta_2) = 15 \times 31.41 \times 30.22 = 14{,}238.15 \text{ W},$$

which is about 19.35 hp. The power required for lifting the mass at a velocity of v is

$$P = mgv,$$

and therefore

$$v = \frac{P}{mg} = \frac{14{,}238.153}{100 \times 9.8} = 14.52 \text{ m/s}.$$

11.5.3 Turbine Degree of Reaction

The degree of reaction was used in Subsection 11.3.5 to define the pressure-drop ratio between the rotor and the entire stage (including the stator). Similarly, the

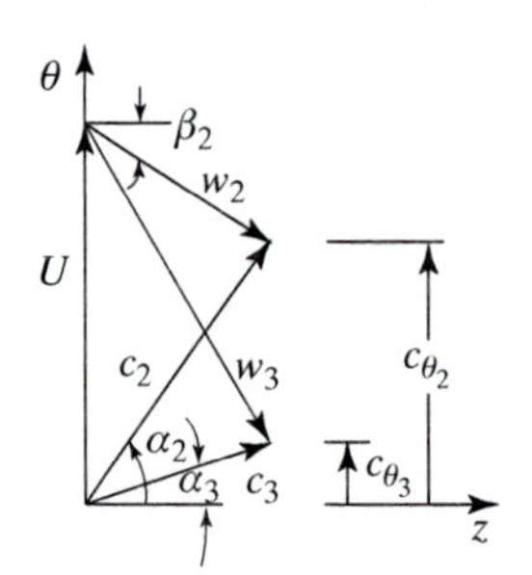

Figure 11.35. Generic velocity triangles for an axial turbine.

degree of reaction R can be defined for a turbine as the enthalpy drop in the rotor versus the stagnation enthalpy change in the whole stage. We approximated the changes by using the pressure changes in Subsection 11.3.5, and using the same approach we get

$$R = \frac{h_2 - h_3}{h_{01} - h_{03}} \approx \frac{p_2 - p_3}{p_{01} - p_{03}}. \tag{11.86}$$

We can estimate the rotor pressure drop across the rotor by using the incompressible Bernoulli equation (note that this is an approximation only!):

$$h_2 - h_3 = \frac{1}{\rho}(p_2 - p_3) = \frac{1}{2}\left(w_3^2 - w_2^2\right),$$

and the total stage stagnation enthalpy change is taken from Eq. (11.80) [e.g., $p_{0_1} - p_{0_3} = U(c_{\theta_2} - c_{\theta_3})$]; therefore

$$R = \frac{w_3^2 - w_2^2}{2U(c_{\theta 2} - c_{\theta 3})}. \tag{11.87}$$

We can rewrite the numerator, similar to the case of the axial compressor [Eq. (11.44)], in terms of the velocity components:

$$w_3^2 - w_2^2 = w_{\theta_3}^2 + w_{z_3}^2 - w_{\theta_2}^2 - w_{z_2}^2 = w_{\theta_3}^2 - w_{\theta_2}^2 = (w_{\theta_3} + w_{\theta_2})(w_{\theta_3} - w_{\theta_2}),$$

but the tangential velocity difference is the same: $(w_{\theta_3} - w_{\theta_2}) = (c_{\theta 2} - c_{\theta 3})$ and therefore the degree of reaction is

$$R = \frac{w_{\theta_3} + w_{\theta_2}}{2U}. \tag{11.88}$$

Observing the velocity triangles in Fig. 11.35, we can write

$$w_{\theta 3} = c_z \tan\beta_3,$$
$$w_{\theta 2} = U - c_z \tan\alpha_2.$$

Consequently the sum in the numerator becomes

$$w_{\theta 3} + w_{\theta 2} = c_z \tan\beta_3 + U - c_z \tan\alpha_2.$$

With this in mind, the expression for the degree of reaction becomes

$$R = \frac{w_{\theta 3} + w_{\theta 2}}{2U} = \frac{c_z \tan\beta_3 + U - c_z \tan\alpha_2}{2U} = \frac{1}{2} - \frac{c_z}{2U}(\tan\alpha_2 - \tan\beta_3). \tag{11.89}$$

This indicates that if a 50% degree-of-reaction design is sought (e.g., $R = 1/2$), then the two angles α_2 and β_3 must be equal (and of course pointing in opposite directions, as shown in Fig. 11.36.

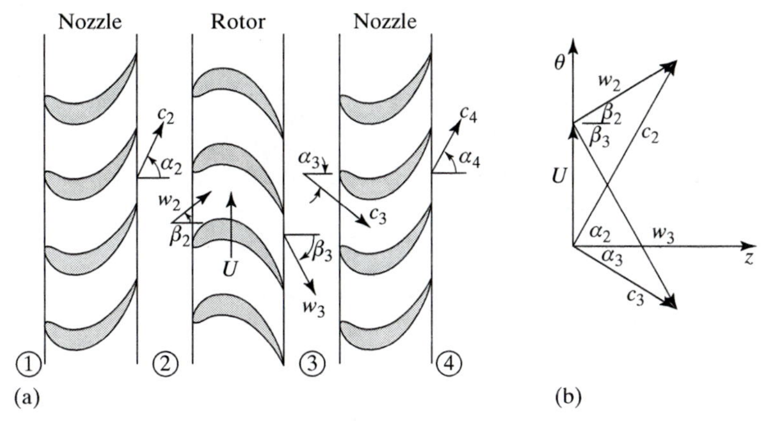

Figure 11.36. (a) 50% degree-of-reaction turbine and (b) corresponding rotor velocity triangle.

Such a design is shown schematically in Fig. 11.36a, and the nozzle (stator) and rotor incoming- and exiting-velocity vectors are the same (but reversed in direction). Consequently the rotor velocity diagram [(Fig. 11.36(b)] is symmetrical about an imaginary horizontal line, as shown in the figure. This implies that the blades of the nozzle and of the rotor are exposed to similar angles of attack and therefore their shape is the same, as well. Figure 11.36(a) depicts this condition in which the nozzle exit at station 2 is the same as the nozzle exit at station 4, and so on. Using the symmetry assumption, we find that the tangential velocities of the rotor are

$$c_{\theta 2} - c_{\theta 3} = 2(c_z \tan\alpha_2 - U) + U.$$

Substituting this into Eq. (11.81), we get

$$\frac{\Delta T_0}{T_{01}} = \frac{U(c_{\theta 2} - c_{\theta 3})}{c_p T_{01}} = \frac{U}{c_p T_{01}}[2(c_z \tan\alpha_2 - U) + U]$$
$$= \frac{U^2}{c_p T_{01}}\left[2\frac{c_z}{U}\tan\alpha_2 - 1\right]. \tag{11.90}$$

This equation indicates that larger nozzle angles α_2 result in more power. Of course, that will reduce the axial velocity, and $\alpha_2 = 70°$ is considered to be the maximum nozzle angle.

EXAMPLE 11.9. MULTISTAGE AXIAL TURBINE. The average radius of the second stage in a multistage axial turbine is $r = 0.3$ m, rotation speed is 7000 RPM, and the hot gas flow leaves the nozzle at a temperature of $T_{01} = 1200$ K (and $\rho_2 = 0.5$ kg/m^3). Assuming that this is a 50% degree-of-reaction design, the rotor inflow angle is $\beta_2 = 30°$, and the exit angle is $\beta_3 = 60°$, calculate the following values:

(a) The power per unit mass generated by this turbine ($\eta = 0.90$, $\gamma = 1.4$, $c_p = 0.24$).
(b) The stagnation pressure drop in the turbine. How much of the static pressure drop takes place in the rotor?

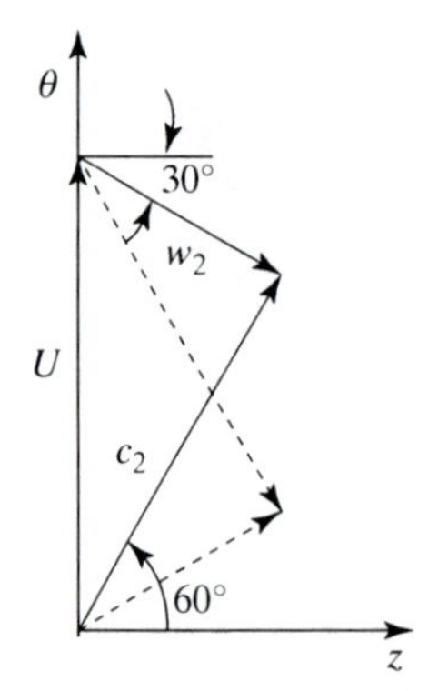

Figure 11.37. Velocity triangles for a 50% degree-of-reaction turbine (in Example 11.9).

Solution: To establish the rotor velocity triangle, we calculate the local blade speed U:

$$U = r\Omega = 0.3\, 2\pi \frac{7000}{60} = 219.91 \text{ m/s}.$$

Using the symmetry of the velocity triangle in Fig. 11.37, we conclude that $\alpha_2 = |\beta_3| = 60°$. In terms of a graphical solution we plot the vector U on the θ axis and then draw a 30° line from the top, representing the direction of w_2. Similarly, a 60° line is drawn at the bottom to represent the direction of the c_2 vector. As the two curves intersect, the axial velocity is found and the velocity triangle for station 2 is complete. We draw the velocity triangle next at the turbine exit (station 3) by using the symmetry principle (dashed lines).

An algebraic solution can also be based on the velocity diagram at station 2. For example, we can write for the tangential component $w_{\theta 2}$

$$U - c_2 \sin 60° = w_2 \sin 30°$$

and for the axial component c_z

$$c_2 \cos 60° = w_2 \cos 30°.$$

Solving these two equations for the unknowns c_2 and w_2 results in

$$w_2 = U/2,$$
$$c_2 = 0.866U.$$

The tangential velocity difference is therefore

$$c_{\theta 2} - c_{\theta 3} = U - 2w_2 \sin 30° = U/2 = 109.95 \text{ m/s}.$$

The power of the turbine (per unit mass) is calculated with Eq. (11.73):

$$w_t = U(c_{\theta 2} - c_{\theta 3}) = 219.91 \times 109.95 = 24{,}179.10 \text{ m}^2/\text{s}^2.$$

Next, the temperature drop is calculated with Eq. (11.93),

$$\frac{\Delta T_0}{T_{01}} = \frac{U(c_{\theta 2} - c_{\theta 3})}{c_p T_{01}} = \frac{219.91 \times 109.95}{0.24 \times 4200 \times 1200} = 0.02,$$

and the pressure ratio with Eq. (11.83),

$$\frac{p_{03}}{p_{01}} = \left(1 - \frac{T_{03} - T_{01}}{\eta_t T_{01}}\right)^{\frac{\gamma}{\gamma-1}} = \left(1 - \frac{0.02}{0.9}\right)^{3.5} = 0.924.$$

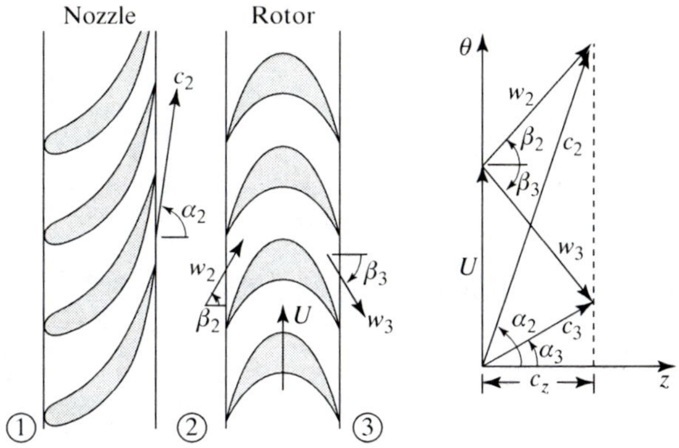

Figure 11.38. Impulse turbine and corresponding rotor velocity triangle.

Because this is a 50% degree-of-reaction turbine, half of the pressure drop takes place in the turbine.

Another interesting example is the *impulse turbine*. In this case, all of the stagnation pressure drop takes place in the nozzle and the pressure energy is converted entirely into velocity. Impulse turbines are usually selected for single-stage designs or the last stage of a multistage turbine. This concept is also the oldest form of turbine used with water jets (see Fig. 11.34). It is clear that the best results could be obtained when all the energy in the water is converted into velocity (or a jet) and then this jet impinges on the turbine blade. Such a design is less sensitive to the delicate airfoil shape of more complex turbine designs. Because all the stagnation pressure drop takes place in the nozzle, the static pressure change in the rotor is zero! Consequently, the degree of reaction, according to Eq. (11.86), is zero as well. For an impulse turbine,

$$R = \frac{p_2 - p_3}{p_{01} - p_{03}} = 0. \tag{11.91}$$

Recalling Eq. (11.87) and equating it to zero yields

$$R = \frac{w_3^2 - w_2^2}{2U(c_{\theta 2} - c_{\theta 3})} = 0,$$

and therefore the incoming- and exiting-velocity vectors are of the same magnitude. So for the case of an impulse turbine:

$$w_2 = w_3. \tag{11.92}$$

As a result the velocity triangles will have the shape depicted in Fig. 11.38. The rotor blade shape will reflect this fore–aft symmetry, as shown schematically in the figure.

From the velocity diagram shown in Fig. 11.38 and assuming that the rotor incoming- and exiting-velocity vectors have the same magnitude, we can write:

$$c_{\theta 2} - c_{\theta 3} = 2(c_z \tan \alpha_2 - U),$$

and after substituting this into Eq. (11.81), we get

$$\frac{\Delta T_0}{T_{01}} = \frac{U(c_{\theta 2} - c_{\theta 3})}{c_p T_{01}} = \frac{2U^2}{c_p T_{01}} \left(\frac{c_z}{U} \tan \alpha_2 - 1 \right). \tag{11.93}$$

Again, this equation indicates that larger nozzles angles α_2 result in more power. And the maximum recommended nozzle angle is $\alpha_2 = 70°$.

EXAMPLE 11.10. IMPULSE TURBINE. An impulse turbine with an average radius of $r_2 = 0.1$ m is operated by a hot gas flow at a rate of $\dot{m} = 1$ kg/s and a temperature of $T_{01} = 1200$ K. The jet leaves the nozzle at an angle of $\alpha_2 = 65°$ and a speed of $c_2 = 330$ m/s ($\rho_2 = 0.5$ kg/m^3).

(a) If the flow at the exit has no swirl, then calculate the power generated by this turbine ($\eta = 0.90$, $\gamma = 1.4$, $c_p = 0.24$).
(b) Calculate the stagnation pressure drop in the turbine. How much of the static pressure drop takes place in the rotor?
(c) Calculate shaft RPM.

Solution: If the exit velocity c_3 has no swirl then $\alpha_3 = 0$, and by observing the velocity diagram in Fig. 11.39(a) for the impulse turbine with $\alpha_3 = 0$ only; we get

$$c_{\theta 2} - c_{\theta 3} = 2U = c_2 \sin 65 = 299.08 \text{ m/s}.$$

The power of the turbine is calculated with Eq. (11.72):

$$P = \dot{m}U(c_{\theta 2} - c_{\theta 3}) = 1.00\frac{299.08}{2}299.08 = 44{,}724.8 \text{ W},$$

which is about 60 hp. The stagnation temperature drop is calculated with Eq. (11.93),

$$\frac{\Delta T_0}{T_{01}} = \frac{U(c_{\theta 2} - c_{\theta 3})}{c_p T_{01}} = \frac{\frac{299.08}{2}299.08}{0.24 \times 4200 \times 1200} = 0.037,$$

and the pressure ratio from Eq. (11.83),

$$\frac{p_{03}}{p_{01}} = \left(1 - \frac{T_{03} - T_{01}}{\eta_t T_{01}}\right)^{\frac{\gamma}{\gamma-1}} = \left(1 - \frac{0.037}{0.9}\right)^{3.5} = 0.863.$$

This is an impulse turbine and all the stagnation pressure drop takes place in the nozzle (e.g., zero pressure drop in the rotor). Because the vector U and the rotor radius are known, the rotation rate can be calculated as

$$\text{RPM} = \frac{U}{2\pi r} = \frac{299.08/2}{2 \times 3.14 \times 0.1}60 = 14{,}280.$$

Usually in the last turbine stage, or if there is only one stage, it is desirable for the exit flow to leave without swirl (or zero tangential velocity). In terms of the velocity diagram this means that $\alpha_3 = 0$. Such a condition is sketched in Fig. 11.39 for a 50% degree-of-reaction and an impulse turbine. Assuming that both have the same radius, rotation speed (same U), and axial velocity, we can see that for the $R = 50\%$ design the tangential velocity difference is equal to the blade speed ($c_{\theta 2} - c_{\theta 3} = U$). For the impulse turbine, the tangential velocity difference is twice the rotation speed ($c_{\theta 2} - c_{\theta 3} = 2U$), suggesting that it generates twice the power of the $R = 50\%$ design. This is true for the conditions shown in Fig. 11.39, but at the same time the nozzle exit speed c_2 is significantly higher for the impulse turbine.

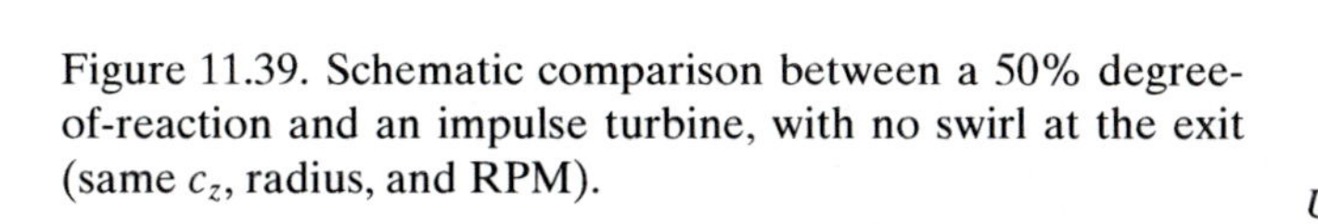

Figure 11.39. Schematic comparison between a 50% degree-of-reaction and an impulse turbine, with no swirl at the exit (same c_z, radius, and RPM).

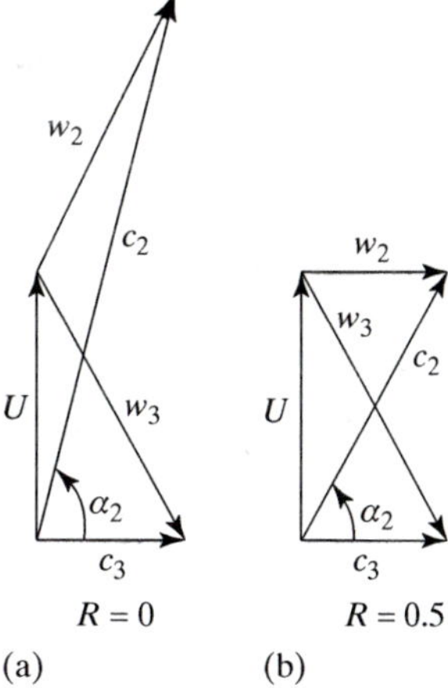

Also the fluid crossing the turbine blades is turned more sharply by the impulse turbine (more camber). In spite of these considerations, the impulse turbine has the potential to generate more power.

EXAMPLE 11.11. THE PELTON WHEEL. One important form of the impulse turbine, called the Pelton wheel (after Ohio-born Lester Allan Pelton, 1829–1908), is used in power-generating plants and is shown schematically in Fig. 11.40(a). In this case all the pressure drop takes place in the nozzle and the turbine blade is simply changing the direction of the jet. This is shown schematically in Fig. 11.40(b).

The velocity diagram in the z–θ coordinate system is shown schematically in Fig. 11.40(c) and the jet velocity c_2 has no velocity component in the z direction! Still, the treatment of this turbine is no different from that of the axial impulse turbine. As an example, consider a Pelton wheel with an average radius of $r =$ 0.1 m, which rotates at 1000 RPM by a water jet with an exit velocity of $c_2 =$ 30 m/s (assume the jet cross section is 2 cm^2). Also, the angle of the flow leaving the blade in this example is $\beta_3 = 60°$. Because this is a Pelton wheel and the nozzle is aligned with the wheel ($\alpha_2 = 90°$), the velocity relative to the

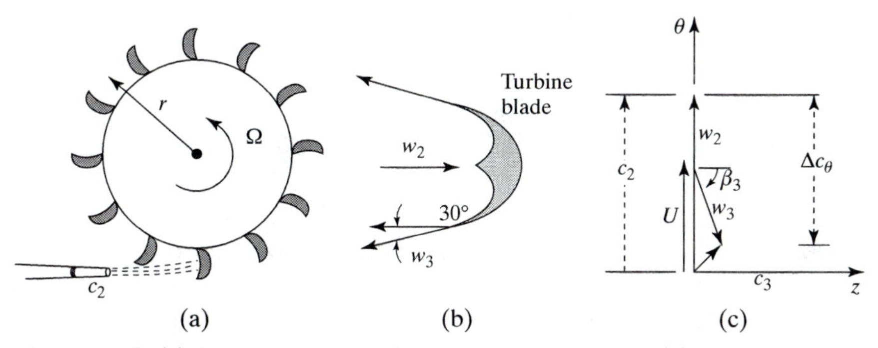

Figure 11.40. (a) Schematic description of the Pelton wheel, (b) the rotor blade geometry, and (c) the velocity diagram.

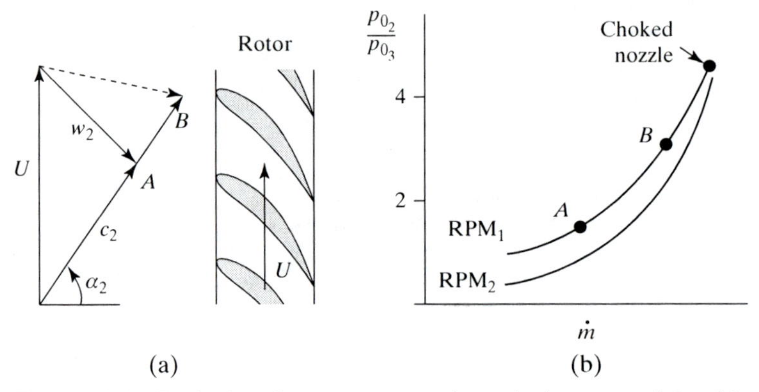

Figure 11.41. Typical performance curves for a single-stage axial turbine.

blade (w_2) is the difference between the jet and the wheel tip speeds, as shown Fig. 11.40(c):

$$w_2 = c_2 - U.$$

The blade tip velocity is

$$U = r\Omega = 0.1 \times 2\pi \frac{1000}{60} = 10.47 \text{ m/s}.$$

and w_2 is

$$w_2 = c_2 - U = 30 - 10.47 = 19.53 \text{ m/s}.$$

Because this is an impulse turbine the exit velocity is unchanged,

$$w_3 = w_2 = 19.53 \text{ m/s}.$$

and the torque and power can be calculated with Eqs. (11.69) and (11.70). The change in tangential velocity is therefore [see Fig. 11.40(c)],

$$\Delta c_\theta = c_{\theta 2} - c_{\theta 3} = 19.53(1 + \sin 60) = 36.44 \text{ m/s}.$$

The mass flow rate is

$$\dot{m} = \rho c_2 S = 1000 \times 30 \times 2 \times 10^{-4} = 6 \text{ kg/s}.$$

The torque and power now can be calculated as

$$Tq = \dot{m} r \Delta c_\theta = 6 \times 0.1 \times 36.44 = 21.86 \text{ Nm},$$
$$P = \dot{m} U \Delta c_\theta = 6 \times 10.47 \times 36.44 = 2289.16 \text{ W}.$$

Now that the relation among turbine geometry, power, and pressure drop is established, we can discuss the generic shape of a typical performance map. Usually, turbine mass flow rate can be controlled, thereby dictating the incoming velocity c_2 and the pressure drop across the stage. Consequently a typical turbine performance map will show the pressure drop versus mass flow rate across the stage (as shown schematically in Fig. 11.41). The rotor blade geometry is depicted in Fig. 11.41(a) and the solid-line velocity triangle represent a lower mass flow rate (point A). If mass flow rate increases at a constant RPM$_1$, so will c_2, as shown by the dashed line (point B). Because the incoming flow angle α_2 (nozzle angle) remains unchanged,

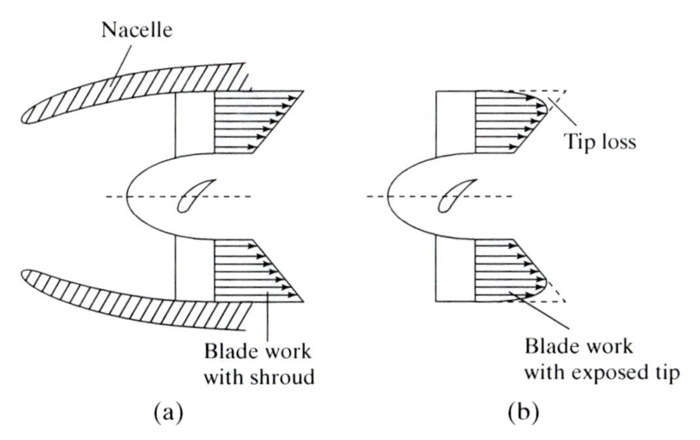

Figure 11.42. The effect of an exposed tip on rotor performance.

c_2 in the second case is longer, as shown. This increases the angle of attack of the rotor blade, resulting in a higher load (e.g., higher lift coefficient) and a higher pressure ratio. Therefore, at the higher mass flow rate, the compression ratio will also increase, as shown by point B on the performance map. Now, if rotation is reduced to RPM_2, then the same trend is observed, but at a slightly lower pressure ratio. If the mass flow rate is further reduced, the incoming flow angle of attack is reduced, up to a point where the turbine cannot generate power (so no airfoil lift – this is where the lines end at the lower mass flow rates). At the higher mass flow rates, a point is reached at which the nozzle is chocked, and this represents the highest mass flow rate point in Fig. 11.41(b).

11.5.4 Remarks on Exposed Tip Rotors (Wind Turbines and Propellers)

The discussion so far has focused on turbomachines in which the rotor is enclosed in an outer casing and the discussion has been limited to simple 2D models (as demonstrated with the rotor velocity diagrams). However, frequently seen, similar devices such as wind turbines, propellers, and many cooling fans have no outer shroud. The question then is this: Why wasn't the simple analysis of this chapter extended to include those cases, as well? The main reason is that with open-tip rotors the flow is highly 3D and more advanced analysis methods are required (such as CFD in Chapter 9). These effects can be grouped into two main categories that can be addressed here briefly, mainly as a prelude for more advanced course work. First, at the tip, no pressure difference between the upper and lower surfaces can be maintained. Therefore the loading drops to zero (at the tip) compared with the shrouded case, in which high loading can be present. This is depicted schematically in Fig. 11.42(a), where the expected pressure rise is shown for a shrouded compressor. On the right-hand side [Fig. 11.42(b)], the same rotor is shown but without the shroud. In this case the tip effect is reducing the loading (or pressure rise) not only at the tip but along the whole blade. So, in principle, ducted (or shrouded) fans can generate more thrust (or pressure rise) but will also require more torque because of higher loading at the larger-radius stations. Therefore the efficiency of open rotors is not necessarily inferior, particularly in airplane applications (because of the added drag and weight of the nacelle).

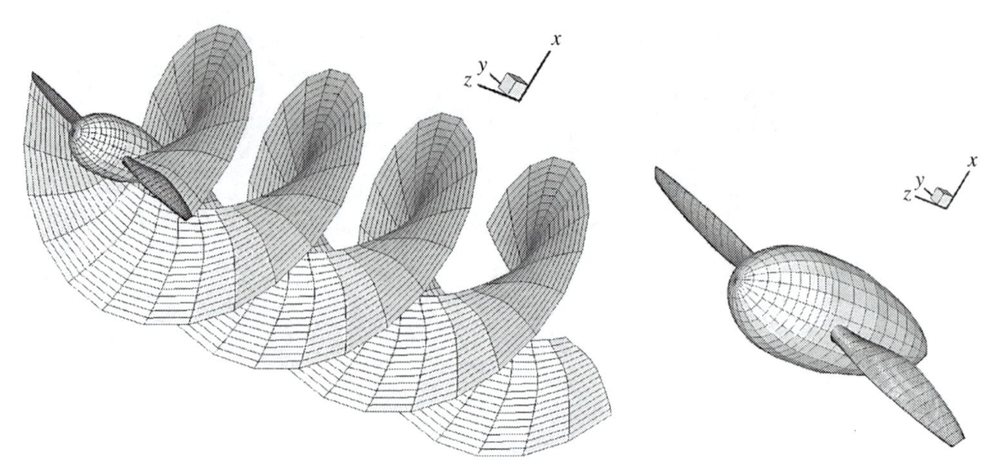

Figure 11.43. Visualization of the spiral wake behind a rotor (rotor geometry is shown on the right).

The second major feature of open-tip rotors is the creation of a strong vortex wake (which effect is negligible in the shrouded case), as shown in Fig. 11.43. This is similar to the tip effect of a finite wing, discussed in reference to Fig. 8.42, in which the pressure difference between the pressure and suction sides of the airfoil results in a rotating flow near the tip. A simple modeling of this wake is shown in Fig. 11.43, and the velocity field induced by this vortex wake has a strong effect as well. For example, when the forward speed is slow compared with the rotor-tip speed (as for a hovering helicopter) this wake-induced velocity may have the same order of magnitude as that of the free stream.

In spite of the previously mentioned modeling difficulties, simple stream-tube-based models were developed in the past to estimate both propeller and wind turbine performance. These approaches fail to connect blade geometry to parameters such as the pressure rise, but at least one, which discusses the maximum efficiency of a wind turbine, is worth mentioning.

This (anecdotal) model of a wind turbine is shown in Fig. 11.44 and is based on the 1D stream-tube model presented earlier in Chapter 2.

The flow enters the stream tube at station 1 at a speed of U_∞ and crosses through the propeller with a disk area of S. As the flow drives the rotor it slows down (between stations a and b) and the stream tube becomes wider at exit station 2

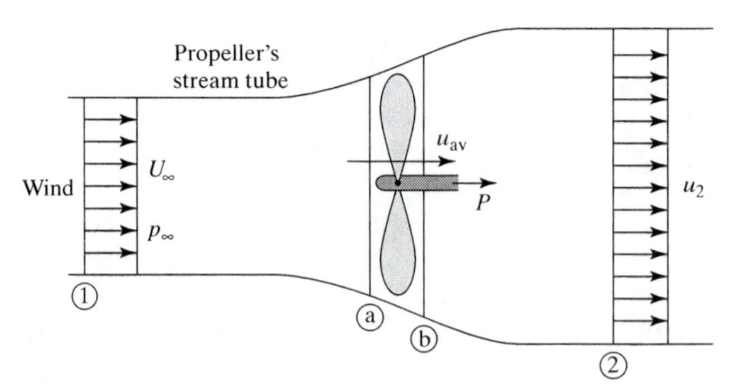

Figure 11.44. Simple stream-tube model for a wind turbine.

(recall the continuity equation), where the velocity is u_2. The force on the turbine must be equal to the momentum loss of the stream-tube flow:

$$F = \dot{m}(U_\infty - u_2),$$

where $\dot{m}$ is the mass flow rate. The power given away by the flow is the force times the average velocity (u_{av}) across the rotor:

$$P = F\, u_{av}.$$

For this 1D model the mass flow rate is

$$\dot{m} = \rho u_{av}\, S,$$

and an effort is made to use the rotor disk area instead of the conditions at station 1. Substituting the mass flow rate and the force term into the power expression yields

$$P = \rho u_{av} S(U_\infty - u_2)u_{av} = \rho u_{av}^2 S(U_\infty - u_2).$$

Next, the average velocity across the rotor disk is estimated as

$$u_{av} = \frac{1}{2}(U_\infty + u_2).$$

With this average velocity the power relation becomes

$$P = \rho\left(\frac{U_\infty + u_2}{2}\right)^2 S(U_\infty - u_2) = \frac{\rho S}{4}(U_\infty + u_2)^2(U_\infty - u_2). \qquad (11.94)$$

The importance of this relation is that it provides a hint about the best wind turbine performance. To find the maximum of this expression, a derivative of the power relative to the exit velocity u_2 yields

$$\frac{dP}{du_2} = \frac{\rho}{4}\left[(U_\infty + u_2)^2(-1) + 2(U_\infty + u_2)(U_\infty - u_2)\right] = 0,$$

and by solving this equation, we find that the condition for the maximum power becomes

$$u_2 = \frac{U_\infty}{3}. \qquad (11.95)$$

Substituting this result into the power relation provides the maximum wind turbine power:

$$P = \frac{\rho S}{4}(U_\infty + u_2)^2(U_\infty - u_2) = \frac{8}{27}\rho U_\infty^3 S. \qquad (11.96)$$

To place this result in perspective, consider the total kinetic energy entering the stream tube at station 1

$$P_{avail} = \frac{1}{2}\dot{m}U_\infty^2 = \frac{1}{2}\rho U_\infty^3 S.$$

Therefore, based on this simple analysis, the maximum energy that a wind turbine can extract from a stream having the same cross-section area S (let us call this the maximum efficiency) is

$$\eta_{max} = \frac{P}{P_{avail}} = \frac{16}{27}. \qquad (11.97)$$

Now, 16/27 is about 59.26%, and this number is usually called the Betz number or Betz efficiency (dating from circa 1926), after Albert Betz (1885–1968) who succeeded Ludwig Prandtl as the head of the Göttingen Center in Germany.

The important part of this discussion is that wind turbine performance is evaluated by a similar nomenclature. Typical values for the efficiency defined by Eq. (11.97) are near 20%–40%, which are significantly less than the Betz value.

EXAMPLE 11.12. SIZING OF A WIND TURBINE. It is proposed to build a small wind turbine capable of producing 5-kw power. If the average wind speed is 30 km/h and the expected design is capable of extracting 70% of the power predicted by the Betz formula ($\eta_p = 0.7$) calculate the wind turbine diameter.

Solution: Based on Eq. (11.96), the power is

$$P = \eta_P \frac{8}{27} \rho U^3 \pi R^2.$$

Solving for the propeller radius, we get

$$R = \sqrt{\frac{P \times 27}{8 \times \eta_p \times \rho \times (U/3.6)^3 \pi}},$$

$$R = \sqrt{\frac{5000 \times 27}{8 \times 0.7 \times 1.22 \times (30/3.6)^3 \pi}} = 3.29\,\text{m}.$$

11.6 Concluding Remarks

The formulation presented in this chapter establishes the relation between turbomachinery component geometry and the expected performance. It allows proper orientation of the airfoil-shaped blade components and clearly explains undesirable conditions such as compressor surge (or blade stall). Furthermore, this method allows the preliminary design of compressors, pumps, and turbines, and the ability to size their components. Power requirements and expected pressure ratios are estimated, based on parameters such as the geometry and rotation speed. Such data for commercially available turbomachinery are provided in performance maps, typically connecting pressure ratio to mass or volumetric flow rate for several rotation speeds. The trends in these performance maps are also explained, relating component geometry to performance for both compressors–pumps and turbines.

REFERENCE

[1] Wankel, F., *Rotary Piston Machines*, Iliffe Books, London, 1965.

PROBLEMS

11.1. It is required to pump 2 L/s water from a pool into a container 7 m above the water level in the pool. If the overall efficiency of the pump is 0.7, estimate the continuous power required for pumping the water.

11.2. A segment of an axial compressor with six identical stages has a compression ratio of 3 (all stages with degree of reaction of $R = 1/2$, and with the same velocity diagrams, compression ratio, average radius, etc.). The shaft rotates at 8000 RPM and the average blade radius is $r_m = 0.3$ m. The axial velocity across these stages is assumed to be constant at $c_z = 120$ m/s. The first-stage inlet temperature is $T_{01} = 360$ K, the efficiency of all stages is $\eta_{st} = 0.95$, and $\gamma = 1.4$.

(a) Calculate the rotor blade angles β_1 and β_2.
(b) Calculate the rotor pressure coefficient C_p.
(c) Calculate power (per unit mass flow rate) required for driving all six stages?

11.3. The axial velocity across the second stage of an axial compressor is $c_z =$ 120 m/s, the average radius is $r_m = 0.4$ m, and the stagnation temperature ahead of the rotor is $T_0 = 300$ K. If the compressor rotates at 5000 RPM, the angle $\alpha_1 =$ 25°, and the pressure coefficient of this rotor stage is $C_p = 0.5$, calculate the following values.

(a) The angles β_1 and β_2.
(b) The stagnation pressure rise p_{03}/p_{01} across this stage ($\eta_{stage} = 0.98$, $c_p =$ 0.24, $\gamma = 1.4$).
(c) Degree of reaction R.
(d) During the starting process the axial velocity is reduced to $c_z =$ 80 m/s but RPM = 5000. How many degrees should the stator blade (ahead of this rotor) be rotated so that the rotor blade angle β_1 will not change?

11.4. The average radius of an axial compressor blade is $r_{av} = 0.6$ m and it rotates at a constant speed of 3500 RPM. The axial velocity through this stage is $c_z = 130$ m/s and the two rotor angles are $\beta_1 = 55°$ and $\beta_2 = 35°$. Also assume that $\gamma = 1.4$, $c_p =$ 0.24, and $T_{01} = 300$ K.

(a) Draw a velocity diagram for the compressor blade.
(b) Calculate the stagnation pressure ratio p_{03}/p_{01} ($\eta_{st} = 1.0$).
(c) Calculate the pressure coefficient (C_p) for this rotor. Does blade stall seem likely?

11.5. A turbofan jet engine operates at a flight speed of $M_a = 0.7$ and the ambient conditions are $T_a = 300$ K and $p_a = 1$ atm. The axial fan has no stator ahead of the rotor, and the average radius of the fan blade is $r_{av} = 0.7$ m and it rotates at 3600 RPM [$\gamma = 1.4$, $c_p = 0.24$, $R = 286.78$ (m/s)^2/K].

(a) Assuming $\eta = 1.0$ and the inlet area ratio is $A_a/A_2 = 0.63$ (A_2 is ahead of the fan and A_a is in the free stream), calculate the axial velocity ahead of the fan blade.
(b) If the pressure coefficient of the fan rotor blade is $C_p = 0.55$, calculate the stagnation pressure ratio p_{03}/p_{0a} for the fan (for simplicity, assume $\eta_{stage} =$ 1.0).

11.6. A single-stage axial compressor rotates at 4000 RPM. The incoming flow speed is $c_z = 40$ m/s, the rotor average radius is $r_m = 0.25$ m, and the rotor blade consists of a symmetric airfoil shape oriented at 60° ($\beta_2 = 60°$). Also, $\eta_{st} = 0.85$, $\gamma = 1.4$, $c_p = 0.24$ kcal/kg K, and $T_{01} = 300$ K.

(a) Draw the velocity diagram for the rotor (assume no stator ahead of this stage, $\alpha_1 = 0$).

(b) Calculate rotor angle β_1.
(c) Calculate stagnation pressure ratio p_{02}/p_{01}.
(d) Calculate torque and power requirements.

11.7. The average radius of an axial compressor blades in the first row of a jet engine compressor is $r_m = 0.7$ m and the local Mach number is $M = 0.45$.

(a) Calculate the axial velocity ahead of the rotor ($T_{01} = 356$ K, $\eta_{st} = 0.95$, $\gamma = 1.4$, $c_p = 0.24$).
(b) Calculate the stagnation pressure rise across this stage at 3300 RPM if the rotor pressure-rise coefficient is $C_p = 0.5$.
(c) Calculate degree of reaction.

11.8. An axial water pump with an average radius of 0.05 m rotates at 1000 RPM and the incoming axial velocity is 1.5 m/s. If the blades turn the flow by $\beta_2 - \beta_1 = 10°$ and the pump efficiency is $\eta = 0.8$, calculate pressure rise and power per unit mass.

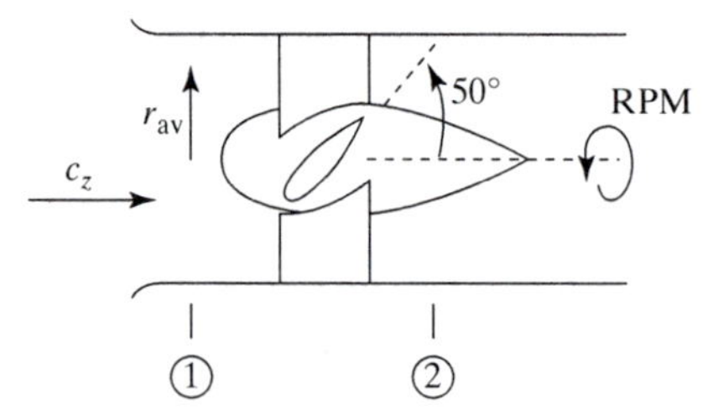

Problem 11.8.

11.9. The axial pump shown in the previous problem is pumping 25 L/s water in a 12-cm-diameter pipeline and the required head is $h_f = 0.8$ m of water. If the rotation rate is set by the motor at 1200 RPM, calculate the symmetric blade angle β_2 (assume that the pump average radius is 0.05 m, the efficiency is 0.75, and the average velocity in the pipe is the same as the axial velocity entering the rotor). Also calculate the pressure-rise coefficient and estimate whether blade stall is likely.

11.10. The axial velocity across an axial compressor blade is $c_z = 43.5$ m/s, the average radius is $r_{av} = 0.2$ m, and the stagnation temperature ahead of the rotor is $T_0 = 300$ K. The rotor blade is a symmetric untwisted airfoil that rotates at 3600 RPM; the angle β_1–$\beta_2 = 10°$ and $\alpha = 0^0$. Calculate the following values:

(a) The angles β_1 and β_2.
(b) The change in the rotor tangential velocity, Δc_θ.
(c) The pressure coefficient of this rotor stage, C_p.
(d) The stagnation pressure rise p_{03}/p_{01} across this stage ($\eta_{stage} = 0.98$, $c_p = 0.24$, $\gamma = 1.4$).
(e) Degree of reaction R.
(f) During the starting process the axial velocity is reduced to $c_z = 20$ m/s but the RPM stayed at 3600. How many degrees should the stator blade (ahead of this rotor) be rotated so that the rotor blade angle β_1 will not change?

11.11. A centrifugal compressor rotates at 7000 RPM and compresses 6-kg/s air. The impeller dimensions are $r_1 = 0.1$ m, $r_2 = 0.35$ m, and the blade height at the exit is $b_2 = 0.015$ m. The tangential velocity at the impeller inlet is negligible, inlet temperature is $T_{01} = 300$ K, and the air density at the impeller exit is 2.5 kg/m^3.

Also, at the exit, the blade is leaning backward at $\beta = 5°$, and $\eta = 0.93$, $c_p = 0.24$, $\gamma = 1.4$.

(a) Calculate the stagnation pressure rise p_{03}/p_{01} across this stage.
(b) Calculate the power required to drive this compressor.

11.12. What is the minimum flow rate for the centrifugal compressor described in the previous problem? Assume air density at the impeller inlet is 1.22 kg/m^3.

11.13. A centrifugal water pump rotates at 750 RPM. The impeller inlet radius is 0.04 m, the exit radius is 0.10 m, and it has straight vanes. The water enters the inlet at an average velocity of 2.5 m/s (without swirl). Calculate pressure rise and the power required for operating the pump ($\eta = 0.95$). Also estimate the angle β_1 at the inlet tip.

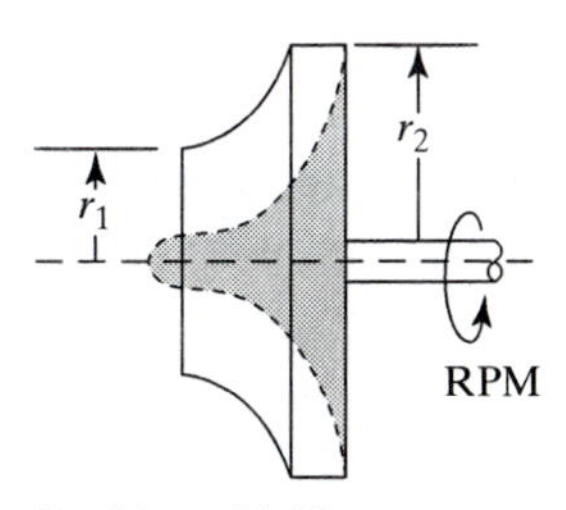

Problem 11.13.

11.14. A 0.1-m-diameter centrifugal pump (straight exit vanes) is used to pump 2-L/s water from a pool into a container 5 m above the water level in the pool. Calculate the required power and RPM if overall pump efficiency is 0.7.

11.15. A centrifugal pump is used to pump water from a lower to a higher reservoir, as shown in the figure. The impeller has simple straight vanes, diameter $D_2 = 12$ cm, and its inlet diameter is $D_1 = 6$ cm. If $z_1 = 2$ m, $z_2 = 10$ m, and the friction losses can be summarized as a head loss of $h_f = 0.5$ m, calculate the required RPM (assume efficiency $\eta = 0.8$). Estimate the power requirement when the flow rate is 12 L/s.

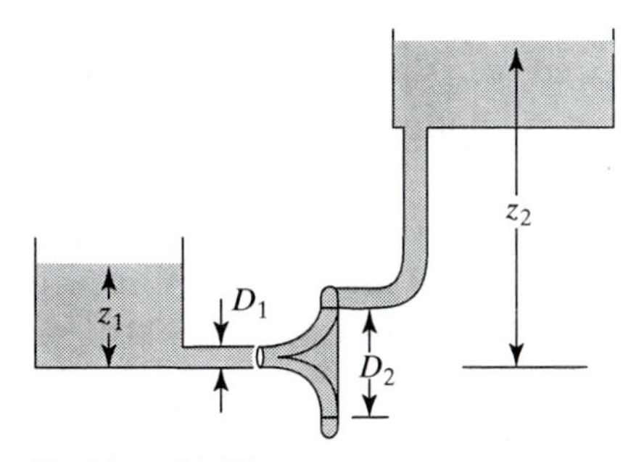

Problem 11.15.

11.16. Calculate the inlet velocity c_1, impeller inlet tip angle β_1, and the pressure coefficient [Eq. (11.58)] for the previous problem. Can the flow rate be reduced to 8 L/s?

11.17. A centrifugal water pump is designed to provide a pressure jump equivalent to a 2-m water column. The motor designated to run the pump provides a constant rotation speed of 800 RPM. If efficiency is estimated at 0.75, calculate the straight-vane impeller diameter.

11.18. Estimate the compression ratio of a centrifugal compressor with straight impeller exit blades operating at 6000 RPM. The inlet radius is 0.05 m and the impeller exit radius is 0.30 m. The compressed fluid is air ($T_{01} = 300$ K, $\eta = 0.90$, $c_p = 0.24$, $\gamma = 1.4$).

11.19. Suppose a pump with the dimensions used in the previous problem rotates at 1500 RPM, but this time the fluid is water. Estimate the pressure rise.

11.20. Estimate the compression ratio of a centrifugal compressor with forward-leaning blades $\beta = 15^\circ$ at the impeller exit, rotating at 5000 RPM. Assume the exit radial velocity $w_{r2} = U_2/2$. The inlet radius is 0.10 m and the impeller exit radius is 0.40 m. The compressed fluid is air ($T_{01} = 300$ K, $\eta = 0.90$, $c_p = 0.24$, $\gamma = 1.4$).

11.21. A centrifugal water pump rotates at 3000 RPM. The impeller inlet radius is 0.04 m, the exit radius is 0.10 m, and it has straight vanes. The water enters at the inlet at an average velocity of 1.5 m/s (without swirl). Calculate mass flow rate, the pressure rise, and the power required for operating the pump ($\eta = 0.95$).

11.22. A small impeller (of the centrifugal compressor shown in Fig. 11.25) with the dimensions $r_1 = 12$ mm and $r_2 = 18$ mm is used to compress air, and its rotation speed is 130,000 RPM. (Assume that $\gamma = 1.4$, $\rho_p = 1.0$ kg/m^3, $c_p = 0.24$, $T_{01} = 300$ K, and compressor efficiency is $\eta = 0.85$.)

(a) Draw the velocity diagrams at the tip of the impeller inlet and impeller exit. Assume that $\beta_1 = 48^\circ$ and $c_{\theta 1} = 0$ at the inlet and the blades are turned back 10° at the exit. (Also, blade height at the exit is $b_2 = 3$ mm).
(b) Calculate mass flow rate and exit velocity w_{r2}.
(c) How much power is needed to drive this impeller and what is the compression ratio of this compressor?

11.23. The impeller of a centrifugal compressor stage with the dimensions $r_1 = 0.08$ m and $r_2 = 0.32$ m rotates at a constant speed of 5000 RPM. The fluid that is being compressed has the following properties: $\gamma = 1.4$, $\rho_1 = 1.2$ kg/m^3, $c_p = 0.24$, $T_{01} = 300$ K, and the compressor efficiency is $\eta = 0.85$.

(a) Draw the velocity diagrams at the tip of the impeller inlet and impeller exit. Assume that $\beta_1 = 45^\circ$ and $c_{\theta 1} = 0$ at the inlet and the blades are turned back 10° at the impeller exit. (Also, blade height at the exit is $b_2 = 0.01$ m, and the fluid density there is $\rho_2 = 1.5$ kg/m^3).
(b) Calculate mass flow rate and exit velocity w_{r2}.
(c) How much power is needed to drive this impeller and what is the compression ratio of this compressor?
(d) Calculate compression ratio and power if the blades lean forward by 10° at the impeller exit (instead of backward).

11.24. A centrifugal compressor rotates at 5000 RPM and compresses 5-kg/s air. The impeller dimensions are $r_1 = 0.1$ m, $r_2 = 0.3$ m, and the blade height at the exit is $b_2 = 0.015$ m. The tangential velocity at the impeller inlet is negligible, inlet temperature is $T_{01} = 300$ K, and the air density at the impeller exit is 2.4 kg/m^3. Also, at the exit, the blade is leaning forward at $\beta = -5°$, and $\eta = 0.90$, $c_p = 0.24$, $\gamma = 1.4$.

(a) Calculate the stagnation pressure rise p_{03}/p_{01} across the compressor.
(b) Calculate the power required for driving this compressor.

11.25. An axial turbine with an average radius of $r_2 = 0.3$ m is operated by a hot gas flow at a temperature of $T_{01} = 1200$ K and $\rho_2 = 1.0$ kg/m^3. The jet leaves the nozzle at an axial velocity of $c_z = 300$ m/s. The nozzle total exit area (normal to c_2) is $A_2 = 0.03$ m^2.

(a) Draw the rotor velocity diagram based on $\beta_2 = 38°$, $\beta_3 = -40°$, and c_3 has *no* tangential component. Based on the velocity diagram, calculate rotor RPM and nozzle angle α_2.
(b) Based on your finding, sketch the turbine blade shape and calculate mass flow rate.
(c) How much power is generated by this turbine? Calculate the degree of reaction.

11.26. The average radius of a 50% degree-of-reaction axial turbine is $r = 0.2$ m and rotation speed is 8000 RPM. The hot gas flow leaves the nozzle at a temperature of $T_{01} = 1200$ K ($\rho_2 = 0.5$kg/m^3) and at an angle of $\beta_2 = 30°$. If the rotor trailing edge is aligned such that $\beta_3 = 60°$, calculate the power per unit mass generated by this turbine ($\eta = 0.90$, $\gamma = 1.4$, $c_p = 0.24$) and the stagnation pressure drop in the turbine.

11.27. An impulse turbine with an average radius of $r_2 = 0.25$ m is operated by a water jet. The nozzle area is 3 cm^2, its jet angle is $\alpha_2 = 60°$, and the water flow rate is 5 kg/s ($\rho = 1000$ kg/m^3).

(a) If the flow at the exit has no swirl, then calculate the power generated by this turbine.
(b) The turbine operates a lift, raising a mass of 75 kg, at sea-level conditions. If the mechanical efficiency of the gear system is assumed to be 1.0, then calculate how fast the mass could be lifted.

11.28. An impulse turbine with an average radius of $r_2 = 0.2$ m is operated by a hot gas flow at a rate of $m = 1$ kg/s and a temperature of $T_{01} = 1200$ K. The jet leaves the nozzle at an angle of $\alpha_2 = 60°$ and speed of $c_2 = 300$ m/s ($\rho_2 = 0.5$ kg/m^3).

(a) If the flow at the exit has no swirl, then calculate the power generated by this turbine ($\eta = 0.90$, $\gamma = 1.4$, $c_p = 0.24$).
(b) Calculate the stagnation pressure drop in the turbine. How much of the static pressure drop takes place in the rotor?
(c) Calculate shaft RPM.

11.29. A Pelton turbine wheel with an average radius of $r = 0.2$ m rotates at 800 RPM; it is operated by a water jet with a cross section of 3 cm^2 and exit velocity $c_2 = 25$ m/s. The nozzle is aligned with the wheel ($\alpha_2 = 90°$) and the blade outflow angle is $\beta_3 = 60°$. Calculate the torque and the power of the turbine.

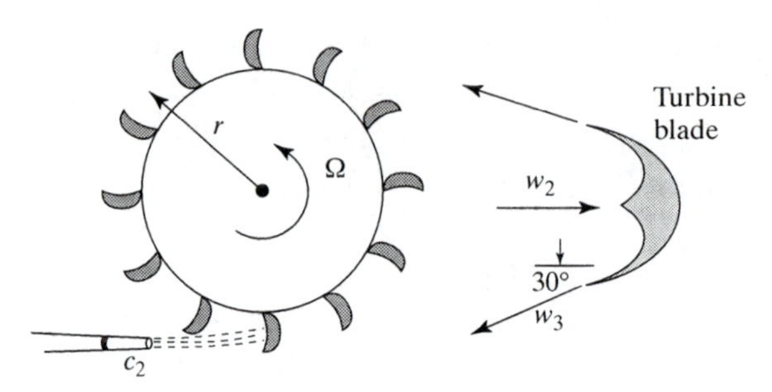

Problem 11.29.

11.30. The turbine wheel shown in Fig. 11.34 rotates at 1000 RPM and is operated by a water jet with a velocity of $c_2 = 30$ m/s. The nozzle angle is $\alpha_2 = 60°$ and the blade inflow and outflow angles are the same ($\beta_2 = -\beta_3$). The average radius of the turbine blades is $r = 0.2$ m. Calculate the blade angle β_2, the pressure drop in the nozzle, and the power per unit mass ($\eta = 0.9$).

11.31. It is proposed to build a small wind turbine capable of lighting up a 100-W lightbulb. If the average wind speed is 30 km/h and the expected design is capable of extracting 80% of the power predicted by the Betz formula, calculate the wind turbine diameter.

APPENDIX A

Conversion Factors

Length	1 m = 3.281 ft
	1 cm = 0.3937 in
	1 km = 3281 ft = 0.6214 mi
	1 in = 2.54 cm = 25.4 mm
	1 mi = 5280 ft = 1.609 km
Mass	1 kg = 2.2046 lbm = 0.06853 slug
	1 lbm = 453.59 gm = 0.45359 kg = 0.03108 slug
	1 slug = 32.17 lbm = 14.59 kg
Density	1 kg/m^3 = 0.06243 lbm/ft^3 = 0.001941 slug/ft^3
	1 slug/ft^3 = 32.17 lbm/ft^3 = 515.36 kg/m^3
Force	1 N = 0.2248 lbf = 10^5 dynes
	1 lbf = 4.448 N
	1 kg$_f$ = 9.81 N
Pressure	1 Pa = 1 N/m^2 = 0.02089 lbf/ft^2
	1 kPa = 0.14504 lbf/in^2
	1 lbf/in^2 (psi) = 6.8948 kPa
	1 lbf/ft^2 (psf) = 47.88 Pa
	1 atm = 101.3 kPa = 14.696 psi = 760 mm Hg @ 0°
Temperature	T(K) = T(°C) + 273.15
	T(K) = T(°R)/1.8
	T(°C) = [T(°F) − 32]/1.8
	T(°F) = 1.8 T(°C) + 32
	T(°R) = T(°F) + 459.67
Velocity	1 m/s = 3.281 ft/s = 3.6 km/h
	1 ft/s = 0.3048 m/s
	1 mi/h = 0.4470 m/s = 1.609 km/h = 1.467 ft/s
	1 RPM = 0.1047 rad/s
	1 rad/s = 9.549 RPM
Energy/Work	1 J = 1 N m = 0.7376 ft-lbf = 0.9478 × 10^{-3} Btu
	1 Btu = 778.17 ft-lbf = 1055 J
Power	1 W = 1 J/s = 0.7376 ft-lbf/s = 1.341 × 10^{-3} horsepower
	1 horsepower = 550 ft-lbf/s = 0.7457 kW
	1 Btu/s = 1.415 horsepower

Viscosity	$1\ \mathrm{Pa \cdot s} = 0.02089\ \mathrm{lbf \cdot s/ft^2}$
	$1\ \mathrm{m^2/s} = 10.764\ \mathrm{ft^2/s}$
	$1\ \mathrm{lbf \cdot s/ft^2} = 47.88\ \mathrm{Pa \cdot s}$
	$1\ \mathrm{ft^2/s} = 0.09290\ \mathrm{m^2/s}$
Angle	$1\ \mathrm{degree} = 0.01745\ \mathrm{rad}$
	$1\ \mathrm{rad} = 57.30\ \mathrm{degrees}$

APPENDIX B

Properties of Compressible Isentropic Flow

M	A/A^*	p/p_0	ρ/ρ_0	T/T_0	M	A/A^*	p/p_0	ρ/ρ_0	T/T_0
0.00		1.000	1.000	1.000	0.66	1.13	0.747	0.812	0.920
0.01	57.87	0.9999	0.9999	0.9999	0.68	1.12	0.734	0.802	0.915
0.02	28.94	0.9997	0.9999	0.9999	0.70	1.09	0.721	0.792	0.911
0.04	14.48	0.999	0.999	0.9996	0.72	1.08	0.708	0.781	0.906
0.06	9.67	0.997	0.998	0.999	0.74	1.07	0.695	5.771	0.901
0.08	7.26	0.996	0.997	0.999	0.76	1.06	0.682	0.761	0.896
0.10	5.82	0.993	0.995	0.998	0.78	1.05	0.669	0.750	0.891
0.12	4.86	0.990	0.993	0.997	0.80	1.04	0.656	0.740	0.886
0.14	4.18	0.986	0.990	0.996	0.82	1.03	0.643	0.729	0.881
0.16	3.67	0.982	0.987	0.995	0.84	1.02	0.630	0.719	0.876
0.18	3.28	0.978	0.984	0.994	0.86	1.02	0.617	0.708	0.871
0.20	2.96	0.973	0.980	0.992	0.88	1.01	0.604	0.698	0.865
0.22	2.71	0.967	0.976	0.990	0.90	1.01	0.591	0.687	0.860
0.24	2.50	0.961	0.972	0.989	0.92	1.01	0.578	0.676	0.855
0.26	2.32	0.954	0.967	0.987	0.94	1.00	0.566	0.666	0.850
0.28	2.17	0.947	0.962	0.985	0.96	1.00	0.553	0.655	0.844
0.30	2.04	0.939	0.956	0.982	0.98	1.00	0.541	0.645	0.839
0.32	1.92	0.932	0.951	0.980	1.00	1.00	0.528	0.632	0.833
0.34	1.82	0.923	0.944	0.977	1.02	1.00	0.516	0.623	0.828
0.36	1.74	0.914	0.938	0.975	1.04	1.00	0.504	0.613	0.822
0.38	1.66	0.905	0.931	0.972	1.06	1.00	0.492	0.602	0.817
0.40	1.59	0.896	0.924	0.969	1.08	1.01	0.480	0.592	0.810
0.42	1.53	0.886	0.917	0.966	1.10	1.01	0.468	0.582	0.805
0.44	1.47	0.876	0.909	0.963	1.12	1.01	0.457	0.571	0.799
0.46	1.42	0.865	0.902	0.959	1.14	1.02	0.445	0.561	0.794
0.48	1.38	0.854	0.893	0.956	1.16	1.02	0.434	0.551	0.788
0.50	1.34	0.843	0.885	0.952	1.18	1.02	0.423	0.541	0.782
0.52	1.30	0.832	0.877	0.949	1.20	1.03	0.412	0.531	0.776
0.54	1.27	0.820	0.868	0.945	1.22	1.04	0.402	0.521	0.771
0.56	1.24	0.808	0.859	0.941	1.24	1.04	0.391	0.512	0.765
0.58	1.21	0.796	0.850	0.937	1.26	1.05	0.381	0.502	0.759
0.60	1.19	0.784	0.840	0.933	1.28	1.06	0.371	0.492	0.753
0.62	1.17	0.772	0.831	0.929	1.30	1.07	0.361	0.483	0.747
0.64	1.16	0.759	0.821	0.924	1.32	1.08	0.351	0.474	0.742

M	A/A^*	p/p_0	ρ/ρ_0	T/T_0	M	A/A^*	p/p_0	ρ/ρ_0	T/T_0
1.34	1.08	0.342	0.464	0.736	2.18	1.97	0.097	0.188	0.513
1.36	1.09	0.332	0.455	0.730	2.20	2.01	0.094	0.184	0.508
1.38	1.10	0.323	0.446	0.724	2.22	2.04	0.091	0.180	0.504
1.40	1.11	0.314	0.437	0.718	2.24	2.08	0.088	0.176	0.499
1.42	1.13	0.305	0.429	0.713	2.26	2.12	0.085	0.172	0.495
1.44	1.14	0.297	0.420	0.707	2.28	2.15	0.083	0.168	0.490
1.46	1.15	0.289	0.412	0.701	2.30	2.19	0.080	0.165	0.486
1.48	1.16	0.280	0.403	0.695	2.32	2.23	0.078	0.161	0.482
1.50	1.18	0.272	0.395	0.690	2.34	2.27	0.075	0.157	0.477
1.52	1.19	0.265	0.387	0.684	2.36	2.32	0.073	0.154	0.473
1.54	1.20	0.257	0.379	0.678	2.38	2.36	0.071	0.150	0.469
1.56	1.22	0.250	0.371	0.672	2.40	2.40	0.068	0.147	0.465
1.58	1.23	0.242	0.363	0.667	2.42	2.45	0.066	0.144	0.461
1.60	1.25	0.235	0.356	0.661	2.44	2.49	0.064	0.141	0.456
1.62	1.27	0.228	0.348	0.656	2.46	2.54	0.062	0.138	0.452
1.64	1.28	0.222	0.341	0.650	2.48	2.59	0.060	0.135	0.448
1.66	1.30	0.215	0.334	0.645	2.50	2.64	0.059	0.132	0.444
1.68	1.32	0.209	0.327	0.639	2.52	2.69	0.057	0.129	0.441
1.70	1.34	0.203	0.320	0.634	2.54	2.74	0.055	0.126	0.437
1.72	1.36	0.197	0.313	0.628	2.56	2.79	0.053	0.123	0.433
1.74	1.38	0.191	0.306	0.623	2.58	2.84	0.052	0.121	0.429
1.76	1.40	0.185	0.300	0.617	2.60	2.90	0.050	0.118	0.425
1.78	1.42	0.179	0.293	0.612	2.62	2.95	0.049	0.115	0.421
1.80	1.44	0.174	0.287	0.607	2.64	3.01	0.047	0.113	0.418
1.82	1.46	0.169	0.281	0.602	2.66	3.06	0.046	0.110	0.414
1.84	1.48	0.164	0.275	0.596	2.68	3.12	0.044	0.108	0.410
1.86	1.51	0.159	0.269	0.591	2.70	3.18	0.043	0.106	0.407
1.88	1.53	0.154	0.263	0.586	2.72	3.24	0.042	0.103	0.403
1.90	1.56	0.149	0.257	0.581	2.74	3.31	0.040	0.101	0.400
1.92	1.58	0.145	0.251	0.576	2.76	3.37	0.039	0.099	0.396
1.94	1.61	0.140	0.246	0.571	2.78	3.43	9.038	0.097	0.393
1.96	1.63	0.136	0.240	0.566	2.80	3.50	0.037	0.095	0.389
1.98	1.66	0.132	0.235	0.561	2.82	3.57	0.036	0.093	0.386
2.00	1.69	0.128	0.230	0.556	2.84	3.64	0.035	0.091	0.383
2.02	1.72	0.124	0.225	0.551	2.86	3.71	0.034	0.089	0.379
2.04	1.75	0.120	0.220	0.546	2.88	3.78	0.033	0.087	0.376
2.06	1.78	0.116	0.215	0.541	2.90	3.85	0.032	0.085	0.373
2.08	1.81	0.113	0.210	0.536	2.92	3.92	0.031	0.083	0.370
2.10	1.84	0.109	0.206	0.531	2.94	4.00	0.030	0.081	0.366
2.12	1.87	0.106	0.201	0.526	2.96	4.08	0.029	0.080	0.363
2.14	1.90	0.103	0.197	0.522	2.98	4.15	0.028	0.078	0.360
2.16	1.94	0.100	0.192	0.517	3.00	4.23	0.027	0.076	0.357

APPENDIX C

Properties of Normal Shock Flow

M_1	M_2	p_2/p_1	T_2/T_1	$(p_0)_2/(p_0)_1$	M_1	M_2	p_2/p_1	T_2/T_1	$(p_0)_2/(p_0)_1$
1.00	1.000	1.000	1.000	1.000	1.68	0.646	3.126	1.444	0.864
1.02	0.980	1.047	1.013	1.000	1.70	0.641	3.205	1.458	0.856
1.04	0.962	1.095	1.026	1.000	1.72	0.635	3.285	1.473	0.847
1.06	0.944	1.144	1.039	1.000	1.74	0.631	3.366	1.487	0.839
1.08	0.928	1.194	1.052	0.999	1.76	0.626	3.447	1.502	0.830
1.10	0.912	1.245	1.065	0.999	1.78	0.621	3.530	1.517	0.821
1.12	0.896	1.297	1.078	0.998	1.80	0.617	3.613	1.532	0.813
1.14	0.882	1.350	1.090	0.997	1.82	0.612	3.698	1.547	0.804
1.16	0.868	1.403	1.103	0.996	1.84	0.608	3.783	1.562	0.795
1.18	0.855	1.458	1.115	0.995	1.86	0.604	3.869	1.577	0.786
1.20	0.842	1.513	1.128	0.993	1.88	0.600	3.957	1.592	0.777
1.22	0.830	1.570	1.140	0.991	1.90	0.596	4.045	1.608	0.767
1.24	0.818	1.627	1.153	0.988	1.92	0.592	4.134	1.624	0.758
1.26	0.807	1.686	1.166	0.986	1.94	0.588	4.224	1.639	0.749
1.28	0.796	1.745	1.178	0.983	1.96	0.584	4.315	1.655	0.740
1.30	0.786	1.805	1.191	0.979	1.98	0.581	4.407	1.671	0.730
1.32	0.776	1.866	1.204	0.976	2.00	0.577	4.500	1.688	0.721
1.34	0.766	1.928	1.216	0.972	2.02	0.574	4.594	1.704	0.711
1.36	0.757	1.991	1.229	0.968	2.04	0.571	4.689	1.720	0.702
1.38	0.748	2.055	1.242	0.963	2.06	0.567	4.784	1.737	0.693
1.40	0.740	2.120	1.255	0.958	2.08	0.564	4.881	1.754	0.683
1.42	0.731	2.186	1.268	0.953	2.10	0.561	4.978	1.770	0.674
1.44	0.723	2.253	1.281	0.948	2.12	0.558	5.077	1.787	0.665
1.46	0.716	2.320	1.294	0.942	2.14	0.555	5.176	1.805	0.656
1.48	0.708	2.389	1.307	0.936	2.16	0.553	5.277	1.822	0.646
1.50	0.701	2.458	1.320	0.930	2.18	0.550	5.378	1.839	0.637
1.52	0.694	2.529	1.334	0.923	2.20	0.547	5.480	1.857	0.628
1.54	0.687	2.600	1.347	0.917	2.22	0.544	5.583	1.875	0.619
1.56	0.681	2.673	1.361	0.910	2.24	0.542	6.687	1.892	0.610
1.58	0.675	2.746	1.374	0.903	2.26	0.539	5.792	1.910	0.601
1.60	0.668	2.820	1.388	0.895	2.28	0.537	5.898	1.929	0.592
1.62	0.663	2.895	1.402	0.888	2.30	0.534	6.005	1.947	0.583
1.64	0.657	2.971	1.416	0.880	2.32	0.532	6.113	1.965	0.575
1.66	0.651	3.048	1.430	0.872	2.34	0.530	6.222	1.984	0.566

M_1	M_2	p_2/p_1	T_2/T_1	$(p_0)_2/(p_0)_1$	M_1	M_2	p_2/p_1	T_2/T_1	$(p_0)_2/(p_0)_1$
2.36	0.527	6.331	2.003	0.557	2.70	0.496	8.338	2.343	0.424
2.38	0.525	6.442	2.021	0.549	2.72	0.494	8.465	2.364	0.417
2.40	0.523	6.553	2.040	0.540	2.74	0.493	8.592	2.396	0.410
2.42	0.521	6.666	2.060	0.532	2.76	0.491	8.721	2.407	0.403
2.44	0.519	6.779	2.079	0.523	2.78	0.490	8.850	2.429	0.396
2.46	0.517	6.894	2.098	0.515	2.80	0.488	8.980	2.451	0.389
2.48	0.515	7.009	2.118	0.507	2.82	0.487	9.111	2.473	0.383
2.50	0.513	7.125	2.139	0.499	2.84	0.485	9.243	2.496	0.376
2.52	0.511	7.242	2.157	0.491	2.86	0.484	9.376	2.518	0.370
2.54	0.509	7.360	2.177	0.483	2.88	0.483	9.510	2.541	0.364
2.56	0.507	7.479	2.198	0.475	2.90	0.481	9.645	2.563	0.358
2.58	0.506	7.599	2.218	0.468	2.92	0.480	9.781	2.586	0.352
2.60	0.504	7.720	2.238	0.460	2.94	0.479	9.918	2.609	0.346
2.62	0.502	7.842	2.260	0.453	2.96	0.478	10.055	2.632	0.340
2.64	0.500	7.965	2.280	0.445	2.98	0.476	10.194	2.656	0.334
2.66	0.499	8.088	2.301	0.438	3.00	0.475	10.333	2.679	0.328
2.68	0.497	8.213	2.322	0.431					

Index

CPSIA information can be obtained at www.ICGtesting.com
Printed in the USA
BVOW06s1016140114

341827BV00010B/413/P